Lecture Notes
in Control and Information Sciences 247

Editor: M. Thoma

Springer-Verlag London Ltd.

K.D. Young and Ü. Özgüner (Eds)

Variable Structure Systems, Sliding Mode and Nonlinear Control

Springer

Editors

K.D. Young, PhD
Applied Technology Center, Hong Kong University of Science and Technology, Clear Water Bay, Kowloon, Hong Kong

Ü. Özgüner, PhD
Department of Electrical Engineering, Ohio State University, Columbus, 43210 USA

ISBN 978-1-85233-197-9

British Library Cataloguing in Publication Data
Variable structure systems, sliding mode and nonlinear control. - (Lectures notes in control and information sciences ; 247)
1.Automatic control - Congresses
I.Young, K.D. II.Ozguner, U.
629.8
ISBN 978-1-85233-197-9 ISBN 978-1-84628-540-0 (eBook)
DOI 10.1007/978-1-84628-540-0

Library of Congress Cataloging-in-Publication Data
A catalog record for this book is available from the Library of Congress

Typesetting: Camera ready by contributors

69/3830-543210 Printed on acid-free paper SPIN 10713312

Foreword

This volume of work has been selected from papers presented at VSS'98 (5th International Workshop on Variable Structure Systems) held in Sarasota, Florida. This workshop was the fifth in a series of VSS international workshops, and the first to be held in the United States.

The initiator of this workshop series, Dr. Asif Sabanovic first organized and chaired VSS'90 in Sarajevo amidst the warmth of the Bosnian people and the spectacular surroundings. Since then an international workshop has been organized and held biennially. The second workshop was in 1992 (Sheffield, UK) and the third in 1994 (Benevento, Italy). VSS'96 (Tokyo, Japan) was the first in the series to be held outside Europe.

Work presented herein on theoretical developments and applications on VSS and Sliding Mode, reflects how trends have advanced beyond the original ideas that are now well documented in a number of books and research monographs. In particular, the concepts of Sliding Sector and Second Order Sliding Mode introduced in this volume, will stimulate discussions and invite further extensions. Also, the focus on Sampled Data systems represents a positive trend towards practical industrial implementations of sliding mode controllers.

K. David Young, Hong Kong
Ümit Özgüner, Columbus, Ohio, USA

Contents

Sliding Sector for Variable Structure System

Katsuhisa Furuta* Yaodong Pan**

* Department of Mechanical and Environmental Informatics
Tokyo Institute of Technology
2-12-1 Oh-okayama, Meguro-ku, Tokyo 152, Japan
** Department of Computers and Systems Engineering
Tokyo Denki University
Hiki-gun, Saitama 350-0394, Japan

Abstract. This paper presents VS controllers with sliding sector. The sliding sector is designed so that some norm of the state is decreasing inside it without any control action. VS control law ensures the norm decreases in the state space and thus yields a quadratic stable VS control system. The discrete-time system described by transfer function is also taken into consideration.

1 Introduction

The variable structure control has started in Russia by many researchers, like Barbashin[1], Utkin[2], Emelyanov[3]. The variable structure control has been mainly considered for continuous time systems in the form of sliding mode control. But when the continuous-time sliding mode is implemented by the digital controller, the unstable phenomena is known to be possibly happened as follows: The stability on the sliding mode

$$s = c^T x \tag{1}$$

for the continuous-time linear system represented by

$$\dot{x} = Ax + Bu \tag{2}$$

is determined by the zeros of the transfer function between the input u and s. The zeros of the transfer function are those of

$$\det \begin{bmatrix} A - sI & B \\ c & 0 \end{bmatrix} = \det \begin{bmatrix} A - sI & B \\ 0 & c^T(A - sI)^{-1}B \end{bmatrix} \tag{3}$$

But when the continuous system is realized by the digital control, not only the chattering around the sliding mode may be generated, but also the stable zeros may be happened unstable as Åstrom pointed out [4].

To overcome this problem discrete-time VS controllers have been proposed in literatures [5], [6], [7], [8] and [9]. Drakunov and Utkin [5] proposed a discrete-time variable structure control based on the contraction mapping. Furuta [6] proposed to use the conventional Lyapunov like approach and has shown that the sliding sector should be considered for the discrete-time system. He also considered the variable structure control using the transfer function for the unknown parameter system [10].

The close observation of the sliding mode for a simple linear system

$$\dot{x} = \begin{bmatrix} 0 & 1 \\ 0 & 0 \end{bmatrix} x + \begin{bmatrix} 0 \\ 1 \end{bmatrix} u$$
$$s = \begin{bmatrix} a & 1 \end{bmatrix} x, \quad a > 0$$

Decompose the state space into three sets by

$$S_+ = \{x|s > 0\} \tag{4}$$
$$S_0 = \{x|s = 0\} \tag{5}$$
$$S_- = \{x|s < 0\} \tag{6}$$

and the control law is given by

$$u = \begin{cases} u_- & \forall x \in S_+ \\ 0 & \forall x \in S_0 \\ u_+ & \forall x \in S_- \end{cases} \tag{7}$$

For this system, the state can not be stay inside the sliding mode for a sampling interval even if the sampling mode is small. So instated of the sliding mode, the set of the state in which the state has the desirable properties without control will be considered. One of such sets may be one in which the norm of the state decreases with zero control input. To define such set, the following P-norm of the state is introduced.

$$||x||_P = (x^T P x)^{(\frac{1}{2})} \tag{8}$$

Letting

$$P = I_2$$

then

$$L = x_1^2 + x_2^2$$

and the derivative is

$$\dot{L} = x^T(A^T P + PA)x = 2x_1x_2$$

where I_2 is the 2×2 identity matrix. This tells that the norm of the zero input system is deceasing inside the sector

$$S = \{x|x_1x_2 < 0\}$$

i.e., inside the second and fourth quadrants, the P-norm of the state is decreasing for zero input. If we restrict the condition in that the derivative of L is less than $-x^T Rx$ for S, then the following set can be defined.

$$S_R = \{x|\dot{L} < -x^T Rx\}$$

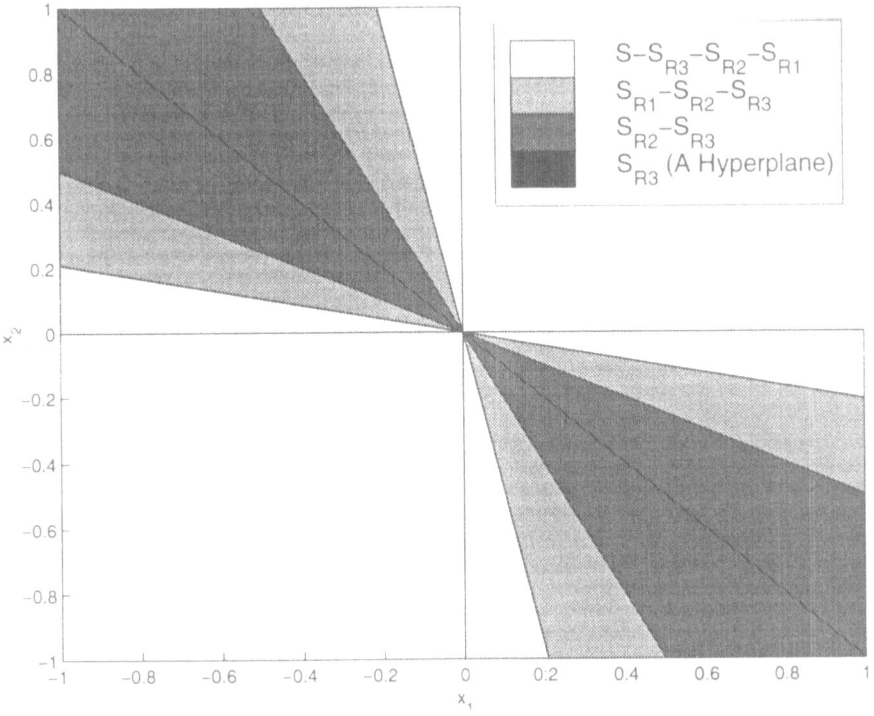

Fig. 1. Sectors $\mathcal{S}$, $\mathcal{S}_{R1}$, $\mathcal{S}_{R2}$, and $\mathcal{S}_{R3}$ for $R = O_2$, $R = 0.2 * I_2$, $R = 0.4I_2$, and $R = 0.5I_2$, respectively

Inside this set the P-norm is decreasing faster and

$$\mathcal{S}_R \subset \mathcal{S}$$

Figure 1 shows the sector where R is chosen to be O_2, $0.2 * I_2$, $0.4 * I_2$, and $0.5 * I_2$, respectively where O_2 is the 2×2 zero matrix. The sector is reduced to a hyperplane if $R = 0.5 * I_2$ and is the origin if $R > 0.5 * I_2$. This sector is named PR-sliding sector and we take the control be zero when the state is in the sector. And when the state is outside of the sector, the control to transfer the state into the sector will be considered. The sector has been shown to be useful in the control design not only for the continuous-time system but also for the discrete-time system [15] and the sampled-data system [14].

In the paper, the definition and design of the PR-sliding sector for continuous-time system is presented and the corresponding VS controller is designed. Then parameter uncertainty and external disturbance are considered, a robust VS controller and an adaptive VS controller are proposed.

For a discrete-time system described by transfer function, another kind of sliding sector is also proposed. Inside the sliding sector, the closed loop system is designed to be stable. The corresponding discrete-time VS control law transfers the state to the inside of the sliding sector and stabilizes the closed loop system.

The organization of the paper is as follows: Section 2 defines the sliding sector for continuous-time systems. Section 3 designs a VS controller with the sliding sectors. Section 4 proposes a robust VS controller and an adaptive VS controller for continuous-time system with parameter uncertainty and external disturbance. Section 5 presents the discrete-time VS controller for discrete-time system described by transfer function. Section 6 gives some simulation results.

2 Design of Sliding Sector for Continuous-time Systems

Consider a single input continuous-time plant:

$$\dot{x}(t) = Ax(t) + Bu(t) \tag{9}$$

where $x(t) \in R^n$ and $u(t) \in R^1$ are state and input vectors, respectively. A and B are constant matrices of appropriate dimensions, pair (A, B) is controllable.

Definition 2. 1 *The P-*Norm *$||\cdot||_P$ of the system state is defined as*

$$||x||_P = (x^T Px)^{\frac{1}{2}}, \quad x \in R^n \tag{10}$$

where $P \in R^{n\times n}$ is a positive definite symmetric matrix.

The square of the P-norm is denoted as

$$L = ||x||_P^2 = x^T Px > 0, \quad \forall x \in R^n, x \neq 0 \tag{11}$$

which will be considered as a Lyapunov function candidate. If the autonomous system of (9) is quadratic stable, then there exist a positive definite symmetric matrix P and a positive semi-definite symmetric matrix $R = C^T C$ such that

$$\dot{L} = x^T(A^T P + PA)x \leq -x^T Rx, \quad \forall x \in R^n$$

where $P \in R^{n\times n}$, $R \in R^{n\times n}$, and $C \in R^{l\times n}$, $l \geq 1$, (C, A) is observable pair. But for an unstable one, this inequality does not hold. It is possible to decompose all state space into two parts such that one satisfies $\dot{L} > -x^T Rx$ for some element $x \in R^n$, and the other one satisfies $\dot{L} \leq -x^T Rx$ for some other element $x \in R^n$. The latter ones form a special subset in which the P-norm $||x||_P$ decreases. Accordingly this special subset is defined as a PR-sliding sector.

Definition 2. 2 *The PR-*Sliding Sector *is a subset of R^n defined as*

$$\mathcal{S} = \{x|\ x^T(A^T P + PA)x \leq -x^T Rx, x \in R^n\ \} \tag{12}$$

where $P \in R^{n\times n}$ is a positive definite symmetric matrix, $R \in R^{n\times n}$ is a positive semi-definite symmetric matrix, and $R = C^T C$, $C \in R^{l\times n}$, $l \geq 1$, (C, A) is observable pair.

Inside the PR-sliding sector, the P-norm $||x||_P$ of the plant (9) without any control action decreases with the rate of

$$\dot{L} \leq -x^T R x \leq 0, \quad \forall x \in \mathcal{S}, x \neq 0. \tag{13}$$

Theorem 2. 1 *[11]For any plant (9), the PR-sliding sector defined in (12) exists for any positive definite symmetric matrix P and any positive semi-definite symmetric matrix R described in Definition 2.2, and can be rewritten as*

$$\mathcal{S} = \{x | \; s^2(x) \leq \delta^2(x) \; \} \tag{14}$$

where

$$s^2(x) = x^T P_1 x \geq 0 \tag{15}$$
$$\delta^2(x) = x^T P_2 x \geq 0, \tag{16}$$

P_1 and P_2 are $n \times n$ positive semi-definite symmetric matrices.

Corollary 2. 1 *Let $n_i = rank(P_i)$ $(i = 1, 2)$ and $n_3 = n - n_1 - n_2$. Then n_1 and n_2 are the numbers of the positive eigenvalues and the negative eigenvalues of Ω, respectively, n_3 is the number of the eigenvalues of Ω in the origin, where $\Omega = A^T P + PA + R$.*

Example 2. 1 *Consider a third order plant represented by*

$$\dot{x} = \begin{bmatrix} -1 & 0 & 0 \\ 0 & -2 & 0 \\ 0 & 0 & -4 \end{bmatrix} x + \begin{bmatrix} 1 \\ 1 \\ 1 \end{bmatrix} u. \tag{17}$$

If R is chosen to be the 3×3 identity matrix I_3, i.e. $R = I_3$, P is chosen to be σI_3 $(\frac{1}{4} < \sigma < \frac{1}{2})$, then the PR-sliding sector is determined by

$$x^T P_1 x \leq x^T P_2 x$$

where $P_1 = diag(1 - 2\sigma, 0, 0) \geq 0$ and $P_2 = diag(0, 4\sigma - 1, 8\sigma - 1) \geq 0$. Thus $n_1 = 1$, $n_2 = 2$, and the PR-sliding sector can be rewritten as

$$\mathcal{S} = \{x : \; |s(x)| \leq \delta(x) \; \}.$$

where $s(x)$ and $\delta(x)$ are a linear function and the square root of a quadratic function on x, respectively, i.e.

$$\begin{cases} s(x) = \quad Sx \quad = \quad \left[\sqrt{1 - 2\sigma} \; 0 \; 0\right] x \\ \delta(x) = \sqrt{x^T P_2 x} = \sqrt{x^T diag(0, 4\sigma - 1, 8\sigma - 1) x}. \end{cases}$$

Figure 2 shows the PR-sliding sector for the plant (17) when $\sigma = \frac{2}{5}$.

In the PR-sliding sector shown in Figure 2, a hyperplane $s(x) = 0$ is included. We call such PR-sliding sector the simplified PR-sliding sector.

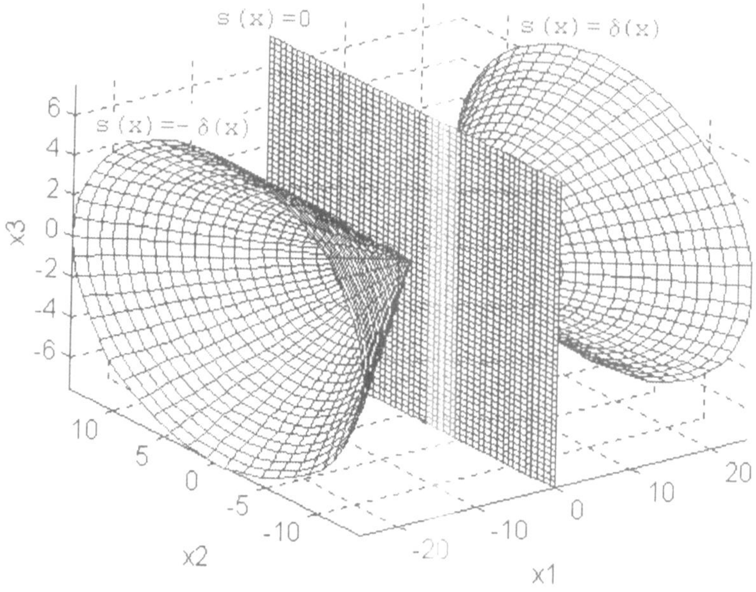

Fig. 2. PR-Sliding Sector : Complementary Set of a Cone

Definition 2. 3 *A* Simplified PR-Sliding Sector *is a subset of* R^n *defined as*

$$\mathcal{S} = \{x|\ |s(x)| \leq \delta(x), x \in R^n\ \} \tag{18}$$

where the linear function $s(x)$ *and the square root* $\delta(x)$ *of the quadratic function* $\delta^2(x)$ *are determined by*

$$s(x) = Sx, \quad S \in R^{1\times n} \tag{19}$$

$$\delta(x) = \sqrt{x^T \Delta x}, \quad \Delta \in R^{n\times n} \text{ and } \Delta \geq 0 \tag{20}$$

Inside the simplified PR*-sliding sector* $\mathcal{S}$ *(18), the* P*-norm decreases with zero input and the derivative of the Lyapunov function candidate* $L(t)$ *in (11) satisfies that*

$$\begin{aligned}\dot{L}(t) &= \frac{d}{dt}(x^T(t)Px(t)) = s^2(x) - \delta^2(x) - x^T(t)Rx(t)\\ &\leq -x^T(t)Rx(t), \quad \forall x(t) \in \mathcal{S}\end{aligned}$$

where the matrices P *and* R *are defined in Definition 2.1.*

Remark 2. 1 *The* PR*-sliding sector designed in Theorem 2.1 is a simplified* PR*-sliding sector with* $S^T S = P_1$ *and* $\Delta = P_2$ *if* $n_1 = 1$ *and* $n_2 \geq 1$.

Theorem 2. 2 *[15]For any controllable plant (9), the PR-sliding sector defined in (14) can be rewritten as*

$$\mathcal{S} = \{x|\ |s(x)| \leq \delta(x)\ \} \tag{21}$$

which is determined by linear function $s(x)$ and quadratic function $\delta^2(x)$

$$s(x) = Sx(t), \quad S = B^T P \tag{22}$$

$$\delta(x) = \sqrt{x^T(t)\Delta x(t)}, \Delta \in R^{n\times n} and \Delta \geq 0 \tag{23}$$

where P is the solution of the following Riccati equation:

$$PA + A^T P - PBB^T P = -Q, \tag{24}$$

$R \in R^{n\times n}$ is a symmetric positive semi-definite matrix which is chosen so that $\Delta = Q - R \geq 0$, $Q \in R^{n\times n}$ is a symmetric positive definite matrix.

In conclusion of the section, steps to design a PR-sliding sector (21) using Riccati equation are given as follows.

1. Choose a $n \times n$ positive definition symmetric matrix as Q,
2. Solve the Riccati equation (24) for the positive definition symmetric solution P,
3. Choose a positive constant r ($0 < r < 1$) and let $\Delta = rQ$ and $R = (1-r)Q$.

3 VSC for Continuous-time Systems

The VS controller with PR-sliding sector for continuous-time systems is designed such that

1. the system state is controlled to move from the outside toward the inside of the PR-sliding sector if the initial state is outside the PR-sliding sector,
2. the P-norm decreases in the state space.

To design a chattering free VS controller based on the PR-sliding sector, an inner sector $\mathcal{S}_i$ and an outer sector $\mathcal{S}_o$ are defined as subsets of the PR-sliding sector (21), i.e.

$$\mathcal{S}_i = \{x|\ |s(x)| \leq \alpha\delta(x)\ \} \tag{25}$$

$$\mathcal{S}_o = \{x|\ \alpha\delta(x) < |s(x)| \leq \delta(x)\ \} \tag{26}$$

where positive constant α satisfies that $0 < \alpha < 1$. It is obvious that $\mathcal{S} = \mathcal{S}_i \bigcup \mathcal{S}_o$ and $\mathcal{S}_i \bigcap \mathcal{S}_o = \Phi$ where Φ is an empty set in R^n.

With the inner sector $\mathcal{S}_i$ (25) and the outer sector $\mathcal{S}_o$ (26) of the PR-sliding sector (21), the VS control law designed in the paper will be active to move the system state from the outside of the PR-sliding sector to the inside of the inner sector, then will become zero as soon as the state enters the inner sector and will not be active until the system state moves out of the PR-sliding sector.

Theorem 3.1 *[15] Corresponding to the inner sector $\mathcal{S}_i$ (25) and the outer sector $\mathcal{S}_o$ (26) of the PR-sliding sector $\mathcal{S}$ (21) with $\Delta = rQ$ and $R = (1-r)Q$ for some positive constants r $(0 < r < 1)$ and α $(0 < \alpha < 1)$, the VS control law*

$$u(t) = -\sigma(x(t), \delta(x))(SB)^{-1}(SAx + ks(x)) \tag{27}$$

ensures the movement of the state from the outside of the PR-sliding sector to the inside of the inner sector and the decreasing of the P-norm at any state if the positive constant k is large enough so that

$$k > \max\{\frac{SB}{2}, k_0\} \tag{28}$$

where the positive constant k_0 satisfies the following quadratic inequality.

$$2k_0 r\alpha Q + S^T SA + A^T S^T S > 0.$$

$\sigma(s(x), \delta(x))$ is a hysteresis dead-zone function on $s(x)$ and $\delta(x)$ and is defined as

$$\sigma(x) = \begin{cases} 0 & x \in \mathcal{S}_i \\ unchange & x \in \mathcal{S}_o \\ 1 & x \bar{\in} \mathcal{S}. \end{cases} \tag{29}$$

SB is equal to $B^T PB$ which is nonsingular according to Theorem 2.2. Therefore the control law (27) results in a quadratic stable VS controller.

4 VSC for Uncertainty Continuous-time Systems

Consider a plant with parameter uncertainty and external disturbance:

$$\dot{x}(t) = Ax(t) + Bh(x,t) + Bd(t) + Bu(t). \tag{30}$$

where (A, B) is controllable pair, $d(t)$ is the external disturbance, $h(x,t)$ is the parameter uncertainty. It is assumed that

$$h(x,t) = f(x,t)\theta, \quad \text{and} \quad |h(x,t)| \leq H(x,t)$$
$$|d(t)| \leq \phi(t).$$

$f(x,t)$, $H(x,t)$, and $\phi(t)$ are known functions, and θ is an unknown parameter matrix of appropriate dimension.

In this section a robust VS control law to stabilize the uncertainty plant (30) with nominal system (9) is proposed, the adaptive control algorithm is used to estimate the unknown parameter matrix θ and an adaptive VS controller is designed.

4.1 Robust VSC

The robust VS control law is designed to move the system state into the inner sector of the PR-sliding sector (21) and to let the P-norm decrease in the state space even if there exist parameter uncertainty and external disturbance.

Theorem 4. 1 *The robust VS control law*

$$u(t) = \sigma(x(t), \delta(x))u_o(t) + (1 - \sigma(x(t), \delta(x)))u_i(t) \tag{31}$$

quadratically stabilizes the uncertainty plant (30) if the pair $(A,\ B)$ in (9) is controllable, where the hysteresis dead-zone function $\sigma(s(x), \delta(x))$ on $s(x)$ and $\delta(x)$ is defined in (29), the PR-sliding sector $\mathcal{S}$ (21) is designed in Theorem 2.2 with $\Delta = rQ$ and $R = (1-r)Q$ for some positive definite symmetric matrix $Q \in R^{n\times n}$ and some positive constant r $(0 < r < 1)$, the inner sector $\mathcal{S}_i$ and the outer sector $\mathcal{S}_i$ of the PR-sliding sector (21) are defined in (25) and (26), respectively, and $u_i(t)$ and $u_o(t)$ are defined as

$$u_i(t) = -(\phi(t) + k_1(x,t))sgn(s(x)) \tag{32}$$
$$u_o(t) = -(SB)^{-1}(SAx(t) + ks(x)) - (\phi(t) + k_2(x,t))sgn(s(x)) \tag{33}$$

k, $k_1(x,t)$ and $k_2(x,t)$ satisfy the following inequalities, respectively.

$$k > \max\{\frac{SB}{2}, k_0\}, \tag{34}$$
$$k_1(x,t) > H(x,t) \quad \forall x \in \mathcal{S} \tag{35}$$
$$k_2(x,t) > H(x,t) + (SB)^{-1}|SAx| \quad \forall x \bar{\in} \mathcal{S}_i \tag{36}$$

and the positive constant k_0 enables the following inequality hold.

$$2k_0 r\alpha Q + (S^TSA + A^TS^TS) > 0.$$

Proof. Because it is assumed that the pair $(A,\ B)$ is controllable, the PR-sliding sector can be designed for the nominal system (9) by using Riccati equation as shown in (21).

Consider the square of the P-norm as a Lyapunov function candidate, i.e.,

$$L(t) = x^T(t)Px(t) > 0, \quad \forall x \in R^n, x \neq 0. \tag{37}$$

It is assumed that the initial state is outside the PR-sliding sector. In this case, the control law $u(t)$ should be $u_o(t)$ (33), i.e.,

$$u(t) = u_o(t) = -(SB)^{-1}(SAx(t) + ks(x)) - (\phi(t) + k_2(x,t))sgn(s(x))$$

which will be active before the system state is moved into the inner sector. With the control law $u_o(t)$ (33), the derivate of $s^2(x)$ satisfies that

$$\begin{aligned}\frac{d}{dt}s^2(x) &= 2s(x)\dot{s}(x) \\ &= 2s(x)(SAx(t) + SBh(x,t) + SBd(t) + SBu_o(t)) \\ &= -2ks^2(x) + B^TPBs(x)(h(x,t) + d(t) + (\phi(t) + k_2(x,t))sgn(s(x))) \\ &\leq -2ks^2(x) < 0\end{aligned}$$

which means that the absolute value of the linear function $s(x)$ will decrease so that the system state moves inside the inner sector in a finite time if the decreasing rate of $\delta(x)$ is slower than the one of $|s(x)|$ for some large enough k.

While the system state is being moved from the outside of the PR-sliding sector to the inside of the inner sector by the VS control law (33), the P-norm is decreasing because according to the definition of the PR-sliding sector (21) with $\Delta = rQ$ and $R = (1-r)Q$, it follows from $|s(x)| > \alpha\delta(x)$ and the inequality (34) that

$$\begin{aligned}
\dot{L}(t) &= (Ax(t) + Bu(t))^T Px(t) + x^T(t)P(Ax(t) + Bf(x,t) + Bd(t) + Bu(t)) \\
&= x^T(t)(A^T P + PA)x(t) + 2x^T(t)PB(f(x,t) + d(t) + u(t)) \\
&= s^2(x) - \delta^2(x) - x^T(t)Rx(t) + 2s(x)(f(x,t) + d(t) + u(t)) \\
&< -(2(SB)^{-1}k - 1)s^2(x) - 2s(x)(SB)^{-1}SAx(t) - \delta^2(x) - x^T(t)Rx(t) \\
&\le -(2(SB)^{-1}k - 1)\alpha\delta^2(x) - 2x^T(t)S^T(SB)^{-1}SAx(t) - \delta^2(x) - x^T(t)Rx(t) \\
&< -2(SB)^{-1}k\alpha\delta^2(x) - 2x^T(t)S^T(SB)^{-1}SAx(t) - x^T(t)Rx(t) \\
&= -(SB)^{-1}x^T(t)(2kr\alpha Q + S^T SA + A^T S^T S)x(t) - x^T(t)Rx(t) \\
&< -x^T(t)Rx(t)
\end{aligned}$$

where $SB = B^T PB > 0$.

After the state is moved into the inner sector and before goes out of the PR-sliding sector, the control input $u(t)$ should be $u_i(t)$ (32), i.e.,

$$u(t) = u_i(t) = -(\phi(t) + k_1(x,t))sgn(s(x))$$

In this case, $|s(x)| \le \delta(x)$ and the derivate of the Lyapunov function $L(t)$ is

$$\begin{aligned}
\dot{L}(t) &= s^2(x) - \delta^2(x) - x^T(t)Rx(t) + 2s(x)(f(x,t) + d(t) + u_i(t)) \\
&\le -x^T(t)Rx(t).
\end{aligned}$$

If the system state goes out of the PR-sliding sector with $u_i(t)$ (32), the control input $u_o(t)$ (33) will let it back to the inside of the inner sector again while the P-norm keeps decreasing.

Thus the system state will be moved from the outside of the PR-sliding sector to the inside of the inner sector and the Lyapunov function $L(t)$ (37) keeps decreasing in the state space with the VS control law (27), i.e.

$$\dot{L}(t) \le -x^T(t)Rx(t) < 0, \quad \forall x \in R^n$$

which means that the VS control law (27) quadratically stabilizes the plant (9).

Remark 4. 1 *The control law $u_i(t)$ inside the PR-sliding sector includes the sign function* sgn$(s(x))$ *, which may result in the chattering for some systems. If the external disturbance is not taken into consideration, i.e., letting $d(t) = 0$ in (30), then $u_i(t)$ does not included $\phi(t)$*sgn$(s(x))$*. If the unknown parameter matrix θ is estimated by some adaptive rule, then a chattering free adaptive VS controller is available for the uncertainty plant (30) without external disturbance. Such adaptive VS control is proposed in the next subsection.*

4.2 Adaptive VSC

In this subsection, the adaptive VS control law to stabilize the uncertainty plant (30) is designed to move the system state into the PR-sliding sector and to let the P-norm decrease in the state space while the adaptive control is used to estimate the unknown parameter θ.

Theorem 4. 2 *If the VS control law $u_i(t)$ (32) and $u_o(t)$ (33) designed in Theorem 4.1 are replaced by*

$$u_i(t) = -f(x,t)\tilde{\theta} - \phi(t)\mathrm{sgn}(s(x)) \tag{38}$$

$$u_o(t) = -f(x,t)\tilde{\theta} - \phi(t)\mathrm{sgn}(s(x)) - (SB)^{-1}(SAx + ks(x)) \tag{39}$$

respectively, where $\tilde{\theta}$ is the estimated matrix of the unknown parameter matrix θ according to

$$\dot{\tilde{\theta}} = \Gamma f^T(x,t)s(x) \tag{40}$$

and Γ is a positive definite symmetric matrix of appropriate dimensions, then the VS control law (31) designed in Theorem 4.1,i.e.

$$u(t) = \sigma(x(t),\delta(x))u_o(t) + (1 - \sigma(x(t),\delta(x)))u_i(t) \tag{41}$$

quadratically stabilizes the uncertainty plant (30).

Proof. Consider two Lyapunov functions $L_1(t)$ and $L_2(t)$, i.e.,

$$L_1(x) = \frac{1}{2}s^2(x) + \frac{1}{2}SB\Delta\theta^T\Gamma_1^{-1}\Delta\theta > 0, \quad \forall x \in R^n, x \neq 0 \tag{42}$$

$$L_2(x) = \frac{1}{2}x^T Px + \frac{1}{2}\Delta\theta^T\Gamma_2^{-1}\Delta\theta > 0, \quad \forall x \in R^n, x \neq 0 \tag{43}$$

where

$$SB = B^T PB > 0.$$
$$\Delta\theta = \tilde{\theta} - \theta$$

With the control law (41) and adaptive rule (40), it can be shown that

1. when $|s(x)| > \alpha\delta(x)$ and the control law $u_o(t)$ (39) is active, the derivatives of $L_1(x)$ and $L_2(t)$ satisfy the following inequalities, respectively.

$$\begin{aligned}
\dot{L}_1(x) &= s(x)\dot{s}(x) + SB\Delta\theta^T\Gamma_1^{-1}\dot{\Delta\theta} \\
&= s(x)(SAx + SBh(x,t) + SBd(t) + SBu_o(t)) + SB\Delta\theta^T\Gamma_1^{-1}\dot{\tilde{\theta}} \\
&= -ks^2(x) - SB(\phi(t) + \mathrm{sgn}(s(x)d(t))|s(x)| \\
&\leq -ks^2(x), \quad \forall x\bar{\in}S_i
\end{aligned}$$

$$\begin{aligned}
\dot{L}_2(t) &= x^T P\dot{x} + \Delta\theta^T\Gamma_2^{-1}\dot{\Delta\theta} \\
&= \frac{1}{2}x^T(A^T P + PA)x + x^T PB(h(x,t) + d(t) + u_o(t)) + \Delta\theta^T\Gamma_2^{-1}\dot{\tilde{\theta}} \\
&= \frac{1}{2}(s^2(x) - \delta^2(x)) - \frac{1}{2}x^T Rx - s(x)(SB)^{-1}(SAx + ks(x)) \\
&\quad -|s(x)|(\phi(t) + d(t)\mathrm{sgn}(s(x)) \\
&\leq -\frac{1}{2}x^T Rx, \quad x\bar{\in}S_i
\end{aligned}$$

2. when $|s(x)| \leq \delta(x)$ and the control law $u_i(t)$ (38) is active,the derivative of $L_2(t)$ satisfies

$$\begin{aligned}
\dot{L}_2(t) &= x^T P\dot{x} + \Delta\theta^T \Gamma_2^{-1} \dot{\Delta\theta} \\
&= \frac{1}{2} x^T (A^T P + PA)x + x^T PB(h(x,t) + d(t) + u_i(t)) + \Delta\theta^T \Gamma_2^{-1} \dot{\tilde{\theta}} \\
&= \frac{1}{2}(s^2(x) - \delta^2(x)) - \frac{1}{2} x^T Rx - |s(x)|(\phi(t) + \dot{d}(t)\mathrm{sgn}(s(x)) \\
&\leq -\frac{1}{2} x^T Rx, \quad \forall x \in \mathcal{S}
\end{aligned}$$

Therefore the Lyapunov function $L_1(t)$ and $L_2(t)$ decrease in the outside of the inner sector and in the state space, respectively. The decreasing of $L_1(t)$ means $|s(x)|$ will decrease and/or the estimation error $\Delta\theta$ will vanish until the system state moves into the inner sector. The decreasing of $L_2(t)$ means that the proposed VS control law (41) quadratically stabilizes the uncertainty plant (30) as

$$\dot{L}_2(t) < -x^T(t)Rx(t) < 0 \quad \forall x(t) \in R^n \text{ and } x(t) \neq 0$$

Remark 4. 2 *The above adaptive VS controller is chattering free if the external disturbance is not taken into consideration and $\phi(t)$ in the control law $u_i(t)$ and $u_o(t)$ is set to be zero.*

5 VSC for Discrete-time Input-Output System

This section considers a single input and single output system. The following discrete relation represents the controlled plant with input u_k, disturbance w_k and output y_k,

$$A(q^{-1})y_k = q^{-d}B(q^{-1})u_k + D(q^{-1})w_k, \tag{44}$$

where $A(q^{-1})$ and $B(q^{-1})$ have no common factors, q denotes the time shift operator defined by

$$q^{-t}y_k = y_{k-t}$$

and q^{-d} is the pure time delay of the system, $d (d \geq 1)$ is an integer. $A(q^{-1})$ and $B(q^{-1})$ are assumed known and representing

$$\begin{aligned}
A(q^{-1}) &= 1 + a_1 q^{-1} + a_2 q^{-2} + \cdots + a_n q^{-n}, \\
B(q^{-1}) &= b_0 + b_1 q^{-1} + b_2 q^{-2} + \cdots + b_m q^{-m} \quad (b_0 \neq 0),
\end{aligned}$$

The objective of the control is that the output y_k tracks the reference r_k in the presence of the disturbance. The polynomial models of the reference and the disturbance are assumed to be

$$\Psi_r(q^{-1})r_k = 0, \quad \Psi_w(q^{-1})w_k = 0.$$

Let $\Psi(q^{-1})$ be the least common multiple of $\Psi_r(q^{-1})$ and $\Psi_w(q^{-1})$ then

$$\Psi(q^{-1})r_k = 0, \quad \Psi(q^{-1})w_k = 0,$$

where $\Psi(q^{-1})$ is coprime with $A(q^{-1}),\ B(q^{-1})$. Subtract $A(q^{-1})r_k$ from (44) and multiplying $\Psi(q^{-1})$ to both sides, the following relation can be obtained

$$A(q^{-1})\Psi(q^{-1})e_k = q^{-d}B(q^{-1})\Psi(q^{-1})u_k, \tag{45}$$

where the error will be defined as

$$e_k = y_k - r_k.$$

5.1 Servo Control of Generalized Minimum Variance Control

The objective of the control in this subsection is to minimize the generalized variance of the controlled variables s_{k+d}, that is, in the deterministic case, to give the control input satisfying

$$s_{k+d} = C(q^{-1})e_{k+d} + Q(q^{-1})\Psi(q^{-1})u_k = 0, \tag{46}$$

where real polynomials $C(q^{-1})$ and $Q(q^{-1})$ are determined so that the error vanishes if the above (46) is satisfied.

At first, the generalized minimum variance control without using VSS will be discussed. The equation (46) is rewritten as

$$\begin{aligned} s_{k+d} &= (E(q^{-1})B(q^{-1}) + Q(q^{-1}))\Psi(q^{-1})u_k + F(q^{-1})e_k \\ &= G(q^{-1})\Psi(q^{-1})u_k + F(q^{-1})e_k, \end{aligned} \tag{47}$$

where $E(q^{-1})$ and $F(q^{-1})$ are polynomials determined to satisfy

$$C(q^{-1}) = A(q^{-1})\Psi(q^{-1})E(q^{-1}) + q^{-d}F(q^{-1}) \tag{48}$$

and $G(q^{-1})$ is defined as

$$G(q^{-1}) = \{E(q^{-1})B(q^{-1}) + Q(q^{-1})\}. \tag{49}$$

The control input to make $s_{k+d} = 0$ is, therefore, given by

$$u_k = -[G(q^{-1})\Psi(q^{-1})]^{-1}[F(q^{-1})e_k]. \tag{50}$$

This subsection considers the use of variable structure control in addition to the above conventional generalized minimum variance control so that the output satisfies $s_k = 0$. $C(q^{-1})$ and $Q(q^{-1})$ should be chosen to satisfy the following lemma.

Lemma 1 . *[10] The necessary and sufficient condition that the output with zero reference making $s_{k+d} = 0$ stable is that all zeros of*

$$A(q^{-1})Q(q^{-1})\Psi(q^{-1}) + B(q^{-1})C(q^{-1}) = 0 \tag{51}$$

are inside the unit disk.

Instead of (50), the following input is considered to be used,

$$u_k = -[G(q^{-1})\Psi(q^{-1})]^{-1}[F(q^{-1})e_k - \beta s_k - v_k], \tag{52}$$

where

$$0 < \beta \le 1.$$

Substituting (52) into (44) yields

$$s_{k+d} = v_k + \beta s_k. \tag{53}$$

The auxiliary control input v_k is chosen as the state feedback with the variable coefficients.

$$v_k = h_0 e_k + h_1 e_{k-1} + \cdots + h_{n-1} e_{k-n+1}. \tag{54}$$

The control input with $v_k = 0$ is called the β equivalent control where $s_{k+d} = \beta s_k$. The control law given as follows gives a stable system.

Theorem 5. 1 *[10] For the plant (44), if the coefficients of the feedback control law are chosen*

$$h_i = \begin{cases} h & s_k e_{k-i} < -\delta_i \\ 0 & | s_k e_{k-i} | \le \delta_i \\ -h & s_k e_{k-i} > \delta_i \end{cases} \quad (i = 0, 1, \cdots, n-1), \tag{55}$$

then the control system becomes stable, where

$$\delta_i = \eta \sum_{j=0}^{n-1} | e_{k-i} || e_{k-j} | h \tag{56}$$

and

$$\eta \ge \frac{\alpha}{2(\alpha\beta - \alpha + 1)} \tag{57}$$

and it is assumed that $\alpha \ge 1$.

5.2 Self-Tuning Servo Control based on VSS

Now it is assumed that the plant (44) is with parameter uncertainty. The given plant is considered to have the known delay d in the plant. Parameters $\{a_i, b_i\}$ of $A(q^{-1})$ and $B(q^{-1})$ are, however, assumed unknown except b_0 ($\neq 0$). In this subsection, the control algorithm is determined based on generalized minimum variance control. Different control laws and parameter identification methods are employed inside and outside the sector. The sector is defined by

$$S_k = \{s_k | \; | s_k | \le \left(\frac{(1-\beta)p + \sqrt{(1-\beta)^2 p^2 + 2p^2\gamma}}{\gamma} \right) (|\phi_k|)\}, \tag{58}$$

where

$$\gamma = \frac{2}{\alpha}(1-\beta) - (1-\beta)^2,$$

$$|\phi_k| = \sum_{j=0}^{n-1} |e_{k-j}| + \sum_{j=1}^{m+d-1} |\Psi(q^{-1})u_{k-j}|$$

and p will be defined later in (66).

In the outside of the sector, control and parameter identification are done simultaneously based on the Lyapunov function. The polynomial s_k is defined by (46). When parameters $A(q^{-1})$, $B(q^{-1})$ are known, s_k is given by (47).

When $A(q^{-1})$ and $B(q^{-1})$ are unknown, $G(q^{-1})$ and $F(q^{-1})$ can not be obtained exactly. In this case, the control input u_k is determined by using the estimate of $G(q^{-1})$ and $F(q^{-1})$, denoted by $\hat{G}(q^{-1})$ and $\hat{F}(q^{-1})$, as follows:

For the outside of the sector:

$$\begin{aligned} u_k &= -\{\hat{G}_k(q^{-1})\Psi(q^{-1})\}^{-1}\left[\hat{F}_k(q^{-1})e_k - \beta s_k \right. \\ &\quad \left. - \sum_{j=0}^{n-1} h_j e_{k-j} - \sum_{j=1}^{m+d-1} w_j \Psi(q^{-1})u_{k-j}\right], \end{aligned} \tag{59}$$

where $\{h_j\}$, and $\{w_j\}$ are nonlinear bang-bang type functions depending on the state outside the sector and take values out of h, 0, $-h$. Since b_0 is assumed known, g_0 can be given. Let the estimate of $G(q^{-1})$ and $F(q^{-1})$ be

$$\hat{\theta}_k = \hat{\theta}_{k-d} + \Gamma^{-1}\phi_k s_k \quad (\Gamma > 0), \tag{60}$$

θ, ϕ_k are defined as

$$\begin{aligned} \theta &= [f_0, f_1, \cdots, f_{n-1}, g_1, \cdots, g_{m+d-1}]^T \\ \phi_k &= \left[e_k, e_{k-1}, \cdots, \Psi(q^{-1})u_{k-1}, \cdots, \Psi(q^{-1})u_{k-m-d+1}\right]^T \end{aligned}$$

For the inside of the sector;

$$u_k = -\{\hat{G}_k(q^{-1})\Psi(q^{-1})\}^{-1}\left[\hat{F}_k(q^{-1})e_k\right]. \tag{61}$$

The following main theorem establishes the stability of the closed loop system.

Theorem 5. 2 *[10] The control system, which is employed for the outside of the sector S_k where the control and the simultaneous parameter estimation are determined by (59) and (60) respectively, brings either the system into the sector or the error to zero. The coefficients of (59), $\{h_i\}$ and $\{w_i\}$ are given by the following relations:*

$$h_i = \begin{cases} h & s_k e_{k-i} < -\delta_i \\ 0 & |s_k e_{k-i}| \le \delta_i \\ -h & s_k e_{k-i} > \delta_i \end{cases} \tag{62}$$
$$(i = 0, 1, \cdots, n-1)$$

$$w_i = \begin{cases} h & s_k\Psi(q^{-1})u_{k-i} < -\sigma_i \\ 0 & |s_k\Psi(q^{-1})u_{k-i}| \le \sigma_i \\ -h & s_k\Psi(q^{-1})u_{k-i} > \sigma_i \end{cases} \tag{63}$$
$$(i = 1, 2, \cdots, m + d - 1)$$

where

$$\delta_i = \eta_a|e_{k-i}| \left(\sum_{j=0}^{n-1}|e_{k-j}| + \sum_{j=1}^{m+d-1}|\Psi(q^{-1})u_{k-j}| + \sum_{j=0}^{n-1}|r_{k-j}|\right) h$$
$$(i = 0, 1, \cdots, n - 1), \tag{64}$$
$$\sigma_i = \eta_a|\Psi(q^{-1})u_{k-i}| \left(\sum_{j=0}^{n-1}|e_{k-j}| + \sum_{j=1}^{m+d-1}|\Psi(q^{-1})u_{k-j}| + \sum_{j=0}^{n-1}|r_{k-j}|\right) h$$
$$(i = 0, 1, \cdots, m + d - 1), \tag{65}$$

where

$$\eta_a \ge \frac{\alpha}{1 - \alpha + \alpha\beta},$$

p is the upper bound of the uncertainty of the parameters defined by

$$\max_i|\theta_i - \hat{\theta}_{ki}| < p, \tag{66}$$

where $\hat{\theta}_{ki}$ denotes the i-th element of $\hat{\theta}_k$ and $\alpha > 1$. The following Lyapunov function

$$V_k = \frac{1}{2}s_k^2 + \frac{1}{2}\tilde{\theta}_{k-d}^T\Gamma\tilde{\theta}_{k-d} \tag{67}$$

decreases outside the sector.

By using the control (61), inside the sector, the stability of the closed loop system is assured.

6 Example

6.1 VSC for Continuous-time System

Consider a second order continuous-time plant represented by:

$$\dot{x} = \begin{bmatrix} 0 & 1 \\ 0 & 0 \end{bmatrix} x + \begin{bmatrix} 0 \\ 1 \end{bmatrix} (x^T \begin{bmatrix} 0.6 \\ -0.1 \end{bmatrix} + 0.1 \sin 10t + u).$$

where the parameter uncertainty and external disturbance are described by

$$f(x,t) = x^T$$
$$\theta = \begin{bmatrix} 0.5 \\ 1.5 \end{bmatrix}$$
$$d(t) = 0.1 \sin 10t.$$

Choose Q to be the 2×2 identity matrix I_2, then the positive definite symmetric solution P of the Riccati equation (24) is

$$P = \begin{bmatrix} 1.73205081 & 1.0 \\ 1.0 & 1.73205081 \end{bmatrix}$$

the PR-sliding sector (21) has a form as shown in Figure 3 with the parameters of

$$C = \begin{bmatrix} 1.0 & 1.73205081 \end{bmatrix}$$
$$\Delta = R = 0.5 I_2, \quad \text{i.e., } r = 0.5$$

The inner sector (25) and the outer sector (26) shown in Figure (3) are designed with $\alpha = 0.1$.

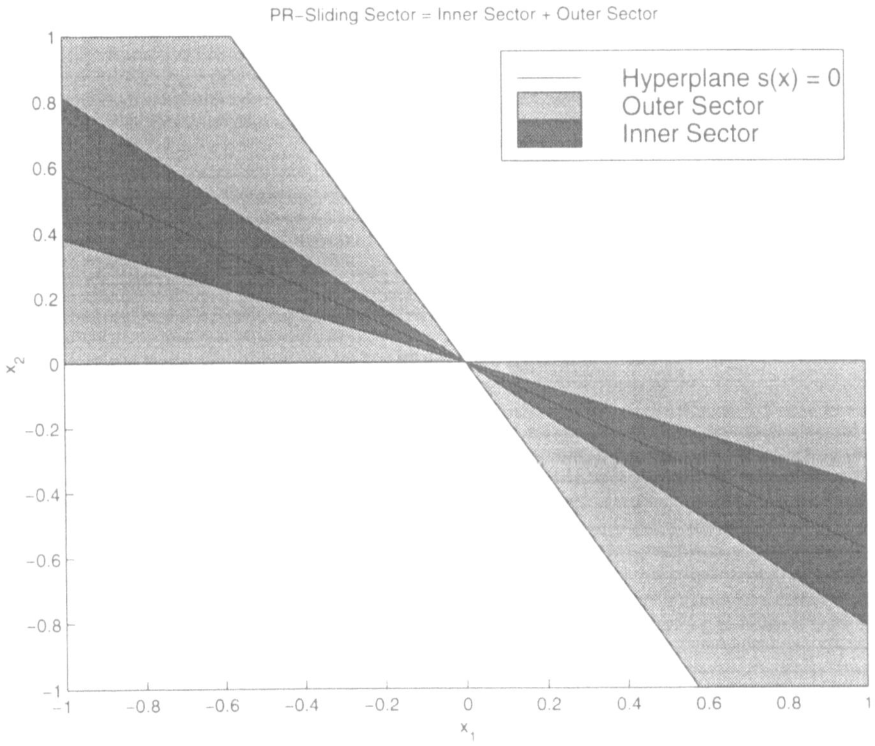

Fig. 3. PR-Sliding Sector

The simulation results with the adaptive VS control law (41) are shown in Figure 4, 5, and 6 where the coefficient k of the nonlinear VS control law, the positive definite symmetric matrix Γ, the sampling interval τ, the initial state $x(0)$ and the initial estimated matrix $\theta(0)$ are respectively chosen as

$$k = 10, \quad \Gamma = I_2, \quad \tau = 0.01 \text{ Second}$$
$$x(0) = \begin{bmatrix} 1.0 \\ 1.0 \end{bmatrix}, \quad \theta(0) = \begin{bmatrix} 0.0 \\ 0.0 \end{bmatrix}$$

The simulation result shows that the proposed adaptive VS control law (41) quadratically stabilizes the uncertainty plant with good control performance. As shown in Figure 6, the chattering may happen because the time-variance external disturbance exists. If the external disturbance $d(t)$ is a constant, the chattering can be avoided. Figure 7, 8, and 9 show the simulation result for the uncertainty plant with constant external disturbance, i.e.

$$d(t) = 0.1$$

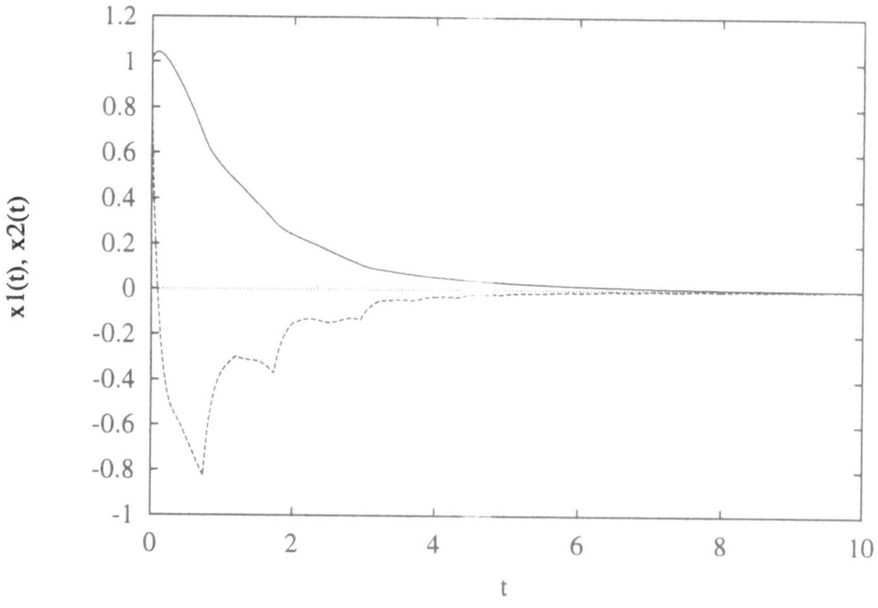

Fig. 4. Response with Adaptive VS Controller ($d(t) = 0.1 \sin 10t$)

6.2 VSC for Discrete-time Input-Output System

In this simulations, the objective is to let a second order plant with external disturbance to track an unit-step reference. The plant is represented by

$$A(q^{-1})y_k = q^{-d}B(q^{-1})u_k + D(q^{-1})w_k,$$

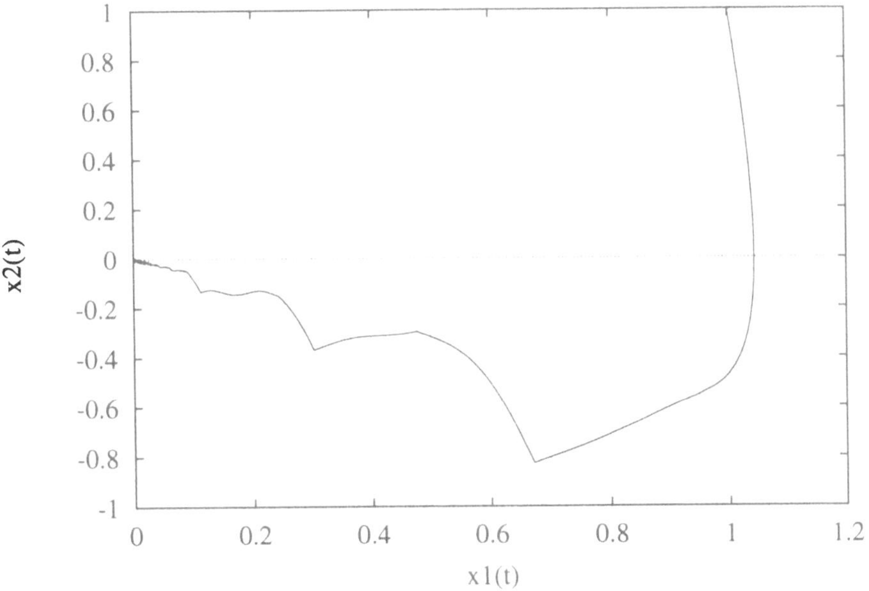

Fig. 5. Phase Plane with Adaptive VS Controller ($d(t) = 0.1 \sin 10t$)

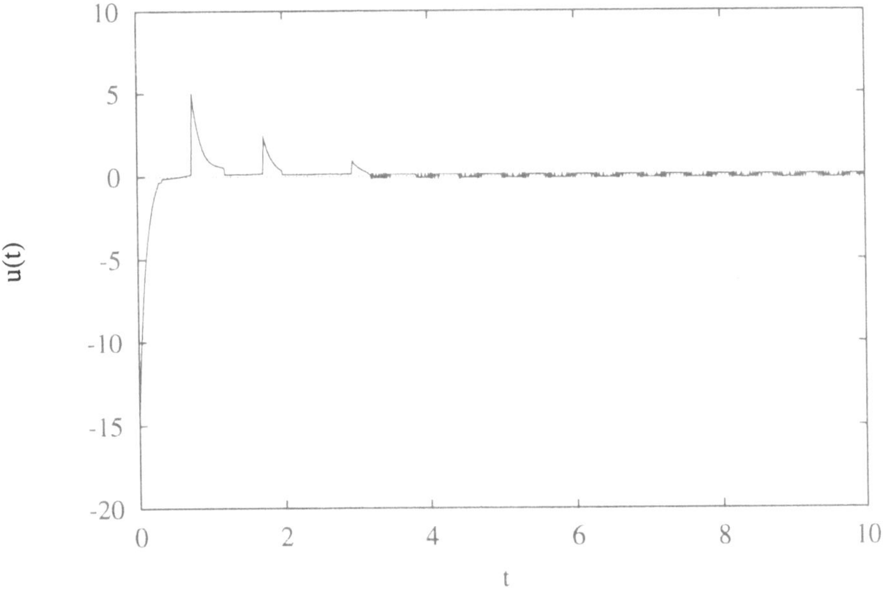

Fig. 6. Input with Adaptive VS Controller ($d(t) = 0.1 \sin 10t$)

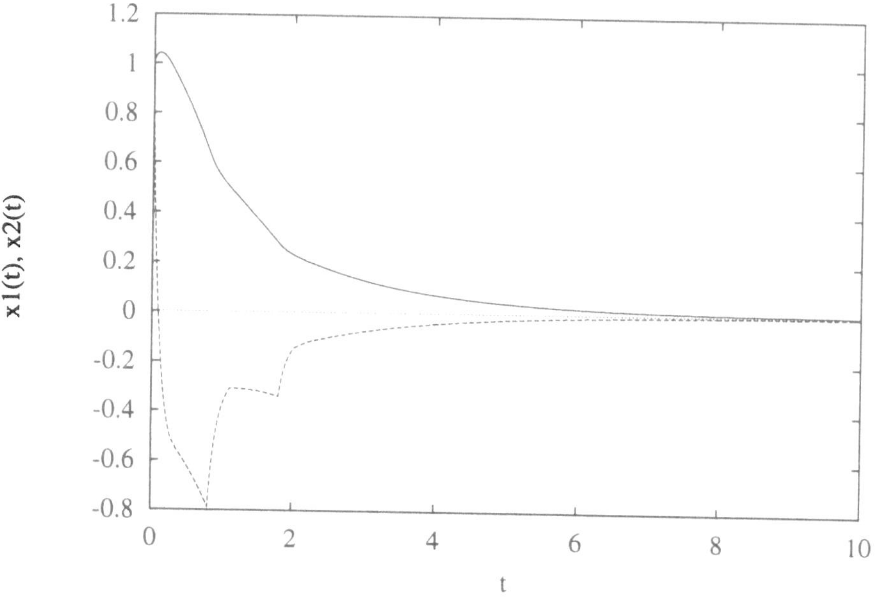

Fig. 7. Response with Adaptive VS Controller ($d(t) = 0.1$)

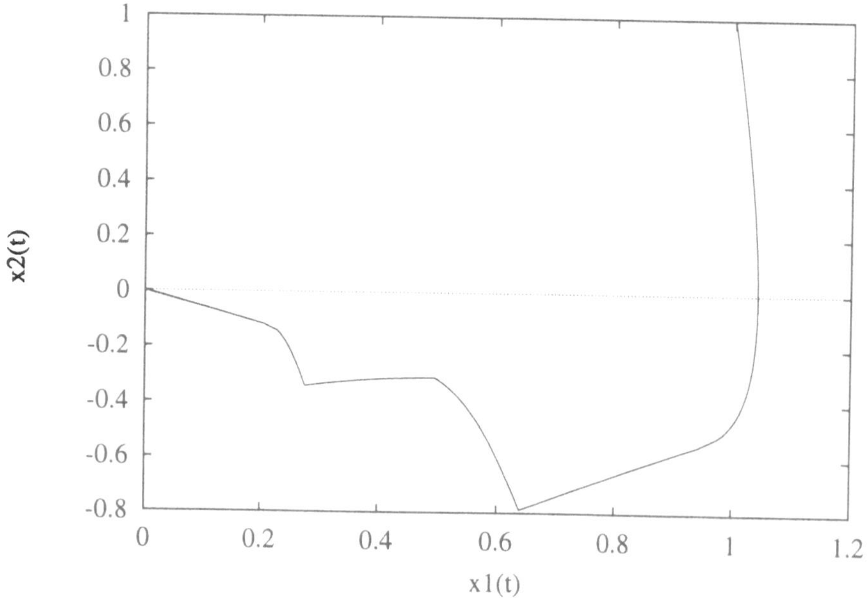

Fig. 8. Phase Plane with Adaptive VS Controller ($d(t) = 0.1$)

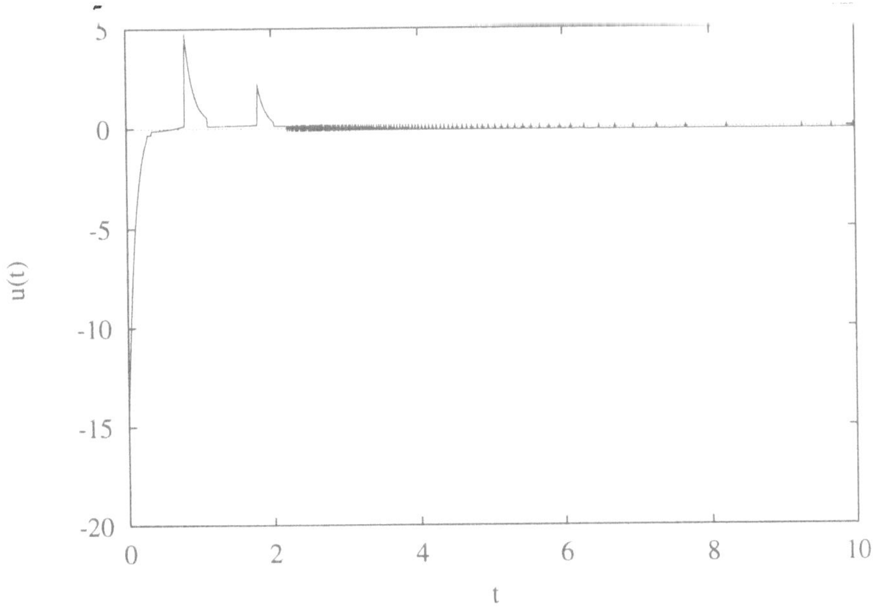

Fig. 9. Input with Adaptive VS Controller ($d(t) = 0.1$)

where

$$A(q^{-1}) = 1 - 1.3q^{-1} + 0.42q^{-2},$$
$$B(q^{-1}) = 1 - 0.8q^{-1},$$
$$D(q^{-1}) = 1,$$
$$d = 1.$$

Assuming that

$$\Psi(q^{-1}) = \Psi_r(q^{-1}) = \Psi_w(q^{-1}) = 1 - q^{-1},$$
$$r_k = 1,$$
$$w_k = 0.2,$$

we choose the polynomials $C(q^{-1})$ and $Q(q^{-1})$ as

$$C(q^{-1}) = 1 + q^{-1} + 0.25q^{-2},$$
$$Q(q^{-1}) = 0.5$$

Considering the parameter uncertainty, the estimated parameters are assumed to be

$$\hat{A}(q^{-1}) = 1 - q^{-1} + 0.5q^{-2},$$
$$\hat{B}(q^{-1}) = 1 - 0.7q^{-1}.$$

The polynomials $E(q^{-1})$ and $F(q^{-1})$ are chosen as

$$E(q^{-1}) = 1,$$
$$F(q^{-1}) = 3 - 1.25q^{-1} + 0.5q^{-2}$$

which satisfy (48).

Then the responses of the generalized minimum variance control (50) and the discrete-type VSS control (52) with $h = 0.2$, $\beta = 0.85$, and $\alpha = 1.0$ are shown in Figure 10 and Figure 11, respectively. The responses of the self-tuning VS control (59) with $h = 0.2$, $p = 0.15$, $\beta = 0.85$, $\alpha = 1.0$, and $\Gamma = 100 * I$ is shown in Figure 12.

The simulation results show that the discrete-time VS control and the discrete-time self-tuning VS control are with better responses than the generalized minimum variance control.

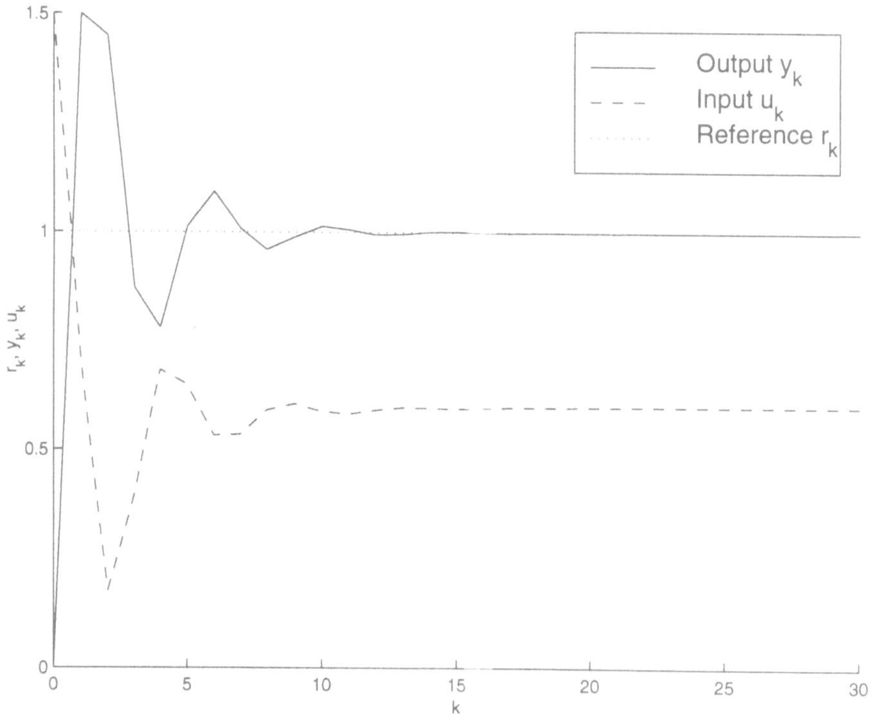

Fig. 10. Response with Minimum Variance Control

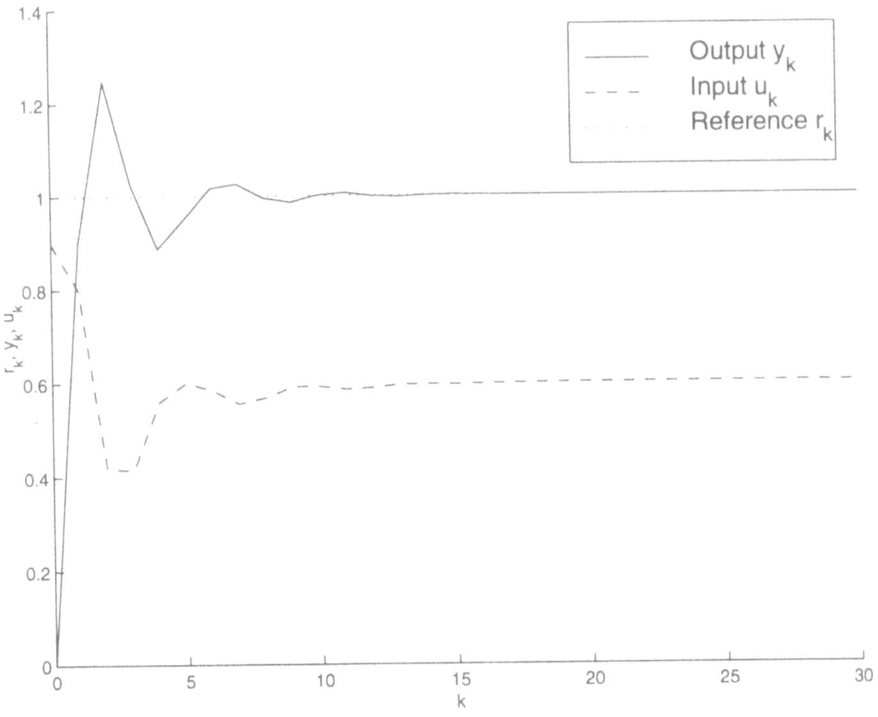

Fig. 11. Response with Discrete-time VS Control

7 Conclusion

This paper studied the sliding sectors for the VS control. With the PR-sliding sector presented in this paper, VS controller for continuous-time systems is designed so that the state is moved from the outside to the inside of the PR-sliding sector and the P-norm decreases in the state space. Thus the resultant VS control system is quadratically stable. The parameter uncertainty and external disturbance are also taken into consideration. An robust VS controller and an adaptive VS controller are proposed. As another kind of sliding sector, the sliding sector for discrete-time input-output system is designed by minimal variance control method. The corresponding VS control law ensures that the system state moves from the outside to the inside of the sliding sector where the closed loop system is stable.

References

1. E.A. Barbashin and E.I. Geraschenko, "On speeding up sliding modes in automatic control systems (in Russian)," *Differentzialniye Uravneniya*, vol. 1, pp. 25–32, 1965.

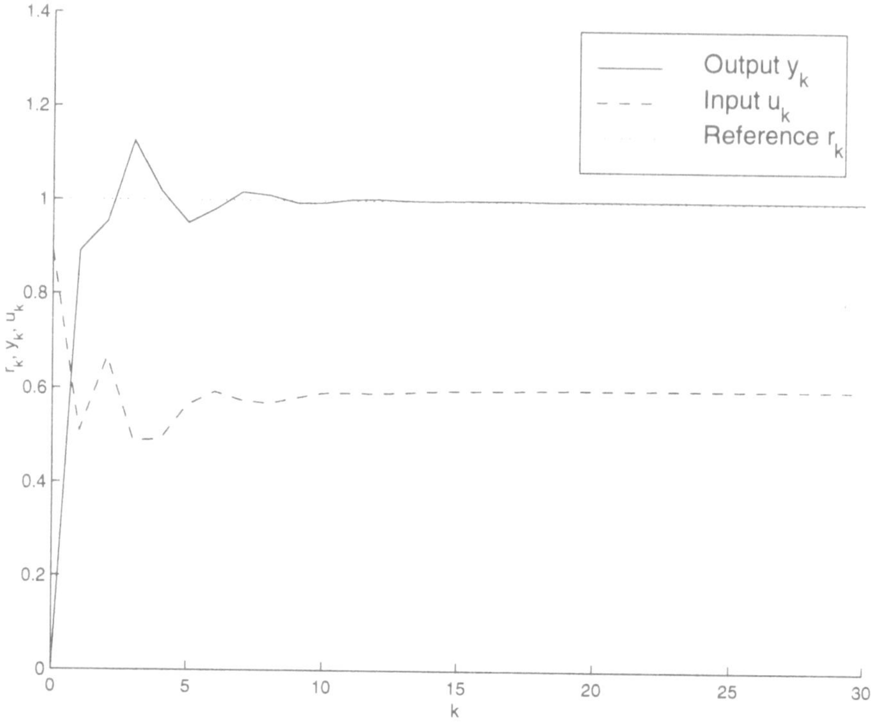

Fig. 12. Response with Discrete-time Self-tuning VS Control

2. U.I. Utkin, *Sliding Modes in Control and Optimization*, Springer-Verlag, 1992.
3. S.V. Emelyanov, "Binary control systems," *International Research Institute for Management Sciences*, vol. Issue 1, 1985.
4. K.J. Åström, P. Hagander, and J. Sternby, "Zeros of sampled systems," *Automatica*, vol. 20, pp. 31–38, 1984.
5. S.V. Drakunov and V.I. Utkin, "On discrete-time sliding mode," in *IFAC Symposium on Nonlinear Control System Design*, 1989, pp. 484–489.
6. K. Furuta, "Sliding mode control of a discrete system," *System & Control Letters*, vol. 14, pp. 145–152, 1990.
7. C.Y. Chan, "Servo-systems with discrete-variable structure control," *System & Control Letters*, vol. 17, pp. 321–325, 1991.
8. X. Yu and P.B. Potts, "A class of discrete variable structure systems," in *Proc. of 30th CDC*, Brighton, UK, 1991, pp. 1367–1372.
9. G. Bartolini, A. Ferrara, and V.I. Utkin, "Design of discrete-time adaptive sliding mode control," in *Proc. of the 31st CDC*, Tucson, Arizona, 1992, pp. 2387–2391.
10. K. Furuta, "VSS type self-tuning control," *IEEE Trans. on Industrial Electronics*, vol. 40, pp. 37–44, 1993.
11. K. Furuta and Y. Pan, "A new approach to design a sliding sector for VSS controller," in *Proc. of the American Control Conference*, Seattle, 1995, pp. 1304–1308.

12. Y. Pan and K. Furuta, "Adaptive VSS controller based on sliding sector," in *IFAC'96 World Congress*, San Francisco, 1996, vol. F, pp. 235–240.
13. K. Furuta and Y. Pan, "Design of discrete-time VSS controller based on sliding sector," in *IFAC'96 World Congress*, San Francisco, 1996, vol. F, pp. 487–492.
14. K. Furuta and Y. Pan, "Variable structure control for sampled-data systems," in *Lecture Notes in Control and Information Sciences: Learning, Control, and Hybrid Systems*, Vol. 241, Springer, 1998
15. K. Furuta and Y. Pan, "Variable structure control with Sliding Sector," *To be published in Automatica*, 1999.
16. M. Hara, K. Furuta, Y. Pan, and T. Hoshino, "Evaluation of discrete-time VSC on inverted pendulum apparatus with additional dynamics," *International Journal of Applied Mathematics and Computer Science*, vol. 7, pp. 101–123, 1998.

On Discrete Variable Structure Control with Switching Sector

Choon Yik Tang[1] and Eduardo A. Misawa[2]

[1] Department of Mechanical Engineering and Applied Mechanics, University of Michigan, Ann Arbor, MI 48109-2125, USA
[2] School of Mechanical and Aerospace Engineering, Oklahoma State University, Stillwater, OK 74078-5016, USA

Abstract. In this chapter, potential instability and performance limitation due to the use of an attractive and non-invariant *switching sector* in a number of discrete variable structure controllers (DVSCs) for linear systems with parametric uncertainties are discussed. Analytical explanation as well as counter examples are given to support the arguments. A state feedback DVSC based on the concept of switching sector and capable of avoiding the potential pitfalls is proposed. It is shown that global uniform asymptotic stability can be achieved despite the non-attractiveness and non-invariance of the switching sector. Conservativeness of the proposed controller is investigated and a numerical example is presented.

1 Introduction

In continuous-time sliding mode control, replacement of the sliding surface by a *boundary layer* is made essentially for the reduction of undesirable chattering [10]. The boundary layer is typically of uniform thickness, and the dynamics inside it are linear. In its discrete-time counterpart, the emergence of a boundary layer is generally unavoidable because finite sampling rate prevents the existence of ideal sliding mode [6]. The shape of the boundary layer in this case depends on the type of uncertainties in the system as well as the version of DVSC.

For systems with only additive uncertainties (exogenous disturbances), boundary layer with uniform thickness is usually adopted [4, 7, 11, 12], whereas for systems with parametric uncertainties, due to the state-dependency of these uncertainties, boundary layer with thickness that grows proportionally with the state norm is usually adopted [3, 9, 15, 1, 13]. For the latter, the boundary layer takes the shape of a cone or pyramid and is called the *sliding sector* in [9], *switching region* in [13], and *switching sector* in this chapter.

Although ideal sliding mode no longer exists, there are switching sector-based DVSCs capable of maintaining a level of robustness not achievable with linear controllers, such as the one shown in the following example[3]:

[3] The notation, equations, and theorems referred to in this example are based on the cited paper.

Example 1 (Wang et al. [14]). Consider a system described by

$$\bar{X}(k+1) = (\bar{A} + \Delta\bar{A})\bar{X}(k) + \bar{B}\bar{U}(k)$$

where $\bar{A} = \begin{bmatrix} 0.9 & 0.6 \\ 0 & 0.2 \end{bmatrix}$, $\bar{B} = [0\ 1]^T$, $\Delta\bar{A} = \bar{B}\bar{D}$, and $\bar{D} = [0.4\ 0.2]$. Since $\bar{B}$ has the form of $[0\ B_2]^T$, a change of coordinates is not necessary and hence $X = \bar{X}$, $A = \bar{A}$, $B = \bar{B}$, $B_2 = 1$, $U = \bar{U}$, and $D = \bar{D}$. Let the control law be given by (7), (8), and (10). Let $C = [C_1\ C_2] = [1\ 1]$ so that the nominal system inside the switching sector, i.e.,

$$X(k+1) = [A + B(C - CA)/(CB)]X(k) = \begin{bmatrix} 0.9 & 0.6 \\ 0.1 & 0.4 \end{bmatrix} X(k) ,$$

has eigenvalues at 0.3 and 1. Then, $L_1 = 0.8571$, $L_2 = 0.1429$, and

$$\left| \frac{2}{C_2 B_2 (E_{n-1}|L_1| + |L_2|)} \right| = 2$$

are obtained. Next, let $\delta = 1.2$ and $\bar{d} = 0.6$ so that (4) and (12) are satisfied. Finally, let $k_d = -0.7$, 0, or 0.7 depending on the conditions in (11a), (11b), and (11c). The simulation result for $X(0) = [0.5\ 0]^T$ is shown in Fig. 1 of this chapter. Note that the perturbed system is robustly stabilized despite the fact that it has an unstable eigenvalue inside the switching sector. ∎

Unfortunately, there are times when these DVSCs lead to stability and performance problems. Section 2 discusses these problems analytically and through counter examples. Section 3 presents, as an attempt to solve these problems, a state feedback switching sector-based DVSC for linear multivariable systems with matched parametric uncertainties and investigates its conservativeness. Section 4 illustrates the proposed controller via a multivariable system while Section 5 concludes the chapter.

2 Problems with Switching Sector

2.1 Potential Instability

The underlying cause of potential instability in existing switching sector-based DVSCs is the attractiveness and non-invariance of the switching sector. Here, attractiveness and non-invariance mean that any state outside the switching sector will be attracted into the sector but will not be confined to stay there. As the state is allowed to switch between two regions, the system may go unstable even though the linear dynamics inside the sector are stable. This fact is illustrated next in a counter example to a theorem stated in Furuta [3][3]:

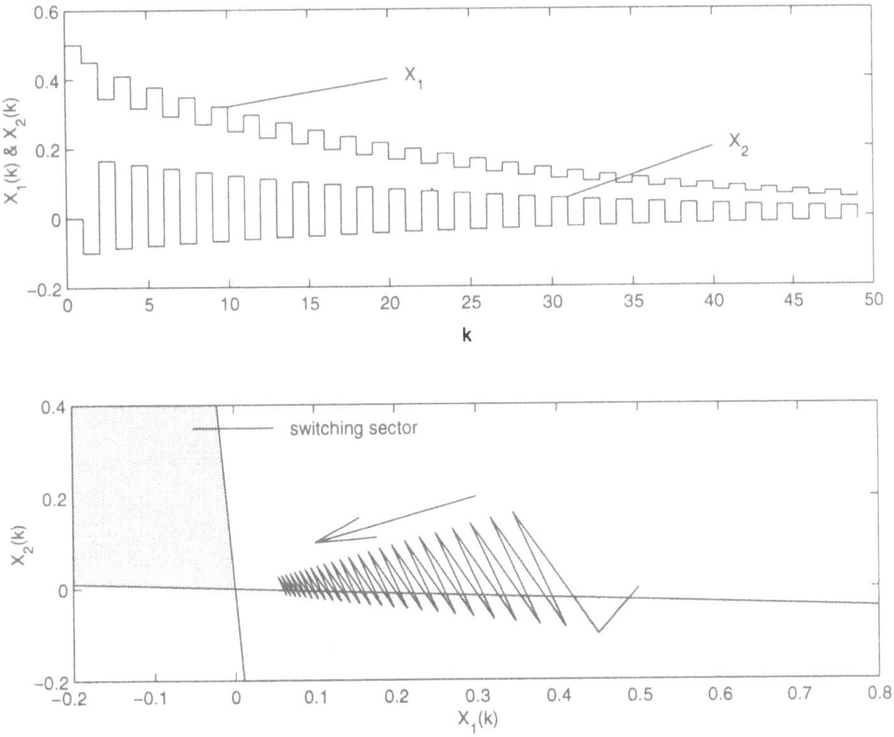

Fig. 1. Simulation result for Example 1

Example 2 (Furuta [3]). Consider a system described by

$$x_{k+1} = \Phi x_k + \Gamma u_k$$

where $\Phi = \Phi_0 + \Delta\Phi$, $\Gamma = [0\ 1]^T$, $\Phi_0 = \begin{bmatrix} 2 & 10 \\ 0 & 1 \end{bmatrix}$, $\Delta\Phi = \Gamma D$, and $D = [-0.09\ 0.03]$. Let the control law be given by (19), (13)', (32), and (33). Let $G = [1\ 4]$ so that both the nominal and perturbed systems are stable inside the switching sector as well as on the sliding hyperplane, i.e.,

$$\begin{aligned}
&\mathrm{eig}\left(\Phi_0 - \Gamma(G\Gamma)^{-1}G(\Phi_0 - I)\right) = \{1, -0.5\}\ , \\
&\mathrm{eig}\left(\Phi_0 - \Gamma(G\Gamma)^{-1}G(\Phi_0 - I) + \Delta\Phi\right) = \{0.2650 \pm 0.6243j\}\ , \\
&\mathrm{eig}\left(\Phi_0 - \Gamma(G\Gamma)^{-1}G\Phi_0\right) = \{0, -0.5\}\ , \\
&\mathrm{eig}\left(\Phi_0 - \Gamma(G\Gamma)^{-1}G\Phi_0 + \Delta\Phi\right) = \{-0.2350 \pm 0.9512j\}\ .
\end{aligned}$$

Next, let $\bar{d} = 0.1$, $f_0 = 0.11$, $t_1 = [0.0588\ 0.2353]^T$, and $t_2 = [-0.9701\ 0.2425]^T$ so that (31b) and (34) are satisfied, $Gt_1 = 1$, and $t_2 \in \ker G$. The simulation result for $x_0 = [1\ 0]^T$ is shown in Fig. 2 of this chapter. Note that the resulting system is unstable although all its eigenvalues are inside the unit circle. This implies that Theorem 3 is incorrect. ∎

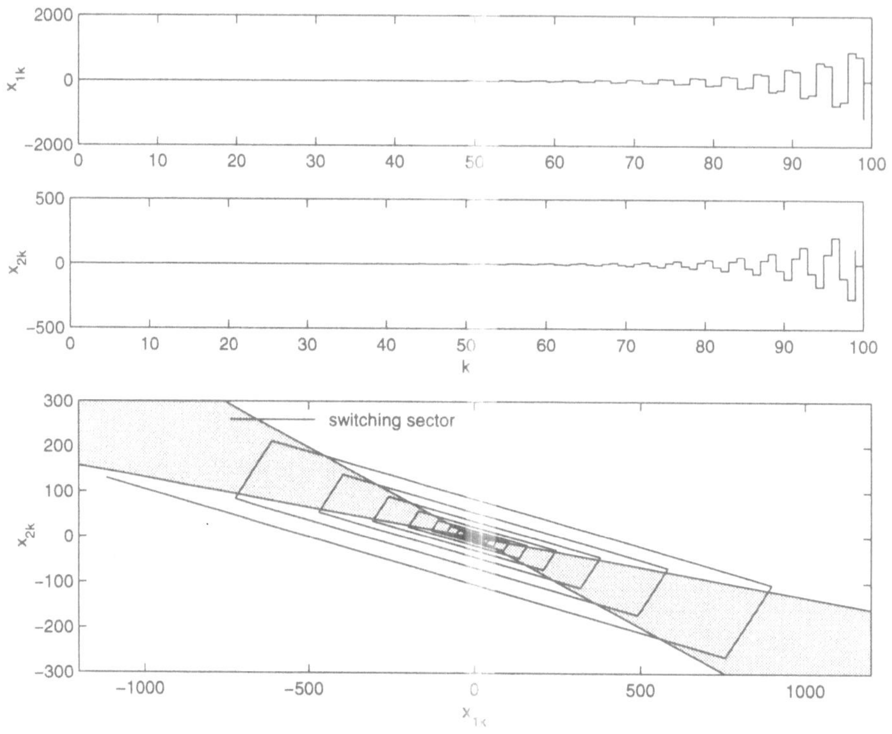

Fig. 2. Simulation result for Example 2

It can be seen from Fig. 2 that for $k > 70$, whenever the state is outside the sector, it is attracted into the sector. After it enters the sector, the system is driven alone by the equivalent control which provides a stable linear dynamics. Due to the non-invariance of the sector and parametric perturbations, the state exits the sector at the subsequent time step. The discontinuous control then comes into effect again and pushes the state back into the sector. The whole process is repeated with the state spiraling away from the origin and going unbounded.

Analytically, this incident can be explained as follows. Since no restriction is imposed on the state outside the sector except that its distance from the sliding hyperplane is guaranteed to decrease, its Euclidean 2-norm may *increase* as it approaches or enters the sector. Since the linear dynamics inside the sector are asymptotically stable, the state inside the sector will move on to the surface of a smaller ellipsoid every subsequent time step, i.e., $x^T(k+1)Px(k+1) <$

$x^T(k)Px(k)$ where $x(k)$ is the state and $P > 0$. For two successive sampling points, this does not imply that its 2-norm is decreasing. In fact, for $x(k+1) = Ax(k)$, it is possible that $||x(k+1)||_p > ||x(k)||_p$ even if A is Schur stable because the spectral radius of any matrix is less than or equal to any of its induced p-norm. As the sector is not invariant, the state may exit the sector with *increasing* 2-norm. It is therefore possible that the state norm increases continually as the state switches between two regions.

An important implication from Example 2 is that to ensure stability, it is *not* sufficient to only show that the switching sector is attractive and the linear dynamics inside the sector are stable.

Observe from Examples 1 and 2 that, unlike linear control, the eigenvalues are not completely crucial in switching sector-based DVSC. The presence of the switching sector has entangled the stability conditions. It possesses the opposing features of ***robustifying*** and ***derobustifying*** the system. This argument is strengthened by the phenomenon of "unstable switchback" [13] displayed in the following example, which is also a counter example to a theorem stated in Wang et al. [14][3]:

Example 3 (Wang et al. [14]). Reconsider Example 1 of this chapter but let $\bar{D} = [0.5\ 0.2]$, which is still well within the bound of $\bar{d} = 0.6$. The simulation result shown in Fig. 3 of this chapter demonstrates the "unstable switchback" phenomenon. This implies that Theorem 2 is incorrect. ∎

Apparently, the analytical explanation of this incident is conceptually similar to that of Example 2.

In addition to the "unstable switchback" phenomenon, there are several other dramatic behaviors which may occur, including but not limited to the one depicted in Fig. 4 as well as limit cycles. The existence of limit cycles is conceivable because the switching between two regions provides opportunities for the state to coincide with the points on its previous path and run into endless loops.

The phenomenon of "unstable switchback" has been successfully avoided by Lee and Wang [5] and Wang et al. [13]. Yet, they did not explicitly take into account the type of instabilities shown in Figs. 2 and 4 as well as the existence of limit cycles.

2.2 Performance Limitation

In existing switching sector-based DVSCs, although the dynamics inside the sector is linear, the applicability of well-established linear control design strategies is limited by the non-invariance of the sector. Frequency domain techniques say little about performance if the state ever exits the sector and causes the response to be nonlinear. Time domain techniques such as the optimal sliding sector design procedure developed in [9] may not be effective because the quadratic performance index is minimized if and only if the state is restricted inside the sector after its first entrance.

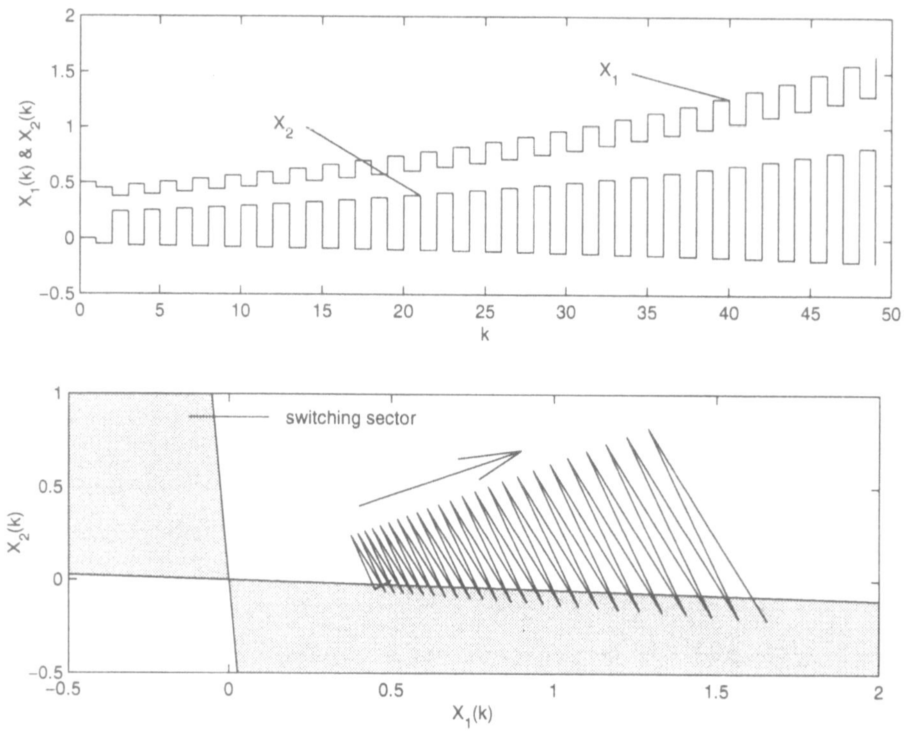

Fig. 3. Simulation result for Example 3

On the other hand, it has been claimed that the switching sector is capable of reducing chattering [3, 14, 5]. Here, it is noted that the contribution of the switching sector towards chattering reduction is questionable. This is because in most existing switching sector-based DVSCs [3, 14, 5, 13], switching is expected to occur along the edges (or boundaries) of the sector *instead of* the sliding surface. As a result, chattering takes place on the edges of the sector (see Figs. 1 and 3). The presence of the switching sector does *not* reduce chattering—it merely changes the region chattering occurs. Furthermore, notice that chattering on the edges of the sector may not be reduced by simply increasing the sector thickness. It may be reduced, as is revealed in the following example, by the uncertainties, which is quite contrary to continuous-time sliding mode control[3]:

Example 4 (Furuta [3]). Reconsider Example 2 of this chapter but let: (i) $D = [0\ 0]$, (ii) $D = [-0.05\ 0]$. The simulation result is shown in Fig. 5 of this chapter. Note that: (i) chattering occurs when there are *no* parametric uncertainties, (ii) chattering is fully eliminated by a particular amount of parametric uncertainties. ∎

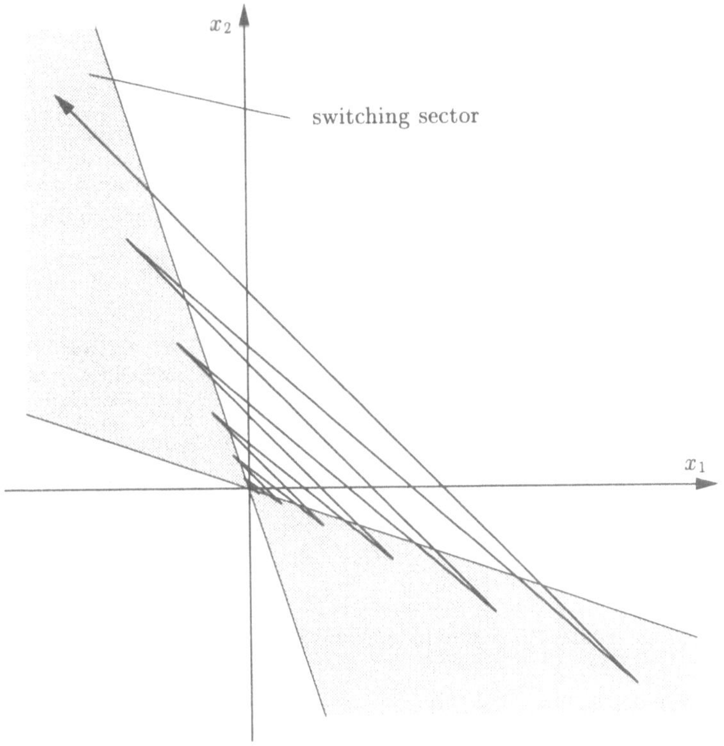

Fig. 4. A possible system behavior with switching sector

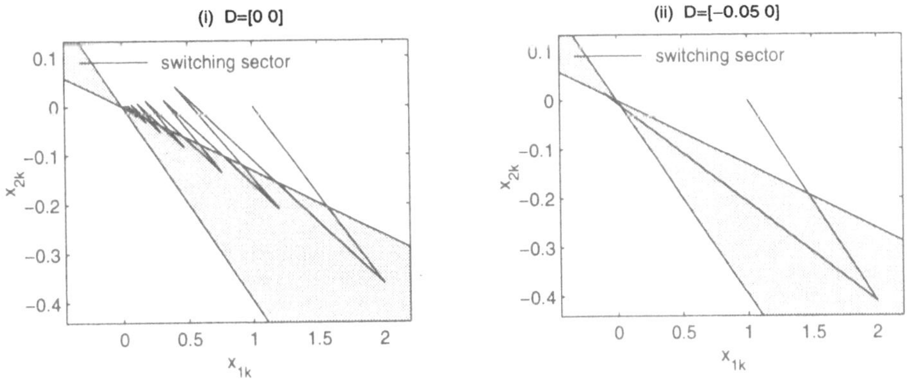

Fig. 5. Simulation result for Example 4

In the upcoming section, a switching sector-based DVSC is developed as an attempt to solve the problems mentioned above.

3 A Robust Stabilization Scheme

Consider the following discrete-time linear multivariable system which may be obtained by discretizing its continuous-time equivalent with sampling period Δt:

$$x(k+1) = (A + \Delta A(k))x(k) + Bu(k) \tag{1}$$

where $x = [x_1\ x_2\ \cdots\ x_n]^T \in \mathbf{R}^n$ is the state vector, $u = [u_1\ u_2\ \cdots\ u_m]^T \in \mathbf{R}^m$ is the input vector, A and B are perfectly known constant matrices with appropriate dimensions, and $\Delta A(k)$ is the time-varying parametric uncertainty matrix satisfying the matching condition $\operatorname{rank}([B \mid \Delta A(k)]) = \operatorname{rank}(B)$. The objective is to find a suitable control input $u(k)$ so that $x(k)$ approaches zero asymptotically in the presence of $\Delta A(k)$.

Assumption 1. The system (1) satisfies the following conditions: $n \geq 2$, $\operatorname{rank}(B) = m$, the pair (A, B) is controllable, and every element of $\Delta A(k)$ is bounded.

Lemma 2. *Let Assumption 1 hold and suppose $\Delta A(k) = 0$. Let $n_1, n_2, \ldots, n_m$ be the Kronecker invariant [8] of the system (1) in which $\sum_{i=1}^{m} n_i = n$ and define $h_1, h_2, \ldots, h_m$ as $h_j = \sum_{i=1}^{j} n_i$. Then, there exists a nonsingular matrix $T \in \mathbf{R}^{n\times n}$ such that with $x = T\bar{x}$, $\bar{x} = [\bar{x}_1\ \bar{x}_2\ \cdots\ \bar{x}_n]^T$, the system (1) can be transformed into the controllable canonical form:*

$$\begin{aligned} \bar{x}_{h_i-n_i+1}(k+1) &= \bar{x}_{h_i-n_i+2}(k) \\ \bar{x}_{h_i-n_i+2}(k+1) &= \bar{x}_{h_i-n_i+3}(k) \\ &\ \vdots \\ \bar{x}_{h_i}(k+1) &= \sum_{j=1}^{n} a_{ij}\bar{x}_j(k) + u_i(k) + \sum_{j=i+1}^{m} b_{ij}u_j \end{aligned} \tag{2}$$

for $i = 1, \ldots, m$. Moreover, there exists a nonsingular matrix $R \in \mathbf{R}^{m\times m}$ given by

$$R = \begin{bmatrix} 1 & b_{12} & b_{13} & \cdots & b_{1m} \\ 0 & 1 & b_{23} & \cdots & b_{2m} \\ 0 & 0 & 1 & \ddots & \vdots \\ \vdots & \vdots & \ddots & \ddots & b_{(m-1)m} \\ 0 & 0 & \cdots & 0 & 1 \end{bmatrix}^{-1}$$

such that with $u = R\bar{u}$, $\bar{u} = [\bar{u}_1\ \bar{u}_2\ \cdots\ \bar{u}_m]^T$, the last equation in (2) can be written as

$$\bar{x}_{h_i}(k+1) = \sum_{j=1}^{n} a_{ij}\bar{x}_j(k) + \bar{u}_i(k)\,. \tag{2'}$$

Proof. This is a standard result. See, e.g., [8] pp. 704. □

Corollary 3. *Let Assumption 1 hold and let* $\Delta\bar{A}(k) := [\Delta a_{ij}(k)] \in \mathbf{R}^{m\times n}$ *be related to* $\Delta A(k)$ *by*

$$\Delta A(k)T = BR\Delta\bar{A}(k) .$$

Then, (2') *can be written as*

$$\bar{x}_{h_i}(k+1) = \sum_{j=1}^{n}(a_{ij} + \Delta a_{ij}(k))\bar{x}_j(k) + \bar{u}_i(k) . \tag{2''}$$

Proof. It is evident from the matching condition and Lemma 2. □

Definition 4. Let the switching sector corresponding to the ith input, $\mathcal{W}_i$, be defined as

$$\mathcal{W}_i = \{\bar{x} : |\bar{x}_{h_i}| \leq \sum_{j=1}^{n} K_{ij}|\bar{x}_j|\}$$

where $K_{ij} > 0$ and let $\mathcal{W}_i^+$ and $\mathcal{W}_i^-$ be defined as

$$\mathcal{W}_i^+ = \{\bar{x} : \bar{x}_{h_i} > \sum_{j=1}^{n} K_{ij}|\bar{x}_j|\} ,$$

$$\mathcal{W}_i^- = \{\bar{x} : \bar{x}_{h_i} < -\sum_{j=1}^{n} K_{ij}|\bar{x}_j|\}$$

so that $\mathcal{W}_i \bigcup \mathcal{W}_i^+ \bigcup \mathcal{W}_i^- = \mathbf{R}^n$.

Theorem 5. *Consider the system* (1) *and let Assumption 1 hold. If the control law is chosen as*

$$u(k) = -RF\bar{x}(k) + Ru_d(k) , \; u_d = [u_{d_1} \; u_{d_2} \; \cdots \; u_{d_m}]^T , \tag{3}$$

$$u_{d_i}(k) = \begin{cases} \bar{x}_{h_i}(k) - \mathrm{sgn}(\bar{x}_{h_i}(k))\sum_{j=1}^{n} K_{ij}|\bar{x}_j(k)|, & \bar{x}(k) \notin \mathcal{W}_i \\ 0, & \bar{x}(k) \in \mathcal{W}_i \end{cases} \tag{4}$$

and if $\Delta a_{ij}(k)$, $F := [f_{ij}]$, *and* $K := [K_{ij}]$ *satisfy*

$$|a_{ij} + \Delta a_{ij}(k) - f_{ij}| + \varepsilon \leq K_{ij} , \; 0 < \varepsilon < 1 , \tag{5}$$

$$\|K\|_\infty \leq 1 \tag{6}$$

for $i = 1, \ldots, m$, $j = 1, \ldots, n$, *then the following hold:*

(i) $\bar{x}(k) \notin \mathcal{W}_i \Rightarrow |\bar{x}_{h_i}(k+1)| \leq (1-\varepsilon)|\bar{x}_{h_i}(k)|$, *for* $i = 1, \ldots, m$.
(ii) $\bar{x}(k) \in \mathcal{W}_i \Rightarrow |\bar{x}_{h_i}(k+1)| \leq (1-n\varepsilon)\|\bar{x}(k)\|_\infty$, *for* $i = 1, \ldots, m$.
(iii) $\|\bar{x}(k+1)\|_\infty \leq \|\bar{x}(k)\|_\infty$.
(iv) $\bar{x}(k) \to 0$ *as* $k \to \infty$, *i.e., the system* (1) *is globally uniformly asymptotically stable.*

Proof. To begin with, observe from Lemma 2, Corollary 3, and (3) that the system (1) can be expressed as

$$\begin{aligned} \bar{x}_{h_i-n_i+1}(k+1) &= \bar{x}_{h_i-n_i+2}(k) \\ \bar{x}_{h_i-n_i+2}(k+1) &= \bar{x}_{h_i-n_i+3}(k) \\ &\vdots \\ \bar{x}_{h_i}(k+1) &= u_{d_i}(k) + \sum_{j=1}^{n}(a_{ij} + \Delta a_{ij}(k) - f_{ij})\bar{x}_j(k) \end{aligned} \tag{7}$$

for $i = 1, \ldots, m$. Besides, since $n \geq 2$ and $K_{ij} > 0$ by Assumption 1 and Definition 4, inequalities (5) and (6) imply that

$$0 < \varepsilon \leq \min(K_{ij}) \leq \frac{1}{n} < 1\,. \tag{8}$$

(i) Suppose $\bar{x}(k) \notin \mathcal{W}_i$. Then, $\bar{x}_{h_i}(k) \neq 0$ by contrapositive and either $\bar{x}(k) \in \mathcal{W}_i^+$ or $\bar{x}(k) \in \mathcal{W}_i^-$, exclusively. For $\bar{x}(k) \in \mathcal{W}_i^+$, note from (4) and (5) that the last equation in (7) can be written as

$$\bar{x}_{h_i}(k+1) = \bar{x}_{h_i}(k) - \sum_{j=1}^{n} K_{ij}|\bar{x}_j(k)| + \sum_{j=1}^{n}(a_{ij} + \Delta a_{ij}(k) - f_{ij})\bar{x}_j(k)$$

and the bounds on $\bar{x}_{h_i}(k+1)$ are given by

$$\begin{aligned} &\bar{x}_{h_i}(k) - 2\sum_{j=1}^{n} K_{ij}|\bar{x}_j(k)| + \varepsilon\sum_{j=1}^{n}|\bar{x}_j(k)| \ \ldots \\ &\qquad\ldots \leq \bar{x}_{h_i}(k+1) \leq \bar{x}_{h_i}(k) - \varepsilon\sum_{j=1}^{n}|\bar{x}_j(k)|\,. \end{aligned} \tag{9}$$

Next, by definition of $\mathcal{W}_i^+$, one can write

$$2\sum_{j=1}^{n} K_{ij}|\bar{x}_j(k)| < 2\bar{x}_{h_i}(k)\,.$$

Adding $\varepsilon\sum_{j=1}^{n}|\bar{x}_j(k)|$ to both sides and rearranging the inequality lead to

$$-\bar{x}_{h_i}(k) + \varepsilon\sum_{j=1}^{n}|\bar{x}_j(k)| < \bar{x}_{h_i}(k) - 2\sum_{j=1}^{n} K_{ij}|\bar{x}_j(k)| + \varepsilon\sum_{j=1}^{n}|\bar{x}_j(k)|\,. \tag{10}$$

Since the left-hand side of (9) is similar to the right-hand side of (10), the inequalities can be combined to give

$$\begin{aligned} &|\bar{x}_{h_i}(k+1)| \leq (1-\varepsilon)|\bar{x}_{h_i}(k)| - \varepsilon\sum_{\substack{j=1 \\ j\neq h_i}}^{n}|\bar{x}_j(k)| \\ &\Rightarrow |\bar{x}_{h_i}(k+1)| \leq (1-\varepsilon)|\bar{x}_{h_i}(k)|\,. \end{aligned} \tag{11}$$

For $\bar{x}(k) \in \mathcal{W}_i^-$, one can proceed in a similar manner to obtain (11). It is obvious that the above deduction is valid for $i = 1, \ldots, m$.

(ii) Suppose $\bar{x}(k) \in \mathcal{W}_i$. Then, from (4), (5), (6), (7), and (8),

$$\begin{aligned} \bar{x}_{h_i}(k+1) &= \sum_{j=1}^{n} (a_{ij} + \Delta a_{ij}(k) - f_{ij}) \bar{x}_j(k) \\ \Rightarrow |\bar{x}_{h_i}(k+1)| &\leq \sum_{j=1}^{n} |a_{ij} + \Delta a_{ij}(k) - f_{ij}| \cdot |\bar{x}_j(k)| \\ &\leq \sum_{j=1}^{n} (K_{ij} - \varepsilon) |\bar{x}_j(k)| \\ &\leq (\|K\|_\infty - n\varepsilon) \|\bar{x}(k)\|_\infty \\ &\leq (1 - n\varepsilon) \|\bar{x}(k)\|_\infty \,. \end{aligned} \tag{12}$$

Again, it is obvious that the above deduction is valid for $i = 1, \ldots, m$.

(iii) Note from (7) that $n - m$ states of $\bar{x}(k+1)$ receive values from $\bar{x}(k)$, i.e., $\bar{x}_i(k+1) = \bar{x}_{i+1}(k)$ for $i \notin \{h_1, h_2, \ldots, h_m\}$. Thus, to ensure that

$$\|\bar{x}(k+1)\|_\infty \leq \|\bar{x}(k)\|_\infty \,, \tag{13}$$

it is sufficient that the remaining m states of $\bar{x}(k+1)$, namely $\bar{x}_i(k+1)$ for $i \in \{h_1, h_2, \ldots, h_m\}$, satisfy

$$\|[\bar{x}_{h_1}(k+1) \; \bar{x}_{h_2}(k+1) \; \cdots \; \bar{x}_{h_m}(k+1)]^T\|_\infty \leq \|\bar{x}(k)\|_\infty \,. \tag{14}$$

Suppose, without loss of generality, that $\bar{x}(k) \notin \mathcal{W}_i$ for $i = 1, \ldots, m'$ and $\bar{x}(k) \in \mathcal{W}_i$ for $i = m'+1, \ldots, m$, with $0 \leq m' \leq m$. It is then clear from (11) and (12) that

$$|\bar{x}_{h_i}(k+1)| \leq \begin{cases} (1-\varepsilon)\|\bar{x}(k)\|_\infty, & i = 1, \ldots, m' \\ (1-n\varepsilon)\|\bar{x}(k)\|_\infty, & i = m'+1, \ldots, m \end{cases} \tag{15}$$

which implies that (14) always holds and so does (13).

(iv) First, observe that (13) and (15) imply

$$|\bar{x}_{h_i}(k+j)| \leq (1-\varepsilon)\|\bar{x}(k)\|_\infty \tag{16}$$

for $i = 1, \ldots, m$, $j \geq 1$. Next, let $N = \max(n_1, n_2, \ldots, n_m)$ and observe from (7) that

$$\bar{x}_{h_i - j}(k+N) = \bar{x}_{h_i}(k+N-j) \tag{17}$$

for $i = 1, \ldots, m$, $j = 1, \ldots, n_i - 1$. It then follows from (16) and (17) that

$$|\bar{x}_i(k+N)| \leq (1-\varepsilon)\|\bar{x}(k)\|_\infty$$

for $i \in \{h_1, h_2, \ldots, h_m\}$ and $i \notin \{h_1, h_2, \ldots, h_m\}$. Thus,

$$\|\bar{x}(k+N)\|_\infty \leq (1-\varepsilon)\|\bar{x}(k)\|_\infty$$

which, together with (13), imply that $\bar{x}(k) \to 0$ as $k \to \infty$.

This completes the proof. □

Theorem 5 has the following corollary:

Corollary 6. *There exists a system and a set of controller parameters such that* (5) *and* (6) *hold and*

(i) $\mathcal{W}_i$ *is not attractive, i.e.,* $\bar{x}(k) \notin \mathcal{W}_i \ \forall k$.
(ii) $\mathcal{W}_i$ *is not invariant, i.e.,* $\exists k$ *such that* $\bar{x}(k) \in \mathcal{W}_i$ *and* $\bar{x}(k+1) \notin \mathcal{W}_i$.

Proof. (i) Consider a system described by $\bar{x}_i(k+1) = u_i(k)$, $i = 1, 2$, which is already in the form of (2). Let $\Delta a_{ij}(k) = f_{ij} = 0$ and $K_{ij} = 0.1$ so that (5) and (6) hold. The solution to this system for $\bar{x}_1(0) = \bar{x}_2(0) = 1$ is $\bar{x}_1(k) = \bar{x}_2(k) = 0.8^k$. Note that $\bar{x}(k) \to 0$ as $k \to \infty$ but $\bar{x}(k) \notin \mathcal{W}_i \ \forall k$, for $i = 1, 2$.
(ii) Shown in the numerical example of Section 4. □

Corollary 6 says that $\mathcal{W}_i$ is, in general, neither attractive nor invariant. Nevertheless, it is proven in Theorem 5 that "unstable switchback", limit cycles, etc., cannot appear—the resulting system is globally uniformly asymptotically stable.

Note from (5) and (6) that if $f_{ij} = a_{ij}$, then the admissible bound on uncertainties reaches the maximum, i.e.,

$$\|\Delta \bar{A}(k)\|_\infty \leq 1 - n\varepsilon \tag{18}$$

where ε can be arbitrarily small. To get a feel on the conservativeness of (18), consider a state feedback linear controller which is, in the notation of this chapter, equivalent to (3) with $u_d(k) = 0$. Again, suppose that $f_{ij} = a_{ij}$. Based on the facts that $\|XY\|_\infty \leq \|X\|_\infty \|Y\|_\infty$ for any matrices X and Y and that the system (2) is in controllable canonical form, one can infer that (18) is *also* the maximum admissible bound on uncertainties for this linear controller. In the case where $\Delta \bar{A}$ is time-independent, keeping in mind that the spectral radius of any matrix never gets larger than its ∞-norm, one can infer that (18) is sufficient but not necessary for the asymptotic stability of linear systems. In this case, the admissible bound on uncertainties under the proposed controller is smaller than that of a linear controller.

Finally, it is pointed out that, like other switching sector-based DVSCs, the proposed controller suffers from the performance limitation mentioned in Section 2.2.

4 Numerical Example

Consider the following pressurized flow box system [2] which has also been adopted in [13]:

$$\begin{bmatrix} \dot{H}(t) \\ \dot{h}(t) \\ \dot{u}_a(t) \end{bmatrix} = \begin{bmatrix} -0.2 & 0.1 & 1 \\ -0.05 & 0 & 0 \\ 0 & 0 & -1 \end{bmatrix} \begin{bmatrix} H(t) \\ h(t) \\ u_a(t) \end{bmatrix} + \begin{bmatrix} 0 & 1 \\ 0 & 0.7 \\ 1 & 0 \end{bmatrix} \begin{bmatrix} u_c(t) \\ u_s(t) \end{bmatrix}$$

Suppose that this is the nominal system. Its discrete-time equivalent, obtained by applying $u_c(t)$ and $u_s(t)$ through a zero-order hold with $\Delta t = 0.2$, is given by (1) where

$$A = \begin{bmatrix} 0.9607 & 0.0196 & 0.1776 \\ -0.0098 & 0.9999 & -0.0009 \\ 0 & 0 & 0.8187 \end{bmatrix}, \quad B = \begin{bmatrix} 0.0185 & 0.1974 \\ -0.0001 & 0.1390 \\ 0.1813 & 0 \end{bmatrix}.$$

By Lemma 2 with $n_1 = 2$, $n_2 = 1$,

$$T = \begin{bmatrix} 0.0168 & 0.0185 & 0.1986 \\ -0.0003 & -0.0001 & 0.1390 \\ -0.1767 & 0.1813 & 0.0116 \end{bmatrix}, \quad R = \begin{bmatrix} 1 & -0.0642 \\ 0 & 1 \end{bmatrix}^{-1},$$

the controllable canonical form is obtained as

$$\bar{x}(k+1) = \begin{bmatrix} 0 & 1 & 0 \\ -0.7979 & 1.7935 & -0.0107 \\ -0.0025 & 0 & 0.9858 \end{bmatrix} \bar{x}(k) + \begin{bmatrix} 0 \\ \bar{u}_1(k) \\ \bar{u}_2(k) \end{bmatrix}.$$

Let

$$F = [f_{ij}] = \begin{bmatrix} -0.4279 & 1.5934 & 0.0147 \\ -0.0020 & 0.0045 & 0.4359 \end{bmatrix}$$

so that the linear portion of the nominal closed-loop system has eigenvalues at 0.55 and $0.1 \pm 0.6j$. Let

$$\Delta\bar{A}(k) = [\Delta a_{ij}(k)] = 0.12 \times \begin{bmatrix} -1 & \cos(2k\Delta t) & 1 \\ \sin(3k\Delta t) & -1 & -\sin(k\Delta t) \end{bmatrix}$$

and note that $\max(|\Delta a_{ij}(k)|) = 0.12$. By selecting $\varepsilon = 0.001$ and

$$K_{ij} = |a_{ij} - f_{ij}| + \max(|\Delta a_{ij}(k)|) + \varepsilon,$$

i.e.,

$$K = [K_{ij}] = \begin{bmatrix} 0.4911 & 0.3211 & 0.1464 \\ 0.1215 & 0.1255 & 0.6709 \end{bmatrix},$$

inequalities (5) and (6) are satisfied. The simulation result for $x(0) = [1\ 1\ 1]^T$ is shown in Fig. 6. By noticing that $u_{d_1}(k) \neq 0$ for some k and $u_{d_2}(k) = 0$ for all k, it is known that $\bar{x}(k)$ enters and exits $\mathcal{W}_1$ several times before converging to zero but never leaves $\mathcal{W}_2$ throughout the simulation. Also, it is seen that $\|\bar{x}(k+1)\|_\infty \leq \|\bar{x}(k)\|_\infty$ regardless of whether $\bar{x}(k)$ is inside or outside the switching sectors and the system is uniformly asymptotically stable in the presence of $\Delta\bar{A}(k)$.

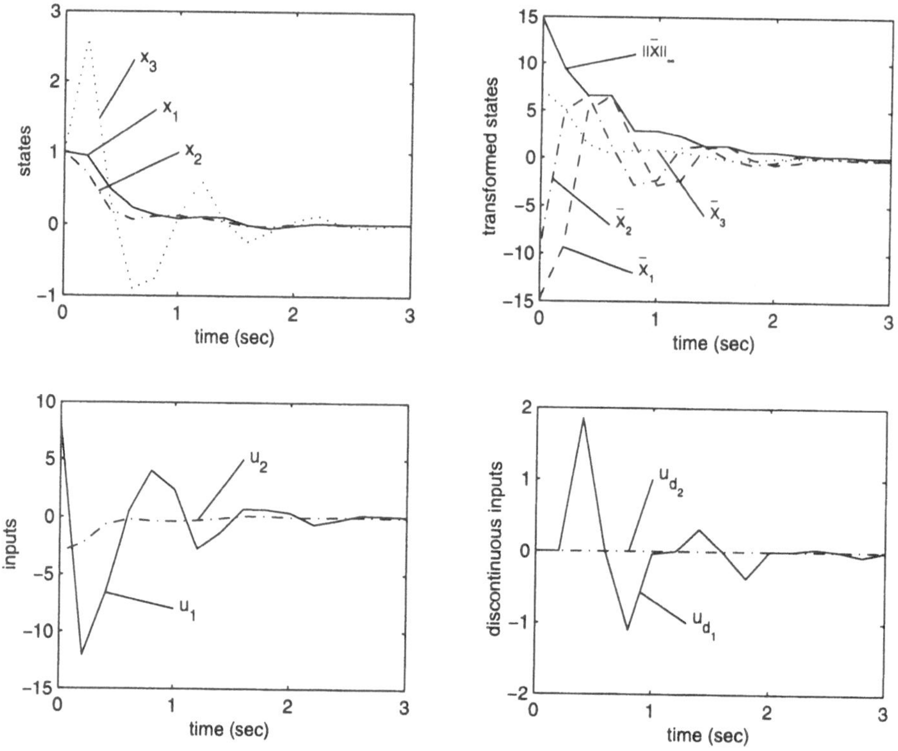

Fig. 6. Time response of the pressurized flow box

5 Concluding Remarks

Potential instability and performance limitation in a number of switching sector-based DVSCs have been detailed in this chapter. It has been shown analytically and by counter examples that the attractiveness of the switching sector and the stability of the linear dynamics inside the sector are not sufficient for overall stability. A state feedback switching sector-based DVSC for linear multivariable systems with matched parametric uncertainties has been developed. The proposed controller guarantees the global uniform asymptotic stability of the uncertain system in spite of the non-attractiveness and non-invariance of the switching sector. The admissible bound on uncertainties, however, is found to be conservative. A numerical example has been given to illustrate the effectiveness of the proposed controller.

References

1. M. L. Corradini and G. Orlando, "Variable Structure Control for Uncertain Sampled Data Systems," in *Proc. IEEE Int. Workshop on Variable Structure Systems,*

(Tokyo, Japan), pp. 117–121, 1996.
2. G. Franklin, J. Powell, and M. Workman, *Digital Control of Dynamic Systems.* Addison-Wesley, 2nd ed., 1990.
3. K. Furuta, "Sliding Mode Control of a Discrete System," *Systems & Control Letters*, vol. 14, pp. 145–152, 1990.
4. W. Gao, Y. Wang, and A. Homaifa, "Discrete-Time Variable Structure Control Systems," *IEEE Trans. Industrial Electronics*, vol. 42, no. 2, pp. 117–122, 1995.
5. R.-C. Lee and W.-J. Wang, "Robust Variable Structure Control Synthesis in Discrete-Time Uncertain Systems," *Control-Theory and Advanced Technology*, vol. 10, no. 4, pp. 1785–1796, 1995.
6. C. Milosavljevic, "General Conditions for the Existence of a Quasisliding Mode on the Switching Hyperplane in Discrete Variable Structure Systems," *Automation and Remote Control*, vol. 46, no. 3, pp. 307–314, 1985.
7. E. A. Misawa, "Discrete-Time Sliding Mode Control: The Linear Case," *ASME Journal of Dynamic Systems, Measurement, and Control*, vol. 119, no. 4, pp. 819–821, 1997.
8. K. Ogata, *Discrete-Time Control Systems.* Prentice-Hall, 2nd ed., 1995.
9. Y. Pan and K. Furuta, "VSS Controller Design for Discrete-Time Systems," *Control-Theory and Advanced Technology*, vol. 10, no. 4, pp. 669–687, 1994.
10. J.-J. E. Slotine, "Sliding Controller Design for Non-linear Systems," *Int. J. Control*, vol. 40, no. 2, pp. 421–434, 1984.
11. C. Y. Tang and E. A. Misawa, "Discrete Variable Structure Control for Linear Multivariable Systems: The State Feedback Case," in *Proc. American Control Conference*, (Philadelphia, Pennsylvania), pp. 114–118, 1998.
12. C. Y. Tang and E. A. Misawa, "Discrete Variable Structure Control for Linear Multivariable Systems: The Output Feedback Case," in *Proc. ASME Dynamic Systems and Control Division*, vol. 64, (Anaheim, California), pp. 355–360, 1998.
13. W.-J. Wang, R.-C. Lee, and D.-C. Yang, "Sliding Mode Control Design in Multi-Input Perturbed Discrete-Time Systems," *ASME Journal of Dynamic Systems, Measurement, and Control*, vol. 118, pp. 322–327, 1996.
14. W.-J. Wang, G.-H. Wu, and D.-C. Yang, "Variable Structure Control Design for Uncertain Discrete-Time Systems," *IEEE Trans. Automat. Contr.*, vol. 39, no. 1, pp. 99–102, 1994.
15. X. Yu, "Digital Variable Structure Control with Pseudo-Sliding Modes," in *Variable Structure and Lyapunov Control* (A. S. I. Zinober, ed.), pp. 133–155, Springer-Verlag, 1994.

Variable Structure Control of Nonlinear Sampled Data Systems by Second Order Sliding Modes

Giorgio Bartolini, Alessandro Pisano, Elio Usai

Department of Electrical and Electronic Engineering, University of Cagliari, Piazza d' Armi, 09123 Cagliari, Italy

1 Introduction

Due to the growing diffusion of computer–controlled systems, discrete time implementation of variable structure controllers is one of the main topics in modern VSS theory. The sample-and-hold effect makes not straightforward a direct extension of the structural properties showed by continuous time sliding modes to the discrete time context.

For this reason, a great research effort has been devoted at giving a consolidated theoretical basis to this particular kind of VSSs. In [13] Milosavljevic analyzed the effect of the discretization of measures in a proximity of the sliding surface, defining the so called "quasi-sliding" motion. Drakunov and Utkin developed a semigroup approach to the analysis of discrete time VSSs [7], and they proposed a synthesis procedure which extends the well known equivalent control concept to discrete–time control systems. This procedure, which can be viewed as an extension of the "dead beat control", requires a mathematical characterization of the system to be controlled in terms of the delay operator, and it is theoretically able to constrain the system state on the sliding surface in one sampling period, even if at the cost of a very high control effort [8, 16, 14]. Nevertheless, an important feature of this approach is that the control effort decreases while the sliding manifold is approached, and the discrete time equivalent control turns out to be smooth within the boundary layer. Then the ringing phenomenon, appearing as a discrete time counterpart of the chattering phenomenon when a direct discretization of continuous-time VSC is performed, is avoided. To achieve this task, the discrete time equivalent control is defined in a slightly different form than the continuous time one, since it provides both the reaching and the sliding phase [16].

Furuta developed the sliding sector approach, a two stages control strategy which consists in a direct discretization of a continuous-time sliding mode control outside a properly chosen sector in the state space including the sliding manifold, and commutes to a smooth discrete time robust control within this sector [10, 11].

In presence of disturbances and/or uncertainties some adaptation or estimation technique should be adopted to counteract them, in order to reduce the size of the boundary layer. In [1] a discrete time MRAC approach is used to improve the performances of the VSC in presence of model uncertainties. By using a predictor of the uncertainties, in [9] Drakunov *et al.* dealt with the sliding mode control of a large class of sampled data nonlinear uncertain systems, attaining a $\mathcal{O}(T^2)$ accuracy, being T the sampling period, and avoiding the ringing phenomenon. This result is an improvement respect to the $\mathcal{O}(T)$ accuracy provided by discretized first order SMC. By resorting to a second order sliding mode control technique, the accuracy can be further improved, and the effectiveness can be extended to a larger class of systems.

Let the "sliding variable" be the state-dependent quantity that vanishes when the manifold is reached. The relative degree q between the sliding variable and the control input, plays a fundamental role in solving the control problem.

The sliding order r is defined as the relative degree between the sliding variable and the discontinuous control signal. In classical SMC the sliding order is equal to one, so that the first derivative of the sliding quantity is discontinuous and its sign changes with theoretically infinite frequency. In second order sliding modes $r = 2$, and both the sliding variable and its first time derivative converge to zero in a finite time.

Obviously, the sliding order cannot be smaller than q, and, for this reason, first order SMC is not effective if $q = 2$, meaning that the control input does not affect the first time derivative of the sliding variable directly.

Moreover, when $q = 1$, second order SMC allows to obtain chattering elimination. The second derivative of the sliding quantity can be properly modified by using, as a discontinuous auxiliary control signal, the derivative of the actual control input. The control input, obtained by integrating the discontinuous derivative, results to be continuous, and chattering is avoided.

The digital realization of the control law is treated by using both the continuous model and a sufficiently accurate discrete model, which allow direct visibility on the intersampling system behavior and preserve bad phenomena such as nonminimum phase effects.

The aim of this Chapter is that of dealing with the sliding mode control of a class of uncertain nonlinear sampled data systems in which, due to the not complete availability of the states, the sliding variable must be chosen such that the control input acts on its second time derivative. This problem is addressed and solved by using the equivalent control concept, properly coupled with a second order sliding mode control scheme.

The Chapter is organized as follows: the next Section is devoted to the problem statement. In Section 3 the discrete model is derived, while in Section 4 a digital VSC, based on a dead-beat like procedure involving the discrete time equivalent control concept, is proposed, providing ringing avoidance and $\mathcal{O}(T^3)$ accuracy. In the subsequent Section 5, a different control strategy, based on a piecewise constant approximation of the continuous time equivalent control, is considered. This procedure guarantees the same accuracy order $\mathcal{O}(T^3)$ with respect to that based on the discrete time equivalent control. Section 6 deals with simulation results, while in the final Section some remarks regarding both the proposed algorithms and future researches are discussed.

2 The Control Problem

Consider the class of uncertain nonlinear feedback linearizable systems whose dynamics is defined by

$$\begin{cases} \dot{x}_i(t) = x_{i+1}(t) & i = 1, 2, \ldots, n-1 \\ \dot{x}_n(t) = f[\mathbf{x}(t), t] + g[\mathbf{x}_\mathrm{M}(t), t]u(t) \end{cases} \tag{1}$$

where $\mathbf{x}(t) \equiv [x_1(t), x_2(t), \ldots, x_n(t)] \in \Re^n$ is the state vector and $\mathbf{x}_\mathrm{M}(t) \equiv [x_1(t), x_2(t), \ldots, x_{n-1}(t)] \in \Re^{n-1}$ is its measurable component, $u(t) \in \Re$ is the bounded system input such that

$$|u(t)| \leq U \tag{2}$$

$f[\mathbf{x}(t), t]$ is an uncertain bounded and smooth function satisfying the constraints

$$\begin{cases} |f[\mathbf{x}(t), t]| \leq F \\ |\dot{f}[\mathbf{x}(t), \dot{\mathbf{x}}(t), t]| \leq F_\mathrm{d} \end{cases} \tag{3}$$

and $g[\mathbf{x}_\mathrm{M}(t), t]$ is a known measurable smooth function such that

$$\begin{cases} 0 < G_1 \leq g[\mathbf{x}_\mathrm{M}(t), t] \leq G_2 \\ |\dot{g}[\mathbf{x}(t), t]| \leq G_\mathrm{d} \end{cases} \tag{4}$$

being F, F_d , G_1, G_2, G_d and U non negative constants.

The control task consists in steering to zero the state vector components in spite of their uncertain dynamics and of their not complete measurability.

According to the VSS approach, a proper manifold in the state space is defined such that any motion constrained on such a manifold fulfills the control objective.

It is common practice in sliding mode control to choose the sliding quantity as a suitable linear combination of the available states. Define

$$s[\mathbf{x}_\mathrm{M}(t)] = x_{n-1} + \sum_{i=1}^{n-2} c_i x_i(t) \tag{5}$$

being c_i proper positive constants such that the polynomial

$$P(q) = q^{n-2} + \sum_{i=1}^{n-2} c_i q^{i-1} \tag{6}$$

is Hurwitz.

If we are able to provide the fulfillment of condition

$$s = \dot{s} = 0 \tag{7}$$

then the states converge asymptotically to zero according to the choice of the c_i constants.

From a different point of view, the sliding variable can be viewed as an output variable of system (1), which is wanted to be steered to zero with its unavailable first derivative.

Double differentiating (5) yields

$$\begin{cases} y_1(t) = s[\mathbf{x}_\mathrm{M}(t)] \\ \dot{y}_1(t) = y_2(t) \\ \dot{y}_2(t) = f[\mathbf{x}(t), t] + \sum_{i=1}^{n-2} c_i x_{i+2}(t) + g[\mathbf{x}_\mathrm{M}(t), t]u(t) \end{cases} \tag{8}$$

The second order sliding mode control problem is reduced to the stabilization of uncertain second order systems with not completely available state. It can be noted that the control input $u(t)$ acts on the second derivative of the sliding variable but not on the first one, and therefore the control problem cannot be solved by classical VSS theory, in which, basically, the first derivative of the sliding variable is discontinuously modified by relay control so as the invariance condition $s\dot{s} < 0$ holds [15].

Steering to zero both the sliding variable s and its first derivative $\dot{s}$ by acting on $\ddot{s}$ is the operating mode for second order sliding mode control.

Second order sliding modes consist in a special class of sliding motions which occur at the intersection of the manifolds $s = 0$ and $\dot{s} = 0$. This extended manifold is called "second order sliding set" [12].

Special contractive behaviours must be activated to generate second order sliding modes. In previous Chapter, a number of second order sliding controllers are presented, and their features are discussed [12, 2, 3]. In the following, the term "second order sliding mode" will be abridged to "2-sliding mode" for the sake of brevity.

In continuous time, the control problem has been solved for a wider class of systems than that expressed by (1), (3) and (4). No assumptions was made about the uncertainties derivatives, and the control gain was assumed unknown and possibly depending on the unavailable state x_n. Assumptions (3) and (4) reduce to

$$\begin{cases} |f[\mathbf{x}(t), t]| + \sum_{i=1}^{n-2} c_i |x_{i+2}(t)| \le \Phi \\ 0 < G_1 \le g[\mathbf{x}(t), t] \le G_2 \end{cases} \quad t \ge 0\,;\ \mathbf{x} \in \mathbf{X} \subseteq \Re^n \tag{9}$$

$\mathbf{X}$ being a bounded region within which the boundedness of the system dynamics is assured, and Φ being a positive constant.

Various continuous time 2-sliding mode control algorithms solving such control problem have been presented in the literature [2, 3, 12]. In the next we will refer to the so-called "suboptimal 2-sliding algorithm", dealt with in previous Chapter [2, 3].

The first step of this treatment consists in analyzing the behaviour of the controlled system when a direct discretization of the continuous time suboptimal 2-sliding control scheme is performed.

The continuous time suboptimal algorithm provides a real sliding regime such that

$$\begin{cases} |s| \leq \mathcal{O}(T^2) \\ |\dot{s}| \leq \mathcal{O}(T) \end{cases} \tag{10}$$

being T the measurement step of the sliding variable.

The direct discretization of the control law preserves such accuracy according to the following statement.

Consider the sequence of the sampled values of the sliding variable $y_1(t)$

$$y_1[k] = y_1(kT)$$

being T the considered sampling period.

Theorem 1. *Given system* (8), *with its uncertain dynamics satisfying* (9), *then the discrete controller*

$$\begin{cases} u(t) = u[k] = -U_M\alpha[k] sign\left\{y_1[k] - \frac{1}{2}\hat{y}_{1M}[k]\right\} \\ \qquad t \in [kT, (k+1)T) \quad k = 0, 1, 2, \ldots \end{cases} \tag{11}$$

$$\alpha[k] = \begin{cases} 1 & if \left\{y_1[k] - \frac{1}{2}\hat{y}_{1M}[k]\right\}\left\{y_1[k] - \hat{y}_{1M}[k]\right\} \leq 0 \\ \alpha^* & otherwise \end{cases} \tag{12}$$

$$\begin{array}{l} \hat{y}_{1M}[k] = \begin{cases} y_1[k-1] & if \Delta[k] \leq 0 \\ \hat{y}_{1M}[k-1] & otherwise \end{cases} \\ \Delta[k] = (y_1[k] - y_1[k-1])(y_1[k-1] - y_1[k-2]) \\ y_1[-1] = \hat{y}_{1M}[-1] = y_1(0) \ ; \ y_1[-2] = 0 \end{array} \tag{13}$$

with the additional constraints

$$U_M \in \left(\frac{F}{\alpha^* G_1}, \infty\right) \cap \left(\frac{4F}{3G_1 - \alpha^* G_2} + M_1 T\ ,\ M_2 T^{-2}\right) \tag{14}$$

$$\alpha^* \in (0, 1] \cap \left(0, \frac{3G_1}{G_2}\right) \tag{15}$$

provides the fulfillment of

$$\begin{array}{ll} |y_1(t)| \leq \rho_1 F T^2 & \\ |y_2(t)| \leq \rho_2 T & t \geq T_{\text{reach}} \end{array} \tag{16}$$

being M_1, M_2, ρ_1, ρ_2 proper positive constants and T_{reach} a finite transient time. In particular, if $G_1 = G_2 = \alpha^* = 1$, then ρ_1 can be minimized by choosing

$$U_M = 6F \tag{17}$$

Proof.

See the Appendix.

Similar result was attained by Drakunov *et al.* for systems with $q = 1$ and known control gain [9]. The use of a predictor of the uncertain dynamics was the main point of the above approach, increasing by one the order of the accuracy provided by first order sliding mode control and reducing the control effort.

Second order SMC technique allows a proper design choice between an improvement of the accuracy, under the same assumptions regarding the uncertainties, and the enlargement of the class of systems for which the controller is effective, with no further contraction of the boundary layer.

The aim of the present Chapter is that of showing that, under the same assumptions regarding the knowledge of the control gain, the application of second order sliding mode control strategies allows to achieve a system motion confined within a $\mathcal{O}(T^3)$ boundary layer of the sliding manifold, that is the same accuracy featured by real third order sliding mode control. Note that, in any case, the attained motion cannot be defined as a third order sliding mode, since $\ddot{s}$ has discontinuous dynamics [12].

This result has less general validity respect to the general case dealt with in previous Theorem 1, but it constitutes an important step in the complete solution of the problem. The relaxation of the assumption of the knowledge of the control gain remains an open problem, and will be the object of future researches.

In next section we derive a sufficiently accurate discrete model of the controlled plant which will be used for the synthesis of two control strategies providing the desired accuracy.

3 A discrete time uncertain model with $\mathcal{O}(T^3)$ accuracy

Our first purpose is to provide a discrete model of the continuous system (8) with an approximation, at any time instant, not worse than the accuracy of the sliding motion that it is wanted to be assured by the proposed digital control, that is $\mathcal{O}(T^3)$. To this end, let us indicate with $a[k] = a(kT)$ the k–th sample of a generic variable a. Assume that the control is applied by means of a ZOH.

Consider system (8), and define

$$\begin{cases} \varphi[\mathbf{x}(t), t] = f[\mathbf{x}(t), t] + \sum_{i=1}^{n-2} c_i x_{i+2}(t) \\ \gamma[\mathbf{x}_{\mathrm{M}}(t), t] = g[\mathbf{x}_{\mathrm{M}}(t), t] \end{cases} \tag{18}$$

By (9), $\varphi[\mathbf{x}(t), t]$ and $\gamma[\mathbf{x}(t), t]$ satisfy the following boundedness assumptions

$$\begin{cases} |\varphi[\mathbf{x}(t), t]| \le \Phi \\ \\ 0 < \Gamma_1 \le \gamma[\mathbf{x}_{\mathrm{M}}(t), t] \le \Gamma_2 \end{cases} \qquad t \ge 0\,;\ \mathbf{x} \in \mathbf{X} \subseteq \Re^n \tag{19}$$

$\mathbf{X}$ being the bounded region within which the boundedness assumptions hold and Γ_1, Γ_2 being the constants previously defined as G_1 and G_2.

By (2),(3) and (4), $\varphi[\mathbf{x}(t), t]$ and $\gamma[\mathbf{x}_M(t), t]$ can be proved to have bounded derivatives according to

$$\begin{cases} |\dot{\varphi}[\mathbf{x}(t), t]| \le \Phi_{\mathrm{d}} \\ \\ |\dot{\gamma}[\mathbf{x}(t), t]| \le \Gamma_{\mathrm{d}} \end{cases} \qquad t \ge 0\,;\ \mathbf{x} \in \mathbf{X} \subseteq \Re^2 \tag{20}$$

for any $\mathbf{x} \in \mathbf{X}$, being Φ_d and Γ_d proper positive constants. A suitable region $\mathbf{X}$ can be made an invariant set by using other control techniques, for example first order SMC.

Two subsequent samples of the sliding variable s satisfy the following relationship

$$s[k+1] = s[k] + \int_{kT}^{(k+1)T} \dot{s}(\tau)d\tau \tag{21}$$

The argument of the integral function, $\dot{s}(\tau)$, can be reduced in Taylor series as follows:

$$\begin{gathered} \dot{s}(\tau) = \dot{s}[k] + \varphi[k](\tau - kT) + \gamma[k]u[k](\tau - kT) + \\ + \tfrac{1}{2}\left(\dot{\varphi}[\mathbf{x}(\xi), \dot{\mathbf{x}}(\xi), \xi] + \tfrac{1}{2}\dot{\gamma}[\mathbf{x}(\xi), \xi]u[k]\right)(\tau - kT)^2 \\ \tau \in [kT\ ;\ (k+1)T] \\ \xi \in (kT\ ;\ (k+1)T) \end{gathered} \tag{22}$$

Considering (22) into (21) yields

$$s[k+1] = s[k] + \dot{s}[k]T + \frac{1}{2}\left[\varphi[k] + \gamma[k]u[k]\right]T^2 + \eta_1(T) \tag{23}$$

where $\eta_1(T)$ is the discretization error due to the Taylor approximation in (22), satisfying, in accordance with (20), the following constraint

$$|\eta_1(T)| \leq \frac{1}{6}\left(\Phi_\mathrm{d} + \Gamma_\mathrm{d} u[k]\right)T^3 \tag{24}$$

In order to obtain a discrete model effective for the synthesis procedure, the unavailable sample $\dot{s}[k]$ in (23) must be eliminated. To this end, consider (23) in two subsequent sampling and subtract one each other, then it follows

$$\begin{gathered} s[k+1] = 2s[k] - s[k-1] + (\dot{s}[k] - \dot{s}[k-1])T + \\ + \tfrac{1}{2}\left(\varphi[k] - \varphi[k-1] + \gamma[k]u[k] - \gamma[k-1]u[k-1]\right)T^2 + \eta_2(T) \end{gathered} \tag{25}$$

being, by (20),

$$|\eta_2(T)| \leq \left[\frac{1}{3}\Phi_\mathrm{d} + \frac{1}{6}\Gamma_\mathrm{d}(u[k] + u[k-1])\right]T^3 \tag{26}$$

By (22) it results

$$\begin{gathered} \dot{s}[k] - \dot{s}[k-1] = \varphi[k-1]T + \gamma[k-1]u[k-1]T + \\ + \tfrac{1}{2}\left(\dot{\varphi}[\mathbf{x}(\xi'), \dot{\mathbf{x}}(\xi'), \xi'] + \dot{\gamma}[\mathbf{x}(\xi'), \xi']u[k-1]\right)T^2 \\ \xi' \in ((k-1)T\ ;\ kT) \end{gathered} \tag{27}$$

and, consequently,

$$\dot{s}[k] - \dot{s}[k-1] = \left(\varphi[k-1] + \gamma[k-1]u[k-1]\right)T + \eta_3(T) \tag{28}$$

with, by (20)

$$|\eta_3(T)| \leq \frac{1}{2}\left(\Phi_{\rm d} + \Gamma_{\rm d} u[k-1]\right) T^2 \tag{29}$$

Taking into account (25) and (28), the discrete time model of the sliding variable dynamics is

$$\begin{aligned} s[k+1] = 2s[k] - s[k-1] + \varphi[k-1]T^2 \\ +\tfrac{1}{2}\left(\gamma[k]u[k] + \gamma[k-1]u[k-1]\right)T^2 + \varepsilon[k] \\ k = 0, 1, 2 \ldots \end{aligned} \tag{30}$$

At any sampling time, by (26) and (29), the discretization error $\varepsilon[k]$ is such that

$$|\varepsilon[k]| \leq \left(\frac{4}{3}\Phi_{\rm d} + \frac{1}{6}\Gamma_{\rm d} u[k] + \frac{2}{3}\Gamma_{\rm d} u[k-1]\right) T^3 \tag{31}$$

The discrete time sliding mode control problem can be therefore re–defined as that of finding, by means of the above approximate model, a control sequence $u[k]$ ($k = 0, 1, \ldots$) such that the sliding variable s is constrained within a $\mathcal{O}(T^3)$ boundary layer of the origin from a finite time instant on.

4 The Dead–Beat Approach

The discrete model (30)-(31) may be used to define a digital control law ensuring that system (8), (19)- (20) is confined within a small vicinity of the sliding set (7).

In [14] the discrete–time equivalent control has been defined as the control sequence $u[k]$ such that $s[k+1] = 0$ ($k = 1, 2, \ldots$). The discrete equivalent control $u^{\rm d}_{\rm eq}[k]$ ($k = 1, 2, \ldots$) is able to drive the system into the sliding surface in one sampling period, and to constrain on the system state in all the subsequent sampling time instants. This approach can be viewed, in some sense, as a dead-beat control for tracking the null sequence.

For the above model, the discrete time equivalent control can be defined as

$$u^{\rm d}_{\rm eq}[k] = -\frac{1}{\gamma[k]}\left(\gamma[k-1]u[k-1] + 2\varphi[k-1] + 2\frac{2s[k] - s[k-1]}{T^2}\right) \tag{32}$$

By direct inspection of (32), it can be noted that the equivalent control is inversely proportional to T^2 unless a $\mathcal{O}(T^2)$ boundary layer of the sliding manifold $s = 0$ is reached.

In order to reduce the amplitude of such signal, an initialization procedure, consisting in the digital suboptimal algorithm presented in Theorem 1, must be performed until a $\mathcal{O}(T^2)$ boundary layer is attained.

From this point on, the sliding variable can be steered to $\mathcal{O}(T^3)$ in one sampling step by means of the discrete equivalent control. Unfortunately, the equivalent control is not directly measurable, due to the uncertain dynamics of the controlled system, and some form of prediction must be implemented to estimate it.

At the actual $k-th$ sampling instant, an estimate of the uncertain sample $\varphi[k-1]$ can be found as follows:

- Rewrite the discrete model (30) with one step delay

$$\begin{aligned} s[k] &= 2s[k-1] - s[k-2] + \varphi[k-2]T^2 \\ &+\tfrac{1}{2}\left(\gamma[k-1]u[k-1] + \gamma[k-2]u[k-2]\right)T^2 + \varepsilon[k-1] \end{aligned} \tag{33}$$

  As the uncertain function $\varphi(\mathbf{x}(t),t)$ has bounded derivative, $\varphi[k-1]$ can be suitably estimated by means of its previous value at time instant $t_{k-2} = (k-2)T$.
- Define $\hat{\varphi}[k-1]$ as

$$\begin{aligned} \hat{\varphi}[k-1] &= (s[k] - 2s[k-1] + s[k-2])T^{-2} - \\ &-\tfrac{1}{2}\left(\gamma[k-1]u[k-1] + \gamma[k-2]u[k-2]\right) \end{aligned} \tag{34}$$

  By (31), the estimation error is upperbounded as follows

$$|\hat{\varphi}[k-1] - \varphi[k-1]| \leq \left(\frac{7}{3}\Phi_{\mathrm{d}} + \frac{2}{3}\Gamma_{\mathrm{d}}u[k-2] + \frac{1}{6}\Gamma_{\mathrm{d}}u[k-1]\right)T \tag{35}$$

- Use the estimated sample $\hat{\varphi}[k-1]$ in place of $\varphi[k-1]$ in (32), and define $\hat{u}^{\mathrm{d}}_{\mathrm{eq}}[k]$ accordingly

$$\hat{u}^{\mathrm{d}}_{\mathrm{eq}}[k] = \frac{1}{\gamma[k]}\left(\gamma[k-2]u[k-2] - 2\left(3s[k] - 3s[k-1] + s[k-2]\right)T^{-2}\right) \tag{36}$$

  The estimate of the discrete-time equivalent control can be shown to be affected by a $\mathcal{O}(T)$ estimation error.

After the initialization phase, the control $\hat{u}^{\mathrm{d}}_{\mathrm{eq}}[k]$, which is available at the beginning of any control interval, is used, and its effect on the reduction of the size of the boundary layer is stated in the following Theorem

Theorem 2: *Given system* (1),(2) *with its uncertain dynamics satisfying* (18)–(20), *then the digital feedback controller*

$$u[k] = \begin{cases} -\frac{U_M}{\gamma[k]}\mathrm{sign}\left(s[k] - \frac{1}{2}\hat{s}_M[k]\right) & \text{if } kT < T_{\mathrm{reach}} \\ \frac{\gamma[k-2]}{\gamma[k]}u[k-2] - 2\frac{3s[k]-3s[k-1]+s[k-2]}{\gamma[k]T^2} & \text{if } kT \geq T_{\mathrm{reach}} \end{cases} \tag{37}$$

$$u[-1] = u[-2] = 0$$

$$\hat{s}_{\mathrm{M}}[k] = \begin{cases} s[k-1] & if \;\; \Delta[k] \le 0 \\ \\ \hat{s}_{\mathrm{M}}[k-1] & if \;\; \Delta[k] > 0 \end{cases}$$
$$\begin{gathered} \Delta[k] = (s[k] - s[k-1])(s[k-1] - s[k-2]) \\ s[-1] = \hat{s}_{\mathrm{M}}[-1] = s(0) \quad s[-2] = 0 \end{gathered} \tag{38}$$

being s the sliding variable defined in (5) *and with the additional constraints*

$$U_{\mathrm{M}} = 6 \frac{\Phi}{1 - 6\frac{\Gamma_{\mathrm{d}}}{\Gamma_1} T} \tag{39}$$

$$T < \frac{\Gamma_1}{6\Gamma_{\mathrm{d}}} \tag{40}$$

guarantees the finite time reaching of a $\mathcal{O}(T^3)$ *vicinity of the sliding manifold* $s = 0$, *being* T *the sampling period and* T_{reach} *the finite reaching time of the boundary layer*

Proof:
Consider the uncertain dynamics (8), and define the auxiliary control signal

$$w(\mathbf{x}_{\mathrm{M}}, t) = \gamma(\mathbf{x}_{\mathrm{M}}, t) u(t) \tag{41}$$

The considered system belongs to the class of systems dealt with in Theorem 1 with $G_1 = G_2 = 1$, and $w(\mathbf{x}_{\mathrm{M}}, t)$ can be chosen according to the digital suboptimal 2-sliding control algorithm with $\alpha^* = 1$. The actual control $u[k]$ ($k = 1, 2, \ldots$) turns out to be defined as

$$u[k] = -\frac{W_{\mathrm{M}}}{\gamma[k]} \mathrm{sign}\left(s[k] - \frac{1}{2}\hat{s}_{\mathrm{M}}[k] \right) \tag{42}$$

where W_{M} is a positive constant to be suitably determined.

Due to sample-and-hold effect, the drift of the gain function $\gamma(\mathbf{x}_{\mathrm{M}}, t)$ in the control interval $T_k \equiv [kT \;;\; (k+1)T)$ must be taken into account, and the resulting dynamics of the sliding variable is

$$\begin{gathered} \ddot{s}(t) = \varphi(\mathbf{x}(t)) - \frac{\eta_4(T)}{\gamma[k]} W_M \mathrm{sign}\left(s[k] - \tfrac{1}{2}\hat{s}_{\mathrm{M}}[k]\right) + \\ + W_{\mathrm{M}} \mathrm{sign}\left(s[k] - \tfrac{1}{2}\hat{s}_{\mathrm{M}}[k]\right); \quad t \in T_k \end{gathered} \tag{43}$$

being, by (4), $|\eta_4(T)| \le \Gamma_{\mathrm{d}} T$.

By collecting the first two terms of the right hand side of (43) into one uncertain function $\bar{\varphi}$, it results

$$|\bar{\varphi}| \le \Phi + T \frac{\Gamma_{\mathrm{d}}}{\Gamma_1} W_M = \bar{\Phi} \tag{44}$$

The control effort W_{M} can be optimally defined, so as to minimize the size of the attained boundary layer, according to Theorem 1.

In particular, W_{M} can be chosen equal to six times the uncertain dynamics upper bound $\bar{\Phi}$, and then (39) and (40) are sufficient conditions for the finite time convergence to the admissible boundary layer, whose size is $\mathcal{O}(T^2)$ [6].

From this time instant T_{reach} on, the second phase of the control algorithm is activated, reducing the boundary layer size. The attained sliding accuracy can be evaluated by means of the discrete model (30) taking into account the discretization error (31) and the estimation error (35).

By the above considerations it results

$$|s[k]| \leq \left\{ \frac{11}{3}\Phi_{\mathrm{d}} + \frac{\Gamma_{\mathrm{d}}}{3}\left(u[k-1] + 4u[k-2]\right) \right\} T^3 \tag{45}$$

As the control effort does not depend on the sampling time T, (45) states that the size of the neighbourhood of the sliding manifold in which the sliding motion occurs is $\mathcal{O}(T^3)$. □

This result is not surprising, and it is related to the effect of the equivalent control predictor, which increases by one the power of T appearing in the size of the boundary layer. In [9] a similar approach improved from $\mathcal{O}(T)$ to $\mathcal{O}(T^2)$ the sliding accuracy for systems with $q = 1$.

5 The Combined Continuous/Discrete Approach

In Sect. 4, by using a feedback control law close to the discrete time equivalent control [14], the ideal control goal has been achieved with $\mathcal{O}(T^3)$ accuracy. Another possible approach is to use some piecewise–constant approximation of the continuous–time equivalent control, which can be generated by means of digital control and measurement devices.

The continuous–time equivalent control has been defined as the continuous control law $u_{\mathrm{eq}}(t)$ such that $\dot{s}(t) = 0$ [15]. Note that it is able to maintain the sliding mode but not to reach it.

In the considered problem the control does not affect directly the first derivative of the sliding variable, and the above definition does not apply. The equivalent control for second order sliding modes has been straightforwardly defined as the continuous control $u^{\mathrm{c}}_{\mathrm{eq}}(t)$ such that $\ddot{s}(t) = 0$ [4], and, from (8) and (18), it results

$$u^{\mathrm{c}}_{\mathrm{eq}}(t) = -\gamma^{-1}[\mathbf{x}_{\mathrm{M}}(t), t]\varphi[\mathbf{x}(t), t] \tag{46}$$

As $\varphi[\mathbf{x}(t), t]$ is not available, the equivalent control must be estimated at each sampling instants. In previous Section, an estimate of the uncertain term $\varphi[k-1]$ has been used to evaluate the discrete time equivalent control $u^{\mathrm{d}}_{\mathrm{eq}}[k]$. By extending this approach, and by estimating the actual value of $\varphi[k]$ with that at time instant $(k-2)T$ in accordance with (34), $u^{\mathrm{c}}_{\mathrm{eq}}(t)$ can be estimated, at the sampling instants, as follows

$$\hat{u}^{\mathrm{c}}_{\mathrm{eq}}[k] = -\frac{\hat{\varphi}[k]}{\gamma[k]} = \frac{1}{\gamma[k]}(\tfrac{1}{2}\gamma[k-1]u[k-1]+ \\ +\gamma[k-2]u[k-2]) - \frac{1}{\gamma[k]}\frac{s[k]-2s[k-1]+s[k-2]}{T^2} \tag{47}$$

Then, it is possible to conceive a two–components piecewise–constant control law, the first one being the equivalent control estimate (47) and the second one to be determined such that the desired manifold is reached and the sliding motion is guaranteed:

$$u[k] = \hat{u}^{\mathrm{c}}_{\mathrm{eq}}[k] + \frac{w[k]}{\gamma[k]} \tag{48}$$

The following Theorem clarifies the proposed strategy and the attained performances

Theorem 3: *Given system* (1),(2) *with its uncertain dynamics satisfying* (18)-(20), *then the digital feedback controller*

$$u[k] = \frac{1}{2\gamma[k]}(\gamma[k-1]u[k-1]+\gamma[k-2]u[k-2]) - \\ -\frac{s[k]-2s[k-1]+s[k-2]}{\gamma[k]T^2} - \frac{W_{\mathrm{M}}}{\gamma[k]}\mathrm{sign}\left(s[k]-\tfrac{1}{2}\hat{s}_{\mathrm{M}}[k]\right) \\ u[-1]=u[-2]=0 \tag{49}$$

$$\hat{s}_{\mathrm{M}}[k] = \begin{cases} s[k-1] & \mathit{if}\ \ \Delta[k]\le 0 \\ \hat{s}_{\mathrm{M}}[k-1] & \mathit{if}\ \ \Delta[k]>0 \end{cases} \\ \Delta[k] = (s[k]-s[k-1])(s[k-1]-s[k-2]) \\ s[-1]=\hat{s}_{\mathrm{M}}[-1]=s(0) \quad s[-2]=0 \tag{50}$$

being s the sliding variable defined in (5) *and with the additional constraints*

$$W_{\mathrm{M}} = 6\left(\frac{26\Phi_{\mathrm{d}}\Gamma_1+11\Gamma\Phi_{\mathrm{d}}+15\Phi_{\mathrm{d}}\Gamma_{\mathrm{d}}T}{6\Gamma_1-71\Gamma_{\mathrm{d}}T}\right)T \tag{51}$$

$$T < \frac{6}{71}\frac{\Gamma_1}{\Gamma_{\mathrm{d}}} \tag{52}$$

guarantees the finite time reaching of a $\mathcal{O}(T^3)$ vicinity of the sliding manifold $s=0$, being T the sampling period.

Proof: By (8), (47) and (48), the dynamics of the sliding variable s can be represented as

$$\ddot{s}(t) = \varphi[k]-\hat{\varphi}[k]+\eta_\varphi(T) \\ -\frac{1}{\gamma[k]}(\hat{\varphi}[k]-w[k])\,\eta_\gamma(T)+w[k] \quad t\in T_k \equiv [kT\ ,\ (k+1)T) \tag{53}$$

η_φ and η_γ being two uncertain terms taking into account the drift of $\varphi(\mathbf{x}(t),t)$ and $\gamma(\mathbf{x}_{\mathrm{M}},t)$ within the control period, which, by (20), are upperbounded in accordance with

$$\begin{aligned} |\eta_\varphi(T)| &\leq \Phi_\mathrm{d} T \\ |\eta_\gamma(T)| &\leq \Gamma_\mathrm{d} T \end{aligned} \tag{54}$$

Let $\Psi_k(T)$ be the uncertain function collecting all the terms in (53) except the control $w[k]$; the sliding variable dynamics can be rewritten as follows

$$\ddot{s}(t) = \Psi_k(T) + w[k] \quad t \in T_k \tag{55}$$

If $|\Psi_k(T)|$ is upper bounded by some non negative constant then the sub-optimal second order sliding mode controller is able to stabilize system (55).

Due to the choice of estimating $\varphi[k]$ by means of (34), it results

$$|\hat{\varphi}[k] - \varphi[k]| \leq \left(\frac{10}{3}\Phi_\mathrm{d} + \frac{2}{3}\Gamma_\mathrm{d} u[k-2] + \frac{1}{6}\Gamma_\mathrm{d} u[k-1]\right) T \tag{56}$$

By (48) it can be derived that, in any control interval T_k, the control amplitude is such that

$$|u[k]| \leq \frac{1}{\Gamma_1}\left\{\Phi + \max(|\hat{\varphi}[k] - \varphi[k]|) + W_M\right\} \tag{57}$$

If (57) is considered into (56), then (56) can be rewritten as

$$|\hat{\varphi}[k] - \varphi[k]| \leq \frac{5T}{6\Gamma_1 - 5\Gamma_\mathrm{d} T}\left[4\Phi_\mathrm{d}\Gamma_1 + (\Phi + W_\mathrm{M})\,\Gamma_\mathrm{d}\right] \tag{58}$$

Therefore, by the definition of $\Psi_k(T)$, and taking into account (19),(20), (54) and (58), the modulus of the uncertain term is upper bounded by the constant

$$\Psi_\mathrm{M} = \left(\frac{26\Phi_\mathrm{d}\Gamma_1 + 11\Phi\Gamma_\mathrm{d} + 15\Phi_\mathrm{d}\Gamma_\mathrm{d} T}{6\Gamma_1 - 5\Gamma_\mathrm{d} T}\right) T + \frac{11\Gamma_\mathrm{d} T}{6\Gamma_1 - 5\Gamma_\mathrm{d} T} W_\mathrm{M} \tag{59}$$

By applying the sufficient conditions for the finite time convergence of the control algorithm defined in Theorem 1, i.e., $W_\mathrm{M} = 6\Psi_\mathrm{M}$, theorem's assumptions (51) and (52) are directly derived.

Furthermore, the size of the boundary layer of the sliding manifold (5), in which the attained real sliding motion occurs, can be evaluated by (16).

Due to the estimation of the uncertainties, in the considered problem the uncertain dynamics $\Psi_k(T)$ is made $\mathcal{O}(T)$ according to (59) and (51), and therefore the corresponding size of the boundary layer, which has been also proved to be reached in a finite time, is $\mathcal{O}(T^3)$ in accordance with

$$|s| \leq \rho^* \left(\frac{26\Phi_\mathrm{d}\Gamma_1 + 11\Phi\Gamma_\mathrm{d} + 15\Phi_\mathrm{d}\Gamma_\mathrm{d} T}{6\Gamma_1 - 71\Gamma_\mathrm{d} T}\right) T^3 \tag{60}$$

□

6 Simulation results

Consider system (1) with $n = 3$ and

$$\begin{cases} f(\mathbf{x}(t), t) = 3 + sin(10t + x_1)cos(x_2^2 + x_3^2) \\ g(\mathbf{x_M}) = 3 + sin(3 + x_1 + x_2) \end{cases} \tag{61}$$

x_3 is assumed to be not available, $T = 10^{-3}$s is the sampling period and the initial conditions are set to $\mathbf{x}(0) = [1, 1, 1]$.

The chosen manifold is

$$s(\mathbf{x_M}) = x_2 + 2x_1 \tag{62}$$

The control task is that of asymptotically steering to zero the states by constraining the controlled system on the sliding set $s = \dot{s} = 0$.

The digital suboptimal control algorithm is first implemented to compare it with the two new control schemes here proposed, one based on the dead beat approach (Sect. 4) and the other on the combined continuous-discrete (Sect. 5). The transient behaviour is the same for all the control strategies until a $\mathcal{O}(T^2)$ vicinity is reached (fig. 1). The further contraction from $\mathcal{O}(T^2)$ to $\mathcal{O}(T^3)$ is performed by the discrete approach in one sampling step, while the combined approach requires a slower transient.

The more evident positive feature of the novel 2-sliding controllers lies in the appearance of the control law. The suboptimal algorithm does not compensate the uncertainties, and the resulting control effort is large with subsequent samples which differ by a finite quantity (Fig. 2). The new proposed strategies resort to the suboptimal algorithm for an initialization procedure, and then they commute to different structures which give rise to practically continuous controls, with subsequent samples differing by $\mathcal{O}(T)$ (Figs. 3,4).

In Figs. 5-7 the accuracy in the sliding constraint maintenance is evidenced for each of the three different controllers. The residual $\mathcal{O}(T)$ switching component in the combined approach control increases the boundary layer size and leads to high frequency ripple around the manifold. The discrete approach, which has not discontinuous feedback component, produces better accuracy and smoothness of the motion within the boundary layer.

The same tests have been conducted by using $T = 0.5 \cdot 10^{-3}$s. The simulation results confirm the expected accuracy with respect to the sampling period (Figs. 8-10).

For the sake of completeness, the convergence to zero of the states x_i, $i = 1, 2, 3$, is investigated in Fig. 11 when the suboptimal algorithm is used.

7 Conclusions

In this paper the discrete time control of uncertain nonlinear systems with uncomplete state availability has been dealt with. The adopted philosophy was that of resorting to the equivalent control principle, in both its discrete-time and continuous-time versions.

The first proposed approach involves a controller synthesis completely developed in the discrete-time domain. It leads to very good performances of the controlled system, due to the $\mathcal{O}(T^3)$ size of the attained boundary layer and to the smoothness of the resulting control law. As a result, output chattering is eliminated in the steady state, since there is no switching component in the control input. Moreover, the above procedure is feasible only after a $\mathcal{O}(T^2)$ vicinity of the manifold is reached.

The combined continuous-discrete time approach preserves the $\mathcal{O}(T^3)$ accuracy of the discrete approach, but the residual control switching component introduces high frequency ripple in the system outputs.

The actual size of the boundary layer depends on the particular case under investigation, as it can be not by direct inspection of (60) and (45).

The combined approach does not involve admissible regions for the smooth control, so that it can be applied from the first sampling instant on. However, in practical applications, it is suitable to activate the smooth control after that some vicinity of the sliding manifold is reached to improve the convergence speed.

Moreover, by taking into account in the synthesis procedure the errors due to the discretization, the choice of the sampling period is not subjected to possible non minimum phase behaviors of the controlled system, and the resulting controllers show, theoretically, better robustness properties.

Future developments of this work will be devoted to deal with multi input systems and to extend the class of systems for which the proposed techniques are effective, in particular removing the knowledge assumption of the control gain. Further extensive simulations will be carried out to show the effectiveness of the proposed controllers in practical applications such as robotics, underwater vehicles etc.

APPENDIX

Proof. of Theorem 1

The proposed algorithm requires the approximate real-time evaluation of the singular points of the available state variable $y_1(t)$, (that is of the values corresponding to the time instants at which its derivative $y_2(t)$ is zero). The approximate peak holder (13) is implemented with this aim.

The *modus operandi* of the Algorithm consists in constraining the state trajectories on the y_1Oy_2 plane between two limiting lines, defined taking into account the extreme constant bounds, $\pm F$, G_1 and G_2, of the uncertain dynamics, both converging to a neighbourhood of the origin (fig. 11).

The proposed feedback law causes subsequent crossings of the state trajectory with the abscissa axis, and the control aim is attained by choosing the controller parameters in order to assure that these subsequent crossings are nearer and nearer to the origin of the state plane, so assuring the convergence property.

Let $y_{1\mathrm{M}_j}$ be the actual $j-th$ singular value of $y_1(t)$, $\hat{y}_{1\mathrm{M}_j}$ its estimate, t_{M_j} the corresponding time instant and t_{c_j} the time instant subsequent t_{M_j} at which a commutation occurs $(j = 1, 2, \ldots)$.

The proof can be splitted into three different parts:

1. Reaching of the first singular value

It is trivial to verify that, if the control amplitude satisfies the dominance condition $U_\mathrm{M} \geq \frac{F}{\alpha^* G_1}$ (such that the sign of $\dot{y}_2(t)$ is directly affected by that of the input $u(t)$), for any initial condition $(y_1(0), y_2(0))$ a point of the abscissa axis is reached in a finite time, and it is the first singular value $y_{1\mathrm{M}_1}$ of the trajectory $y_1(t)$.

2. Contraction property

U_M and α^* are chosen such that the contractive behavior defined by condition

$$|y_{1\mathrm{M}_{j+1}}| < |y_{1\mathrm{M}_j}| \qquad j = 1, 2, \ldots \tag{63}$$

takes place.

Suppose, without loss of generality, that the actual value of the $j-th$ singular value is such that $y_{1\mathrm{M}_j} > 0$, i.e. it lies on the right side of the abscissa axis. Due to the symmetry of the problem with respect to the origin of the state plane, analogous consideration are also valid if $y_{1\mathrm{M}_j} < 0$.

Due to the sampled nature of the measures, the updating of the gain coefficient $\alpha[k]$ and the switching of the control can occur with a delay at most equal to T with respect to the ideal ones in $t = t_{\mathrm{M}_j}$ and in $y_1[k] = \frac{1}{2}\hat{y}_{1\mathrm{M}}[k]$ respectively.

Consequently, at the actual switching time instant $t = t_{\mathrm{c}_j}$, the states satisfy the following conditions

$$\begin{aligned} y_1(t_{\mathrm{c}_j}) \in [&\tfrac{1}{2}y_{1\mathrm{M}_j} - \tfrac{1}{16}(F + G_2U_\mathrm{M})T^2 - \tfrac{1}{2}(F + \alpha^* G_2 U_\mathrm{M})T^2 \\ &-T\sqrt{y_{1\mathrm{M}_j}(F + \alpha^* G_2 U_\mathrm{M}) + aT^2}\ ,\ \tfrac{1}{2}y_{1\mathrm{M}_j}] \\ y_2(t_{\mathrm{c}_j}) \in [&-(F + \alpha^* G_2 U_\mathrm{M})T - \sqrt{y_{1\mathrm{M}_j}(F + \alpha^* G_2 U_\mathrm{M}) + aT^2}\ , \\ &-\sqrt{y_{1\mathrm{M}_j}(F - \alpha^* G_1 U_\mathrm{M})}] \end{aligned} \tag{64}$$

$$a = \frac{1}{8}(F + G_2U_\mathrm{M})[(F + \alpha^* G_2 U_\mathrm{M}) + 18G_2U_\mathrm{M}(1 - \alpha^*)] \tag{65}$$

As the system trajectory is constrained between the limit lines in fig. 1, the subsequent crossing of the abscissa axis belongs to the interval

$$\begin{aligned} y_{1\mathrm{M}_{j+1}} \in [&-\tfrac{1}{2}\tfrac{(\alpha^* G_2 - G_1)U_\mathrm{M} + 2F}{G_1U_\mathrm{M} - F}y_{1\mathrm{M}_j} - bT^2 \\ &-\tfrac{G_1 + \alpha^* G_2}{G_1U_\mathrm{M} - F}U_\mathrm{M}T\sqrt{y_{1\mathrm{M}_j}(F + \alpha^* G_2 U_\mathrm{M}) + aT^2}\ , \\ &\tfrac{1}{2}\tfrac{(G_2 - \alpha^* G_1)U_\mathrm{M} + 2F}{G_2U_\mathrm{M} + F}y_{1\mathrm{M}_j}] \end{aligned} \tag{66}$$

$$\begin{aligned} b = &\tfrac{1}{16}(F + G_2U_\mathrm{M})\left[\tfrac{(G_1 + \alpha^* G_2)}{G_1U_\mathrm{M} - F}U_\mathrm{M}\right] + \tfrac{9}{8}(F + G_2U_\mathrm{M})(1 - \alpha^*)G_2U_\mathrm{M} + \\ &+\tfrac{1}{2}(F + \alpha^* G_2 U_\mathrm{M})\left[\tfrac{(G_1 + \alpha^* G_2)}{G_1U_\mathrm{M} - F}U_\mathrm{M}\right] \end{aligned} \tag{67}$$

Define the following normalized non negative variables

$$z = \frac{U_M}{F} \tag{68}$$

$$\rho = \frac{|y_{1M_j}|}{FT^2} \tag{69}$$

Sufficient condition for the fulfillment of the contraction condition (63) is represented by the following system of inequalities

$$\begin{cases} \rho \geq 0 \\ z \geq \frac{1}{\alpha^* G_1} \\ (3G_1 - \alpha^* G_2)z - 4 > (G_1 + \alpha^* G_2)(1 + \alpha^* G_2 z)\frac{z}{\rho} + \\ \quad + \frac{1}{8}[G_1 + (18 - 17\alpha^*)G_2](1 + G_2 z)\frac{z}{\rho} + \\ \quad\quad + 2(G_1 + \alpha^* G_2)\frac{z}{\rho}\sqrt{(1 + \alpha^* G_2 z)\rho + \frac{1}{8}(1 + G_2 z)[1 + (18 - 17\alpha^*)G_2 z]} \end{cases} \tag{70}$$

The second inequality represents the control's dominance condition, ensuring that the sign of the control $u(t)$ sets the sign of $\dot{y}_2(t)$. The third inequality in (70) defines the set $\mathcal{Z} \subseteq \mathcal{R}$ such that, $\forall z \in \mathcal{Z}$, the points of the straight line defined by the left-hand side term $w_1(z) = (3G_1 - \alpha^* G_2)z - 4$ lie above the points of the parametric function

$$\begin{aligned} w_2(z) = {} & (G_1 + \alpha^* G_2)(1 + \alpha^* G_2 z)\frac{z}{\rho} + \frac{1}{8}[G_1 + (18 - 17\alpha^*)G_2](1 + G_2 z)\frac{z}{\rho} + \\ & + 2(G_1 + \alpha^* G_2)\frac{z}{\rho}\sqrt{(1 + \alpha^* G_2 z)\rho + \frac{1}{8}(1 + G_2 z)[1 + (18 - 17\alpha^*)G_2 z]} \end{aligned} \tag{71}$$

The function $w_2(z;\rho)$ crosses the origin of the cartesian plane zOw_2, and it has a negative local minimum for $z < 0$ and positive first and second derivatives with respect to z if $z > 0$. Moreover, the value of the local minimum is an increasing function of the parameter ρ, and the parametric function $w_2(z;\rho)$ degenerates into the abscissa axis as the parameter ρ goes to infinity. By means of the above considerations it is possible to claim that the two lines defined by the functions $w_1(z)$ and $w_2(z;\rho)$ have at most two cross points, which degenerate into a double contact point for a specific lower value of the parameter ρ called ρ^*. For values of $\rho < \rho^*$ there is no intersection between the two lines and system (70) has not solutions.

This fact imply that the convergence of the sliding variable and of its time derivative to zero is assured only if the the control amplitude is chosen within a proper open set.

The limits of the admissible set depend on ρ, nevertheless the control amplitude could be chosen such that the non negative normalized variable ρ, defining the size of the boundary layer, can reach its minimum at ρ^*. In this case, the admissible set collapses into a single point, which represent a sort of *optimal value* of the control amplitude, which minimizes the theoretical size of the corresponding boundary layer.

The ρ^* value, and the corresponding point $z = z^*$, have been calculated in [6] under the assumption that no gain affects the control input, that is $G_1 = G_2 = \alpha^* = 1$, leading to the following approximate solution

$$\begin{cases} \rho^* = 85 \\ z^* = 6 \end{cases} \tag{72}$$

This means that, if $G_1 = G_2 = \alpha^* = 1$, then $U_M = 6F$ is the control effort that minimizes the size of the boundary layer.

By re–considering the general case, it can be noted that, as ρ^* does not depend on T, then, by (64), (65) and (69),Theorem's statement (16) is directly derived.

The dependence of the bounds of the admissible set $\mathcal{Z}$ by the sampling period T can be investigated by analyzing the limit behaviour of system (70) for $T \to 0$.

Since $\rho = \mathcal{O}(T^{-2})$, an intersection between $w_1(z, \rho)$ and $w_2(z)$ can occur iff $z = \frac{4}{3G_1 - \alpha^* G_2} + \mathcal{O}(T)$ or $z = \mathcal{O}(T^{-2})$.

By these considerations, (14) is directly justified. So the Theorem is proved.

3. Finite time reaching of the boundary layer

As the time interval between two subsequent singular values of $y_1(t)$ is finite, the finite time convergence of the system to the residual set is a straightforward consequence of the contraction condition. □

References

1. G. Bartolini, A. Ferrara, V.I. Utkin, "Adaptive sliding mode control in discrete time systems", *Automatica*, vol. 31, no. 6, pp. 769–773, 1995
2. G. Bartolini, A. Ferrara, E. Usai, "Applications of a sub–optimal discontinuous control algorithm for uncertain second order systems", *Int. J. of Robust and Nonlinear Control*, vol. 7, no. 4, pp. 299–319, 1997
3. G. Bartolini, A. Ferrara, E. Usai, "Output Tracking Control of Uncertain Nonlinear Second-Order Systems", *Automatica*, vol. 33, no. 12, pp. 2203–2212, 1997
4. G. Bartolini, A. Ferrara, A. Pisano, E. Usai, "Adaptive reduction of the control effort in chattering free sliding mode control of uncertain nonlinear systems", *App. Math. and Computer Science*, vol. 8, no. 1, pp. 51–71, 1998
5. G. Bartolini, A. Ferrara, E. Usai, "Chattering Avoidance by Second Order Sliding Mode Control", *IEEE Trans. Automat. Contr.*, vol. 43, no. 2, pp. 241–246, 1998
6. G. Bartolini, A. Pisano, E. Usai, "Digital Second Order Sliding Mode Control of SISO Uncertain Nonlinear Systems", *Proc. of the 1998 American Control Conference ACC'98*, vol. 1, pp. 119–124, Philadelfia, Pensylvania, June 1998
7. S.V. Drakunov, V.I. Utkin, "Sliding mode in dynamic systems", *Int. Journal of Control*, vol. 55, pp. 1029–1037, 1990
8. S.V. Drakunov, U. Ozguner, W.C. Su, K.D. Young, "Sliding Mode with Chattering Reduction in Sampled Data Systems", *Proc. of the 32th Conf. on Decision and Control - CDC'93*, pp. 2452–2457,

9. S.V. Drakunov, U. Ozguner, W.C. Su, "Implementation of Variable Structure Control for Sampled–Data Systems", in *Robust Control via variable structure and Lyapunov techniques.* F.Garofalo and L.Glielmo eds., Lecture Notes in Control and Information Science no. 217, pp. 87–106, Springer-Verlag, London, 1996
10. K. Furuta, "Sliding mode control of a discrete system", *System and Control Letters*, vol. 14, no. 2, pp. 145–152, 1990
11. K. Furuta, Y. Pan, "VSS controller design for discrete time systems", *Control Theory and Advanced Technology*, vol. 10, no. 4/1, pp. 669–687, 1994
12. A. Levant (Levantovsky L.V.), "Sliding order and sliding accuracy in sliding mode control", *Int. Journal of Control*, vol. 58, no. 6, pp. 1247–1263, 1993
13. C. Milosavljevic, "General conditions for the existence of a quasisliding mode on the switching hyperplane in discrete variable systems", *Automation Remote Control*, vol. 43, no. 1, pp. 307–314, 1985
14. V.I. Utkin, S.V. Drakunov, "On Discrete–Time Sliding Mode Control", *Proc. of the IFAC Symposium on Nonlinear Control Systems - NOLCOS*, pp. 484–489, Capri, Italy, 1989, San Antonio, Texas, December 1993
15. V.I. Utkin, *Sliding Modes In Control And Optimization*, Springer Verlag, Berlin, 1992
16. V.I. Utkin, "Sliding Mode Control in Discrete–Time and Difference Systems", in *Variable Structure and Lyapunov Control.* A.S.I. Zinober ed., pp. 83–102, Springer-Verlag, London, 1993

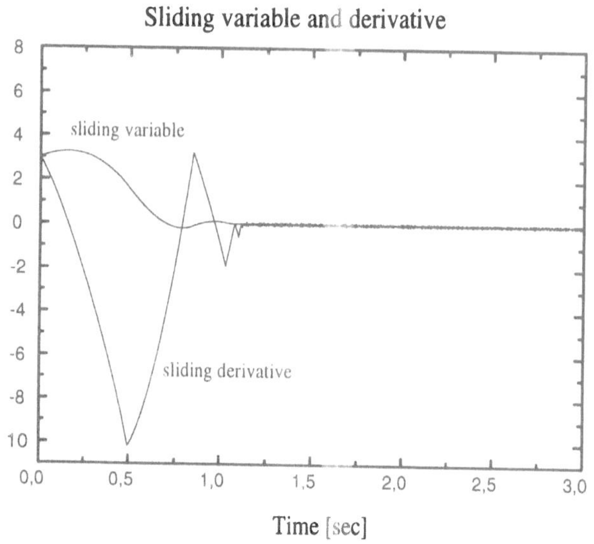

Fig. 1. Suboptimal algorithm. Sliding variable and its derivative ($T = 10^{-3}$s).

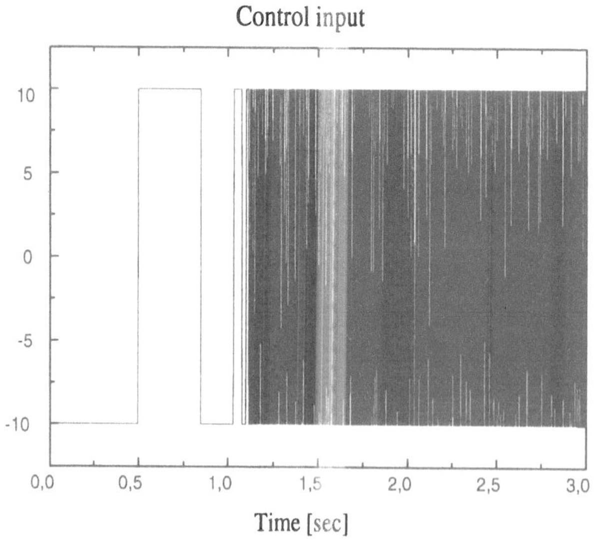

Fig. 2. Suboptimal algorithm. Control input ($T = 10^{-3}$s).

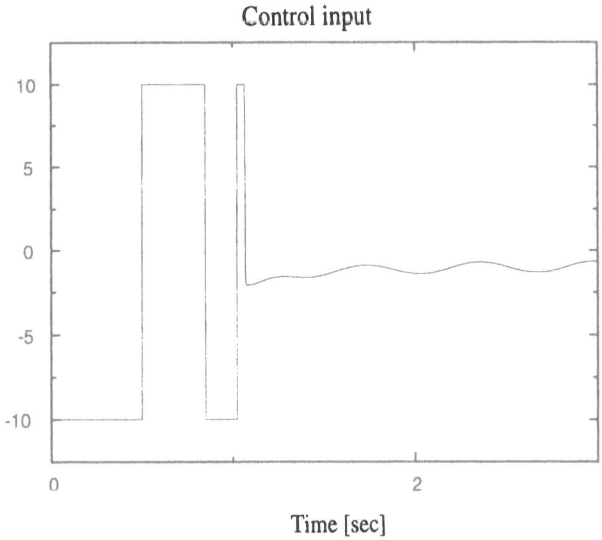

Fig. 3. Combined approach. Control input ($T = 10^{-3}$s).

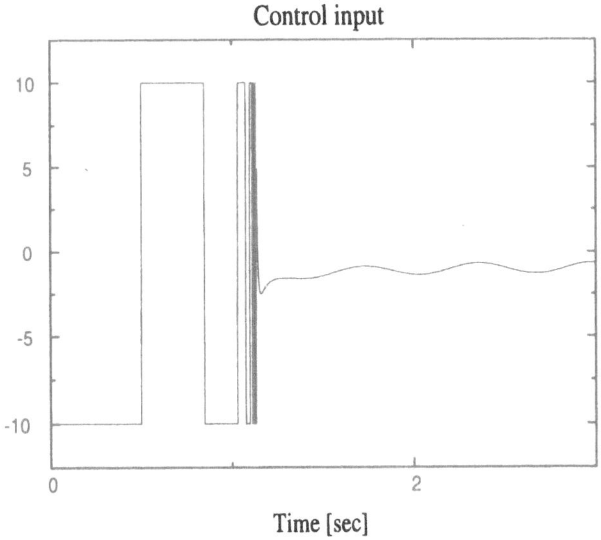

Fig. 4. Discrete approach. Control input ($T = 10^{-3}$s).

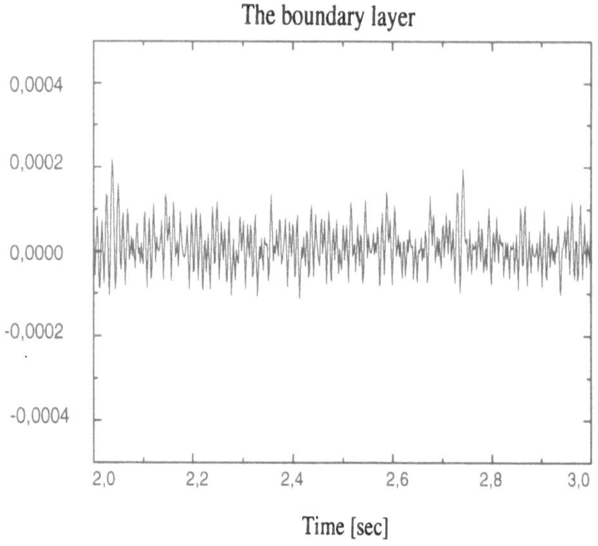

Fig. 5. Suboptimal algorithm. Boundary layer ($T = 10^{-3}$s).

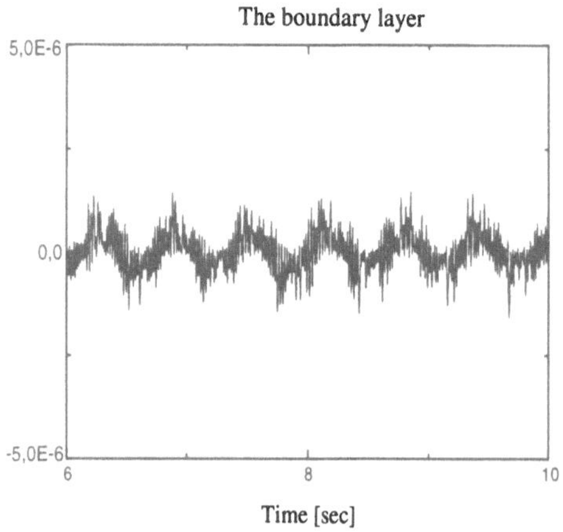

Fig. 6. Combined approach. Boundary layer ($T = 10^{-3}$s).

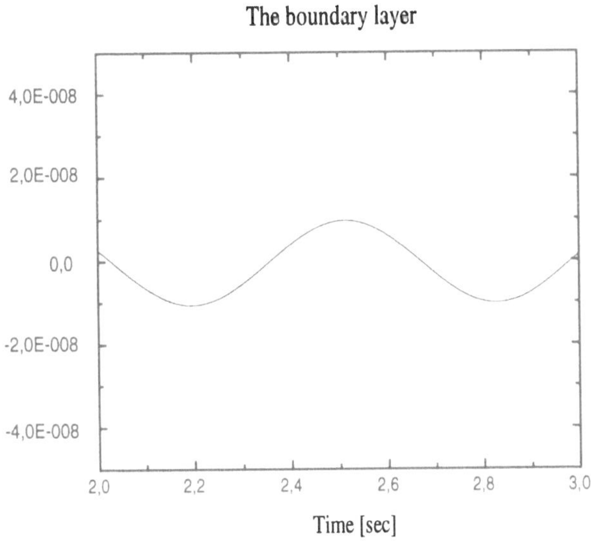

Fig. 7. Discrete approach. Boundary layer ($T = 10^{-3}$s).

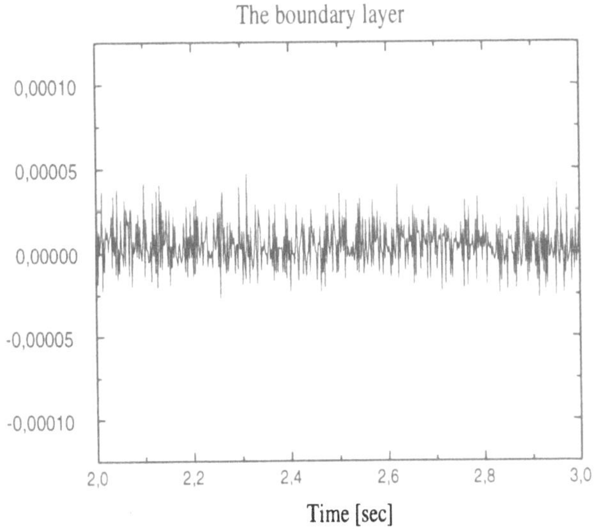

Fig. 8. Suboptimal algorithm. Boundary layer ($T = 0.5 \cdot 10^{-4}$s).

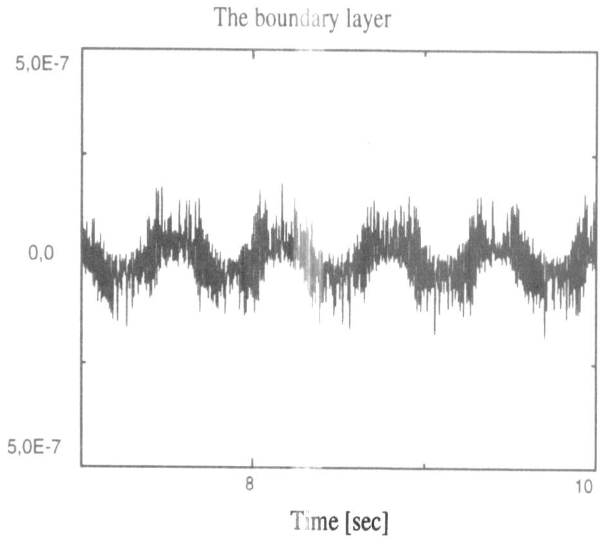

Fig. 9. Combined approach. Boundary layer ($T = 0.5 \cdot 10^{-3}$s).

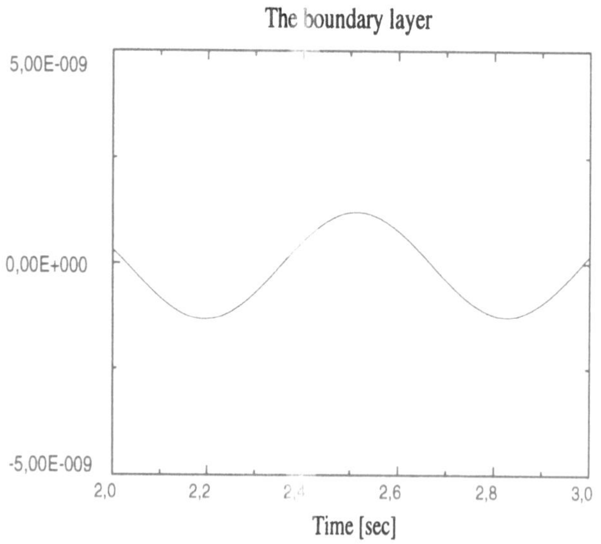

Fig. 10. Discrete approach. Boundary layer ($T = 0.5 \cdot 10^{-3}$s).

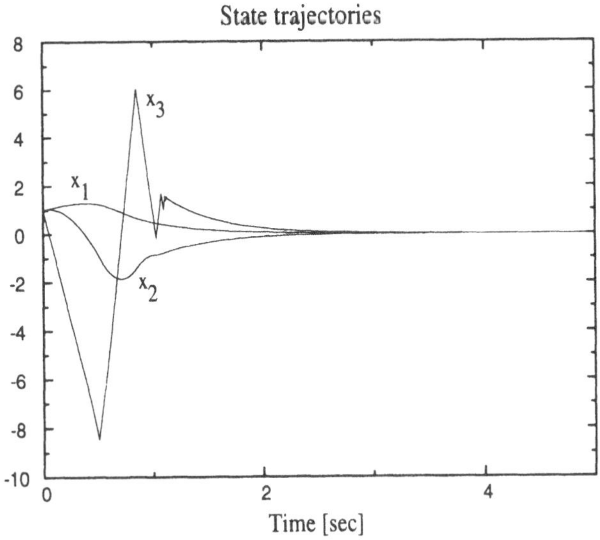

Fig. 11. Suboptimal Algorithm. State behaviour.

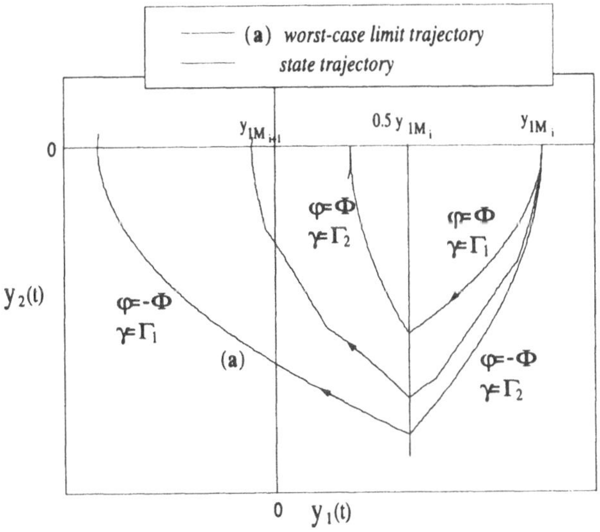

Fig. 12. The limit trajectories.

On Sampled Data Variable Structure Control Systems

Jian-Xin Xu, Feng Zheng and Tong Heng Lee

Department of Electrical Engineering
National University of Singapore
Singapore 119260

Abstract. It is shown that the motion of sampled-data variable structure control systems can be divided into three phases, namely reaching phase, switching phase and chattering phase. The definitions of the three phases are given and the characteristics of the three phases are described. It is shown that in the reaching phase the switching function decreases monotonically, and in the switching phase the switching function stops decreasing but the magnitude of position error decreases monotonically, while in the chattering phase all variables stop decreasing and the chattering is dominated by the velocity. The bound or limit values of corresponding variables are given.

1 Introduction

It is well known that the characteristic feature of continuous variable structure control (VSC) systems is that sliding mode occurs on a prescribed switching surface, which has drawn many attentions from both theoretical and industrial communities (see e.g. [4, 9, 18, 19]). For continuous VSC systems, there are only two phases in ideal case: reaching phase and sliding phase. In sampled-data system, the probability of the existence of ideal sliding mode is zero due to limited sampling rate, sample/hold effect and discretization. As a result, the well-known chattering phenomenon occurs once the states of the concerned systems reach the vicinity of the switching surface.

Since Dote and Hoft [6] first investigated discrete-time VSC systems, many contributions (see e.g. [1, 2, 3, 5, 7, 8, 10, 11, 12, 13, 14, 15, 16, 17, 20, 21, 22, 23]) have been devoted to the study of discrete-time VSC systems. In the above references, the concept of quasi-sliding mode was presented, which is one of the characteristics of discrete-time systems, and various kinds of reaching conditions were given. In [8], some attributes of discrete-time VSC systems were critically presented. It is worthy to notice that these developments is not straight and there are still some essential nature of discrete-time VSC systems to be investigated.

In this paper, we will show that, for the sampled-data VSC systems, the motion of their states can be divided into three phases, namely reaching phase, switching phase and chattering phase. In the *reaching phase*, the system states are forced to reach the prescribed switching surface. Once hitting the switching surface, the system will move, zigzag though, towards the equilibrium point along the switch surface. This is defined to be *switching phase*. In this phase, the switching control ensures the robustness, i.e. forces the system states to stay on or nearby the prescribed switching surface, though impossible to slide on it exactly due to finite sampling rate. Finally when the system states reach a neighborhood of the equilibrium, they will oscillate or chatter and the tracking errors will no longer decrease. The size of this neighborhood depends on the sampling rate, coefficients of switching function and the controller gain.

It will also be shown in this paper that the chattering is dominated by the velocity of a second-order system, when the system is described by phase variables. This is not surprised intuitively since the control is directly applied to the internal channel of the highest order state. By this property, the magnitude of the tracking errors of the position error for second-order systems can be derived, which provides useful guidelines for the design of sliding mode control systems with limited sampling rate.

The organization of this paper is as follows. In Section 2, the problem is formulated. Basic concepts, definitions, assumptions and some derivations are provided in Section 3 to give a basis for the later analysis and discussions. Three phases for sampled-data VSC systems are identified and the characteristics of the three phases are discussed in Sections 4. In Section 5, simulation results are presented to show the correctness of the analysis in the previous sections. Finally concluding remarks are given in Section 6.

2 Problem Formulation

For simplicity of later discussion and analysis, consider the following second-order uncertain system:

$$\begin{cases} \dot{x}_1 = x_2 \\ \dot{x}_2 = f(x_1, x_2, t) + b(x_1, x_2, t)u + d(x_1, x_2, t), \end{cases} \tag{1}$$

where $x_1, x_2 \in R$ are the state variables, $u \in R$ is the control variable, $d \in R$ represents a lumped system uncertainty. Suppose that f and b are known functions of x_1, x_2 and t, $b(x_1, x_2, t) \neq 0$ for all x_1, x_2 and t. and d is bounded with a known bound, i.e.

$$|d(x_1, x_2, t)| \leq \gamma,$$

where γ is a known positive number.

Our objective is to design a variable structure controller to steer the position, x_1, of the system (1) to a constant set point r from any initial states $(x_1(0), x_2(0)) \in R^2$. Hence the velocity, x_2, of the system (1) is expected to settle down to zero.

Define the tracking errors as

$$\begin{cases} e_1 = x_1 - r \\ e_2 = x_2 - \dot{r} = \dot{x}_1. \end{cases} \tag{2}$$

The system (1) reads as in the new coordinate:

$$\begin{cases} \dot{e}_1 &= e_2 \\ \dot{e}_2 &= f(e_1 + r, e_2, t) + b(e_1 + r, e_2, t)u + d(e_1 + r, e_2, t) \\ &\overset{\text{def}}{=} \bar{f}(e_1, e_2, r, t) + \bar{b}(e_1, e_2, r, t)u + \bar{d}(e_1, e_2, r, t). \end{cases} \tag{3}$$

We need only consider the system (3) thereafter. Choose a switching function for the system (3) as

$$\sigma = e_2 + ce_1, \tag{4}$$

where c is a positive number which decides the switching slope for the system. It is easy to verify that, when the following control law is used:

$$u = -\frac{(\bar{f} + ce_2) + (\gamma + \eta)\text{sgn}(\sigma)}{\bar{b}}, \qquad \eta > 0, \tag{5}$$

the following inequality holds

$$\sigma\dot{\sigma} \leq -\eta\,|\sigma|, \tag{6}$$

which is just the well known reaching condition for continuous VSC systems.

Now consider the discrete-time version of the above continuous VSC system. Suppose that the sampling interval is T and a zero-order hold is adopted for the controller.

Write the equations (3) and (4) in a concise form:

$$\begin{cases} \dot{e} = F(e,t) + B(e,t)u + D(e,t) \\ \sigma = Ce \\ u = -\dfrac{(\bar{f} + ce_2) + (\gamma + \eta)\text{sgn}(\sigma)}{\bar{b}}, \end{cases} \tag{7}$$

where $e = [e_1, \quad e_2]^T$, $F(e,t) = [0, \quad \bar{f}(e_1, e_2, t)]^T$, $B(e,t) = [0, \quad \bar{b}(e_1, e_2, t)]^T$, $D(e,t) = [0, \quad \bar{d}(e_1, e_2, t)]^T$, and $C = [c, \quad 1]^T$.

Denote a sampled variable at $t = kT$ by a subscript k, e.g.

$$e_k \overset{\text{def}}{=} e(kT), \quad u_k \overset{\text{def}}{=} u(kT), \quad \sigma_k \overset{\text{def}}{=} \sigma(kT).$$

Then the discrete-time version of the continuous VSC system (7) reads as:

$$\begin{cases} e_{k+1} = e_k + \displaystyle\int_{kT}^{(k+1)T} F(e(t),t)dt + \int_{kT}^{(k+1)T} B(e(t),t)dt\; u_k \\ \qquad + \displaystyle\int_{kT}^{(k+1)T} D(e(t),t)dt \\ \sigma_k = Ce_k \\ u_k = -\dfrac{(\bar{f}_k + ce_{2,k}) + (\gamma + \eta)\text{sgn}(\sigma_k)}{\bar{b}_k}. \end{cases} \tag{8}$$

Our goal is to analyze the behavior and asymptotical properties of the sampled-data system (8). To reach this end, we would like to introduce the following definition and assumptions.

3 Preliminaries

To facilitate the later analysis, it is necessary to introduce the following definitions.

Definition 1. The magnitude of a variable v is said to be of order $O(T^n)$ if and only if

$$\lim_{T\to 0}\frac{v}{T^n}\neq 0 \quad \text{and} \quad \lim_{T\to 0}\frac{v}{T^{n-1}}=0,$$

where n is an integer. We define that $O(T^0)=O(1)$.

Associated with the above definition and the system (3), the following assumptions are made.

Assumption 1 *The sampling internal T is sufficiently small such that for any variables $v_1\in O(T^n)$ and $v_2\in O(T^{n+1})$, we have that $v_1\gg v_2$. Therefore we have the following relations about the effective approximation:*

$$\begin{aligned} O(T^n)+O(T^{n+1})&\approx O(T^n) \qquad \forall n\in \mathcal{Z},\\ O(T^n)\cdot O(1)&\approx O(T^n) \qquad \forall n\in \mathcal{Z}, \end{aligned} \tag{9}$$

where $\approx$ stands for the effective approximation and $\mathcal{Z}$ is the set consisting of all integers.

Remark. Notice that $O(1)$ is not necessary to be of magnitude 1. The order of $O(1)$ is defined relatively to T, e.g. $O(1)$ can be of value 100 if $T=10^{-5}$.

Assumption 2 *The disturbance $\bar{d}$ is sufficiently small such that $\bar{d}$ is at most of the order $O(T)$, i.e., $d\in O(T)$.*

We will discuss the case of large disturbances in our future work.

Assumption 3 *The function $\bar{b}(e,t)$ does not vanish for all e and t. Furthermore, we suppose that*

$$\bar{b}(e,r,t)\geq b_{\min}>0 \qquad \text{and} \quad b_{\min}\in O(1),$$

where $b_{\min}$ is a constant.

Assumption 4 *The functions f, b and d are smooth with respect to their arguments and satisfy the following conditions:*

$$F_{diff}\stackrel{\text{def}}{=}\sup(\left|\frac{\mathrm{d}\,\bar{f}(e_1,e_2,r,t)}{dt}\right|)\in O(1);$$

$$B_{diff}\stackrel{\text{def}}{=}\sup(\left|\frac{\mathrm{d}\,\bar{b}(e_1,e_2,r,t)}{dt}\right|)\in O(1);$$

$$D_{diff}\stackrel{\text{def}}{=}\sup(\left|\frac{\mathrm{d}\,\bar{d}(e_1,e_2,r,t)}{dt}\right|)\in O(1).$$

In addition, $e_1(0),e_2(0),r\in O(1)$, and $f,b\in O(1)$ if $e_1,e_2,r\in O(1)$.

Remark. Assumption 4 implies that the dynamic system (3) should not vary too fast to be controllable by a sampled-data controller with limited sampling rate. This fact, together with Assumption 3, ensures a reasonable bound of the control input.

To make the analysis in Lemma 3 more concise, we need the following lemma.

Lemma 2. *Consider the function $g(v(t), t)$, where t is the time variable and v represents a vector-valued function of t.*

(i) If g is bounded and satisfies

$$|g(v,t)| \le M_g \in O(1), \qquad \forall t \in (0, \infty), \tag{10}$$

where M_g is a constant, then for any numbers $\zeta_1 > 0$ and $\zeta_2 > 0$, $|\zeta_1 - \zeta_2| \in O(T)$, we have

$$\int_{\zeta_1}^{\zeta_2} g(v,t)dt \in O(T).$$

(ii) If g is continuously differentiable with respect to its all arguments and satisfies

$$G_{diff} \stackrel{\text{def}}{=} \sup(\left|\frac{\mathrm{d}\, g(v,t)}{\mathrm{d}t}\right|) \in O(1),$$

then for any numbers $\zeta_1 > 0$ and $\zeta_2 > 0$, $|\zeta_1 - \zeta_2| \in O(T)$, we have

$$g(v(\zeta_2), \zeta_2) - g(v(\zeta_1), \zeta_1) \in O(T).$$

The proof is omitted since it is quite straightforward.

Lemma 3. *Suppose Assumptions 1, 2, 3 and 4 hold, then we have*

$$e_{2,k+1} = (1 - cT)e_{2,k} - (\gamma + \eta)T\mathrm{sgn}\sigma_k + O(T^2), \tag{11}$$

$$e_{1,k+1} = e_{1,k} + e_{2,k}T + O(T^2), \tag{12}$$

$$\sigma_{k+1} = \sigma_k - (\gamma + \eta)T\mathrm{sgn}\sigma_k + O(T^2). \tag{13}$$

Proof: By the formulae shown in the equation (8), we can obtain that

$$\begin{aligned}
&e_{2,k+1} \\
&= e_{2,k} + \int_{kT}^{(k+1)T} [\bar{f}(e_1, e_2, r, t) - \bar{b}(e_1, e_2, r, t) \\
&\quad \cdot \frac{\bar{f}_k + ce_{2,k} + (\gamma+\eta)\mathrm{sgn}(\sigma_k)}{\bar{b}_k} + \bar{d}(e_1, e_2, r, t)]dt \\
&= e_{2,k} + \int_{kT}^{(k+1)T} [-ce_{2,k} - (\gamma+\eta)\mathrm{sgn}(\sigma_k)]\, dt + \int_{kT}^{(k+1)T} \bar{d}(e_1, e_2, r, t)dt \\
&\quad + \int_{kT}^{(k+1)T} \left[\bar{f}(e_1, e_2, r, t) - \frac{\bar{b}(e_1, e_2, r, t)\bar{f}_k}{\bar{b}_k}\right] dt \\
&\quad - \int_{kT}^{(k+1)T} \frac{\bar{b}(e_1, e_2, r, t) - \bar{b}_k}{\bar{b}_k} [ce_{2,k} + (\gamma+\eta)\mathrm{sgn}(\sigma_k)]\, dt.
\end{aligned}$$

Consider the integral terms of the above equation. We can write

$$\int_{kT}^{(k+1)T} \left[\bar{f}(e_1, e_2, r, t) - \frac{\bar{b}(e_1, e_2, r, t)\bar{f}_k}{\bar{b}_k}\right] dt$$

$$= \int_{kT}^{(k+1)T} \left[(\bar{f}(e_1, e_2, r, t) - \bar{f}_k) - \frac{\bar{f}_k}{\bar{b}_k}(\bar{b}(e_1, e_2, r, t) - \bar{b}_k)\right] dt.$$

By virtue of Lemma 2 and Assumptions 3 and 4, we have that

$$\bar{f}(e_1, e_2, r, t) - \bar{f}_k \in O(T) \quad \forall t \in [kT, (k+1)T],$$

and

$$\bar{b}(e_1, e_2, r, t) - \bar{b}_k \in O(T) \quad \forall t \in [kT, (k+1)T].$$

Thus we can conclude that

$$\int_{kT}^{(k+1)T} \left[\bar{f}(e_1, e_2, r, t) - \frac{\bar{b}(e_1, e_2, r, t)\bar{f}_k}{\bar{b}_k}\right] dt \in O(T^2).$$

Similarly, we can show that

$$\int_{kT}^{(k+1)T} \frac{\bar{b}(e_1, e_2, r, t) - \bar{b}_k}{\bar{b}_k} [ce_{2,k} + (\gamma + \eta)\mathrm{sgn}(\sigma_k)]\, dt$$

$$= \frac{ce_{2,k} + (\gamma + \eta)\mathrm{sgn}(\sigma_k)}{\bar{b}_k} \int_{kT}^{(k+1)T} (\bar{b}(e_1, e_2, r, t) - \bar{b}_k) dt.$$

$$\in O(T^2)$$

By Assumptions 2 and 4 we have

$$\int_{kT}^{(k+1)T} \bar{d}(e_1, e_2, r, t) dt \in O(T^2).$$

Combining the above equations yields

$$\begin{aligned} e_{2,k+1} &= e_{2,k} + \int_{kT}^{(k+1)T} [-ce_{2,k} - (\gamma + \eta)\mathrm{sgn}(\sigma_k)]\, dt + O(T^2) \\ &= e_{2,k} - [ce_{2,k} + (\gamma + \eta)\mathrm{sgn}(\sigma_k)]\, T + O(T^2) \\ &= (1 - cT)e_{2,k} - (\gamma + \eta)T\mathrm{sgn}\sigma_k + O(T^2). \end{aligned} \tag{14}$$

Note that $e_2(t)$ is continuous. From equation (3) we can write

$$e_{1,k+1} = e_{1,k} + \int_{kT}^{(k+1)T} e_2(t) dt = e_{1,k} + e_2(\xi)T = e_{1,k} + e_{2,k}T + (e_2(\xi) - e_{2,k})T,$$

where $kT \leq \xi \leq (k+1)T$. Since the system is stabilized by the variable structure controller , in terms of Assumption 4 we have that $|\dot{e}_2| \in O(1)$. According to Lemma 2 we obtain

$$e_2(\xi) - e_{2,k} = \int_{kT}^{\xi} \dot{e}_2 dt \in O(T).$$

Consequently,

$$e_{1,k+1} = e_{1,k} + e_{2,k}T + O(T^2). \tag{15}$$

From equations (4), (14) and (15) we have

$$\begin{aligned}\sigma_{k+1} - \sigma_k &= [e_{2,k+1} - e_{2,k}] + c[e_{1,k+1} - e_{1,k}] \\ &= -cTe_{2,k} - (\gamma + \eta)T\text{sgn}(\sigma_k) + cTe_{2,k} + O(T^2) \\ &= -(\gamma + \eta)T\text{sgn}(\sigma_k) + O(T^2).\end{aligned}$$

This completes the proof.

To simplify analysis, we neglect the terms $O(T^2)$ in the above equations. Thus we have

$$e_{2,k+1} = (1 - cT)e_{2,k} - (\gamma + \eta)T\text{sgn}\sigma_k, \tag{16}$$
$$e_{1,k+1} = e_{1,k} + e_{2,k}T, \tag{17}$$
$$\sigma_{k+1} = \sigma_k - (\gamma + \eta)T\text{sgn}\sigma_k. \tag{18}$$

In the sequel, the main analysis will be based on the equations (16)–(18).

Remark. The equations (11)–(13) hold accurately, whereas the equations (16)–(18) hold approximately, i.e. up to the order of $O(T^2)$. Therefore, the later analysis using (16)–(18) will be only accurate up to the order of $O(T^2)$.

Remark. From equation (16) we can see that the following condition

$$|1 - cT| < 1 \tag{19}$$

must be satisfied to guarantee the sampled-data system (8) to be stable. Generally, $T \ll 1$. Therefore one should choose a large c ($c \gg 1$) to expedite the convergence of the system.

4 Analysis of the System Motion

In this section, the sampled-data VSC system (8) is analyzed. It is shown that the system undergoes three phases: reaching phase, switching phase and chattering phase. The characteristics of each phase will be investigated.

Definition 4. The sampled-data system is said to be in reaching phase if the following equation

$$\text{sgn}\sigma_{k+1} = \text{sgn}\sigma_k$$

holds for any two consecutive instants in some period from $t = 0$ to $t = k_0T$.

Theorem 5. *If the system (8) is in its reaching phase, the magnitude of the switching function will decreases monotonically, i.e.,*

$$|\sigma_{k+1}| < |\sigma_k|.$$

Proof: From Assumption 1 and equation (13), we obtain that if $\sigma_k > 0$,

$$\sigma_{k+1} = \sigma_k - (\gamma + \eta)T + O(T^2) < \sigma_k;$$

and if $\sigma_k < 0$,

$$\sigma_{k+1} = \sigma_k + (\gamma + \eta)T + O(T^2) > \sigma_k.$$

Since $\mathrm{sgn}\sigma_{k+1} = \mathrm{sgn}\sigma_k$, we can conclude that

$$|\sigma_{k+1}| < |\sigma_k|.$$

Generally, the states of the system are far from the switching surface in the beginning, so the system undergoes a reaching phase in the initial stage.

Definition 6. The sampled-data system (8) is said to be in switching phase if the following equations

$$\begin{aligned} \mathrm{sgn}\sigma_{k+1} &= -\mathrm{sgn}\sigma_k \\ \mathrm{sgn}e_{2,k+1} &= \mathrm{sgn}e_{2,k} \end{aligned}$$

hold for any two consecutive instants in some period from $t = k_0T$ to $t = k_1T$.

Theorem 7. *If the system (8) is in its switching phase, and the magnitudes of the state variables $e_{1,k}$ and $e_{2,k}$ are all in $O(1)$, then we have:*
(i) $\quad \mathrm{sgn}e_{1,k+1} = \mathrm{sgn}e_{1,k} \quad \text{for} \quad k = k_0, k_0+1, \ldots, k_1',$
(ii) $\quad |e_{1,k+1}| < |e_{1,k}| \quad \text{for} \quad k = k_0, k_0+1, \ldots, k_1',$
where k_1' is some integer which may be smaller than k_1.

Proof:
(i) By equation (12), we obtain

$$e_{1,k+1} = e_{1,k} + e_{2,k}T + O(T^2) = e_{1,k} + O(T). \tag{20}$$

This clearly shows that $\mathrm{sgn}e_{1,k+1} = \mathrm{sgn}e_{1,k}$ for $k = k_0, k_0+1, \ldots, k_1'$, where k_1' represents the instant at which all the accumulated $O(T)$'s caused by $e_{2,k}$ surpass the quantity e_{1,k_0}.
(ii) Notice that the following equation holds

$$\mathrm{sgn}e_{1,k} = -\mathrm{sgn}e_{2,k} \quad \text{for} \quad k = k_0, k_0+1, \ldots, k_1', \tag{21}$$

otherwise from (i) and the second condition of Definition 6 we would have

$$\mathrm{sgn}\sigma_{k+1} = \mathrm{sgn}\sigma_k = \mathrm{sgn}e_{2,k}, \quad \text{for} \quad k = k_0, k_0+1, \ldots, k_1'.$$

This contradicts the first condition of Definition 6. The second conclusion of the theorem follows immediately from equations (20) and (21).

It is shown by equation (20) that the state variable $e_{1,k}$ cannot change its value, and hence its sign dramatically. This is due to the effect of the integrator.

Remark. Generally, the position error e_1 will keep decreasing during reaching and switching phases, and the magnitude of e_1 will drop from $O(1)$ level at the beginning of reaching phase to $O(T)$ level at the end of switching phase.

Notice that the conclusions (i) and (ii) in Theorem 7 may not be valid in the end of switching phase. This is because when $e_{1,k}$ approaches the level $O(T)$, the neglected term $O(T)$ in the equation (20) may surpass $e_{1,k}$. Thus $|e_{1,k}|$ may cease to decrease in the end of switching phase. However, the fact that $e_{1,k}$ approaches the level $O(T)$ implies that the system is approaching its equilibrium, falling in the phase of chattering, which will be described as below.

Definition 8. The sampled-data system (8) is said to be in chattering phase if the following equations

$$\begin{aligned} \mathrm{sgn}\sigma_{k+1} &= -\mathrm{sgn}\sigma_k \\ \mathrm{sgn}e_{2,k+1} &= -\mathrm{sgn}e_{2,k} \end{aligned}$$

hold for any two consecutive instants after some time $t = k_1 T$.

Theorem 9. *Suppose Assumptions 1, 2, 3 and 4 hold. If the system (8) is in chattering phase, then there exists a positive number k_2 such that the following inequality*

$$c|e_{1,k}| \leq |e_{2,k}| \tag{22}$$

holds for all $k > k_2$ and "=" holds only when $\sigma_k = 0$.

Without loss of generality, we choose c such that

$$0 < 1 - cT < 1, \tag{23}$$

according to Remark 3. For the case $-1 < 1 - cT < 0$ the theorem can be proved in the same way.

Proof of Theorem 9: The proof consists of several Observations.

Observation 1 *Let $W = (\gamma + \eta)T$. Suppose that the initial time be $k = 0$. If $|\sigma_0| \geq W$, then there is a positive number k_3 such that*

$$|\sigma_{k_3}| < W.$$

Proof of Observation 1: Let

$$k_3 = \mathrm{Int}\left[\frac{|\sigma_0|}{W}\right] + 1.$$

where $\mathrm{Int}[\alpha]$ represents the maximal integer which is less than α. Then the number k_3 defined above is just the number needed in the observation.

Now consider the motion of σ_k when $k > k_3$. Let

$$\sigma_{k_3} \stackrel{\mathrm{def}}{=} \beta. \tag{24}$$

Consider the following two cases:

(i) Case 1: $\beta = 0$. In this case, we need to define $\text{sgn}(0) = 0$. Notice that this definition causes no contradiction. From theoretical point of view, there is no ideal sliding modes, and hence no need of defining equivalent control for sampled-data VSC systems. Thus from equation (18) we obtain

$$\sigma_k = 0 \quad \forall k > k_3. \tag{25}$$

From practical point of view, the probability that the equation (25) holds is zero. In fact, the equation (25) cannot hold for some period of time because of the existence of disturbance, the limit on the word length of computer, etc.. We consider this case only for the reason of theoretical completeness.

(ii) Case 2: $\beta \neq 0$. Without loss of generality, we suppose that $\beta > 0$. For the case of $\beta < 0$, the same conclusion can be reached by the similar argument. From equation (18), we obtain

$$\sigma_{k_3+1} = \beta - W\text{sgn}\beta = \beta - W < 0 \tag{26}$$

$$\sigma_{k_3+2} = (\beta - W) - W\text{sgn}(\sigma_{k_3+1}) = \beta > 0. \tag{27}$$

By induction this leads to

$$\sigma_{k_3+2j+1} = -(W - \beta) \quad \text{for} \quad j = 0, 1, 2, \ldots \tag{28}$$

$$\sigma_{k_3+2j+2} = \beta \quad \text{for} \quad j = 0, 1, 2, \ldots . \tag{29}$$

Therefore the switching function will jump at two values, β and $-(W - \beta)$ when $k > k_3$. We summarize the above result as the following observation.

Observation 2 *Once the states of the system (8) drop into the band defined by* $\mathcal{B}_W = \{(e_1, e_2) \in R^2 : |\sigma| < W\}$, *the switching function* σ *will jump at two values,* β *and* $-(W - \beta)$ *thereafter.*

Now consider the motion of e_2 when $k > k_3$. From equation (16) we have

$$\begin{aligned} e_{2,k} &= (1 - cT)e_{2,k-1} - W\text{sgn}\sigma_{k-1} \\ &= (1 - cT)^2 e_{2,k-2} - (1 - cT)W\text{sgn}\sigma_{k-2} - W\text{sgn}\sigma_{k-1} \\ &= \cdots\cdots \\ &= (1 - cT)^{k-k_3} e_{2,k_3} - \sum_{j=k_3}^{k-1} (1 - cT)^{k-1-j} W\text{sgn}\sigma_j \\ &= (1 - cT)^{k-k_3} e_{2,k_3} - \sum_{j=0}^{k-k_3-1} (1 - cT)^j W\text{sgn}\sigma_{k-j-1} \quad \forall k > k_3. \end{aligned} \tag{30}$$

Again, let us consider the following two cases:

(i) Case 1: $\beta = 0$. In this case, in terms of the above discussion we will have

$$e_{2,k} = (1 - cT)^{k-k_3} e_{2,k_3}.$$

Hence we obtain

$$\lim_{k\to\infty} e_{2,k} = 0.$$

(ii) Case 2: $\beta \neq 0$. Again, we suppose that $\beta > 0$. From Observation 2 and equation (30) we have:

$$\begin{aligned} & e_{2,k} \\ &= (1-cT)^{k-k_3} e_{2,k_3} - W \sum_{j=0}^{k-k_3-1} (1-cT)^j (-1)^{j-(k-1-k_3)} \\ &= (1-cT)^{k-k_3} e_{2,k_3} - (-1)^{k-1-k_3}\, W \sum_{j=0}^{k-k_3-1} (-1)^j (1-cT)^j \\ &= (1-cT)^{k-k_3} e_{2,k_3} - (-1)^{k-1-k_3}\, W \,\frac{1-(cT-1)^{k-k_3}}{2-cT} \qquad \forall k > k_3. \qquad (31) \end{aligned}$$

Because $|1-cT| < 1$, we have

$$\lim_{k\to\infty} (1-cT)^{k-k_3} = 0.$$

Thus it is clear that $e_{2,k}$ has two limit values when $k \to \infty$, i.e.,

$$\lim_{j\to\infty} e_{2,k_3+2j+1} \stackrel{\text{def}}{=} E_{2+} = \frac{W}{2-cT}, \qquad (32)$$

$$\lim_{j\to\infty} e_{2,k_3+2j+2} \stackrel{\text{def}}{=} E_{2-} = -\frac{W}{2-cT}. \qquad (33)$$

From (31) it can also be obtained that

$$\lim_{k\to\infty} |e_{2,k}| = \frac{W}{2-cT}. \qquad (34)$$

We summarize the above results as the following observation.

Observation 3 *For any arbitrarily small $\epsilon > 0$, there exists an integer k_ϵ such that when $k > k_\epsilon$, we have*

$$\left| |e_{2,k}| - \frac{W}{2-cT} \right| < \epsilon,$$

i.e.,

$$\frac{W}{2-cT} - \epsilon < |e_{2,k}| < \frac{W}{2-cT} + \epsilon. \qquad (35)$$

Finally consider the behavior of $e_1(k)$. First we have the following observation.

Observation 4 *When the system (8) is in chattering phase, the following equation*

$$\mathrm{sgn}\sigma_k = \mathrm{sgn}e_{2,k}$$

must hold.

Proof of Observation 4: Denote by k_1 the instant at which the system (8) enters into chattering. Suppose there were some integer $k_4 > k_1$ such that

$$\mathrm{sgn}\sigma_k = -\mathrm{sgn}e_{2,k} \qquad \forall k \geq k_4.$$

From equation (16) we would have

$$e_{2,k_4+1} = (1 - cT)e_{2,k_4} + (\gamma + \eta)T\mathrm{sgn}e_{2,k_4}.$$

Clearly, the above equation implies that

$$\mathrm{sgn}e_{2,k_4+1} = \mathrm{sgn}e_{2,k_4}.$$

This contradicts the Definition 8.

Now we are ready to prove the main theorem. We first show that when $|\sigma_k| \geq \frac{W}{2}$, the following inequality holds

$$|ce_{1,k}| < |e_{2,k}| \qquad \forall k \geq k_{\epsilon_1} \text{ and when } |\sigma_k| \geq \frac{W}{2}, \tag{36}$$

where k_{ϵ_1} is some positive integer to be defined.

Let

$$\epsilon_1 = \min\{\frac{1}{2}\left(\frac{W}{2-cT} - \frac{1}{2}W\right),\ \frac{1}{2}\left(W - \frac{W}{2-cT}\right)\}.$$

Noticing that $\frac{1}{2}W < \frac{W}{2-cT} < W$, we have that $\epsilon_1 > 0$. For this ϵ_1, from Observation 3 we know that there exists a positive integer k_{ϵ_1} such that

$$\begin{aligned}|e_{2,k}| &> \frac{W}{2-cT} - \epsilon_1 \\ &\geq \frac{W}{2-cT} - \frac{1}{2}\left(\frac{W}{2-cT} - \frac{1}{2}W\right) \\ &= \frac{1}{2}W + \frac{cT}{4(2-cT)}W \\ &> \frac{1}{2}W \qquad \forall k \geq k_{\epsilon_1}.\end{aligned} \tag{37}$$

Similarly

$$\begin{aligned}|e_{2,k}| &< \frac{W}{2-cT} + \epsilon_1 \\ &\leq \frac{W}{2-cT} + \frac{1}{2}\left(W - \frac{W}{2-cT}\right) \\ &= W - \frac{1-cT}{2(2-cT)}W \qquad \forall k \geq k_{\epsilon_1}.\end{aligned} \tag{38}$$

Thus from Observation 4 we have that

$$\begin{aligned}
|ce_{1,k}| &= |\sigma_k - e_{2,k}| \\
&= |\ |\sigma_k| - |e_{2,k}|\ | \\
&\leq \max\{\sup|\sigma_k| - \inf|e_{2,k}|,\ \ \sup|e_{2,k}| - \inf|\sigma_k|\} \\
&< \max\{W - \left(\frac{1}{2}W + \frac{cT}{4(2-cT)}W\right),\ \left(W - \frac{1-cT}{2(2-cT)}W\right) - \frac{1}{2}W\} \\
&= \max\{\frac{1}{2}W - \frac{cT}{4(2-cT)}W,\ \frac{1}{2}W - \frac{1-cT}{2(2-cT)}W\} \\
&< \frac{1}{2}W \qquad \forall k \geq k_{\epsilon_1} \text{ and when } |\sigma_k| \geq \frac{W}{2}.
\end{aligned} \tag{39}$$

The inequality (36) comes from the inequalities (37) and (39) immediately.

Now consider the chattering phase from the instant $k = k_{\epsilon_1}$. Again denote by β the value of $\sigma_{k_{\epsilon_1}}$. This causes no confusion with the definition of β in the equation (24) since by Observation 2 we can simply let $\beta = \sigma_{k_{\epsilon_1}}$ if $\sigma_{k_{\epsilon_1}} > 0$; otherwise, if $\sigma_{k_{\epsilon_1}} < 0$, we can let $\beta = \sigma_{k_{\epsilon_1}+1}$.

We need only consider the case $0 \leq \beta < W$. Again we divide the proof into three cases: (i) $\beta = 0$; (ii) $0 < \beta \leq \frac{W}{2}$; and (iii) $\frac{W}{2} < \beta < W$.

Case (i): $\beta = 0$. In this case, we have that $\sigma_k = 0$ for all $k > k_{\epsilon_1}$ from the previous discussion. Therefore it can be easily obtained in terms of equation (4) that

$$|ce_{1,k}| = |e_{2,k}| \quad \forall k > k_{\epsilon_1}.$$

Case (ii): $0 < \beta \leq \frac{W}{2}$. From Observation 2 we will have $\sigma_{k_{\epsilon_1}+2j+1} = \beta - W \leq -\frac{W}{2}$ for all $j > 0$. By the inequality (36) we can see that we need only to prove the inequality (22) for the case of $k = k_{\epsilon_1} + 2j$, $j = 0, 1, 2, \ldots$.

From Observation 4 and Definition 8 it can be concluded that $e_{2,k_{\epsilon_1}+2j} > 0$ for all $j > 0$. Let $\epsilon_2 = \frac{1}{2}\beta$. From Observation 3 we know that there exists a positive integer $k_{\epsilon_2} > k_{\epsilon_1}$ such that

$$e_{2,k_{\epsilon_2}+2j} > \frac{W}{2-cT} - \epsilon_2 = \frac{W}{2-cT} - \frac{1}{2}\beta \qquad \forall j > 0, \tag{40}$$

and

$$e_{2,k_{\epsilon_2}+2j} < \frac{W}{2-cT} + \epsilon_1 = \frac{W}{2-cT} + \frac{1}{2}\beta \qquad \forall j > 0, \tag{41}$$

where we suppose that k_{ϵ_2} and k_{ϵ_1} are of the same parity. Thus from equation (4) we obtain that

$$\begin{aligned}
ce_{1,k_{\epsilon_2}+2j} &= \sigma_{k_{\epsilon_2}+2j} - e_{2,k_{\epsilon_2}+2j} \\
&< \beta - (\frac{W}{2-cT} - \frac{1}{2}\beta)
\end{aligned}$$

$$
\begin{aligned}
&= \frac{W}{2-cT} - \frac{1}{2}\beta - 2\left(\frac{W}{2-cT} - \beta\right) \\
&\leq \frac{W}{2-cT} - \frac{1}{2}\beta - 2\left(\frac{W}{2-cT} - \frac{W}{2}\right) \\
&< \frac{W}{2-cT} - \frac{1}{2}\beta \qquad \forall j > 0,
\end{aligned} \tag{42}
$$

and

$$
\begin{aligned}
ce_{1,k_{\epsilon_2}+2j} &= \sigma_{k_{\epsilon_2}+2j} - e_{2,k_{\epsilon_2}+2j} \\
&> \beta - (\frac{W}{2-cT} + \frac{1}{2}\beta) \\
&= -(\frac{W}{2-cT} - \frac{1}{2}\beta) \qquad \forall j > 0.
\end{aligned} \tag{43}
$$

From (40), (42) and (43) we have

$$
|ce_{1,k_{\epsilon_1}+2j}| < e_{2,k_{\epsilon_1}+2j} = |e_{2,k_{\epsilon_1}+2j}| \qquad \forall j > 0.
$$

Case (iii): $\frac{W}{2} < \beta < W$. In this case, we will have that $0 > \sigma_{k_{\epsilon_1}+2j+1} = \beta - W > -\frac{W}{2}$ for all $j > 0$. Therefor we need only to prove the inequality (22) for the case of $k = k_{\epsilon_1} + 2j + 1$, $j = 0, 1, 2, \ldots$.

From Observation 4 and Definition 8 it can be concluded that $e_{2,k_{\epsilon_1}+2j+1} < 0$ for all $j > 0$. Let $\epsilon_3 = \frac{1}{2}(W - \beta)$. From Observation 3 we know that there exists a positive integer $k_{\epsilon_3} > k_{\epsilon_1}$ such that

$$
-e_{2,k_{\epsilon_3}+2j+1} > \frac{W}{2-cT} - \epsilon_3 = \frac{W}{2-cT} - \frac{1}{2}(W - \beta) \qquad \forall j > 0, \tag{44}
$$

and

$$
-e_{2,k_{\epsilon_3}+2j+1} < \frac{W}{2-cT} + \epsilon_3 = \frac{W}{2-cT} + \frac{1}{2}(W - \beta) \qquad \forall j > 0, \tag{45}
$$

where we suppose that k_{ϵ_3} and k_{ϵ_1} are of the same parity. Thus from equation (4) we obtain that

$$
\begin{aligned}
ce_{1,k_{\epsilon_3}+2j+1} &= \sigma_{k_{\epsilon_3}+2j+1} - e_{2,k_{\epsilon_3}+2j+1} \\
&= \beta - W - e_{2,k_{\epsilon_3}+2j+1} \\
&> \beta - W + \frac{W}{2-cT} - \frac{1}{2}(W - \beta) \\
&= -\frac{W}{2-cT} + \frac{1}{2}(W - \beta) + 2\left(\frac{W}{2-cT} + \beta - W\right) \\
&> -\frac{W}{2-cT} + \frac{1}{2}(W - \beta) + 2\left(\frac{W}{2-cT} + \frac{W}{2} - W\right) \\
&> -\left[\frac{W}{2-cT} - \frac{1}{2}(W - \beta)\right] \qquad \forall j > 0,
\end{aligned} \tag{46}
$$

and

$$\begin{aligned} ce_{1,k_{e_3}+2j+1} &= \sigma_{k_{e_3}+2j+1} - e_{2,k_{e_3}+2j+1} \\ &= \beta - W - e_{2,k_{e_3}+2j+1} \\ &< \beta - W + \frac{W}{2-cT} + \frac{1}{2}(W-\beta) \\ &= \frac{W}{2-cT} - \frac{1}{2}(W-\beta) \qquad \forall j > 0. \end{aligned} \tag{47}$$

From (44), (46) and (47) we have

$$|ce_{1,k_{e_3}+2j+1}| < -e_{2,k_{e_3}+2j+1} = |e_{2,k_{e_3}+2j+1}| \qquad \forall j > 0.$$

This completes the proof.

It is shown by Theorem 9 that the chattering process is dominated by the velocity variable.

Remark. According to Observation 4 we have

$$|e_{1,k}| = \frac{1}{c}|\,|\sigma_k| - |e_{2,k}|\,|.$$

Since $|\sigma_k|$ has two limit values, one being β and the other being $W-\beta$, $|e_{1,k}|$ also has two limit values, being

$$\frac{1}{c}\left|\beta - \frac{W}{2-cT}\right| \qquad \text{and} \qquad \frac{1}{c}\left|(W-\beta) - \frac{W}{2-cT}\right|$$

respectively, where $\beta \in (0, W)$. Therefore we have

$$\lim_{k\to\infty} |e_{1,k}| \le \frac{W}{c(2-cT)}.$$

5 Simulation Studies

Consider a set point control problem for the following system:

$$\begin{aligned} \dot{x}_1 &= x_2 \\ \dot{x}_2 &= -x_2 + u + d(t), \end{aligned}$$

where d is an unknown external disturbance with a known upper bound γ. The initial state variables are $x_1(0) = x_2(0) = 0$. The set point is $r = 1$.

The switching function is chosen as $\sigma = e_2 + ce_1$, where $e_1 = x_1 - r$, $e_2 = x_2$, and c is the switching slope. From the equation (5) we obtain the control as

$$u = (1-c)e_2 - (\gamma + \eta)\mathrm{sgn}\sigma,$$

where we chose $\eta = 1$ and $\gamma = 10$.

Case (i): There is no disturbance.

In this case we have $d = 0$. However we still chose $\gamma = 10$ in this case to compare the results here to the ones in the sequel. The switching slope and sampling interval are chosen to be $c = 5$ and $T = 0.02$, respectively. The simulation results are illustrated in Fig. 1 and Fig. 2.

Fig. 1 shows the overall time history of the system variables, while Fig. 2 shows the final part of the time history of the system variables. It is shown by Fig. 1 that before $t \doteq 0.5$, the system is in reaching phase; from $t \doteq 0.5$ to $t \doteq 1.0$, the system is in switching phase; and after $t \doteq 1.0$, the system undergoes chattering phase. From Fig. 2 we can see that all the relations indicated by Observations 2, 3, 4 and Remark 4 are met by e_1, e_2 and σ when the system undergoes chattering phase. Also we can see from Fig. 1 that all the relations demonstrated in Theorems 5, 7 and 9 are met.

Case (ii): There exists small external disturbance.

In this case, we suppose $d(t)$ has the following form:

$$d(t) = 0.1\sin(10t) + 0.05\cos(20t) - 0.05\sin(30t).$$

It can be shown that the maximal disturbance is $d_{max} = 0.2$. So we have that $d_{max} = 10T$. The controller is the same as that for case (i). The corresponding simulation results are illustrated in Fig. 3 and Fig. 4 respectively. It is shown by Fig. 3 and Fig. 4 that all the relationship indicated by Observations 2, 3, 4 and Remark 4 is maintained fairly well.

Case (iii): There exists large external disturbance.

In this case, we suppose $d(t)$ has the following form:

$$d(t) = 5\sin(10t) + 2.5\cos(20t) - 2.5\sin(30t).$$

It can be shown that the maximal disturbance is $d_{max} = 10$. Therefore we have that $d_{max} = 500T$. The controller is the same as that for case (i). The corresponding simulation results are illustrated in Fig. 5 and Fig. 6 respectively.

It is shown by Fig. 5 and Fig. 6 that the relationship indicated by Observations 2, 3, 4 and Remark 4 is not valid now due to the effect of large external disturbance. Nevertheless the tendency is still preserved.

6 Concluding Remarks

In this paper we have shown that the motion of sampled-data VSC systems can be divided into three phases, namely reaching phase, switching phase and chattering phase. The definitions of the three phases are given and the characteristics of the three phases are described. It is shown that in the reaching phase the switching function decreases monotonically; it stops decreasing in the switching phase and the signs of all the state variables except for the highest order state remain invariant in the most part of switching phase; while in the chattering phase the signs of both the switching function and the highest order state change dramatically and the chattering is dominated by the highest order state. The bound or limit values of corresponding variables are given, which may be useful for the design of practical control systems. All the theoretical results are verified by simulation studies.

References

1. Bartolini G., A. Ferrara and V. I. Utkin, "Adaptive sliding mode control in discrete-time systems", *Automatica*, Vol.31, no.5, pp.769-773, 1995.
2. Bartoszewicz, A., "Discrete-time quasi-sliding-mode control strategies", *IEEE Trans. on Industrial Electronics*, Vol. IE-45, no.4, 633-637, 1998.
3. Chen X. and T. Fukuda, "Computer-controlled continuous-time variable structure systems with sliding modes", *Int. J. Control*, Vol.67, no.4, pp.619-639, 1997.
4. DeCarlo R. A., S. H. Zak and G. P. Matthews, "Variable structure control of nonlinear multivariable systems: a tutorial", *Proc. IEEE*, Vol.76, pp.212-232, 1988.
5. Dogruel M., U. Ozguner and S. Drakunov, "Sliding-mode control in discrete-state and hybrid systems", *IEEE Trans. on Automatic Control*, Vol.41, no.3, pp.414-419, 1996.
6. Dote, Y. and R. G. Hoft, "Microprocessor based sliding mode controller for dc motor drives", presented at the Ind. Applicat. Soc. Annual Meeting, Cincinnati, OH, 1980.
7. Furuta, K., "Sliding mode control of a discrete systems", *Systems & Control Letters*, Vol.14, 145-152, 1990.
8. Gao, W.B., Y. Wang, and A. Homaifa, "Discrete-time variable structure control systems", *IEEE Trans. on Industrial Electronics*, Vol.IE-42, no.2, pp.117-122, 1995.
9. Hung J. Y., W. B. Gao and J. C. Hung, "Variable structure control: a survey", *IEEE Trans. on Industrial Electronics*, Vol.IE-40, no.1, pp.2-22, 1993.
10. Kaynak, O. and A. Denker, "Discrete-time sliding mode control in the presence of system uncertainty", *Int. J. Control*, Vol.57, no.5, pp.1171-1189, 1993.
11. Kotta, U., "Comments on the stability of discrete-time sliding mode control systems", *IEEE Trans. on Automatic Control*, Vol.34, pp.1021-1022, 1989.
12. Milosavljevic, C., "General conditions for the existence of a quasisliding mode on the switching hyperplane in discrete variable structure systems", *Automat. Remote Control*, Vol.46, pp.307-314, 1985.
13. Sira-Ramirez, H., "Non-linear discrete variable structure systems in quasi-sliding mode", *Int. J. Control*, Vol.54, no.5, pp.1171-1187, 1991.
14. Sarpturk, S.Z., Istefanopulos, Y. and Kaynak, O., "On the stability of discrete-time sliding mode control systems", *TEEE Trans. on Automatic Control*, Vol.32, no.10, pp.930-932, 1987.
15. Spurgeon, S.K., "Choice of discontinuous control component for robust sliding mode performance", *Int. J. Control*, Vol.53, no.1, 163-179, 1991.
16. Su, W. C., S. V. Drakunov, and U. Ozguner, Implementation of variable structure control for sampled-data systems, *Proc. of the Workshop on Robust Control via Variable Structure and Lyapunov Techniques*, pp.166-173, Benevauro, Italy, 1994.
17. Takahashi, R. H. C., B. R. Menezes and B. J. Cardoso, A geometric approach to sampled data quasi-sliding modes analysis, in *Proc. 30th Conf. on Decision and Control*, Brighton, U.K., pp.1379-1381, 1991.
18. Utkin, V.I., Variable structure systems with sliding mode, *IEEE Trans. on Automatic Control*, vol.22, no.2, pp.212-222, 1977.
19. Utkin, V. I., Discontinuous control systems: state of the art in theory and applications, in *Proc. IFAC 10th World Congress on Automatic Control*, Munich, vol.1, pp.25-44, 1987.
20. Utkin, V. I. and S. V. Drakunov, On discrete-time sliding modes control, in *Preprints IFAC Conf. on Nonlinear Control*, Capri, Italy, pp.484-489, 1989.

21. Yu. X. and R. B. Potts, Analysis of discrete variable structure systems with pseudo-sliding modes, in *Int. J. Systems Science*, vol.23, pp503-516, 1992.
22. Yu, X., Bifurcation and chaotic behaviors in a discrete variable structure system with unbounded control magnitude, *Internation Journal of Bifurcation and Chaos*, vol.7, no.10, 1997.
23. Zheng, F., M. Cheng, and W. B. Gao, Variable Structure Control of Discrete-Time Stochastic Systems, *Proc. 31st IEEE Conference on Decision and Control*, pp.1830–1835, Tucson, Arizona, USA, December 16–18, 1992.

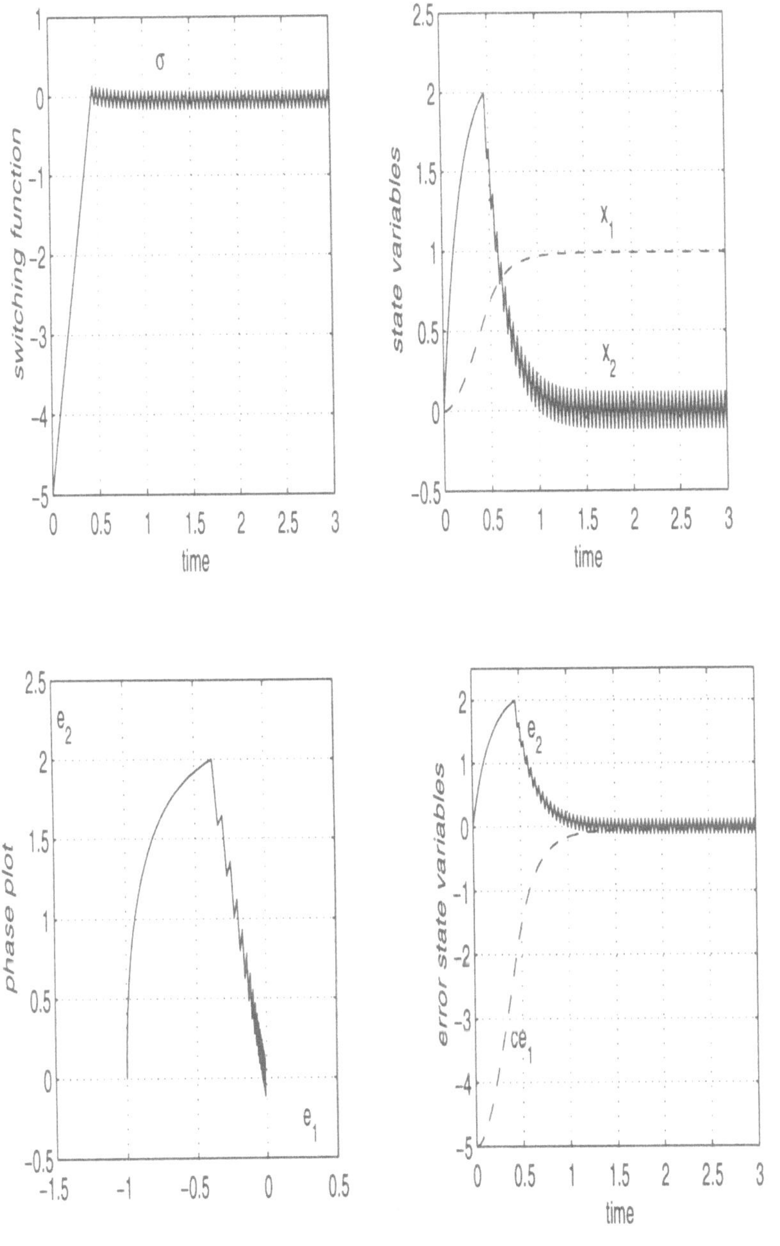

Fig. 1. System response with no external disturbance

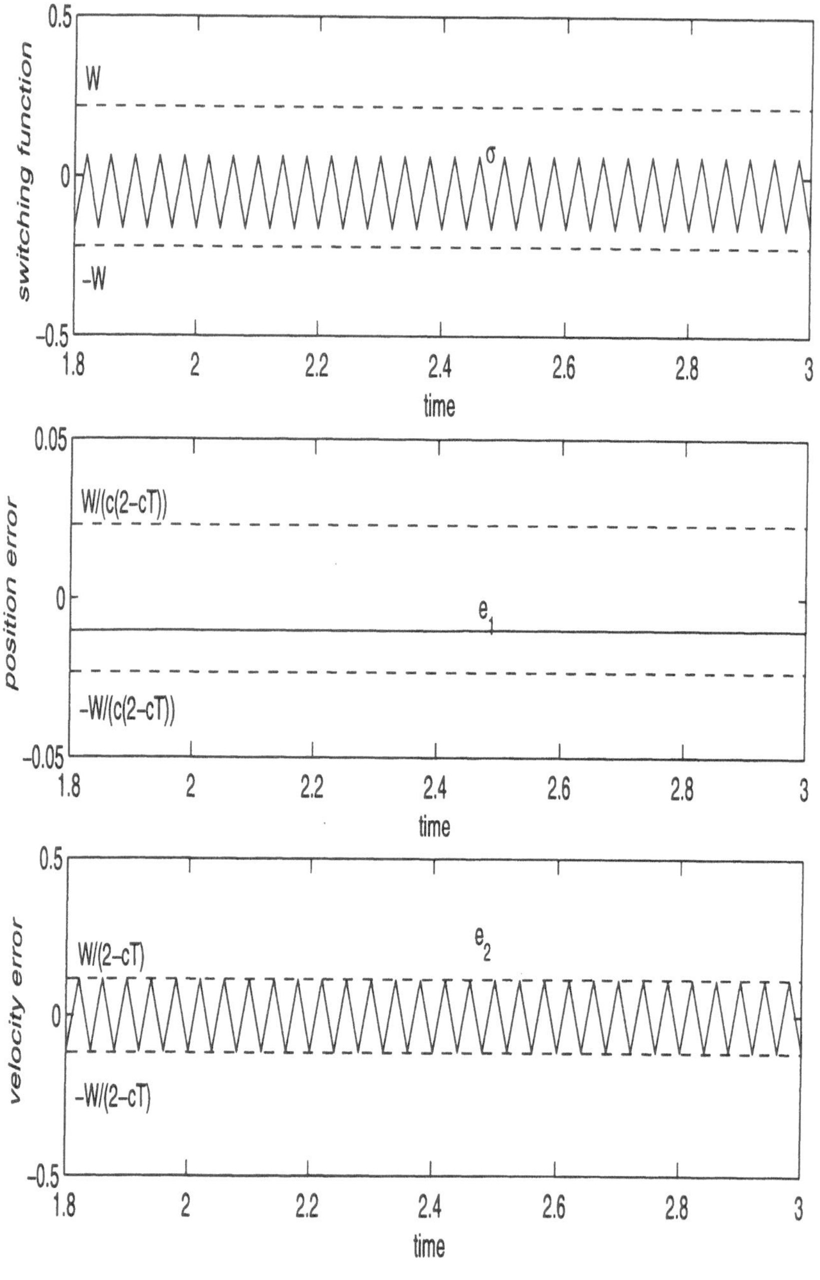

Fig. 2. System steady response with no external disturbance

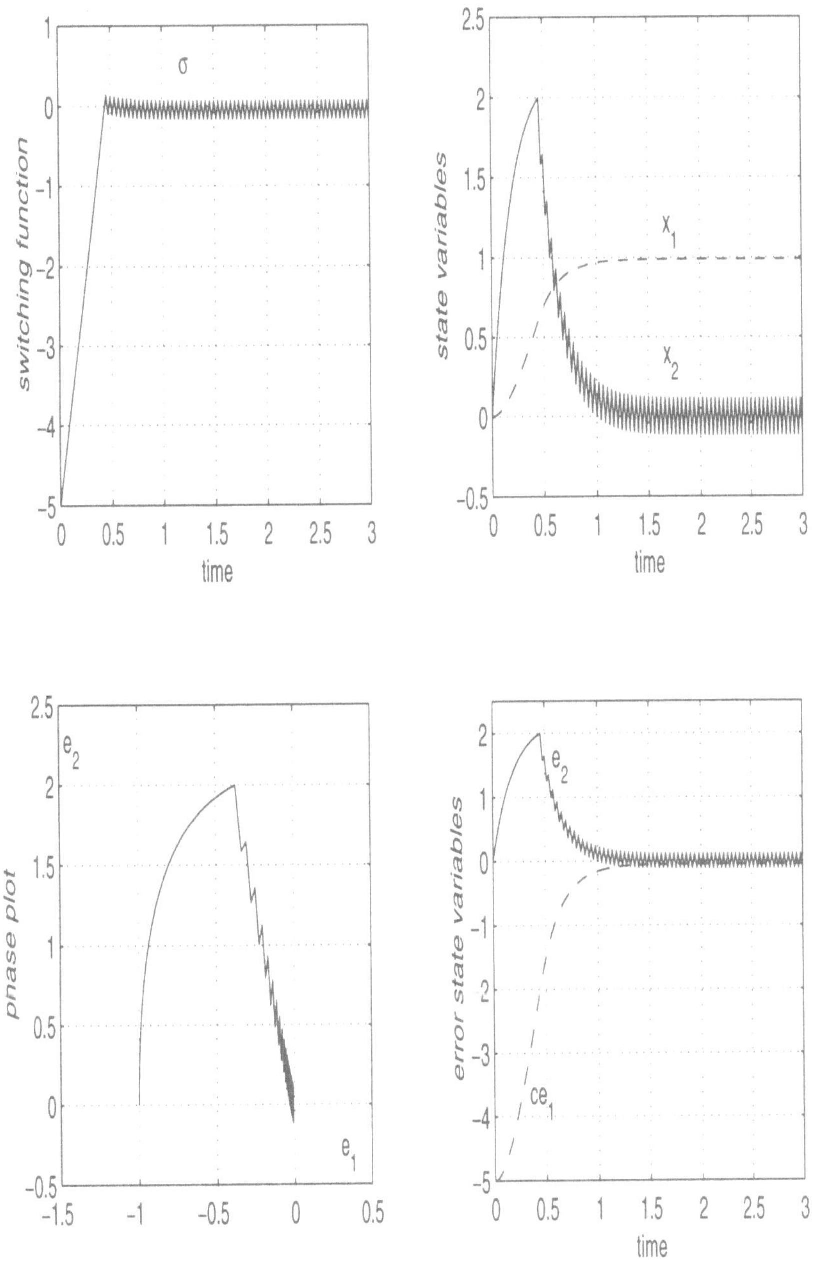

Fig. 3. System response with small external disturbance

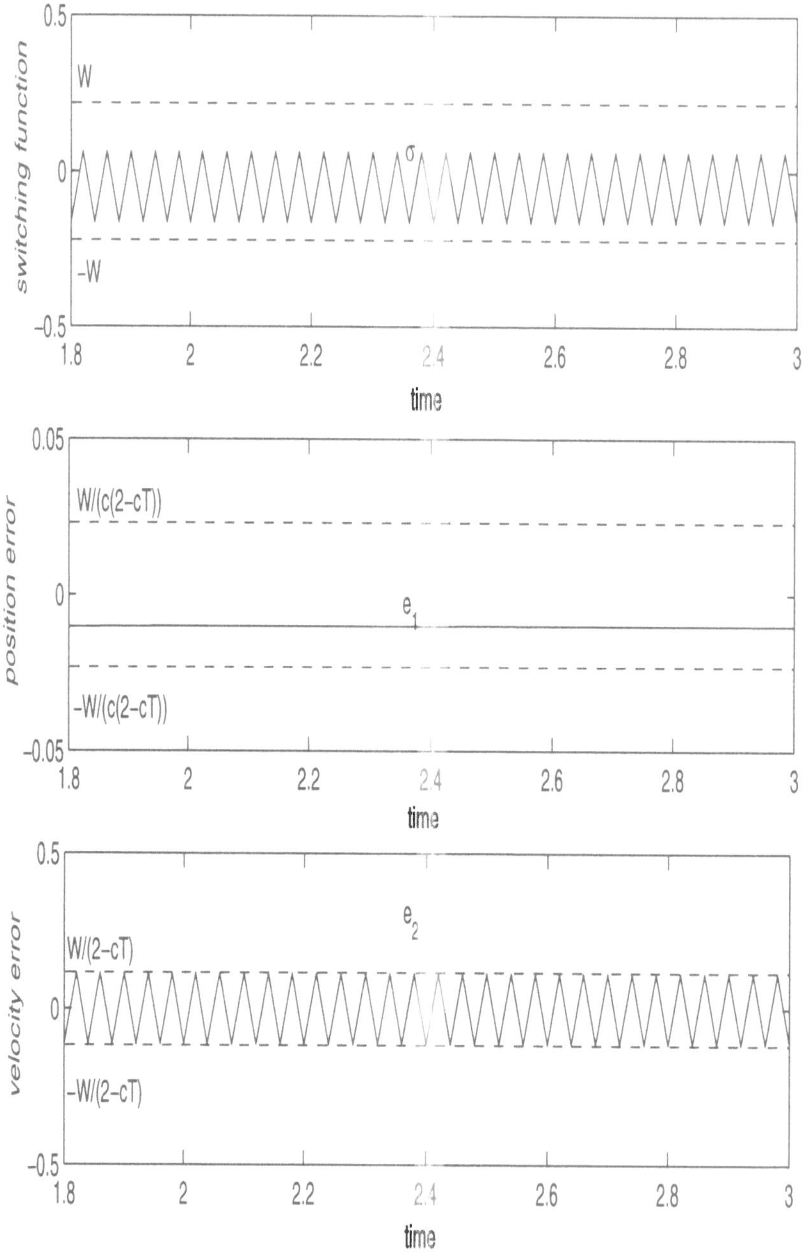

Fig. 4. System steady response with small external disturbance

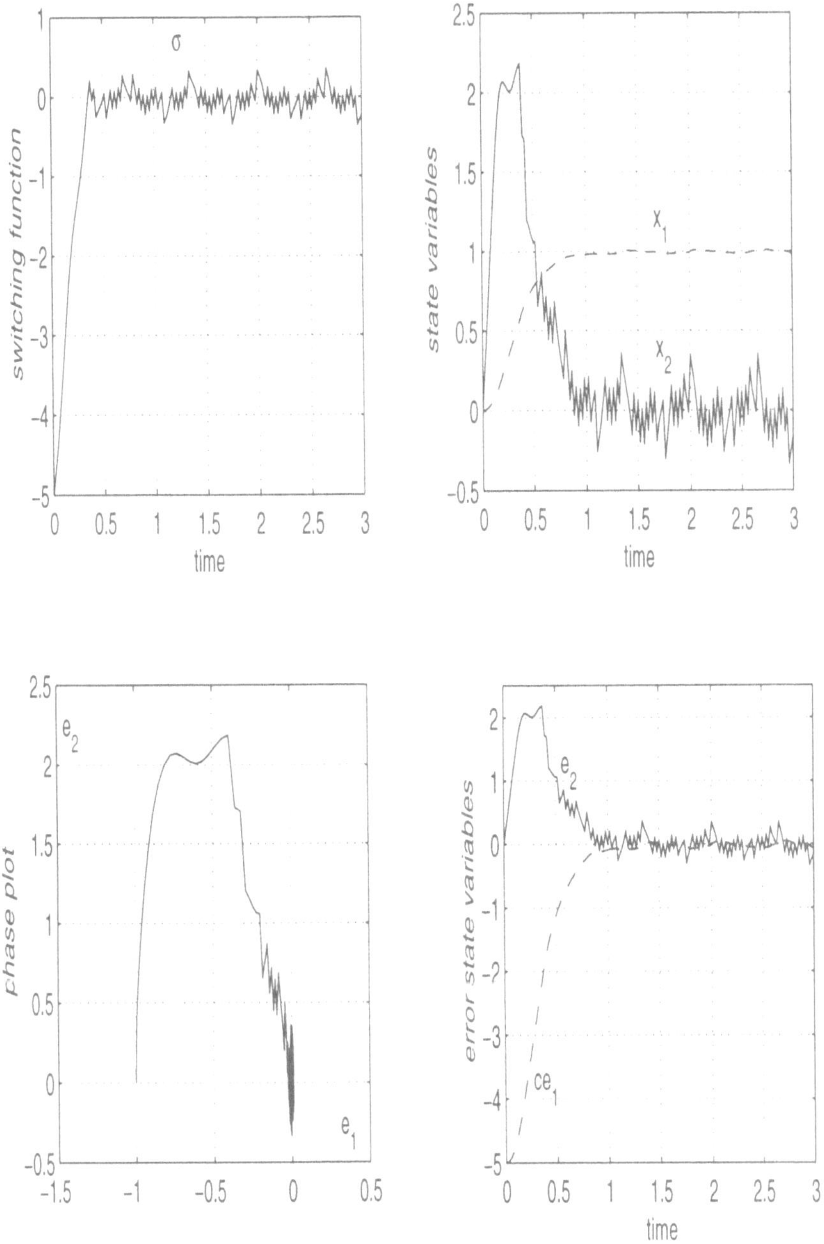

Fig. 5. System response with large external disturbance

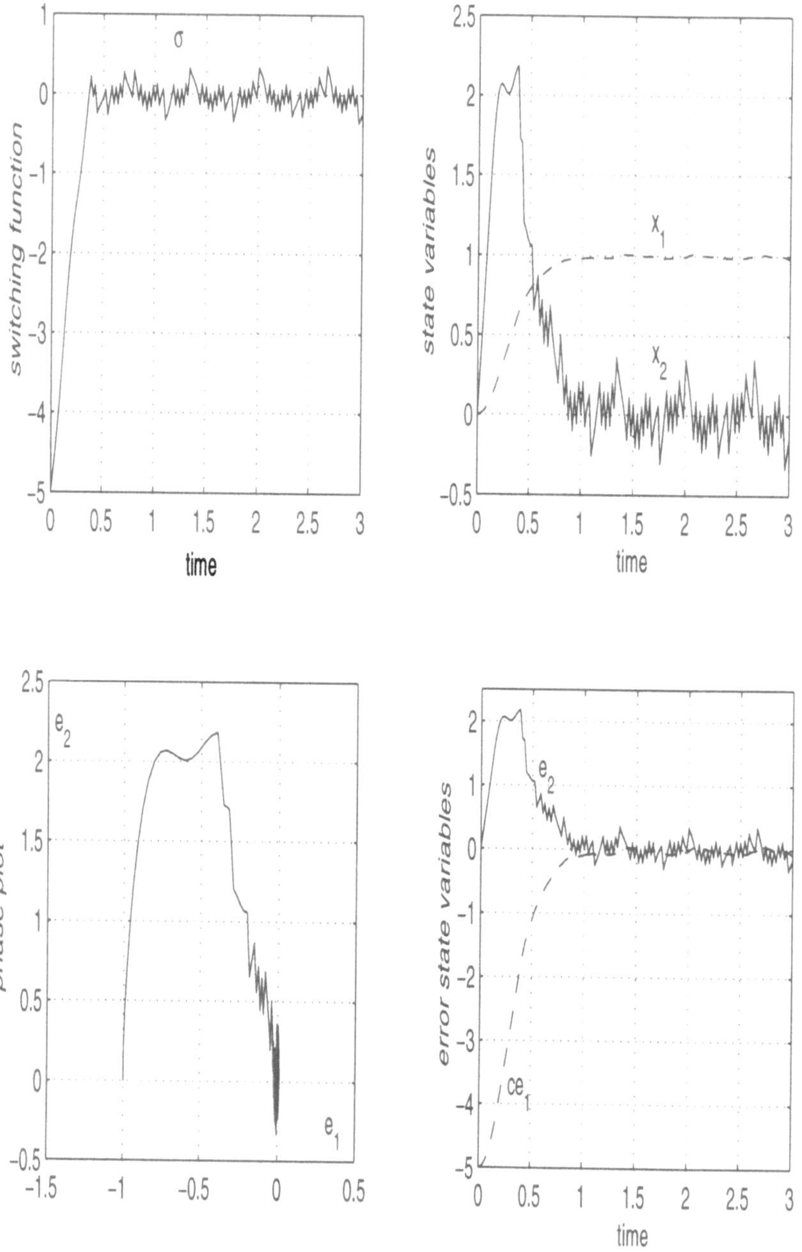

Fig. 6. System steady response with large external disturbance

Sliding Mode Control of Systems with Delayed States and Controls

Xiaoqiu Li Stephen Yurkovich

Center for Automotive Research
Department of Electrical Engineering
The Ohio State University
2015 Neil Avenue
Columbus, OH 43210 USA

Abstract. A sliding mode control (SMC) design method is proposed for a class of point-delayed systems. A linear transformation is applied to convert the delayed system to delay-free system whose spectra embed all the unstable poles of the original system with given stability margin. It is proved that the delayed system is asymptotically stable with SMC based on the delay-free system. The method is applied to internal combustion (IC) engine idle speed control, not for stabilization, but for compensation with the effects of the time delay in the system states.

1 Introduction

SMC is a robust nonlinear control algorithm which employs discontinuous control to enforce the system state trajectory on some prescribed sliding surface [11]. It has been widely used for its robustness to model parameter uncertainties and external disturbances.

Research on SMC has been mostly focused on systems without time delay. While relatively little attention has been given in the open literature to systems with delays, it remains an important issue in practical applications. Several good works have appeared, however, which address some of the difficulties associated with systems with inherent delays. For example, Jafarov [7] gives the necessary and sufficient conditions of the existence of sliding modes for a multidimensional system, and proves the asymptotic stability of the system on the sliding surface using a Lyapunov-Krasovsky functional. Drakunov and Utkin [1] give the two-stage design procedure for dynamic systems which can be written in differential-difference equations in the block form. Hu *et al.* [6] use a transformation for systems with delay only in the control and design a sliding mode controller for the transformed systems; the advantages over state feedback control are also given. Sinha *et al.* [9] give the sufficient conditions for sliding mode control of systems with uncertain, continuously differentiable, and bounded time delay in states using an invariant cone of positive initial functions. El-Khazali [2] proposes an output feedback sliding mode control design method for similar uncertain time-delay systems.

In this paper, we employ an idea which appeared in [3, 4] for transforming delayed systems. Dividing the unstable poles of the delayed system, with given stability margin, into N sets, where each set contains n unstable poles, N characteristic matrices can be found which inherit the nN unstable poles. A linear transformation can then be applied to convert the original delayed system to a delay-free system with state matrix being the direct sum of the N characteristic matrices. SMC can then be designed based on the delay-free system, and asymptotic stability can be shown for both delay-free and delayed systems on certain sliding surfaces. The reaching condition is proved in the case of bounded external disturbance, and the sliding surface is chosen by the means of a Lyapunov function. It should be noted that the sliding surface is a function of previous system states and controls, which obviously makes sense for time-delay systems.

Control of idle speed has become an important problem for IC engine control due to the demands for higher fuel economy and lower exhaust emissions. The control goal is to lower the idle speed while rejecting load disturbance and reducing engine speed variations. Both mass air flow and spark timing are used to control the idle speed. The main difficulty in controller design is with the so-called induction-to-power delay which degrades the control performance. Although the plant is stable itself, we applied the method to compensate the effects of the time delay. An overview of the control scheme, described next, is shown in Figure 1. Simulation results for a realistic engine model are presented in section 4.

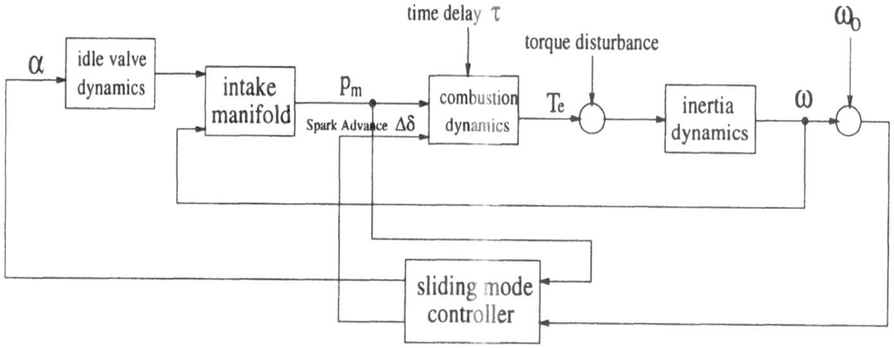

Fig. 1. Diagram of the control structure

2 Problem Formulation

We consider a system with point delay in both states and control, described by,

$$\Sigma: \; \dot{x}(t) = A_0 x(t) + A_1 x(t-r) + B_0 u(t) + B_1 u(t-h) + Df(x,t), \qquad (1)$$

where, r and h are the delays with the system states and control respectively, $x \in \mathcal{R}^n$ represents the system states, $u \in \mathcal{R}^m$ represents the control, $f \in \mathcal{R}^k$ is the disturbance, A_0 and A_1 are $n \times n$ matrices, B_0 and B_1 are $n \times m$ matrices, and D is an $n \times k$ matrix.

Denote $x_t(\theta) = x(t+\theta)$ for $\theta \in [-r, 0]$, and $u_t(\tau) = u(t+\tau)$ for $\tau \in [-h, 0]$. The initial conditions are given as $x(\theta) = x_0(\theta)$, $\theta \in [-r, 0]$, and $u(\tau) = u_0(\tau)$, $\tau \in [-h, 0]$, where, $x_0 \in \mathcal{C}([-r, 0]; \mathcal{R}^n)$, $u_0 \in \mathcal{C}([-h, 0]; \mathcal{R}^m)$, that is, x_0 is a real continuous function of $\theta \in [-r, 0]$, and u_0 is a real continuous function of $\tau \in [-h, 0]$.

3 Controller Design

Denote a nonnegative real number ν_0 as the required "stability margin", and define the set of "unstable" poles [3] of Σ as

$$\sigma_u(\Sigma) = \left\{ s \in C : \det(sI - A_0 - A_1 e^{-rs}) = 0, Re(s) \geq -\nu_0 \right\}. \tag{2}$$

Because the number of the "unstable" poles is finite, generally, one can find ways to divide $\sigma_u(\Sigma)$ as

$$\sigma_u(\Sigma) = \Lambda_c = \Lambda_1 \bigcup \Lambda_2 \bigcup ... \bigcup \Lambda_N \ , \tag{3}$$

where, Λ_i, $i = 1, 2, ..., N$, includes n "unstable" poles. Assuming, for each Λ_i, there exists A^i satisfying the characteristic matrix equation (7), and $\sigma(A^i) = \Lambda_i$, one can obtain,

$$A_c = \bigoplus_{i=1}^{N} A^i, \qquad \sigma(A_c) = \Lambda_c \ , \tag{4}$$

where $\bigoplus$ denotes the direct sum.

Defining a linear transformation

$$\begin{aligned} z_c(t) = (\mathcal{T}_{AC}(x, u))(t) &= \Sigma_{i=1}^{N} (e_i \bigotimes I_n) x(t) \\ &+ \int_{-r}^{0} e^{A_c \theta} \Sigma_{i=1}^{N} (e_i \bigotimes I_n) A_1 x(t - r - \theta) d\theta \\ &+ \int_{-h}^{0} e^{A_c \tau} \Sigma_{i=1}^{N} (e_i \bigotimes I_n) B_1 u(t - h - \tau) d\tau \ , \end{aligned} \tag{5}$$

where, e_i is an N-dimensional unit column vector whose entries are 0 except the ith entry which is 1, $e_i \bigotimes I_n = [0_{n\times n}, ..., 0_{n\times n}, I_{n\times n}, 0_{n\times n}, ..., 0_{n\times n}]^T$, with $I_{n\times n}$ as the ith entry, one can obtain a delay-free system,

$$\Sigma' \qquad \dot{z}(t) = A_c z(t) + B_c u(t) + Df(x, t) \ , \tag{6}$$

where, $B_c = \Sigma_{i=1}^{N} e_i \bigotimes (B_0 + e^{-hA^i} B_1)$, A_c is defined in (4), and A^i, for $i = 1, 2, ...N$, satisfy

$$A^i = A_0 + e^{-rA^i} A_1 \ , \tag{7}$$

which is called the characteristic matrix equation.

Assumption 1 The characteristic matrices A^i, $i = 1, 2, ..., N$, exist, i.e., the n left eigenvectors of Σ corresponding to Λ_i are linearly independent. [3].

Assumption 2 Σ is spectrally controllable, i.e. [8],

$$rank\left[sI - A_0 - e^{-rs}A_1 \mid B_0 + e^{-hs}B_1\right] = n, \quad \forall s \in \{s \in \mathcal{C}, Re(s) \geq -\nu_0\} . \tag{8}$$

Lemma 1 (A_c, B_c) is controllable if Assumption 2 holds and the spectra of A^i, $i = 1, 2, ...N$do not overlap [3].

Sliding mode controller design generally consists of two steps, described in the following.

3.1 Constructing the sliding surface

Step 1 A certain sliding surface is constructed so that on the sliding surface the system has desired behavior. We have the following lemma for constructing the surface.

Lemma 2 The system (6), in the absence of disturbance, is asymptotically stable on the sliding surface

$$S(t) = B_c^T P z(t) = 0 \quad , \tag{9}$$

if there exists a positive definite matrix P satisfying the Ricatti inequality

$$PA_c + A_c^T P < 0 \quad . \tag{10}$$

The proof follows from [10].

Note that the sliding surface designed using this Lyapunov method can not reduce the system order in the sliding mode.

Write the sliding surface as

$$S = B_c^T P z = Cz \quad , \tag{11}$$

where, $C = B_c^T P \in \mathcal{R}^{m \times nN}$.

Substituting (5) into (11), one can get the switching functional

$$\begin{aligned} S(t) = C[&\Sigma_{i=1}^N (e_i \otimes I_n) x(t) + \textstyle\int_{-r}^0 e^{A_c\theta} \Sigma_{i=1}^N (e_i \otimes I_n) A_1 x(t-r-\theta) d\theta \\ &+ \textstyle\int_{-h}^0 e^{A_c\tau} \Sigma_{i=1}^N (e_i \otimes I_n) B_1 u(t-h-\tau) d\tau] \quad . \end{aligned} \tag{12}$$

Partitioning C as $C = [C^1, C^2, ..., C^N]$, where, $C^i \in \mathcal{R}^{m \times n}, i = 1, 2, ..., N$, rewrite S as

$$\begin{aligned} S(t) = &\left(\Sigma_{i=1}^N C^i\right) x(t) + \textstyle\int_{-r}^0 \left(\Sigma_{i=1}^N C^i e^{A^i\theta}\right) A_1 x(t-r-\theta) d\theta \\ &+ \textstyle\int_{-h}^0 \left(\Sigma_{i=1}^N C^i e^{A^i\tau}\right) B_1 u(t-h-\tau) d\tau \quad . \end{aligned} \tag{13}$$

If the system is initially on the switching surface at t_r, then $S(t) = 0$, $\dot{S}(t) = 0$, for $t \geq t_r$. Furthermore,

$$\begin{aligned}\dot{S}(t) &= C\dot{z}(t)\\ &= CA_c z(t) + CB_c u(t)\\ &= CA_c\{\Sigma_{i=1}^N (e_i \otimes I_n)x(t) + \int_{-r}^0 e^{A_c\theta}\Sigma_{i=1}^N (e_i \otimes I_n)A_1 x(t-r-\theta)d\theta\\ &+ \int_{-h}^0 e^{A_c\tau}\Sigma_{i=1}^N (e_i \otimes I_n)B_1 u(t-h-\tau)d\tau\} + CB_c u(t) \quad .\end{aligned} \tag{14}$$

Assuming $(CB_c)^{-1}$ exists, one can obtain the equivalent control by letting $\dot{S}(t) = 0$,

$$\begin{aligned}u_{eq}(t) = -(CB_c)^{-1}CA_c\{&\Sigma_{i=1}^N (e_i \otimes I_n)x(t)\\ &+ \int_{-r}^0 e^{A_c\theta}\Sigma_{i=1}^N (e_i \otimes I_n)A_1 x(t-r-\theta)d\theta\\ &+ \int_{-h}^0 e^{A_c\tau}\Sigma_{i=1}^N (e_i \otimes I_n)B_1 u_{eq}(t-h-\tau)d\tau\} \quad .\end{aligned} \tag{15}$$

Again, partitioning $CA_c \in \mathcal{R}^{m\times nN}$ as $CA_c = [\bar{C}^1, \bar{C}^2, ..., \bar{C}^N]$, where, $\bar{C}^i \in \mathcal{R}^{m\times n}$.

$$\begin{aligned}u_{eq}(t) = -(CB_c)^{-1}\{&\left(\Sigma_{i=1}^N \bar{C}^i\right)x(t) + \int_{-r}^0 \left(\Sigma_{i=1}^N \bar{C}^i e^{A^i\theta}\right)A_1 x(t-r-\theta)d\theta\\ &+ \int_{-h}^0 \left(\Sigma_{i=1}^N \bar{C}^i e^{A^i\tau}\right)B_1 u_{eq}(t-h-\tau)d\tau\} \quad .\end{aligned} \tag{16}$$

The equations for sliding over the manifold $S(t) = 0$ become

$$\dot{x}(t) = A_0 x(t) + A_1 x(t-r) + B_0 u_{eq}(t) + B_1 u_{eq}(t-h) + Df \ , \tag{17}$$

$$\begin{aligned}&\left(\Sigma_{i=1}^N C^i\right)x(t) + \int_{-r}^0 \left(\Sigma_{i=1}^N C^i e^{A^i\theta}\right)A_1 x(t-r-\theta)d\theta\\ &+ \int_{-h}^0 \left(\Sigma_{i=1}^N C^i e^{A^i\tau}\right)B_1 u_{eq}(t-h-\tau)d\tau = 0 \quad .\end{aligned} \tag{18}$$

3.2 Control law design

Step 2 A suitable control law should be chosen to satisfy the reaching condition, i.e., the global stability of the switching surface, $S = 0$. Specifically, for a Lyapunov function, $V = S^T S/2$, satisfaction of the inequality $\dot{V} = S^T\dot{S} < 0$ should be guaranteed.

With the choice of

$$\dot{S} = -(CB_c)^{-1}d\ sgn(S) \ , \tag{19}$$

where, $d = \begin{pmatrix} d_1 & 0 & 0 & 0\\ 0 & d_2 & 0 & 0\\ & \cdots & & \\ 0 & 0 & 0 & d_m\end{pmatrix}$,

$d_i > 0$, $i = 1, 2, ..., m$, $\text{sgn}(S) = [\text{sgn}(S_1), ..., \text{sgn}(S_m)]^T$, one can obtain the control

$$\begin{aligned}u(t) = -(CB_c)^{-1}\{&\left(\Sigma_{i=1}^N \bar{C}^i\right)x(t) + \int_{-r}^0 \left(\Sigma_{i=1}^N \bar{C}^i e^{A^i\theta}\right)A_1 x(t-r-\theta)d\theta\\ &+ \int_{-h}^0 \left(\Sigma_{i=1}^N \bar{C}^i e^{A^i\tau}\right)B_1 u_{eq}(t-h-\tau)d\tau\} - (CB_c)^{-1}d\ \text{sgn}(S) \quad .\end{aligned} \tag{20}$$

3.3 Stability analysis

Now, considering the system with disturbance, we have the following theorems.

Theorem 1 Suppose the disturbance is bounded and $||f(x,t)||_2 \leq \beta$, the system (6) with control (20) is asymptotically stable if $d_0 > ||CD||_2\beta$, where, $d_0 = min(d_1, d_2, ..., d_m)$.

Proof We first prove that the sliding manifold $S = Cz = 0$ is reachable. Choose Lyapunov function $V = S^T S/2 > 0$, then

$$\begin{aligned} \dot{S} &= CA_c z + CB_c u + CDf \\ &= -d\, sgn(s) + CDf \quad . \end{aligned} \tag{21}$$

Hence,

$$\begin{aligned} \dot{V} = S^T\dot{S} &= -d||S||_1 + S^T CDf \\ &\leq -d_0||S||_1 + ||S||_2||CD||_2\beta \quad , \end{aligned} \tag{22}$$

where $||S||_1 = \Sigma_{i=1}^m |S|$. With the fact $||S||_2 \leq ||S||_1$, it is obvious that $\dot{V} < 0$ if $d_0 > ||CD||_2\beta$.

If t_r is the reaching time, then when $t > t_r$, the system reaches the sliding manifold and remains there, and the system dynamics are described by (17), (18). It is shown, by Lemma 2, that the system (6) is asymptotically stable on the sliding manifold, provided the positive P exists, hence, $\lim_{t\to\infty} z(t) = 0$. ■

Furthermore, we can see that from

$$\begin{aligned} \dot{V} &\leq -(d_0 - ||CD||_2\beta)||S||_1 \\ &\leq -(d_0 - ||CD||_2\beta)\sqrt{2V} \quad , \end{aligned} \tag{23}$$

one can obtain

$$t_r \leq \frac{||S_0||_2}{d_0 - ||CD||_2\beta} \quad , \tag{24}$$

where, t_r is the reaching time.

Theorem 2 System (1) with control (20) is asymptotically stable with a stability margin of ν_0 if $\sigma_u(\Sigma) \subseteq \bigcup_{i=1}^{N} \sigma(A^i)$ and $\sigma(A^i) \bigcap \sigma(A^j) = \emptyset$ for $i \neq j$, where $\emptyset$ denotes the void set.

Proof Laplace transforming (5) gives

$$Z(s) = \Gamma_1 X(s) + \Gamma_2 U(s) \tag{25}$$

where,

$$\Gamma_1 = \Sigma_{i=1}^N (e_i \bigotimes I_n) + e^{-rs} \int_{-r}^{0} e^{(A_c - sI)\theta} \Sigma_{i=1}^N (e_i \bigotimes I_n) A_1 d\theta \tag{26}$$

$$\Gamma_2 = e^{-hs} \int_{-h}^{0} e^{(A_c - sI)\tau} \Sigma_{i=1}^{N} (e_i \bigotimes I_n) B_1 d\tau \ . \tag{27}$$

When $t > t_r$, $u(t) = -CA_c z(t)$, and

$$U(s) = -CA_c Z(s) \ . \tag{28}$$

Substituting (28) into (25), one can obtain

$$X(s) = \Gamma_1^{-1}(I_{nN} + \Gamma_2 CA_c) Z(s) \ . \tag{29}$$

After further computation, we have

$$\Gamma_1 = (sI - A_c)^{-1} \Sigma_{i=1}^{N} (e_i \bigotimes I_n)[sI - A_0 - e^{-rs} A_1]. \tag{30}$$

Γ_1^{-1} is a stable term with stability margin of ν_0 because $\sigma_u(\Sigma) \subseteq \sigma(A_c) = \bigcup_{i=1}^{N} \sigma(A^i)$. Hence, the transfer function from $z(t)$ to $x(t)$ is a BIBO stable function, $\lim_{t \to 0} x(t) = 0$ if $\lim_{t \to 0} z(t) = 0$. Moreover, $|x_i(t)|$, $i = 1, 2, ..., n$, would be bounded provided $|z_i(t)|$ is bounded, $i = 1, 2, ...nN$ for $t > t_r$.

Next, we would like to show that the system states, $|x_i(t)|$, are bounded in the reaching process, i.e., for $t < t_r$. We use the mathematical induction method.

It is obvious that $z(t)$ is bounded for $t < t_r$. Hence, $S(t) = Cz(t)$ is bounded. Also, $u(t) = -CA_c z(t) - (CB_c)^{-1} d \ sgn(S)$ is bounded for bounded d; we will use the notation $|u(t)| < \bar{u}$.

Solving (1), in the absence of disturbance, we have

$$x(t + \tau) = e^{A_0 \tau} x(t) + \int_{t}^{t+\tau} e^{A_0 (t-\theta)} [A_1 x(\theta - \tau) + B_0 u(\theta) + B_1 u(\theta - h)] d\theta \tag{31}$$

Obviously, $|x_i(t)|$ is bounded for $t < \tau$.

Suppose $|x_i(t)|$ is bounded and $|x(t)| < \bar{X}$ for $t < t_0$. From (31), we have

$$|x(t + \tau)| < e^{A_0 \tau} |x(t)| + \int_{t}^{t+\tau} e^{A_0 (t-\theta)} |[A_1 \bar{X} + B_0 \bar{u} + B_1 \bar{u}]| d\theta \tag{32}$$

which is bounded. Hence, one can conclude that the right-hand side of (1) is Lipschitz, and the system states are bounded in the reaching process. ■

Remark 1 Real-valued solutions to the characteristic matrix equation (7) are given [3] in the form of $A^i = (Q^i)^{-1} J^i Q^i$, $i = 1, 2, ..., N$, provided A^i exists, where, J is the Jordan matrix, $Q = [Re(v_1^{'})|Im(v_1^{'})|Re(v_2^{'})|Im(v_2^{'})|...$ $|Re(v_p^{'})|Im(v_p^{'})|v_{2p+1}^{'}|...|v_n^{'}|]^{'}$ with v_i, $i = 1, 2, ..., p$, denoting the linearly independent left eigenvectors of Σ corresponding to the p eigenvalues with positive imaginary parts and v_i, $i = 2p + 1, ..., n$, denoting the linearly independent left eigenvectors of Σ corresponding to the $n - 2p$ real eigenvalues of Σ.

Remark 2 Consider the following additional assumptions:

Assumption 3 The number of unstable poles in $\sigma_u(\Sigma) \leq n$.

Assumption 4 $\sigma_u(\Sigma) \subseteq \sigma(A)$, that is, the unstable poles of Σ is a subset of the poles of the characteristic matrix A.

Under these assumptions, a simpler transformation

$$z(t) = x(t) + \int_{-r}^{0} e^{A\theta} A_1 x(t-r-\theta)d\theta + \int_{-h}^{0} e^{A\tau} B_1 u(t-h-\tau)d\tau \tag{33}$$

can be applied to get the delay-free system

$$\dot{z}(t) = Az(t) + Bu(t) + Df(x,t) \quad , \tag{34}$$

where, $A = A_0 + e^{-hA}A_1$, $B = B_0 + e^{-hA}B_1$. In this case, the control would be in the form of

$$\begin{aligned} u(t) = &-(CB)^{-1}CA\{x(t) + \textstyle\int_{-r}^{0} e^{A\theta} A_1 x(t-r-\theta)d\theta \\ &+ \textstyle\int_{-h}^{0} e^{A\tau} B_1 u(t-h-\tau)d\tau\} - (CB)^{-1} d\ sgn\ (S) \quad . \end{aligned} \tag{35}$$

Remark 3 Consider another special case, $r = 0$, which is the same as that discussed in [6]. The controller design procedure would be further simplified. Choosing A_0 as the characteristic matrix, one could apply the transformation,

$$z(t) = x(t) + \int_{-h}^{0} e^{A_0\tau} B_1 u(t-h-\tau)d\tau \quad , \tag{36}$$

to obtain the corresponding delay-free system equation

$$\dot{z}(t) = A_0 z(t) + Bu(t) + Df(x,t) \quad , \tag{37}$$

where, $B = B_0 + e^{-hA_0}B_1$.

Remark 4 It would be straightforward to extend the proposed SMC design method to the distributed time-delay system

$$\Sigma_d : \quad \dot{x}(t) = \int_{-r}^{0} d\alpha(\theta)x(t+\theta) + \int_{-h}^{0} d\beta(\tau)u(t+\tau) + \gamma(t)f(x,t) \tag{38}$$

where α, β are integrable functions.

An effective linear transformation could be defined as [4]

$$\begin{aligned} z(t) = &\Sigma_{i=1}^{N}(e_i \otimes I_n)x(t) + \textstyle\int_{-r}^{0}\int_{t+\theta}^{t} e^{A_c(t+\theta-\tau)}\Sigma_{i=1}^{N}(e_i \otimes I_n)d\alpha(\theta)x(\tau)d\tau \\ &+ \textstyle\int_{-h}^{0}\int_{t+\theta}^{t} e^{A_c(t+\theta-\tau)}\Sigma_{i=1}^{N}(e_i \otimes I_n)d\beta(\theta)u(\tau)d\tau \quad . \end{aligned} \tag{39}$$

The corresponding delay-free system is

$$\dot{z}(t) = A_c z(t) + B_c u(t) + \gamma(t)f(x,t) \quad , \tag{40}$$

where A_c is defined in (4) with $A^i = \int_{-r}^{0} e^{A^i\theta} d\alpha(\theta)$, $B_c = \begin{pmatrix} B^1 \\ B^2 \\ \dots \\ B^N \end{pmatrix}$ with $B^i = \int_{-h}^{0} e^{A^i\theta} d\beta(\theta)$ for $i = 1, 2, ..., N$.

4 Application to Engine Idle Speed Control

The importance of idle speed control partly comes from the fact that vehicles spend a large percentage of fuel during idling. A satisfactory idle speed control algorithm should let the engine operate at a low speed while effectively rejecting typical torque disturbances due to the accessory loads such as the power steering pump, the air conditioning compressor, the various electrical loads, the engagement of the automatic transmission, etc. A production idle speed controller generally consists of an anticipatory term using the information from the accessories, a proportional-integral-derivative (PID) feedback control term for the air mass, a proportional feedback term for the spark timing, and possibly a compensation term for varying engine conditions and environmental conditions. A survey of idle speed model and control methodologies is given by Hrovat *et al.* [5].

A highly simplified two-input (idle by-pass valve opening and spark advance) two-output (engine speed and intake manifold pressure) idle speed control model for IC engines was developed in [13] and used in this work. The model includes intake manifold dynamics, induction-to-power delay, and engine rotational dynamics, encompassed in the equations

$$\frac{dP_m(t)}{dt} + \frac{\eta_v V_d}{4\pi V_m}\omega(t)P_m(t) = K_1\alpha(t) \quad , \tag{41}$$

$$J\frac{d\omega(t)}{dt} + B\omega(t) = K_\tau P_m(t-r) + K_\delta \Delta\delta(t) + \tau_f(t) \quad , \tag{42}$$

where,
P_m is intake manifold pressure,
η_v is volumetric efficiency,
V_d is engine displacement,
V_m is manifold volume,
ω is engine speed,
K_1 is the coefficient relating throttle opening to delayed manifold pressure,
J is engine inertia,
B is lumped damping coefficient,
K_τ is the coefficient relating engine torque to manifold pressure,
K_δ is the coefficient relating engine torque to sparking timing,
$\Delta\delta$ is spark advance,
r is induction-to-power delay, and
τ_f is disturbance torque.
The parameters for a Ford V8 engine are shown in Table 1.

Although simple in nature, we emphasize that this model encompasses the essential dynamics needed for control design. Indeed, successful idle speed control designs have been constructed using this representation, and subsequently applied in an engine test cell [12].

Table 1. V8 Engine Parameters

Parameter	Value	Units
η_v	0.55	*dimensionless*
V_d	0.0046	m^3
V_m	0.0029	m^3
J	0.0843	$Nm - sec^2/rad$
K_τ	5.7143	Nm/kPa
K_1	110	$(kPa/sec)/deg$
B	0.592	$Nm - sec$
ω_0	63.98	rad/sec

Letting $K_2 = \frac{\eta_v V_d}{4\pi V_m}$ and linearizing (41), (42) about the nominal operating point (ω_0, p_{m_0}), one can get

$$\frac{d\Delta p_m(t)}{dt} + K_2\omega_0 \Delta p_m(t) = -K_2 p_{m_0}\Delta\omega(t) + K_1\Delta\alpha(t) \quad , \tag{43}$$

$$J\frac{d\Delta\omega(t)}{dt} + B\Delta\omega(t) = K_\tau \Delta p_m(t-r) + K_\delta \Delta\delta(t) + \tau_f(t) \quad , \tag{44}$$

Converting to a state space form, with $x(t) = [\Delta\omega(t), \Delta p_m(t)]^T$, $u(t) = [\Delta\alpha(t), \Delta\delta(t)]^T$, $f(t) = \tau_f(t)$, we have

$$\dot{x}(t) = A_0 x(t) + A_1 x(t-r) + B_0 u(t) + Df(t) \quad , \tag{45}$$

with $A_0 = \begin{pmatrix} -B/J & 0 \\ -K_2 p_{m_0} & -K_2\omega_0 \end{pmatrix}$, $A_1 = \begin{pmatrix} 0 & K_\tau/J \\ 0 & 0 \end{pmatrix}$, $B_0 = \begin{pmatrix} 0 & K_\delta \\ K_1 & 0 \end{pmatrix}$, $D = \begin{pmatrix} 1/J \\ 0 \end{pmatrix}$

The characteristic polynomial is $(s + B/J)(s + K_2\omega_0) + K_2 K_\tau p_{m_0} e^{-rs}/J$. Choosing $\nu_0 = 4$, one can find two poles, $s = -3.6326 \pm j6.2539$, which violate the stability margin ("unstable poles"). Hence, $N = 1$. Furthermore, $\text{rank}[sI - A_0 - A_1 e^{-rs} | B_0] = 2$ for all $s \in \{s \in C, Re(s) \geq -\nu_0\}$, the system is spectrally controllable.

For $s = -3.6326 + j6.2539$, one can find $Q = \begin{pmatrix} -0.0308 & 1 \\ 0.0569 & 0 \end{pmatrix}$, $J = \begin{pmatrix} -3.6326 & -6.2539 \\ 6.2539 & -3.6326 \end{pmatrix}$, $A_c = Q^{-1}JQ = \begin{pmatrix} -7.0178 & 109.9104 \\ -0.4601 & -0.2474 \end{pmatrix}$, $B_c = B_0$.

Using a linear transformation

$$z(t) = x(t) + \int_{-r}^{0} e^{A_c\theta} A_1 x(t - r - \theta) d\theta \quad , \tag{46}$$

we obtain the delay-free system

$$\dot{z}(t) = A_c z(t) + B_c u(t) + Df(t) \tag{47}$$

To reduce the steady state error, choose the sliding surface as

$$\begin{aligned} S &= z(t) + C\int_0^t z(t')dt' \\ &= [x + \int_{-r}^{0} e^{A_c\theta} A_1 x(t-r-\theta)d\theta] + C\int_0^t [x + \int_{-r}^{0} e^{A_c\theta} A_1 x(t'-r-\theta)d\theta]dt' \end{aligned} \tag{48}$$

where, $C = \begin{pmatrix} c1 & 0 \\ 0 & c2 \end{pmatrix}$, $c1, c2 > 0$.

In order to increase the reaching speed, we choose,

$$\dot{S}(t) = -kS(t) - d\ sgn(S(t)), \quad k > 0, \tag{49}$$

Hence, the control is

$$\begin{aligned} u(t) = &-B_c^{-1}(A_c + C)[x + \int_{-r}^{0} e^{A_c\theta} A_1 x(t-r-\theta)d\theta] \\ &-B_c^{-1}kS(t) - B_c^{-1}d\ sgn(S(t)) \end{aligned} \tag{50}$$

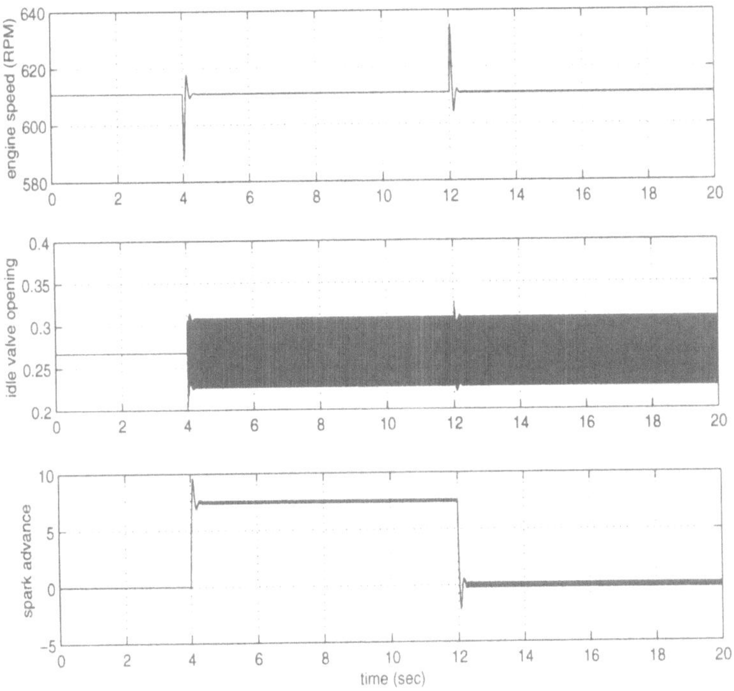

Fig. 2. Engine speed variations and control inputs for linearized model

Simulation results are shown in Figures 2 - 6. A disturbance of $15Nm$ is loaded to the system when $t = 4\ sec$ and unloaded when $t = 12\ sec$. Figure 2 gives the engine speed variations and control inputs for the linearized model. One can see that the controller maintains the idle speed at $611RPM$ and there is no steady-state error. The controller thus has the capability of disturbance rejection for a typical "bad" disturbance. Figure 3 gives the engine speed variations and control inputs for the nonlinear engine idle speed control model. Slightly different responses are evident when the nonlinear simulation model is used, but the controller performance is still quite good. These responses compare quite favorably to previous simulation exercises for linear control designs [13] and for nonlinear fuzzy control designs [12], all of which were later implemented in the engine test cell. Note that for both linearized and nonlinear models, the spark advance shown in degrees plays an important role. Although the engine speed does not oscillate significantly, both of the control inputs have chattering characteristics, which could burden the actuators, and should be avoided if possible in the actual applications. Boundary layers can be used to smooth the control discontinuities in a small neighborhood of the sliding surfaces.

Figures 4, 5, and 6 show the robustness of the controller to system parameter uncertainties. Variations in K_δ, K_1, and K_τ are considered because they are crucial to controller design. Figure 4 gives the engine speed variations to a $15Nm$ disturbance at $t = 4\ sec$ for K_δ varying from 0.25 to 8. One can see that the controller can stabilize the system and reject the disturbance for all the cases. However, as K_δ decreases, the engine speed deviation from the nominal idle speed increases, whereas the chattering becomes significant when K_δ increases. Further increasing K_δ would lead to instability. Figure 5 gives the engine speed for $K_1 = 11$ and 220. We can find the same result as in Figure 4. Increasing K_1 would give higher chattering and even instability, while decreasing K_1 would increase engine speed variations. Figure 6 shows the engine speed variations for $K_\tau = 0.57143, 5.7143$, and 57.143. One can conclude that K_τ does not have as much effect on the engine speed as does K_δ.

5 Conclusions

A new SMC design algorithm is presented in this paper for general point delay systems on the basis of a linear transformation which reduces the delay system to a delay-free system. It is applied in simulation to the IC engine idle speed control problem. Simulation results show the controller is robust to system parameter uncertainties, and has the ability to maintain a low idle speed ($611RPM$) while rejecting the load disturbance. We are in the process of implementing the control algorithm on a 4.6L Ford V8 engine at The Ohio State University.

References

1. S. V. Drakunov and V. I. Utkin, "Sliding mode control in dynamic systems," *Int. J. Control*, Vol.55, pp. 1029-1037, 1992.
2. R. El-Khazali, "Variable structure robust control of uncertain time-delay systems," *Automatica*, Vol.34, pp. 327-332, 1998.

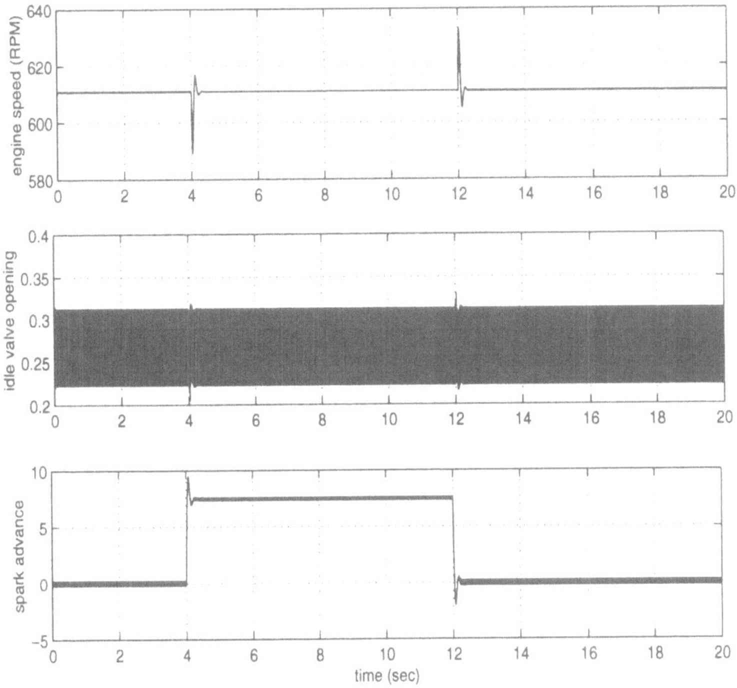

Fig. 3. Engine speed variations and control inputs for nonlinear model

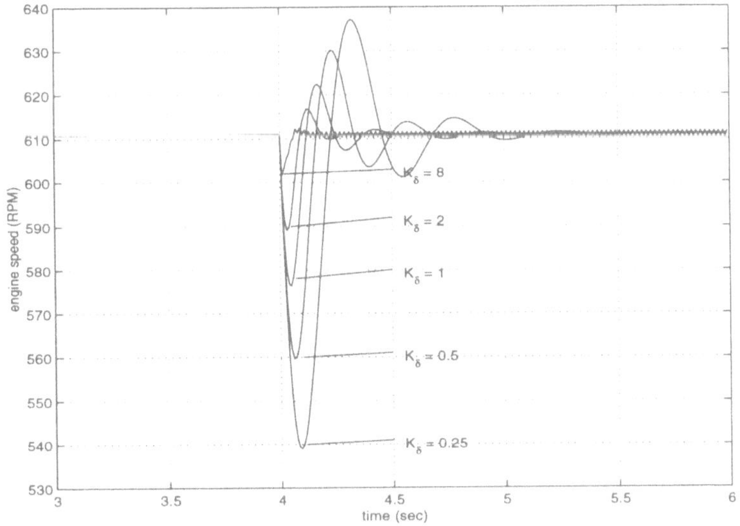

Fig. 4. Engine speed variations with varying K_δ

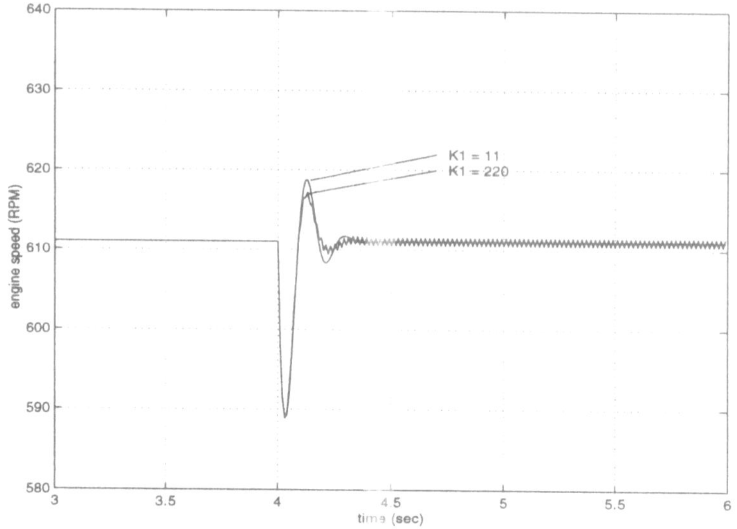

Fig. 5. Engine speed variations with varying K_1

3. Y. A. Fiagbedzi and A. E. Pearson, "Feedback stabilization of linear autonomous time lag systems," *IEEE Trans. Auto. Control*, AC-31, pp. 847-855, 1986.
4. Y. A. Fiagbedzi and A. E. Pearson, "A multistage reduction technique for feedback stabilizing distributed time-lag systems," *Automatica*, Vol.23, pp. 311-326, 1987.
5. D. Hrovat and J. Sun, "Models and control methodologies for IC engine idle speed control design," *Proc. 13th IFAC World Congress*, pp. 243-248, 1996.
6. K. J. Hu, V. R. Basker and O. D. Crisalle, "Sliding mode control of uncertain input-delay systems," *Proc. American Control Conference*, pp. 564-568, 1998.
7. E. M. Jafarov, "Analysis and synthesis of multidimensional SVS with delays in sliding-modes," *Proc. 11th IFAC World Congress*, pp. 46-49, 1990.
8. A. W. Olbrot, "Stabilizability, detectability and spectrum assignment for linear autonomous systems with general time delays," *IEEE Trans. Auto. Control*, AC-23, pp. 887-890, 1978.
9. A. S. C. Sinha, M. El-Sharkawy and M. Rizkalla, "Sliding mode control of uncertain delay differential systems," *Nonlinear Analysis, Theory, Methods & Applications*, Vol. 30, pp. 1075-1086, 1997.
10. W. C. Su, S. V. Drakunov and U. Ozguner, "Constructing discontinuity surfaces for variable structure systems: A Lyapunov approach," *Automatica*, Vol 32, pp. 925-928, 1996.
11. V. I. Utkin, *Sliding modes in control and optimization*, Springer, 1992.
12. J. Wills, S. Yurkovich and G. Rizzoni, "Direct fuzzy control of idle speed in an internal combustion engine," *IFAC Workshop: Advances in Automotive Control*, Mohican State Park, OH, Feb. 1998.
13. S. Yurkovich and M. Simpson, "Comparative analysis for idle speed control: A crank-angle domain viewpoint," *Proc. American Control Conference*, pp. 278-283, 1997.

This article was processed using the LaTeX macro package with LLNCS style

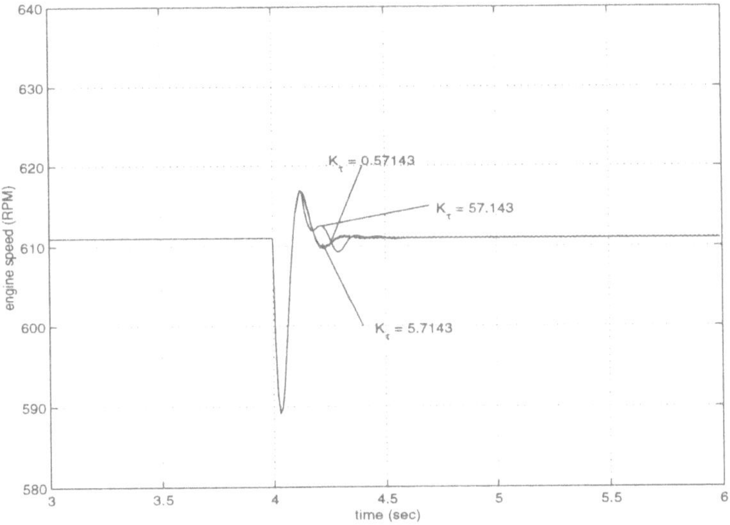

Fig. 6. Engine speed variations with varying K_τ

On Global Stabilization of Nonlinear Dynamical Systems

Xinghuo Yu[1], Yuqiang Wu[2], Man Zhihong[3]

[1]Faculty of Informatics and Communication, Central Queensland University, Rockhampton 4702, Australia
[2]Institute of Automation, Qufu Normal University, Qufu, PR China
[3]Department of Electrical Engineering and Computer Science, University of Tasmania, Hobart 7001, Australia

Abstract. The global stabilization of nonlinear minimum phase systems with partially linear strict nonminimum phase composite dynamics is discussed in this chapter using only the state variables of the linear composite part. The concept of the terminal sliding mode is employed for the control design. The advantage of the approach is that the finite time convergence of the proposed control strategy enables elimination of the effect of asymptotic convergence on the nonlinear systems, hence the peaking phenomenon does not occur. The global stabilization of the nonlinear systems under the developed controller is guaranteed.

1 Introduction

The asymptotic stabilization of minimum phase nonlinear systems with either a well-defined relative degree in a neighbourhood of the equilibrium or augmented state space have been well studied via smooth state feedback control [2, 14, 16, 9, 18, 13]. One major methodology makes use of the nonlinear Byrnes–Isidori normal form derived by an appropriate diffeomorphism and a state feedback transformation [6]. This normal form includes a controllable (or stabilizable) linear composite part and a nonlinear part which is zero input asymptotically stable.

Local stabilizability of the systems in the Byrnes–Isidori normal form has been well understood using the central manifold theory as a linear feedback control can stabilize the linear composite dynamics [1, 5]. For global stabilizability, a counter–example has shown that the global stabilization cannot be guaranteed if the control only keeps the linear composite dynamics exponentially stable, whilst the zero dynamics is globally stable [15]. This is due to the trade–off between the peaking phenomenon caused by the linear state feedback control of the linear composite part and the growth properties of the nonlinear part [14]. To address this problem, extensive research has been focused on exploring special properties of the linear part of the system, such as the passivity properties or the positive real condition [8, 13] so that the dynamic feedback control and passivity analysis tools can be used to realize the asymptotic stabilization [11]. Saberi et al [16] have investigated some special forms for both the nonlinear subsystem and the linear composite system for multivariable dynamic systems. The main difficulty for the global stabilization is that even though the linear smooth feedback control makes the linear composite system globally stable, the dynamics of the nonlinear subsystem may tend to infinity in finite time for some initial conditions when the nonlinear subsystem has more than one equilibriums.

In this chapter, we deal with the global stabilization problem using the terminal sliding mode (TSM) concept [18, 19]. The central problem of the global stabilization problem of the nonlinear minimum phase systems is the asymptotic stability of the linear composite part. Because the state never reaches the equilibrium exactly (or in infinite time), the small decaying value of the state during the transient process may result in the peaking phenomenon. The advantage of our control is that it is independent of the nonlinear part of the system, and only the states of the linear composite dynamics are needed. This implies that the global stabilization is only related to the linear composite part without influence from the nonlinear dynamics by the use of TSM principle.

This chapter is organised as follows. Section 2 presents the problem statement. Section 3 introduces the TSM concept. The TSM controller is designed in Section 4 for the global stabilization. The singularity and global stability issued are dealt with in Section 5. Two examples for illustrating the application of the proposed TSM strategy are given in Section 6. Finally, some conclusions and discussions are drawn in Section 7.

2 System Description

Consider the following nonlinear system with partially linear composite subsystem [16, 14].

$$\left.\begin{aligned} \dot{x} &= f(x,\xi,t),\ x \in R^n,\ \xi \in R^l, \\ \dot{\xi} &= A\xi + Bu,\ u \in R^m, \end{aligned}\right\}, \tag{1}$$

where $f(x, \xi)$ is a smooth vector function, and A and B are constant matrices. Here the nonlinear term $f(x,\xi,t)$ is time-varying. For nonlinear function $f(x,\xi,t)$, in order to obtain a smooth linear feedback controller to globally stabilise the system (1), some special forms were discussed [16, 14]. For example, $f(x,\xi,t)$ is required to be linearly bounded about the norm of dynamics x; $f(x,\xi,t)$ is globally Lipschitz, or $f(x,0,t)$ is globally Lipschitz and $f(x,\xi,t)$ satisfies a growth condition

$$\|f(x,\xi,t) - f(x,0,t)\| \leq L(x,\xi,t)\|\xi\|, \tag{2}$$

where $L(x,\xi,t)/\|x\|$ approaches zero uniformly when $x \to 0$ for bounded ξ [14]. For such nonlinear function $f(x,\xi,t)$, the peaking phenomenon is avoided. For the nonlinear dynamics x in system (1), the following assumptions are made.

H_1 : $f(0,0,t) = 0$ for all t. The origin of the zero dynamics

$$\dot{x} = f(x,0,t) \tag{3}$$

is globally asymptotically stable.

H_2 : the pair (A, B) is controllable.

The control task is to find a control $u(t)$ such that system (1) is asymptotically stable given the assumptions H_1 and H_2. There have been two important methods for the design of such a controller. One uses the linear composite dynamics ξ to design the linear feedback controller to guarantee that the dynamics $\xi(t)$ is exponentially stable and its effect on the nonlinear dynamics $x(t)$ is reduced to zero exponentially. This sometimes may result in the peaking phenomenon for certain nonlinear systems [14]. The other uses all the system states x and ξ for the design of a smooth nonlinear feedback controller. One common requirement in existing research works is that there exists a known (or unknown) positive definite Lyapunov function for (3) such that its derivative along the zero dynamics is negative definite [16] (or negative K-class function). function $f(\bullet)$ must satisfy a particular condition for the global stability or semiglobal stability [3, 13, 17]. This is due to the fact that, although the linear subsystem of (1) is controllable or stabilizable, a smooth linear feedback control can only guarantee the exponential stability of the linear composite system, i.e. ξ is exponentially stable. This exponential stability of the states, however, may result in an unstable closed-loop system. Indeed, let us consider the following example (x is a scalar)

$$\left.\begin{aligned} \dot{x} &= -\tfrac{1}{2}x^3 - \frac{\xi_2^2}{e^{-t^2}+\xi_2^2}\phi(x) \\ \dot{\xi}_1 &= \xi_2 \\ \dot{\xi}_2 &= u \end{aligned}\right\} \tag{4}$$

If $\phi(x) = -x$, we can say that this system can not be globablly stabilised by a linear feedback control. In fact, for any linear feedback control $u = F\xi$ that guarantees the stability of $\xi(t)$, due to the eigenvalues of the closed-loop linear system, the resulting solution $\xi_2(t)$ of the linear system tends to zero exponentially with the following possible forms

$$\xi_2(t) = c_1 e^{-\alpha_1 t} + c_2 e^{-\alpha_2 t}, \tag{5}$$

where c_1, c_2 are constants and α_1, α_2 are positive constants.

$$\xi_2(t) = c_3 t e^{-\alpha_3 t} + c_4 e^{-\alpha_3 t}, \tag{6}$$

where c_3, c_4 are constants and α_3 is positive constant;

$$\xi_2(t) = c_5 e^{-\alpha_4 t}\sin(\alpha_5 t) + c_6 e^{-\alpha_4 t}\cos(\alpha_5 t), \tag{7}$$

where c_5, c_6 are constants and α_4, α_5 are positive constants.

For cases (5) and (6), the dynamics $\xi_2^2(t)$ will be greater than zero and decreases slower than e^{-t^2} when t tends to infinity so that the term $\xi_2^2(t)/(\exp(-t^2) + \xi_2^2(t))$ tends to 1. This shows the first order nonlinear system

$$\dot{x} = -\frac{1}{2}x^3 + \frac{\xi_2^2}{e^{-t^2} + \xi_2^2}x \tag{8}$$

is not stable. As to case (7), we can always choose an initial value $\xi_2(0)$ and $\xi_1(0)$ such that

$$\xi_2(t) = \xi_2(0)e^{-\alpha_4 t}cos(\alpha_5 t) \tag{9}$$

is a solution. The first equation of (4) becomes

$$\dot{x} = -\frac{1}{2}x^3 + \frac{(\xi_2(0)e^{-\alpha_4 t}cos(\alpha_5 t))^2}{e^{-t^2} + (\xi_2(0)e^{-\alpha_4 t}cos(\alpha_5 t))^2}x. \tag{10}$$

And further we obtain

$$\dot{y} = 1 - \frac{2(\xi_2(0)e^{-\alpha_4 t}cos(\alpha_5 t))^2}{e^{-t^2} + (\xi_2(0)e^{-\alpha_4 t}cos(\alpha_5 t))^2}y, \tag{11}$$

where $y = x^{-2}$. Since e^{-t^2} converges fast than e^{-at} for any $a > 0$, It can be seen that although in some time instants $(k\pi + \pi/2)/\alpha_5$ $(k = 0, 1, \cdots)$, $(\xi_2(0)e^{-\alpha_4 t}cos(\alpha_5 t))^2$ equals zero, when k (or t) is large and $\mid y(t) \mid > M > 1$ (M is certain constant), $y(t)$ will strictly decrease in interval $[k\pi+\delta, (k+1)\pi-\delta]$ with slope more than one while in interval $[(k\pi - \delta, k\pi + \delta]$ at most increase in slope one where $0 < \delta < \pi/4$. This implies that $y(t)$ is bounded in time interval $[0, \infty)$ so that the solution of equation (10), $x(t)$, does not tend to zero.

The above analysis shows that the system (4) cannot be stabilised by using only the linear feedback control $u = F\xi$. How to construct a smooth nonlinear feedback $u(t, x, \xi)$ to globally stabilise system (4) is a difficult task. It is obvious that, from this example, the nonlinear dynamics does not meet the Lipschitz condition or the growth condition (2) given in [14]. This example also shows that even the zero dynamics $\dot{x} = -(1/2)x^3$ is asymptotically stable and the linear composite system is exponential stable, the entire nonlinear system may not be globally asymptotically stable. To overcome this problem, some restrictions may be imposed on function $f(\bullet)$ which in turn limits the applicability. However, if a controller can make the dynamics $\xi_1(t)$ and $\xi_2(t)$ of the linear subsystem (4) reach zero in finite time (which is tunable) and remain at zero forever (which means that after the finite time, dynamics $\xi_2(t)$ does not affect the nonlinear dynamics $x(t)$), then the nonlinear subsystem of (4) becomes $\dot{x} = -(1/2)x^3$ exactly after the finite time and is globally asymptotically stable. The goal of this chapter is to find such a controller.

In this chapter, we propose to use the TSM approach [21] for the global stabilization problem. By using the TSM approach, the state $\xi(t)$ will reach zero in a finite time T_0 and stay at zero forever. When $\xi(t)$ remains at zero, according to assumption H_1, we can conclude that $x(t)$ will globally asymptotically tend to zero since the zero dynamics (3) is assumed to be globally asymptotically stable. Hence the globally asymptotic stabilization of the system (1) is realized. In addition, if the system (3) is not asymptotic stable, but Lyapunov stable, that is, the nonlinear subsystem of (1) is weakly minimum phase, our TSM based controller will still guarantee that the closed-loop system is Lyapunov stable.

In the next section, we shall introduce the TSM concept for control design.

3 The Terminal Sliding Mode Concept

The TSM concept can be described by the following first order dynamics

$$s = \dot{x} + \beta\, x^{q/p} = 0 \tag{12}$$

where x is a scalar variable, $\beta > 0$, and p, q $(p > q)$ are positive integers. Note that the parameter p must be an odd integer and only real solution is considered so that for any real number x, $x^{q/p}$ is always a real number. The equation (12) then becomes

$$\dot{x} = -\beta x^{q/p} \tag{13}$$

Given an initial state $x(0) = 0$ the dynamics (12) will reach $x = 0$ in finite time. In fact, the solution of equation (12) is

$$\frac{p}{p-q}[x(t)^{(p-q)/p} - x(0)^{(p-q)/p}] = -\beta t. \tag{14}$$

The time taken from the initial state $x(0)$ to 0, t^s, is determined by

$$t^s = \frac{p}{\beta(p-q)}|x(0)|^{(p-q)/p} \tag{15}$$

Indeed, from the solution (14)

$$\frac{p}{p-q}[x(t^s)^{(p-q)/p} - x(0)^{(p-q)/p}] = -\beta t^s \tag{16}$$

one can easily get (15) for $x(0) > 0$. Since p, q are positive odd integers, then $p - q$ is an even number. Hence we have (15) for $x(t^s) = 0$. It can be proved that the equilibrium 0 is an attractor, i.e. when the state x reaches zero, it will stay at zero forever. Indeed, if we take a Lyapunov function $v = \frac{1}{2}x^2$, the time derivative of v is

$$\dot{v} = x\dot{x} = -\beta x\dot{x}^{q/p} = -\beta x^{(p+q)/p}$$

since $(p+q)$ is even, then $\dot{v}$ is negative definite, so $x = 0$ is "terminally" stable (not necessarily asymptotically stable).

The reaching time t^s which is determined by (15) depends on the parameters p, q, β, and the initial value $x(0)$. As $x(0)$ is fixed or in a known bounded region, one can choose β such that t^s is very small. This property will be used in the following analysis of stability.

The problem with (12) is that while the convergence when near the equilibrium is much improved, the convergence when far away from the equilibrium is worsened since the magnitude of $x^{(p+q)/p}$ is smaller than x^2 for asymptotical convergence. To improve the convergence for both situations, we now introduce the following dynamics,

$$s = \dot{x} + \alpha\, x + \beta\, x^{q/p} = 0 \tag{17}$$

where $x \in R^1$, α, $\beta > 0$ which should take care of both situations. By doing so, we have

$$\dot{x} = -\alpha x - \beta x^{q/p} \tag{18}$$

For properly chosen q, p, given an initial state $x(0) = 0$ the dynamics (18) will reach $x = 0$ in finite time. The physical interpretation is obvious. When x is far away from zero, the approximate dynamics becomes $\dot{x} = -\alpha x$ whose fast convergence when far away from the equilibrium was well understood. When close to $x = 0$, the approximate dynamics becomes $\dot{x} = -\beta x^{q/p}$ whose finite time convergence was shown above.

More precisely, we can solve the differential equation (18) analytically. Let $z = x^{1/p}$, then $x = z^p$. Substitute this into (18) leads to

$$pz^{p-1}\dot{z} = -\alpha z^p - \beta z^q$$

Dividing both sides by z^q yields

$$pz^{p-q-1}\dot{z} = -\alpha z^{p-q} - \beta$$

which can be easily rewritten as

$$\frac{p}{p-q}\frac{d(z^{p-q})}{dt} = -\alpha z^{p-q} - \beta$$

Its solution is

$$\alpha z^{p-q} + \beta = c\exp(-\alpha(p-q)t/p) \tag{19}$$

for $t = 0$, $x(0) = 0$. Hence from (19) and the transformation $x = z^p$,

$$c = \alpha x(0)^{(p-q)/p} + \beta$$

The time to reach zero, t^s, is determined by

$$t^s = \frac{p}{\alpha(p-q)}(\ln c - \ln \beta) \tag{20}$$

Substituting c into (20) gives

$$t^s = \frac{p}{\alpha(p-q)}(\ln(\alpha x(0)^{(p-q)/p} + \beta) - \ln \beta) \tag{21}$$

and as shown above, the equilibrium 0 is a terminal attractor.

To extend the scalar TSM concept to MIMO systems, we now introduce the following TSM vector

$$s = C_1\xi_1 + C_2\xi_2 + C_3\xi_1(\xi_1^T\xi_1)^{-q_0/p_0} \tag{22}$$

where $s \in R^m$, $p_0 > 4q_0$, $m \times (l-m)$, $m \times m$, $m \times (l-m)$ matrices, respectively. by looking at the term $\xi_1(\xi_1^T\xi_1)^{-q_0/p_0}$ in (22). Since

$$\|\xi_1(\xi_1^T\xi_1)^{-q_0/p_0}\| = \|\xi_1^T\xi_1\|^{(p_0-2q_0)/p_0}$$

Thus as $\|\xi_1\|$ becomes sufficiently small, s does not tend to infinity, and s is smooth in ξ.

Now we apply this TSM mechanism to the global stabilization problem. Without loss of generality, assume that the linear composite part of the nonlinear system (1) is in the following form:

$$\dot{\xi}_1 = A_{11}\xi_1 + A_{12}\xi_2 \tag{23}$$
$$\dot{\xi}_2 = A_{21}\xi_1 + A_{22}\xi_2 + B_2 u \tag{24}$$

where $\xi_1 \in R^{l-m}$, $\xi_2 \in R^m$ are system states, A_{11}, A_{12}, A_{21}, A_{22}, B_2 are $(q-m)\times(l-m)$, $(l-m)\times m$, $m\times(l-m)$, $m\times m$, $m\times m$ matrices, respectively. Since the pair (A, B) is controllable, one can always take a coordinates transformation to make the pair (A_{11}, A_{12}) controllable and B_2 nonsingular. The TSM vector (22) will be used for the MIMO control design.

We now present the first result on global stabilization.

Theorem 1. *For the dynamical system (24), If the control law is designed as*

$$\begin{aligned} u = &-(C_2B_2)^{-1}\{[C_1 + C_3(\xi_1^T\xi_1)^{-q_0/p_0} \\ &-2\frac{q_0}{p_0}C_3\xi_1(\xi_1^T\xi_1)^{-1-(q_0/p_0)}\xi_1^T](A_{11}\xi_1 + A_{12}\xi_2) \\ &+C_2(A_{21}\xi_1 + A_{22}\xi_2) + s + Ks(s^Ts)^{-q_0/p_0}\} \end{aligned} \tag{25}$$

where K is positive constant, then the manifold $s = 0$ defined in (22) will be reached in finite time and ξ_1 remains at zero forever.

Proof. Choose the Lyapunov function

$$V = \frac{1}{2}s^Ts$$

The time derivative of V along the system dynamics (22) and (24) is

$$\begin{aligned} \dot{V} &= s^T\dot{s} \\ &= s^T[C_1\dot{\xi}_1 + C_2\dot{\xi}_2 + C_3(\xi_1^T\xi_1)^{-q_0/p_0}\dot{\xi}_1 - 2\frac{q_0}{p_0}C_3\xi_1(\xi_1^T\xi_1)^{-1-(q_0/p_0)}\xi_1^T\dot{\xi}_1] \\ &= s^T[C_1 + C_3(\xi_1^T\xi_1)^{-q_0/p_0} - 2\frac{q_0}{p_0}C_3\xi_1(\xi_1^T\xi_1)^{-1-(q_0/p_0)}\xi_1^T](A_{11}\xi_1 + A_{12}\xi_2) \\ &\quad + s^TC_2(A_{21}\xi_1 + A_{22}\xi_2) + s^TC_2B_2u \end{aligned} \tag{26}$$

Substituting (25) into (26) yields

$$\begin{aligned} \dot{V} &= -s^Ts - K(s^Ts)(s^Ts)^{-q_0/p_0} \\ &= -2V - 2^{(p_0-q_0)/p_0}KV^{(p_0-q_0)/p_0} \end{aligned} \tag{27}$$

According to the analysis given in Section 3 (by letting $p = p_0$ and $q = p_0 - q_0$), $V(t)$ will reach zero in finite time and the finite time t^s is able to be determined by the parameters K, p_0, q_0, and the initial value of $V(0)$ which is calculated by $s(0)$ or $\xi_1(0)$ and $\xi_2(0)$. $V(t)$ reaches zero in finite time t^s means that $s = 0$ will be reached in finite time. It is worth noting that since $V(t)$ is nonnegative, here the requirement of odd integers q and p is not necessary.

Since from (27), $\dot{V}(t)$ is negative, $V(t)$ decreases with time t so that when $V(t)$ reaches zero at time t^s, it will keep zero forever for $t > t^s$. This implies that once the trajectory $\xi(t) = (\xi_1(t), \xi_2(t))$ reaches surfaces $s = 0$, it remains in $s = 0$.

When the sliding mode $s = 0$ is reached, from (22) we have

$$\xi_2 = -C_2^{-1}C_1\xi_1 - C_2^{-1}C_3\xi_1(\xi_1^T\xi_1)^{-q_0/p_0} \tag{28}$$

Substituting (28) into (23) leads to

$$\dot{\xi}_1 = (A_{11} - A_{12}C_2^{-1}C_1)\xi_1 - A_{12}C_2^{-1}C_3\xi_1(\xi_1^T\xi_1)^{-q_0/p_0} \tag{29}$$

When the system state ξ_1 is far away from the equilibrium $\xi_1 = 0$, the dynamics is dominated by

$$\dot{\xi}_1 = (A_{11} - A_{12}C_2^{-1}C_1)\xi_1 \tag{30}$$

When ξ_1 is near the equilibrium $\xi_1 = 0$, using the same reason as above, the dynamics is dominated by

$$\dot{\xi}_1 = -A_{12}C_2^{-1}C_3\xi_1(\xi_1^T\xi_1)^{-q_0/p_0}. \tag{31}$$

in which the term $(\xi_1^T\xi_1)^{-q_0/p_0}$ will become very large when ξ_1 is very small.

We now look at how to choose the constant matrices C_1, C_2, C_3 so that the finite time reachability is achieved. We have the following theorem.

Theorem 2. *Let L_1 and L_2 be*

$$L_1 = (A_{11} - A_{12}C_2^{-1}C_1),\ L_2 = A_{12}C_2^{-1}C_3.$$

If the following conditions hold

1. *The matrices C_1 and C_2 are chosen such that*

$$Re\{\lambda(L_1 + L_1^T)\} < 0 \tag{32}$$

2. *The matrices C_2 and C_3 are chosen such that*

$$Re\{\lambda(L_2 + L_2^T)\} > 0 \tag{33}$$

where $\lambda(\bullet)$ represents the eigenvalues, then the equilibrium $\xi_1 = 0$ is globally stable and can be reached in finite time.

Proof. Consider the Lyapunov function

$$V = \frac{1}{2}\xi_1^T\xi_1$$

Differentiating it along the dynamics (30) leads to

$$\begin{aligned}\dot{V} &= \xi_1^T\dot{\xi}_1 \\ &= \xi_1^T(L_1+L_1^T)\xi_1 - \xi_1^T(L_2^T+L_2)\xi_1(\xi_1^T\xi_1)^{-q_0/p_0} \\ &\leq -\min\{|\ \lambda(L_1^T+L_1)\ |\}\xi_1^T\xi_1 - \min\{|\ \lambda(L_1^T+L_1)\ |\}\xi_1^T\xi_1(\xi_1^T\xi_1)^{-q_0/p_0} \\ &\leq -2\min\{|\ \lambda(L_1^T+L_1)\ |\}V \\ &\quad -2^{-(p_0-q_0)/p_0}\min\{|\ \lambda(L_2^T+L_2)\ |\}V^{(p_0-q_0)/p_0}. \end{aligned} \tag{34}$$

if the conditions (32) and (33) hold. Then using the same analysis as Theorem 1, $V(t)$ as well as the equilibrium ξ_1 is globally stable and can reach zero in finite time. And since $\dot{V} < 0$ for $\xi_1 = 0$, $V(t)$ and $\xi_1(t)$ will remain at zero forever once they reach zero.

It is worth mentioning that in the proof of this theorem, we use the inequality

$$\dot{V} \leq -\alpha V - \beta V^{q/p}$$

where $\alpha = \min\{|\ \lambda(L_1^T+L_1)\ |\}$, $\beta = \min\{|\ \lambda(L_1^T+L_1)\ |\}$, $p = p_0$, $q = p_0 - q_0$ to replace the equation (27). This inequality also guarantees the finite time convergence.

Following is a direct corollary from Theorem 2.

Corollary 3. *If the matrices C_1, C_2, C_3 are chosen such that*

$$A_{11} - A_{12}C_2^{-1}C_1 = -\mathrm{diag}(\eta_1, \cdots, \eta_{n-m})\ (\eta_i > 0) \tag{35}$$

$$A_{12}C_2^{-1}C_3 = -\mathrm{diag}(\rho_1, \cdots, \rho_{n-m})\ (\rho_i > 0) \tag{36}$$

then the equilibrium $\xi_1 = 0$ is globally stable and each entry of ξ_1 can reach zeri in a fixed time specified as

$$t_f(i) = \frac{p}{\eta_i(p-q)}(\ln(\eta_i\xi_i(0)^{(p-q)/p} + \beta) - \ln\rho_i) \tag{37}$$

Proof. The proof directly follows from the proof of Theorem 2 and Section 2.

4 Singularity and Global Stability Issues

Because of introduction of term $(\xi_1^T\xi_1)^{-q_0/p_0}$ in the sliding manifold s, there are a few problems that have to be resolved before it is used for control.

One problem is, as $\xi_1 \to 0$, a singularity may occur in controller (25) which means $u \to \infty$. The following theorem overcomes this problem.

Theorem 4. *If*

$$p_0 - 4q_0 > 0 \tag{38}$$

then when the dynamics $\xi_1(t)$ and $\xi_2(t)$ reach zero on the surface $s = 0$, the control u defined by (25) will be bounded forever.

Proof. Because of controller (25), the terms that would cause the singularity are

$$S_1 = (\xi_1^T \xi_1)^{-q_0/p_0} \times A_{12}\xi_2, \tag{39}$$

and the term

$$S_2 = \xi_1(\xi_1^T \xi_1)^{-1-q_0/p_0} \xi_1^T \times A_{12}\xi_2. \tag{40}$$

When the trajectory moves on the surface $s = 0$, it meets

$$\xi_2 = -C_2^{-1}C_1\xi_1 - C_2^{-1}C_3\xi_1(\xi_1^T\xi_1)^{-q_0/p_0} \tag{41}$$

Substituting (41) into (40) yields

$$\begin{aligned} \|S_2\| &= \|\xi_1(\xi_1^T\xi_1)^{-1-q_0/p_0}\xi_1^T \times [-C_2^{-1}C_1\xi_1 - C_2^{-1}C_3\xi_1(\xi_1^T\xi_1)^{-q_0/p_0}]\| \\ &\le M\|\xi_1\|^{1-2q_0/p_0} + N\|\xi_1\|^{1-4q_0/p_0} \end{aligned} \tag{42}$$

where M, N are positive constant numbers. According to the condition $p_0 - 4q_0 > 0$ given in the theorem, S_2 is bounded. Same argument can be used to prove the boundedness of S_1. Therefore we can conclude that when $\xi_1 \to 0$, the terms S_1 and S_2 tend to zero, and the singularity is avoided.

Another problem is that there may be a case that at an initial stage, $\xi_1 = 0$ while other states are not. this incurs another singularity problem. This problem can be overcome by choosing a pre–TSM controller which is a linear control that takes the state to reach $s = 0$ [20]. The TSM controller then switches on so that the finite time reachability is achieved. This is implementable since in (23) and (24), (A_{11}, A_{12}) is controllable and B_2 is nonsingular and the pair (A, B) is controllable, the state of linear system can reach any place (apart from the equilibrium) by the following linear control.

$$u(t) = B^T \exp(-A^T t)G_c^{-1}(0, t_f)[\exp(At_f)\xi(t_f) - \xi(0)] \tag{43}$$

where $G_c^{-1}(0, t_f)$ is

$$G_c(0, t_f) \triangleq \int_0^{t_f} \exp(-A\tau)BB^T \exp(-A^T\tau)d\tau, \tag{44}$$

where t_f is a specified time for ξ to arrive on the sliding surface.

We now consider the stability issue. As shown in [14], there exist some systems whose dynamics with certain initial values may escape to infinity in a very short time. For such a class of systems, if our TSM control can be used so that the state ξ_1 can reach zero in a finite time which is much shorter than the time the state escapes to infinity, then the nonlinear system (1) is globally asymptotically stabilised.

Theorem 5. *For system (1), if assumptions H_1 and H_2 are satisfied and the conditions of Theorem 2 and Theorem 3 hold, then for any fixed constants R, we can always use pre-TSM controller (43) and TSM controller (25) with appropriate parameters setting such that for any initial value $\xi(0)$ and $x(0)$ which meet $\|x(0)\| \le R$ and $\|\xi(0)\| \le R$, the trajectory $x(t)$ tends to zero asymptotically and $\xi(t)$ reaches zero in finite time.*

Proof: Define M as

$$M = \max\{\|f(x,\xi,t)\| : \|x\| \le 2R, \|\xi\| \le 2R, t \in [0,R]\}.$$

Since $f(x,\xi,t)$ is a continuous function in x, ξ, t, M is a constant. ¿From the first equation of (1), we have

$$x(t) = x(0) + \int_0^t f(x(\tau),\xi(\tau),\tau)d\tau. \tag{45}$$

Further, when t is very small it can be obtained that

$$\begin{aligned} \|x(t)\| &\le \|x(0)\| + \int_0^t \|f(x(\tau),\xi(\tau),\tau)\|d\tau \\ &\le R + Mt. \end{aligned} \tag{46}$$

In addition, under the pre-TSM control, $\xi(t)$ is as

$$\xi(t) = \exp(At)\xi(0) + \exp(At)G_c(0,t)G_c^{-1}(0,t_f)[\exp(At_f)\xi(t_f) - \xi(0)]. \tag{47}$$

It can be seen that as t_f as well as $\|\xi(t_f)\|$ is very small and $t < t_f$, $\xi(t) \le 2R$. If let

$$t \in [0, \min\{t_f, R/M\}],$$

then we obtain from (46)

$$\|x(t)\| \le R + R = 2R.$$

This implies that in $t \in [0, R/M]$, (46) is still valid which means in this time interval $x(t)$ is bounded.

In the pre-TSM controller (43), we choose $t_f = R/2M$, and let $\xi(t_f)$ meets $\|\xi(t_f)\| \le 1$ and $s(\xi(t_f)) = 0$. According to the definition of the TSM control and Theorem 2, the time to arrive on surface $s = 0$, t^s should satisfy

$$\begin{aligned} t^s &\le \frac{p_0 - q_0}{\beta q_0 2^{(p_0-q_0)/p_0} \min\{|\lambda(L_2^T + L_2)|\}} V(t_f)^{(p_0-q_0)/p_0} \\ &\le \frac{p_0 - q_0}{\beta q_0 2^{(p_0-q_0)/p_0} \min\{|\lambda(L_2^T + L_2)|\}} \end{aligned} \tag{48}$$

If we choose a properly large β such that $t^s \le R/2M$, then after $t > R/M$, $\xi(t)$ is zero. Consequently, after $t > R/M$, $x(t)$ meets

$$\dot{x}(t) = f(x(t), 0, t).$$

This implies that $x(t)$ will tend to zero asymptotically since assumption H_1 holds. This completes the proof of this Theorem.

5 Examples

In order to illustrate the application of the proposed control design, we analyse the following two examples.

Example 1. For $c_2 > 0, c_1 < 0$, consider the system

$$\left.\begin{array}{l} \dot{x} = -x^3 - x^3 y,\ \dot{\xi}_1 = \xi_2,\ \dot{\xi}_2 = u \\ y = c_1\xi_1 + c_2\xi_2 \end{array}\right\}. \tag{49}$$

It was shown that since the linear part is not weak minimum phase, as long as the initial condition $(1 - c_1\xi_{10})\xi_0^2 > c_1/2c_2$ holds, the system (49) can not be smoothly stabilised [16]. This is because that for any smooth feedback $u(t)$ and its resulting output $y(t)$, the system (49) has some solutions which escape to infinity in finite time. However, by using our TSM control strategy, it can be globally stabilised by first using a pre–TSM linear control to bring the state ξ_1 to $s = 0$ in a tunable time t_f, and then bring the state ξ_1 to zero in a tunable t^s such that $t_f + t^s$ is much less than the time period x tends to infinity. Now let us look at how to deal with it. It is easy to see that $y(t)$ satisfies

$$\dot{y} = c_1\xi_2 + c_2 u. \tag{50}$$

Let R be a fixed constant. For $| x |> R$, let

$$c_1\xi_2 + c_2 u = -y - x^4,$$

that is

$$u(t) = \frac{1}{c_2}[-c_1\xi_2 - y - x^4]. \tag{51}$$

Under this control law, we have

$$\dot{y} = -y - x^4. \tag{52}$$

In the region $| x |> R$, choose $V(x, y)$ meets

$$V(x, y) = x^2 + y^2.$$

Then it can be obtained that

$$\dot{V}(x, y) = -2x^4 - 2y^2 \leq 0.$$

This shows that the solution $y(t)$ and $x(t)$ of the system (49) for any initial value $| x(0) |> R$ and $y(0)$ or $(\xi_1(0)$ and $\xi_2(0))$, will enter the region

$$R_y = \{y :| y |\leq R\},\ R_x = \{x :| x |\leq R\},$$

respectively after some finite time. Upon the trajectories $x(t)$ and $y(t)$ entering these regions, the control law is switched to the pre-TSM control and then TSM control. By the Theorem 4 we get to know that $x(t)$ and $\xi_1(t)$ and $\xi_2(t)$ then tend to zero.

This example shows that one can first use a nonlinear control law to suppress the dynamics into the known bounded region and then the pre-TSM control and TSM control is used. And finally the system is globally asymptotically stabilised.

Example 2. Let us consider the following system

$$\left.\begin{aligned} \dot{x} &= -\tfrac{1}{2}x^3 + \tfrac{\xi^T\xi}{e^{-t^n}+\xi^T\xi}x \\ \dot{\xi}_1 &= A_{11}\xi_1 + A_{12}\xi_2 \\ \dot{\xi}_2 &= u \end{aligned}\right\} \tag{53}$$

where $n > 1$, $\xi_1 = [\xi_{11}, \xi_{12}]^T$, $\xi_2 = [\xi_{21}, \xi_{22}]^T$, $u = [u_1, u_2]^T$, and A_{11} and A_{12} are as

$$A_{11} = \begin{bmatrix} -1 & 0.2 \\ 0 & -2 \end{bmatrix},\ A_{12} = \begin{bmatrix} -1 & 0 \\ 0 & -1 \end{bmatrix}.$$

Let $C_1 = C_2 = C_3 = A_{12}$ in the design of the TSM controller. It can be seen easily that the conditions of Theorem 2 hold. If $p_0 = 5$ and $q_0 = 1$, then the condition of Theorem 3 holds. For the first equation of system (53), let $y = (1/2)x^{-2}$, it becomes

$$\dot{y} = 1 - \frac{\xi^T\xi}{e^{-t^n} + \xi^T\xi}y.$$

the whole system (53) can be globally asymptotically stabilised.

6 Conclusion

The global stabilization of nonlinear systems with partially linear composite dynamics has been discussed in this chapter using the terminal sliding modes. The proposed control strategy enables the equilibrium of the linear composite subsystem to be reached in finite time, resulting in the effect of asymptotic convergence on the nonlinear system being removed. Under some mild conditions for nonlinear function $f(\bullet)$, the globally asymptotic stabilization is realized.

References

1. Aeyels, D., "Stabilization of a Class of Nonlinear Systems by a Smooth Feedback," *Systems & Control Letters*, Vol. 5, pp. 181-191, 1985.
2. Byrnes, C. I. and A. Isidori, "Asymptotic Stabilization of Minimum Phase Nonlinear Systems," *IEEE Transactions on Automatic Control*, Vol. 36, pp.1122-1137, 1991.
3. Byrnes, C. I. and A. Isidori, "Local Stabilization of Minimum Phase Nonlinear Systems," *Systems & Control Letters*, Vol. 11, pp.7-9, 1988.
4. Byrnes, C. I. and A. Isidori, "A Frequency Domain Philosophy for Nonlinear Systems with Application to Stabilization and to Adaptive control," *23rd IEEE Conference on Decision and Control*, pp.1031-1037, 1985.
5. Carr, J., Application of Centre Manifold Theory, New York: Springer-Verlag, 1981.

6. Isidori, A., Nonlinear Control Systems. 2nd Edition, New York: Springer-Verlag, 1989.
7. Khalil, H. K. and A. Saberi, "Adaptive Stabilization of a Class of Nonlinear Systems using High-gain Feedback," *IEEE Transactions on Automatic Control*, Vol. 32, pp.1031-1035, 1987.
8. Kokotovic, P. V. and H. J. Sussmann, "A Positive Real Condition for Global Stabilization of Nonlinear Systems," *Systems & Control Letters*, Vol. 12, pp. 125-134, 1989.
9. Lin, Z. and A. Saberi, "Semi-global Stabilization of Partially Linear Composite Systems via Feedback of the State of the Linear Part," *Systems & Control Letters*, Vol. 20, pp.199-207, 1993.
10. Lin, Z. and A. Saberi, "Robust Semiglobal Stabilization of Minimum-phase Input-output Linearizable Systems via Partial State and Output Feedback," *IEEE Transactions on Automatic Control*, Vol. 40, pp.1092-1041, 1995.
11. Lozano, R., B. Brogliato and I. D. Landau, "Passivity and Global Stabilization of Cascaded Nonlinear Systems," *IEEE Transactions on Automatic Control*, Vol. 37, pp.1386-1388, 1992.
12. Marino, R., "High-gain Feedback in Nonlinear Control Systems," *International Journal of Control*, Vol. 42, pp.1369-1385, 1985.
13. Ortega, R., "Passivity Properties for Stabilization of Cascaded Nonlinear Systems," *Automatica*, Vol. 27, pp.423–424, 1991.
14. Sussmann, H. J. and P. V. Kokotovic, "The Peaking Phenomenon and the Global Stabilization of Nonlinear Systems," *IEEE Transactions on Automatic Control*, Vol. 36, pp.424-439, 1991.
15. Sussmann, H. J., "Limitations on the Stabilizability of Globally Minimum Phase Systems," *IEEE Transactions Automat Control*, Vol. 35, pp. 117-119, 1990.
16. Saberi, A., P. V. Kokotovic and H. J. Sussmann, "Global Stabilization of Partially Linear Composite Systems," *SIAM Journal of Control and Optimization*, Vol. 28, pp.1491-1503, 1990. Tsinias, J., "Sufficient Lyapunov-like Conditions for Stabilization," *Mathematics Control, Signals Systems*, Vol. 2, pp.343-357, 1989.
17. Teel, A. and L. Praly, " Tools for Semiglobal Stabilization by Partial State and Output Feedback," *SIAM Journal of Control and Optimization*, Vol. 33, pp. 1443-1488, 1995.
18. Yu, X. and Man, Z., "Model Reference Adaptive Control Systems with Terminal Sliding Modes," *International Journal of Control*, Vol. 64, pp. 1165-1176, 1996.
19. Man, Z. and X. Yu, "Terminal Sliding Mode Control of MIMO Systems", *IEEE Transactions on Circuits and Systems - Part I*, Vol. 44, pp. 1065–1070, 1997.
20. Wu, Y., X. Yu and Z. Man, "Terminal Sliding Mode Control Design for Uncertain Dynamic Systems," *Systems and Control Letters*, Vol. 34, No.5, pp.281-288, 1998.

A General Canonical Form for Sliding Mode Control of Nonlinear Systems *

Hebertt Sira-Ramírez **

Centro de Investigación y Estudios Avanzados
CINVESTAV-IPN
Departamento de Ingeniería Eléctrica
Programa Departamental de Mecatrónica
Avenida IPN # 2508, Col. San Pedro Zacatenco A.P. 14-740
07300 México D.F., México

Abstract. A new canonical form of the Generalized Hamiltonian type, including "dissipation" terms, is proposed for nonlinear single input dynamical systems whose state trajectories are required to slide on a given submanifold of the state space. The canonical form clearly depicts the conservative, the dissipative and the destabilizing vector fields thus allowing for a more efficient feedback controller design achieving the reachability of the sliding surface.

1 Introduction

In this chapter, a general canonical form is derived for systems undergoing sliding motions on a submanifold of the state space. The canonical form is based on the use of *projection operators* associated with the sliding surface coordinate function and the control input vector field. A natural decomposition is obtained for the system's drift forces which ranks them as : *workless* or conservative forces, i.e., those yielding invariance of the surface coordinate along the motions of the uncontrolled system; the *attracting* forces, which are those naturally making the sliding surface an attractive manifold and try to drive the surface coordinate function evolutoin towards lower absolute values and, finally, the *rejecting* forces, which locally drive the system trajectories to achieve higher absolute values of the sliding surface coordinate. These two last forces change their nature depending on the local sign of the sliding surface coordinate function i.e., attracting forces "*above*" the surface become repelling forces "*below*" the surface and viceversa. By suitably respecting the local beneficial non-linearities, on each side of the sliding manifold, an autonomous non-divergence from the sliding surface is guaranteed and thus, the variable structure feedback controller yielding convergence towards

* This research was supported by the Centro de Investigación y Estudios Avanzados del Instituto Politécnico Nacional of México and by the Consejo Nacional de Ciencia y Tecnología of Venezuela (CONICIT) under Research Grant S1-95-000886

** On leave of absence from the Departamento de Sistemas de Control, Universidad de Los Andes, Mérida-Venezuela the sliding coordinate zero level set can be designed in a more efficient manner, i.e., with substantially less control authority. Thus, the feedback controller design simply consists in injecting suitable *dissipation*, or attractivity terms, which complement the local beneficial non-linearities of the system while, at the same time, neutralizing those forces which locally increase the magnitude of the sliding surface coordinate.

The results in this chapter constitue an extension of those found in [5] and [6], for feedback passivity of nonlinear systems. In fact, the results here presented give further geometric insight into the long suspected connections and parallelisms between passivity based control (see Ortega *et al* [4]), Generalized Hamiltonian systems (see [3], [10]) and Sliding Mode control [9].

In Section 2 a canonical form for sliding mode control is derived. In Section 3 a sliding mode controller is proposed which exploits the natural structure of the system with respect to the sliding surface. Section 4 presents some illustrative design examples in a tutorial fashion. Section 5 is devoted to some conclusions and suggestions for further research.

2 A Canonical Form for Sliding Mode Control

2.1 Fundamental assumptions, definitions and results

Consider the class of nonlinear single-input single-output systems described by

$$\begin{aligned} \dot{x}(t) &= f(x) + g(x)u \;\; ; \;\; x \in \mathcal{X} \subset \mathbf{R}^n \;\; ; \; u \in \mathcal{U} \subset \mathbf{R} \\ y &= \sigma(x) \;\; ; \;\; y \in \mathcal{Y} \subset \mathbf{R} \end{aligned} \qquad (1)$$

where $\mathcal{X}$ denotes the *operating region* of the system, constituted by a sufficiently large open set containing a continuum of equilibrium points, possibly parametrized by a constant control input value $u = U \in \mathcal{U}$, of the form $x = \overline{x}(U)$ and given by the solution of $f(\overline{x}) + g(\overline{x})U = 0$. In particular, for $u = 0$, we assume that $f(\overline{x}) = 0$ implies $\overline{x} = 0$. However, motivated by a large class of real life systems, we are specifically interested in *nonzero* constant state equilibrium points $x = \overline{x}$, obtained by nonzero constant control inputs $u = U$. Thus, we specifically assume that the operating region $\mathcal{X}$ of the system does not contain the origin of coordinates. The output function $y = \sigma(x)$ is assumed to be zero at the equilibrium point, i.e., $\sigma(\overline{x}) = 0$.

We assume that $\sigma(x)$ is a C^∞ scalar function, called the *sliding surface function* $\sigma : \mathbf{R}^n \to \mathbf{R}$ such that when the state trajectories are confined to its zero level set $\mathcal{S}_0 = \{x \in \mathcal{X} \; : \; \sigma(x) = 0\}$, the behaviour of the system is as desired (for instance, asymptotically stable towards an equilibrium point of interest located at $\mathcal{S}_0$). It is assumed that the sliding surface function $\sigma(x)$ is given, possibly as the result of a previously solved silding surface design problem.

By $\partial\sigma/\partial x$ we denote the *column* vector field with components $\partial\sigma/\partial x_i$ $i = 1, \ldots, n$. The transpose of this gradient field, $(\partial\sigma/\partial x)^T$, is denoted by the *row* vector $\partial\sigma/\partial x^T$. Let $L_g\sigma(x)$ denote the directional derivative of the scalar function $\sigma(x)$ with respect to the control input vector field $g(x)$ at the point $x \in \mathcal{X}$.

We assume throughout the entire chapter that the following assumption holds valid:

$$L_g\sigma(x) = \frac{\partial \sigma}{\partial x^T} g(x) \neq 0 \quad \forall\, x \in \mathcal{X} \tag{2}$$

This last condition is usually known as the ***transversality condition*** and simply establishes that the vector field $g(x)$ is not orthogonal to the gradient of $\sigma(x)$ at any point x in $\mathcal{X}$. In other words, the control vector field $g(x)$ is not tangential, at each x, to the sliding surface function level sets, defined in the state space of the system as, $\mathcal{S}_k = \{x \in \mathcal{X} \ : \ \sigma(x) = constant = k\}$. This condition is quite familiar in sliding mode control of nonlinear systems (see [7]) and it amounts to having a sliding surface function which is locally *relative degree* one in $\mathcal{X}$. The *zero dynamics* corresponding to the ideal sliding condition $y = \sigma(x) = 0$ is assumed to be asymptotically stable towards the isolated equilibrium point $\overline{x} \in \mathcal{S}$. In other words the system is *minimum phase* with respect to the output $y = \sigma(x)$. According to the results in [2], the sliding surface function is a *passive* output.

For each $x \in \mathcal{X}$, we define a *projection operator, along the span of the control vector field* $g(x)$ *onto the tangent space to the constant level sets of the sliding surface function* $\sigma(x)$, as the matrix $M(x)$ given by

$$M(x) = \left[I - \frac{1}{L_g\sigma(x)} g(x) \frac{\partial \sigma}{\partial x^T} \right] \tag{3}$$

The following proposition points out some properties of the matrix $M(x)$ which further justify the given name of "projection operator" (see Figure 1)

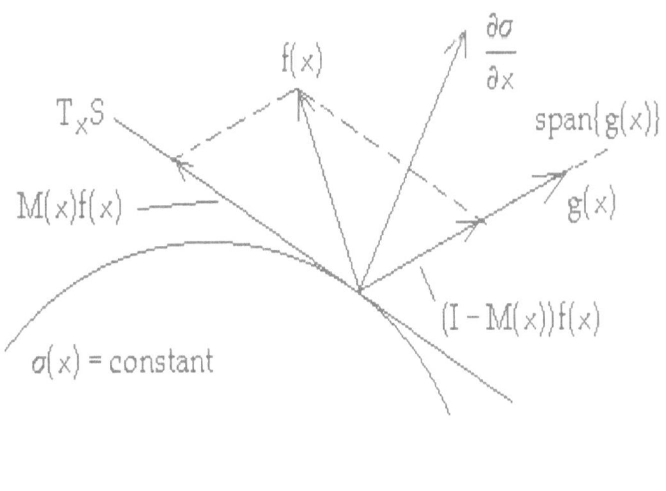

Fig. 1. Decomposition of drift vector field along *span* $\{g(x)\}$, onto T_xS

Proposition 1. *The matrix $M(x)$ enjoys the following properties:*

$$g(x) \in \mathcal{K}er\, M(x)$$

$$\frac{\partial \sigma}{\partial x} \in \mathcal{K}er\, M^T(x)$$

$$M(x)(I - M(x)) = 0 \tag{4}$$

Proof

The first property establishes that, locally, $M(x)g(x) = 0$. Indeed, using the definition of $M(x)$ one has

$$\begin{aligned} &\left[I - \frac{1}{L_g\sigma(x)} g(x) \frac{\partial \sigma}{\partial x^T}\right] g(x) \\ &= g(x) - \frac{1}{L_g\sigma(x)} g(x) \frac{\partial \sigma}{\partial x^T} g(x) \\ &= g(x) - \frac{1}{L_g\sigma(x)} g(x) L_g\sigma(x) = g(x) - g(x) \\ &= 0 \end{aligned} \tag{5}$$

The second property is equivalent to $\partial\sigma/\partial x^T M(x) = 0$.

$$\begin{aligned} &\frac{\partial \sigma}{\partial x^T}\left[I - \frac{1}{L_g\sigma(x)} g(x) \frac{\partial \sigma}{\partial x^T}\right] \\ &= \frac{\partial \sigma}{\partial x^T} - \frac{1}{L_g\sigma(x)} \frac{\partial \sigma}{\partial x^T} g(x) \frac{\partial \sigma}{\partial x^T} \\ &= \frac{\partial \sigma}{\partial x^T} - \frac{\partial \sigma}{\partial x^T} \\ &= 0 \end{aligned} \tag{6}$$

The last property follows immediately from the fact that the columns of the matrix $(I - M(x))$ are all in the subspace *span* $\{g(x)\}$. Indeed,

$$\begin{aligned} I - M(x) &= \frac{1}{L_g\sigma(x)} g(x) \frac{\partial \sigma}{\partial x^T} \\ &= \frac{1}{L_g\sigma(x)} \left[g(x)\frac{\partial \sigma}{\partial x_1}; \cdots ; g(x) \frac{\partial \sigma}{\partial x_n}\right] \end{aligned} \tag{7}$$

This last fact and the use of the first property yields the result.

The following proposition depicts further properties of the projection matrix $M(x)$

Proposition 2. *Let $f(x)$ be a smooth vector field, then the vector $M(x)f(x)$ can be written as*

$$M(x)f(x) = \tilde{\mathcal{J}}(x)\frac{\partial \sigma}{\partial x}$$

where $\tilde{\mathcal{J}}(x)$ is a skew-symmetric matrix, i.e., $\tilde{\mathcal{J}}(x) + \tilde{\mathcal{J}}^T(x) = 0$.

On the other hand, the vector field $(I - M(x))f(x)$ can be written as

$$(I - M(x))f(x) = -\frac{1}{2}\tilde{\mathcal{J}}(x)\frac{\partial \sigma}{\partial x} + \mathcal{S}(x)\frac{\partial \sigma}{\partial x}$$

where $\mathcal{S}(x)$ is a symmetric matrix, i.e., $\mathcal{S}(x) = \mathcal{S}^T(x)$

Proof

The first part of the proposition easily follows from the following string of algebraic manipulations

$$\begin{aligned}
M(x)f(x) &= \left[I - \frac{1}{L_g\sigma(x)}g(x)\frac{\partial \sigma}{\partial x^T}\right]f(x) \\
&= \frac{1}{L_g\sigma(x)}\left[(L_g\sigma(x))\, f(x) - g(x)L_f\sigma(x)\right] \\
&= \frac{1}{L_g\sigma(x)}\left[\frac{\partial \sigma}{\partial x^T}g(x)f(x) - g(x)\frac{\partial \sigma}{\partial x^T}f(x)\right] \\
&= \frac{1}{L_g\sigma}\left[f(x)g^T(x)\frac{\partial \sigma}{\partial x} - g(x)f^T(x)\frac{\partial \sigma}{\partial x}\right] \\
&= \frac{1}{L_g\sigma}\left[f(x)g^T(x) - g(x)f^T(x)\right]\frac{\partial \sigma}{\partial x} \\
&= \tilde{\mathcal{J}}(x)\frac{\partial \sigma}{\partial x}
\end{aligned} \tag{8}$$

For the proof of the second part of the proposition note that,

$$\begin{aligned}
(I - M(x))f(x) &= \frac{1}{L_g\sigma}g(x)\frac{\partial \sigma}{\partial x^T}f(x) \\
&= \frac{1}{L_g\sigma}\left[g(x)f^T(x)\right]\frac{\partial \sigma}{\partial x}
\end{aligned} \tag{9}$$

The result follows from the fact that *any* square matrix $N(x)$ and, in particular,

$$N(x) = (1/L_g\sigma(x))\left[g(x)f^T(x)\right]$$

can always be written as

$$N(x) = (1/2)(N(x) - N^T(x)) + 1/2(N(x) + N^T(x))$$

The first summand, which is written as,

$$\begin{aligned}
&\frac{1}{2}(N(x) - N^T(x)) = \\
&\frac{1}{2L_g\sigma}\left[g(x)f^T(x) - f(x)g^T(x)\right] \\
&= -\frac{1}{2}\tilde{\mathcal{J}}(x)
\end{aligned} \tag{10}$$

is clearly skew-symmetric, while the second summand $(1/2)(N(x) + N^T(x))$ is symmetric. For the purposes of further reference we define the matrix $\mathcal{S}(x)$ as follows

$$\begin{aligned}\mathcal{S}(x) &= \frac{1}{2}\left[N(x) + N^T(x)\right] \\ &= \frac{1}{2L_g\sigma}\left[g(x)f^T(x) + f(x)g^T(x)\right]\end{aligned}$$

and the matrix $\mathcal{J}(x)$ as

$$\mathcal{J}(x) = \frac{1}{2}\tilde{\mathcal{J}}(x)$$

2.2 Vector field decompositions through projection operators

As a consequence of the above propositions and definitions we have the following result.

Proposition 3. *A drift vector field $f(x(t))$ can be naturally decomposed in the following sum,*

$$f(x) = M(x)f(x) + (I - M(x))f(x) = \mathcal{J}(x)\frac{\partial\sigma}{\partial x} + \mathcal{S}(x)\frac{\partial\sigma}{\partial x}$$

Proof
Indeed,

$$M(x)f(x) = \tilde{\mathcal{J}}(x)\frac{\partial\sigma}{\partial x}$$

and

$$(I - M(x))f(x) = -\frac{1}{2}\tilde{\mathcal{J}}(x)\frac{\partial\sigma}{\partial x} + \mathcal{S}(x)\frac{\partial\sigma}{\partial x}$$

then,

$$\begin{aligned}f(x) &= M(x)f(x) + (I - M(x))f(x) \\ &= \tilde{\mathcal{J}}(x)\frac{\partial\sigma}{\partial x} - \frac{1}{2}\tilde{\mathcal{J}}(x)\frac{\partial\sigma}{\partial x} + \mathcal{S}(x)\frac{\partial\sigma}{\partial x} \\ &= \frac{1}{2}\tilde{\mathcal{J}}(x)\frac{\partial\sigma}{\partial x} + \mathcal{S}(x)\frac{\partial\sigma}{\partial x} \\ &= \mathcal{J}(x)\frac{\partial\sigma}{\partial x} + \mathcal{S}(x)\frac{\partial\sigma}{\partial x}\end{aligned} \tag{11}$$

The following lemma is well known,

Lemma 4. *Let $\mathcal{S}(x)$ be a symmetric matrix, then $\mathcal{S}(x)$ can be nonuniquely decomposed as the sum of a positive semi-definite matrix $\mathcal{S}_p(x)$ and a negative semi-definite matrix $\mathcal{S}_n(x)$. If the above decomposition is not possible, then either $\mathcal{S}(x)$ is positive definite or, at least, positive semi-definite or, else, it is negative definite or, at least, negative semi-definite.*

2.3 A canonical form for sliding mode controlled nonlinear systems

As a corollary to the above results, a nonlinear system of the form (1), with a sliding surface function $\sigma(x)$, satisfying the transversality condition $L_g\sigma(x) \neq 0$, can always be rewritten as

$$\dot{x}(t) = \mathcal{J}(x)\frac{\partial \sigma}{\partial x} + \mathcal{S}_p(x)\frac{\partial \sigma}{\partial x} + \mathcal{S}_n(x)\frac{\partial \sigma}{\partial x} + g(x)u \tag{12}$$

with $\mathcal{J}(x)$ being skew-symmetric, and $\mathcal{S}_p(x)$ being positive semi-definite and $\mathcal{S}_n(x)$ negative semi-definite. However, if $\mathcal{S}_p(x)$ is positive definite, then $\mathcal{S}_n(x)$ is zero and conversely if $\mathcal{S}_n(x)$ is negative definite then $\mathcal{S}_p(x)$ is zero.

2.4 Feedback sliding mode control for systems in canonical form

Consider a nonlinear system, given in the following form,

$$\begin{aligned} \dot{x} &= \mathcal{J}(x)\frac{\partial \sigma}{\partial x} + \mathcal{S}(x)\frac{\partial \sigma}{\partial x} + g(x)u \\ y &= \sigma(x) \end{aligned} \tag{13}$$

where the symmetric matrix $\mathcal{S}(x)$ is assumed to be decomposed as the sum of two symmetric matrices $\mathcal{S}_p(x) + \mathcal{S}_n(x)$, as explained above.

Along the solutions of the system, the time derivative of the "energy function"

$$V(x) = \frac{1}{2}\sigma^2(x) \tag{14}$$

is evaluated as $\dot{V} = \sigma(x)\dot{\sigma}(x)$.

For $\sigma(x) > 0$ we have,

$$\begin{aligned} \sigma\dot{\sigma} &= \sigma\Big[\frac{\partial \sigma}{\partial x^T}\mathcal{J}(x)\frac{\partial \sigma}{\partial x} + \frac{\partial \sigma}{\partial x^T}\mathcal{S}_p(x)\frac{\partial \sigma}{\partial x} \\ &\quad + \frac{\partial \sigma}{\partial x^T}\mathcal{S}_n(x)\frac{\partial \sigma}{\partial x} + L_g\sigma(x)u\Big] \\ &= \sigma\Big[\frac{\partial \sigma}{\partial x^T}\mathcal{S}_p(x)\frac{\partial \sigma}{\partial x} + \frac{\partial \sigma}{\partial x^T}\mathcal{S}_n(x)\frac{\partial \sigma}{\partial x} + L_g\sigma(x)u\Big] \\ &\leq \sigma\Big[\frac{\partial \sigma}{\partial x^T}\mathcal{S}_p(x)\frac{\partial \sigma}{\partial x} + L_g\sigma(x)u\Big] \end{aligned} \tag{15}$$

while for $\sigma(x) < 0$ we obtain

$$\begin{aligned} \sigma\dot{\sigma} &= \sigma\Big[\frac{\partial \sigma}{\partial x^T}\mathcal{J}(x)\frac{\partial \sigma}{\partial x} + \frac{\partial \sigma}{\partial x^T}\mathcal{S}_p(x)\frac{\partial \sigma}{\partial x} \\ &\quad + \frac{\partial \sigma}{\partial x^T}\mathcal{S}_n(x)\frac{\partial \sigma}{\partial x} + L_g\sigma(x)u\Big] \\ &= \sigma\Big[\frac{\partial \sigma}{\partial x^T}\mathcal{S}_p(x)\frac{\partial \sigma}{\partial x} + \frac{\partial \sigma}{\partial x^T}\mathcal{S}_n(x)\frac{\partial \sigma}{\partial x} + L_g\sigma(x)u\Big] \\ &\leq \sigma\Big[\frac{\partial \sigma}{\partial x^T}\mathcal{S}_n(x)\frac{\partial \sigma}{\partial x} + L_g\sigma(x)u\Big] \end{aligned} \tag{16}$$

Consider then the following variable structure input coordinate transformation, with v denoting a new external independent control input,

For $\sigma(x) > 0$,

$$u = \frac{1}{L_g\sigma}\left[v - \frac{\partial\sigma}{\partial x^T}\mathcal{S}_p(x)\frac{\partial\sigma}{\partial x}\right] \tag{17}$$

For $\sigma(x) < 0$.

$$u = \frac{1}{L_g\sigma}\left[v - \frac{\partial\sigma}{\partial x^T}\mathcal{S}_n(x)\frac{\partial\sigma}{\partial x}\right] \tag{18}$$

It is clear that the transformed system is given by the following variable structure system:

For $\sigma(x) > 0$

$$\begin{aligned} \dot{x} &= \mathcal{J}(x)\frac{\partial\sigma}{\partial x} + \mathcal{S}_n(x)\frac{\partial\sigma}{\partial x} \\ &+ \left(I - \frac{1}{L_g\sigma(x)}g(x)\frac{\partial\sigma}{\partial x^T}\right)\mathcal{S}_p(x)\frac{\partial\sigma}{\partial x} + \frac{1}{L_g\sigma}g(x)v \\ y &= \sigma(x) \end{aligned} \tag{19}$$

while, for $\sigma(x) < 0$

$$\begin{aligned} \dot{x} &= \mathcal{J}(x)\frac{\partial\sigma}{\partial x} + \mathcal{S}_p(x)\frac{\partial\sigma}{\partial x} \\ &+ \left(I - \frac{1}{L_g\sigma(x)}g(x)\frac{\partial\sigma}{\partial x^T}\right)\mathcal{S}_n(x)\frac{\partial\sigma}{\partial x} + \frac{1}{L_g\sigma}g(x)v \\ y &= \sigma(x) \end{aligned} \tag{20}$$

Notice that, as shown in the previous section, the projected vector field given by either

$$\left(I - \frac{1}{L_g\sigma(x)}g(x)\frac{\partial\sigma}{\partial x^T}\right)\mathcal{S}_p(x)\frac{\partial\sigma}{\partial x}$$

or

$$\left(I - \frac{1}{L_g\sigma(x)}g(x)\frac{\partial\sigma}{\partial x^T}\right)\mathcal{S}_n(x)\frac{\partial\sigma}{\partial x}$$

can be rewritten as,

$$\frac{1}{L_g\sigma}\left[\mathcal{S}_p(x)\frac{\partial\sigma}{\partial x}g^T(x) - g(x)\frac{\partial\sigma}{\partial x^T}\mathcal{S}_p(x)\right]\frac{\partial\sigma}{\partial x} = \mathcal{K}_p(x)\frac{\partial\sigma}{\partial x}$$

and

$$\frac{1}{L_g\sigma}\left[\mathcal{S}_n(x)\frac{\partial\sigma}{\partial x}g^T(x) - g(x)\frac{\partial\sigma}{\partial x^T}\mathcal{S}_n(x)\right]\frac{\partial\sigma}{\partial x} = \mathcal{K}_n(x)\frac{\partial\sigma}{\partial x}$$

with $\mathcal{K}_p(x)$ and $\mathcal{K}_n(x)$ being skew-symmetric matrices. In other words, the transformed system is of the form,

For $\sigma(x) > 0$,

$$\dot{x} = \mathcal{I}_p(x)\frac{\partial\sigma}{\partial x} + \mathcal{S}_n(x)\frac{\partial\sigma}{\partial x} + \frac{1}{L_g\sigma}g(x)v$$
$$y = \sigma(x) \tag{21}$$

with $\mathcal{I}_p(x) = \mathcal{J}(x) + \mathcal{K}_p(x)$ being skew symmetric and,

For $\sigma(x) < 0$,

$$\dot{x} = \mathcal{I}_n(x)\frac{\partial\sigma}{\partial x} + \mathcal{S}_p(x)\frac{\partial\sigma}{\partial x} + \frac{1}{L_g\sigma}g(x)v$$
$$y = \sigma(x) \tag{22}$$

with $\mathcal{I}_n(x) = \mathcal{J}(x) + \mathcal{K}_n(x)$ being skew-symmetric.

The input coordinate transformations, viewed as a partial variable structure feedback, has achieved *neutralization* of the non-beneficial nonlinearities in the system. Notice that this is far less demanding than the usual practise of *elimination* of the non-beneficial nonlinearities. The partial variable sturcture feedback has also achieved passivity for the variable structure system with respect to the sliding surface function viewed as a "degenerate" storage fucntion, as the following proposition establishes,

Proposition 5. *The system (13) is passive with respect to the storage function $h(x) = 1/2\sigma^2(x)$, viewed as a positive semidefinite (i.e., degenerate) storage function, whenever $S_n(x)$, (respectively $S_p(x)$) is negative semidefinite (resp. positive semidefinite) and it is strictly passive if $S_n(x)$ is strictly negative definite (resp. strictly positive definite).*

Proof

The time derivatives of $h(x)$ along the solutions of the transformed system, away from the sliding surface $\mathcal{S}_0$ are given by:

For $\sigma(x) > 0$

$$\begin{aligned}\dot{V}(x) &= \sigma(x)\left[\frac{\partial\sigma}{\partial x^T}\mathcal{I}_p(x)\frac{\partial\sigma}{\partial x} + \frac{\partial\sigma}{\partial x^T}\mathcal{S}_n(x)\frac{\partial\sigma}{\partial x}\right.\\ &\qquad\left. + \frac{\partial\sigma}{\partial x^T}\frac{1}{L_g\sigma}g(x)v\right]\\ &= \sigma(x)\left[\frac{\partial\sigma}{\partial x^T}\mathcal{S}_n(x)\frac{\partial\sigma}{\partial x}\right] + \sigma(x)v\\ &\leq \sigma(x)v = yv\end{aligned} \tag{23}$$

For $\sigma(x) < 0$,

$$\begin{aligned}\dot{V}(x) = \sigma&\left[\frac{\partial\sigma}{\partial x^T}\mathcal{I}_n(x)\frac{\partial\sigma}{\partial x} + \frac{\partial\sigma}{\partial x^T}\mathcal{S}_p(x)\frac{\partial\sigma}{\partial x}\right.\\ &\left. + \frac{\partial\sigma}{\partial x^T}\frac{1}{L_g\sigma}g(x)v\right]\end{aligned}$$

$$
\begin{aligned}
&= \sigma(x)\left[\frac{\partial\sigma}{\partial x^T}\mathcal{S}_p(x)\frac{\partial\sigma}{\partial x}\right] + \sigma(x)v \\
&\leq \sigma(x)v = yv
\end{aligned}
\tag{24}
$$

Notice that if we let $\tilde{g}(x)$ denote the transformed control input vector field $\frac{\sigma(x)}{L_g\sigma}g(x)$ then the variable structure system may, in fact, be written as:

For $\sigma(x) > 0$

$$
\begin{aligned}
\dot{x} &= \mathcal{I}_p(x)\frac{\partial\sigma}{\partial x} + \mathcal{S}_n(x)\frac{\partial\sigma}{\partial x} + \tilde{g}(x)\vartheta \\
y &= \tilde{g}^T(x)\frac{\partial\sigma}{\partial x}
\end{aligned}
\tag{25}
$$

For $\sigma(x) < 0$

$$
\begin{aligned}
\dot{x} &= \mathcal{I}_n(x)\frac{\partial\sigma}{\partial x} + \mathcal{S}_p(x)\frac{\partial\sigma}{\partial x} + \tilde{g}(x)\vartheta \\
y &= \tilde{g}^T(x)\frac{\partial\sigma}{\partial x}
\end{aligned}
\tag{26}
$$

which, except for the "damping" terms $\mathcal{S}_n(x)\partial\sigma/\partial x$ and $\mathcal{S}_p(x)\partial\sigma/\partial x$ are, each one, in the same form as the *Generalized Hamiltonian systems*, widely studied in the literature (see [3],[10]). Notice that the state-dependent input coordinate transformation $\vartheta = v/\sigma(x)$ is defined away from the sliding surface $\mathcal{S}_0$.

The sliding mode controller for the input v may now be obtained by simply injecting complementary "damping" to the natural beneficial nonlinearities acting on each side of the sliding manifold. Let $\mathcal{S}_{nI}(x)$ be a symmetric negative semidefinite matrix such that $\mathcal{S}_n(x) + \mathcal{S}_{nI}(x)$ is negative definite. Simlarly, let $\mathcal{S}_{pI}(x)$ be a symmetric positive semidefinite matrix such that $\mathcal{S}_p(x) + \mathcal{S}_{pI}(x)$ is positive definite. The following variable structure controller achieves the reaching of the sliding surface and the creation of a local sliding regime on such a surface

$$
v = \begin{cases} \dfrac{\partial\sigma}{\partial x^T}\mathcal{S}_{nI}(x)\dfrac{\partial\sigma}{\partial x} & \text{for } \sigma(x) > 0 \\[2ex] \dfrac{\partial\sigma}{\partial x^T}\mathcal{S}_{pI}(x)\dfrac{\partial\sigma}{\partial x} & \text{for } \sigma(x) < 0 \end{cases}
\tag{27}
$$

The *ideal sliding dynamics* is readily obtained from the invariance condition of the sliding surface coordinate $\dot{\sigma} = 0$. Consider the system canonical form before any state feedback precompensation

$$
\begin{aligned}
\dot{x} &= \mathcal{J}(x)\frac{\partial\sigma}{\partial x} + \mathcal{S}(x)\frac{\partial\sigma}{\partial x} + g(x)u \\
y &= \sigma(x)
\end{aligned}
\tag{28}
$$

The ideal control input, or equivalent control, achieving surface coordinate invariance for all motions starting on the sliding surface $\mathcal{S}_0$ is obtained as

$$
u = -\frac{1}{L_g\sigma}\frac{\partial\sigma}{\partial x^T}\mathcal{S}(x)\frac{\partial\sigma}{\partial x}
$$

The *ideal sliding dynamics* is governed by a projected vector field of the form

$$\hat{\mathcal{J}}(x)\frac{\partial \sigma}{\partial x} = \mathcal{J}(x)\frac{\partial \sigma}{\partial x} + \left[I - \frac{1}{L_g\sigma} g(x)\frac{\partial \sigma}{\partial x^T}\right] \mathcal{S}(x)\frac{\partial \sigma}{\partial x}$$

This vector field evidently belongs to the tangent subspace to the sliding manifold. The equivalent control neutralizes *all working forces* in the system (beneficial and non-beneficial forces) and it evidently bestows a workless, or conservative, character to closed loop ideal sliding dynamics.

$$\dot{x} = \hat{\mathcal{J}}(x)\frac{\partial \sigma}{\partial x} \; ; \; \hat{\mathcal{J}}(x) + \hat{\mathcal{J}}^T(x) = 0$$

2.5 Removing the transversality condition limitation

The transversality condition (2) plays an essential role in all our previous developments and provisions should be taken for those cases in which it is not immediately satisfied.

Suppose that, in $\mathcal{X}$, the sliding surface function $\sigma(x)$ is not *relative degree* equals to one with respect to the control input u, i.e., the transversality condition $L_g\sigma(x) \neq 0$ is not satisfied on the operating region $\mathcal{X}$. It is intuitively clear that, in such a case, the sliding surface function $\sigma(x)$ must have *some* relative degree on a subset of $\mathcal{X}$. For if not, then the sliding surface function cannot be modified by any control action whatsoever. In order to avoid needless specifications we make the following assumption:

Assume the sliding surface function $\sigma(x)$ is relative degree $r > 1$, in the operating region $\mathcal{X}$ of the state space, i.e.,

$$\begin{aligned} L_g L_f^j \sigma(x) &= 0 \; ; \; j = 0, 1, \ldots, r-2 \; \forall \, x \in \mathcal{X} \\ L_g L_f^{r-1} \sigma(x) &\neq 0 \; ; \; \forall \, x \in \mathcal{X} \end{aligned} \tag{29}$$

From the above assumption it should also be clear that, in the operating region $\mathcal{X}$, the condition $L_f^{r-1}\sigma(x) \neq 0$ is also trivially satisfied, for, otherwise, the relative degree of $\sigma(x)$ is not r in $\mathcal{X}$, as assumed.

Let α be a nonzero scalar constant. Consider next the following sliding surface function,

$$\omega(x) = \sigma(x) + \alpha L_f^{r-1}\sigma(x)$$

Then it is obviously true that $W(x)$ does satisfy the transversality condition in all of $\mathcal{X}$.

$$\begin{aligned} L_g\omega(x) &= L_g\sigma(x) + \alpha L_g L_f^{r-1}\sigma(x) \\ &= \alpha L_g L_f^{r-1}\sigma(x) \neq 0 \end{aligned} \tag{30}$$

A possibly more suggestive modified sliding surface function $W(x)$ may be taken to be

$$\omega(x) = \alpha_0\sigma(x) + \left[\sum_{k=1}^{r-1} \alpha_k \, L_f^k\sigma(x)\right]$$

with $\alpha_k \neq 0 \; \forall \; k$ being appropriate *Hurwitz coefficients*.

3 Some Illustrative Examples

3.1 Example 1

Consider the following simplified representation of a Cayley-Rodrigues single axis spacecraft model.

$$\begin{aligned}\dot{x}_1 &= \frac{1}{2}\left(1+x_1^2\right)x_2 \\ \dot{x}_2 &= u\end{aligned} \tag{31}$$

where x_1 is the attitude angle measured with respect to a skewed axis and x_2 is the angular velocity with respect to a principal axis.

In this case one has

$$f(x) = \begin{bmatrix}\frac{1}{2}\left(1+x_1^2\right)x_2 \\ 0\end{bmatrix} \quad ; \quad g(x) = \begin{bmatrix}0\\1\end{bmatrix} \tag{32}$$

The linearizing sliding surface, achieving global exponential stabilization towards a prespecified attitude angular position X, is given, in this case, by

$$\sigma = \frac{1}{2}(1+x_1^2)x_2 + \lambda(x_1 - X) \tag{33}$$

with $\lambda > 0$ being a design parameter. The transversality condition is clearly satisfied everywhere in R^2, as

$$L_g\sigma = \frac{\partial\sigma}{\partial x^T}\,g(x) = 0.5(1+x_1^2) \neq 0 \;\; \forall\, x \in R^2 \tag{34}$$

The sliding surface, depicted in Figure 2 is given by the graph of the function

$$x_2 = -\frac{2\lambda(x_1 - X)}{1+x_1^2} \tag{35}$$

in the (x_1, x_2) plane.

The gradient vector $\partial\sigma/\partial x$ is given in this case by

$$\frac{\partial\sigma}{\partial x} = \begin{bmatrix}\dfrac{\partial\sigma}{\partial x_1}\\ \dfrac{\partial\sigma}{\partial x_2}\end{bmatrix} = \begin{bmatrix}x_1x_2+\lambda \\ \dfrac{1}{2}(1+x_1^2)\end{bmatrix} \tag{36}$$

The input coordinate transformation

$$u = v - x_1x_2^2 - \lambda x_2 \tag{37}$$

yields a precompensated system in Hamiltonian form given by

$$\begin{bmatrix}\dot{x}_1\\ \dot{x}_2\end{bmatrix} = \begin{bmatrix}0 & x_2\\ -x_2 & 0\end{bmatrix}\begin{bmatrix}\dfrac{\partial\sigma}{\partial x_1}\\ \dfrac{\partial\sigma}{\partial x_2}\end{bmatrix} + \begin{bmatrix}0\\1\end{bmatrix}v \tag{38}$$

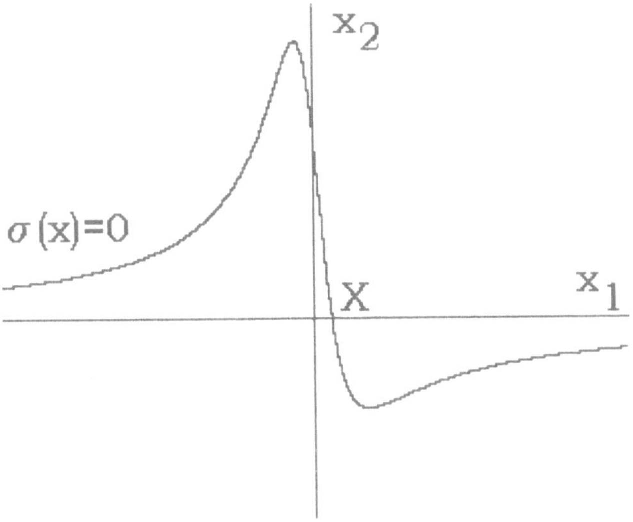

Fig. 2. Sliding Surface in the (x_1, x_2) plane

As evidenced from the previous calculations, the precompensated system exhibits no dissipation nor destabilizing forces. In fact, all remaining forces are of the conservative type meaning that if the control input v is shut-off, at any time, the evolution of the system occurs along leaves of the *foliation*: $\sigma = constant$. The complementary bang-bang control law

$$v = -K\text{sign}\ \sigma \tag{39}$$

wth $K > 0$, yields reachability of the sliding manifold, $\sigma(x) = 0$, in finite time, with the subsequent exponential stabilization of the angular position x_1 to the prespecified value $x = X$, as it is ideally governed, on the sliding surface, by the linear asymptotically stable dynamics

$$\dot{x}_1 = -\lambda\,(x - X) \tag{40}$$

The traditional sliding mode feedback controller, which does not take into account the dissipation structure of thes system is then given by

$$u = v - x_1 x_2^2 - \lambda x_2 - K\text{sign}\ \sigma \tag{41}$$

Evidently, the state dependent input coordinate transformation, which is actually based on the equivalent control, gets rid of all possible dissipation and destabilizing terms without advantageously exploiting the local dissipation and destabilizing forces present in the uncompensated system. Thus, in order to obtain a variable structure control law that locally takes advantage of the beneficial nonlinearities, we must rewrite the original, uncompensated, system in canonical form. We first write the projection operator matrix corresponding to the system field $g(x)$ and the proposed sliding surface.

$$M(x) = \left(I - \frac{1}{L_g\sigma} g(x) \frac{\partial \sigma}{\partial x^T}\right) = \begin{bmatrix} 1 & 0 \\ -\frac{2(x_1x_2+\lambda)}{1+x_1^2} & 0 \end{bmatrix} \tag{42}$$

while

$$(I - M(x)) = \frac{1}{L_g\sigma} g(x) \frac{\partial \sigma}{\partial x} = \begin{bmatrix} 0 & 0 \\ \frac{2(x_1x_2+\lambda)}{1+x_1^2} & 1 \end{bmatrix} \tag{43}$$

The decomposition of the drift vector field, written in terms of the gradient vector field is obtained as

$$f(x) = \begin{bmatrix} \frac{1}{2}\left(1+x_1^2\right)x_2 \\ 0 \end{bmatrix} = \begin{bmatrix} 0 & x_2 \\ -x_2 & 0 \end{bmatrix} \begin{bmatrix} \frac{\partial \sigma}{\partial x_1} \\ \frac{\partial \sigma}{\partial x_2} \end{bmatrix} + \begin{bmatrix} 0 & 0 \\ 0 & \frac{2(x_1x_2+\lambda)x_2}{1+x_1^2} \end{bmatrix} \begin{bmatrix} \frac{\partial \sigma}{\partial x_1} \\ \frac{\partial \sigma}{\partial x_2} \end{bmatrix} \tag{44}$$

The given system is then rewritten in Hamiltonian form as

$$\begin{bmatrix} \dot{x}_1 \\ \dot{x}_2 \end{bmatrix} = \begin{bmatrix} 0 & x_2 \\ -x_2 & 0 \end{bmatrix} \begin{bmatrix} \frac{\partial \sigma}{\partial x_1} \\ \frac{\partial \sigma}{\partial x_2} \end{bmatrix} + \begin{bmatrix} 0 & 0 \\ 0 & \frac{2(x_1x_2+\lambda)x_2}{1+x_1^2} \end{bmatrix} \begin{bmatrix} \frac{\partial \sigma}{\partial x_1} \\ \frac{\partial \sigma}{\partial x_2} \end{bmatrix} + \begin{bmatrix} 0 \\ 1 \end{bmatrix} u \tag{45}$$

The nature of the contribution of the non-conservative forces to the sliding surface coordinate rate of change, $\dot{\sigma}$, is readily evaluated from the expression representing the only eigenvalue of the symmetric matrix in the Hamiltonian model of the system. The sign of this eigenvalue depends only on the sign of the product: $x_2(x_1x_2 + \lambda)$. Therefore, if this product is negative in the region $\sigma > 0$ one should set the control input to the simplest form which guarantees surface reachability in finite time, say $u = -K\text{sign}\ \sigma$. If, on the contrary, the product $x_2(x_1x_2 + \lambda)$ is positive, one should use the controller (37)-(39), since it overrides the locally destabilizing character of this term. In the region $\sigma < 0$ one shold follow precisely the opposite policy. This is summarized in the following prescription for the variable structure feedback control law,

$$u = \begin{cases} -(x_1x_2+\lambda)x_2 - K \text{ sign } \sigma & \text{for } x_2(x_1x_2+\lambda)\sigma > 0 \\ -K \text{ sign } \sigma & \text{for } x_2(x_1x_2+\lambda)\sigma < 0 \end{cases} \tag{46}$$

Figure 3 shows the results of some digital computer simulations demonstrating how the proposed controller exerts far less effort during the regulation of the system towards the sliding surface from the same set of initial conditions. In order to be able to appreciate the fundamental difference in the control effort between the controller (41) and the proposed controller (46), we used a smoothing of the sliding mode controller by means of the usual "smooth" approximation to the switch function.

$$\text{sign } \sigma \approx \frac{\sigma}{\epsilon + |\sigma|} \tag{47}$$

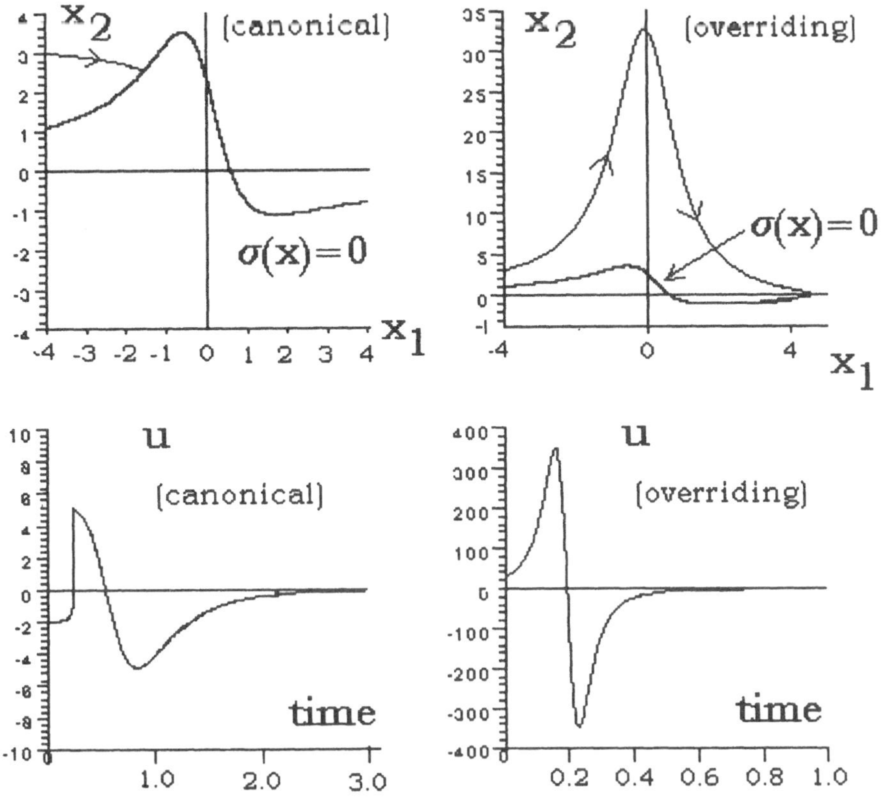

Fig. 3. Comparison of control efforts and sliding mode controlled trajectories.

While the traditional controller, (41), uses a control amplitude of, roughly speaking 400 units, the proposed controller (46) exhibits an amplitude only of 6 units for the same initial conditions. The state trajectories are shown to reach the surface through a more direct path than the corresponding of the traditional sliding mode controller eventhough in a slower fashion.

3.2 Example 2

Consider the following nonlinear system

$$\begin{aligned} \dot{x}_1 &= x_2 + x_1(x_1^2 + x_2^2 - u) \\ \dot{x}_2 &= -x_1 + x_2(x_1^2 + x_2^2 - u) \end{aligned} \tag{48}$$

The vector fields associated with the system dynamics are given by

$$f(x) = \begin{bmatrix} x_2 + x_1(x_1^2 + x_2^2) \\ -x_1 + x_2(x_1^2 + x_2^2) \end{bmatrix} \quad ; \quad g(x) = \begin{bmatrix} -x_1 \\ -x_2 \end{bmatrix} \tag{49}$$

Let K be a strictly positive constant. Consider the sliding surface

$$\sigma = \frac{1}{2}\left(x_1^2 + x_2^2 - R^2\right) \tag{50}$$

Ideal sliding motions on this surface correspond to a sustained oscillatory motion in a circle around the origin of radius K, as governed by the ideal sliding dynamics

$$\dot{x}_1 = x_2 \;\; ; \;\; \dot{x}_2 = -x_1 \tag{51}$$

The transverality condition $L_g\sigma$ is evidently satisfied everywhere on the sliding surface as

$$L_g\sigma = -x_1^2 - x_2^2 = -R^2 \neq 0 \tag{52}$$

The gradient vector of the sliding surface corresponds with the state of the system

$$\frac{\partial \sigma}{\partial x} = \begin{bmatrix} x_1 \\ x_2 \end{bmatrix} \tag{53}$$

The system in canonical form is thus given by

$$\begin{bmatrix} \dot{x}_1 \\ \dot{x}_2 \end{bmatrix} = \begin{bmatrix} 0 & 1 \\ -1 & 0 \end{bmatrix} \begin{bmatrix} x_1 \\ x_2 \end{bmatrix} + \begin{bmatrix} x_1^2 + x_2^2 & 0 \\ 0 & x_1^2 + x_2^2 \end{bmatrix} \begin{bmatrix} x_1 \\ x_2 \end{bmatrix} + \begin{bmatrix} -x_1 \\ -x_2 \end{bmatrix} u \tag{54}$$

The time derivative of the sliding surface coordinates is given by

$$\dot{\sigma} = (x_1^2 + x_2^2)\left[(x_1^2 + x_2^2) - u\right] \tag{55}$$

which immediately suggest the following overriding sliding mode feedback control action

$$u = x_1^2 + x_2^2 - \frac{v}{x_1^2 + x_2^2} \tag{56}$$

leading to the following controlled sliding surface coordinate dynamics,

$$\dot{\sigma} = v \tag{57}$$

In other words, the feedback control action

$$u = x_1^2 + x_2^2 + \frac{K \text{sign}\, \sigma}{x_1^2 + x_2^2} \;\; ; \;\; K > 0 \tag{58}$$

yields sliding surface reachability and the creation of a sliding mode on the sliding surface $\sigma(x) = 0$.

However, since the contribution of the nonconservative part is always postive, from within the circle ($\sigma < 0$) these forces are actually helping in the reaching of the sliding surface and yet the previous controller is overriding these beneficial forces. Naturally from outside the sliding circle ($\sigma > 0$) these forces are in oposition to our desire to reach the sliding surface and they should be overriden. The following controller represents a reasonable choice which exploits the structure of the system with respect to the sliding surface.

$$u = \begin{cases} x_1^2 + x_2^2 + \dfrac{K}{x_1^2 + x_2^2} & \text{for } \sigma > 0 \\ -K(x_1^2 + x_2^2) & \text{for } \sigma < 0 \end{cases} \tag{59}$$

This controller suitably complements the beneficial nonlinearities in the region where the system naturally tends to reach the sliding surface (i.e., on $\sigma < 0$) and neutralizes the destabilizing nonlinearities from the region where the system naturally tends to drift away from the sliding surface (i.e., on $\sigma > 0$).

Figure 4 compares the performance of the canonical form based sliding mode controller versus the overriding controller. Eventhoug the overriding controller reaches the sliding surface first, its control effort is significanly larger. From outside the sliding circle both controllers have exactly the same form and hence the same performance.

3.3 Example 3

Consider the following nonlinear system

$$\begin{aligned} \dot{x}_1 &= x_2 + x_2 u \\ \dot{x}_2 &= -x_2 + u \end{aligned} \tag{60}$$

Let λ be a strictly positive design constant. The sliding surface

$$\sigma = x_2 + x_2^2 + \lambda\left(x_1 - \frac{1}{2}x_2^2 - X\right) \tag{61}$$

locally stabilizes the system state to the equilibrium point $(x_1, x_2) = (X, 0)$.

The system vector fields are, in this case

$$f(x) = \begin{bmatrix} x_2 \\ -x_2 \end{bmatrix} \quad ; \quad g(x) = \begin{bmatrix} x_2 \\ 1 \end{bmatrix} \tag{62}$$

The transversality condition,

$$L_g \sigma = \frac{\partial \sigma}{x^T} g(x) = 1 + 2x_2 \neq 0 \tag{63}$$

implies that the line $x_2 = -1/2$ is a singularity line where the canonical form will not be valid.

The surface coordinate gradient vector is obtained as

$$\frac{\partial \sigma}{\partial x} = \begin{bmatrix} \lambda \\ 1 + (2 - \lambda)x_2 \end{bmatrix} \tag{64}$$

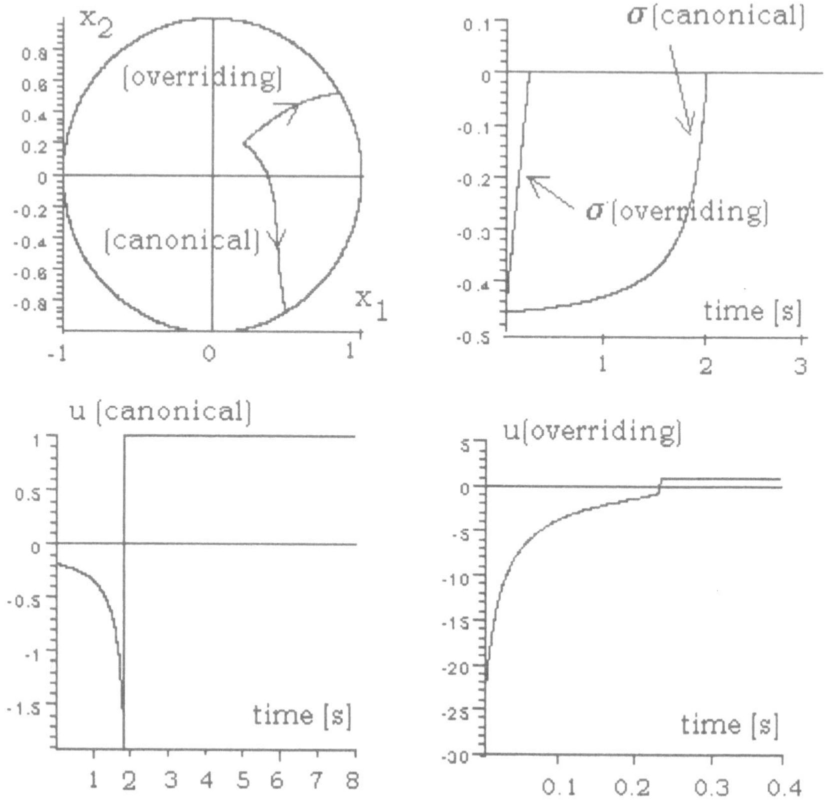

Fig. 4. Comparison between a canonical form based sliding mode control performance and the overriding controller

A state dependent input coordinate transformation which neutralizes all the non-conservative forces acting on the surface coordinate evolution in the system is given by

$$u = \frac{1}{1+2x_2}\left[v + x_2(1-\lambda) + x_2^2(2-\lambda)\right] \tag{65}$$

yields a controlled system in canonical form given by

$$\begin{bmatrix} \dot{x}_1 \\ \dot{x}_2 \end{bmatrix} = \frac{1}{1+2x_2}\begin{bmatrix} 0 & x_2 + x_2^2 \\ -x_2 - x_2^2 & 0 \end{bmatrix}\begin{bmatrix} \dfrac{\partial \sigma}{\partial x_1} \\ \dfrac{\partial \sigma}{\partial x_2} \end{bmatrix} + \frac{1}{1+2x_2}\begin{bmatrix} x_2 \\ 1 \end{bmatrix} v \tag{66}$$

The corresponding controlled rate of change of the sliding surface coordinate σ is given in this case by

$$\dot{\sigma} = v \tag{67}$$

A variable structure control of the form

$$v = -K \text{sign}\, \sigma \quad ; \quad K > 0 \tag{68}$$

achieves reachability of the sliding surface from regions of the state space which yield state trajectories which are bounded away from the singularity line $1 + 2x_2 = 0$. The complete feedback control law is thus given by

$$u = \frac{1}{1 + 2x_2} \left[-K \text{sign}\, \sigma + x_2(1 - \lambda) + x_2^2(2 - \lambda) \right] \tag{69}$$

The complete canonical form of the uncompensated system which exhibits the entire structure of the system with respect to the surface coordinate is readily obtained as

$$\begin{aligned} \begin{bmatrix} \dot{x}_1 \\ \dot{x}_2 \end{bmatrix} &= \frac{1}{2(1 + 2x_2)} \begin{bmatrix} 0 & x_2 + x_2^2 \\ -x_2 - x_2^2 & 0 \end{bmatrix} \begin{bmatrix} \dfrac{\partial \sigma}{\partial x_1} \\ \dfrac{\partial \sigma}{\partial x_2} \end{bmatrix} \\ &+ \frac{1}{2(1 + 2x_2)} \begin{bmatrix} 2x_2^2 & x_2 - x_2^2 \\ x_2 - x_2^2 & -2x_2 \end{bmatrix} \begin{bmatrix} \dfrac{\partial \sigma}{\partial x_1} \\ \dfrac{\partial \sigma}{\partial x_2} \end{bmatrix} + \begin{bmatrix} x_2 \\ 1 \end{bmatrix} u \end{aligned} \tag{70}$$

The contribution of the nonconservative forces to the rate of change of the sliding surface coordinate is sign undefined, as it may be positive, or negative, depending on the region of the state space in which the trajectory happens to be located, at a particular instant of time. This contribution is given by

$$\frac{1}{2(1 + 2x_2)} \frac{\partial \sigma}{\partial x^T} \begin{bmatrix} 2x_2^2 & x_2 - x_2^2 \\ x_2 - x_2^2 & -2x_2 \end{bmatrix} \begin{bmatrix} \dfrac{\partial \sigma}{\partial x_1} \\ \dfrac{\partial \sigma}{\partial x_2} \end{bmatrix} = x_2(\lambda - 1) + x_2^2(\lambda - 2) \tag{71}$$

Chosing $\lambda > 2$ one obtains the following regions of positivity and negativity of the nonconservative forces

$$\begin{aligned} x_2(\lambda - 1) + x_2^2(\lambda - 2) > 0 \text{ for } x_2 &\in \left(-\infty, -\frac{\lambda - 1}{\lambda - 2} \right) \cup (0, \infty) \\ x_2(\lambda - 1) + x_2^2(\lambda - 2) < 0 \text{ for } x_2 &\in \left(-\frac{\lambda - 1}{\lambda - 2}, 0 \right) \end{aligned} \tag{72}$$

It is evident that the contribution of the conservative part of the system helps in reaching the sliding surface only when its sign is opposite to that of the surface. We then may propose the following more efficient variable structure feedback control law

$$u = \begin{cases} u = \dfrac{1}{1 + 2x_2} \left[-K \text{sign}\, \sigma + x_2(1 - \lambda) + x_2^2(2 - \lambda) \right] \\ \qquad\qquad\qquad \text{for } \sigma \left(x_2(\lambda - 1) + x_2^2(\lambda - 2) \right) > 0 \\ -\dfrac{K}{1 + 2x_2} \text{sign}\, \sigma \qquad \text{for } \sigma \left(x_2(\lambda - 1) + x_2^2(\lambda - 2) \right) < 0 \end{cases} \tag{73}$$

4 Conclusions

We have proposed a natural canonical form for nonlinear single input systems for which sliding motions are to be created on a given sliding surface. The canonical form is largely motivated from passivity based considerations on the same class of nonlinear systems. As a result, a clear and natural decomposition of the internal system forces is revealed which definitely helps in designing a more efficient variable structure feedback controller. The approach leads to a controller design characterized by the following facts: 1) It respects the useful nonlinearities of the system which help in locally reaching the sliding surface. 2) It does not eliminate but, rather, *neutralizes* those non-benefical vector fields of the system which tend to make the trajectories move away from the sliding manifold and 3) It simply *complements* the useful nonlinearities of the system, on each side of the sliding surface, so as to locally achieve sliding surface reachability. These three characteristics make the feedback controller signals more naturally tunned to the system structure and control amplitude limitations while achieving the desired control objective.

The results can be extended, with some difficulty, to the sliding mode control of multivariable nonlinear systems (see [8]).

References

1. Bastin, G., and Dochain, D. *On Line Estimation and Adaptive Control of Bioreactors.* Elsevier Science, New York, 1990.
2. Byrnes, Ch. I., Isidori, A., and Willems, J. C. "Passivity, feedback equivalence and the global stabilization of minimum phase nonlinear systems" *IEEE Transactions on Automatic Control* Vol. 36, pp. 1228-1240, 1991
3. Crouch, P., and van der Schaft, A. *Variational and Hamiltonian Control Systems*, Lecture Notes in Control and Information Sciences, Vol. 101, Springer-Verlag, Berlin, 1987.
4. Ortega, R., Loria, A., Nicklasson, P. and Sira-Ramírez, H. *Passivity Based Control of Euler-Lagrange Systems: Applications to Electrical, Mechanical and Electromechanical Systems*, Springer-Verlag, London 1998.
5. Sira-Ramírez, H. "A General Canonical Form for Feedback Passivity" *International Journal of Control*, Special Issue: Recent Advances in the Control of Nonlinear Systems, F. Lamnabhi-Lagarrigue (Editor). Vol 71, No. 5, November 1998.
6. Sira-Ramírez, H. and Angulo-Núñez, M. I. "Passivity based control of nonlinear chemical processes" *International Journal of Control*, Vol. 68, pp. 971-996, 1997.
7. Sira-Ramírez, H. "Differential geometric methods in variable sturcture control" *International Journal of Control* Vol. 48, pp 1359-1390, 1988.
8. Sira-Ramírez, H., and Rios-Bolívar, M. " Feedback Passivity of Nonlinear Multivariable Systems" in *Proc. of the IFAC World Congress*, Beiging, China. July 1999. (to appear).
9. Utkin, V. I. *Sliding Regimes in the Theory of Variable Structure Systems*, MIR Editors, Moscow 1978.
10. van der Schaft, A. *L_2 and Passivity Techniques in Nonlinear Control*, Lecture Notes in Control and Information Systems, Vol. 218, Springer-Verlag, London 1996.

Handling Stiction with Variable Structure Control

Cem Hatipoğlu[1] and Ümit Özgüner[2]

[1] AlliedSignal TBS, 901 Cleveland Street, Elyria, OH 44035, USA
[2] The Ohio State University, Columbus, OH 43210-1272, USA

Abstract. When the control system falls in a class obeying non-Lipschitzian (or "non-smooth") dynamics, the conventional nonlinear theory can not be readily applied. Among the common non-smooth nonlinearities are those which are discontinuous in the state variables. Existence of such inherent right hand side discontinuities may induce undesirable stiction while hardening the control task during reference tracking. In this work, we analyze the "stiction" phenomenon in depth using analogies from the sliding mode control theory and propose a multi layer variable structure reference tracking controller for a class of systems in their companion forms. Results for an interesting physical example is provided to clarify the concepts.

1 Introduction

1.1 Control Theory

Control literature has been one of the most productive and fastest growing literatures in the past few decades. Richness of available approaches helps design various different controllers to meet certain specifications. Today, control engineers possess in their arsenal the tools to attack any given challenge from many different directions. Moreover, the immense growth in the electronics/computer industry created various opportunities to employ high level sophisticated control schemes to life. Not until long ago, control theory was considered mostly theoretical work growing apart from Industry. In process control nothing but PID controllers were used. This was mainly because, (i) they were capable of controlling the process -maybe not as optimally as one would have ideally liked, but were nominally doing the job-, (ii) they were easy to implement, integrate, service and replace and (iii) more importantly they were cost-effective. As the result of recent advances in the electronics sector, we are now capable of implementing high-level logical operations on microchips and have our intelligent controllers do the "control job" in a more efficient and optimal manner.

Today, there are companies that manufacture servo-actuators that are precise to one millionth of an inch which are mostly used in the aerospace industry. Some sensors can do even better and are commonly used in biomedical applications. However, if they were to be be utilized in most applications, they would at least double the cost of any conventional product. "*Marketing needs*" force engineers to use rather less sensitive and less precisely manufactured components in implementation. The encouraging fact -however- is that advanced control techniques have proven to be capable of compensating for such imperfections in most products and thus improving their performances. Advances in electronics technology help those techniques be implemented in a cost effective way by by-passing various problems caused by inherent non-smooth nonlinearities. *Control* is now more like an integral part of manufacturing process (due to needs for compensation against nonlinearities in the product) as well as its well known utilization for function automation.

1.2 Stick-Slip Behavior

Handling non-smooth nonlinearities in control systems (e.g. imperfections in the actuators & sensors, natural frictional forces in mechanical systems) involves the identification of the discontinuities, analysis of the system behavior, the design of a robust controller and the testing of the closed loop performance. For decades, existence of various such nonlinearities in systems caused the toughest of challenges to the control design engineers. All of the Laplace domain and most of the state space control design techniques have been developed exclusively for linear systems. Hence, by many, the existence of actuator, sensor and system nonlinearities were neglected and the controllers were designed on a nominal linearized plant model and later modified to work on the actual plant by means of heuristic approaches. These methods encountered significant difficulties in all of the analysis, model fitting (and hence simulation) and control design stages. Especially it was extremely hard, if not impossible, to prove stability of such systems. Also the lack of available straight-forward, well-defined techniques to identify such nonlinearities emerged as a tough problem by itself.

Identification problem of nonlinearities in system description has been very attractive for researchers for decades both from physical and control stand points. Especially stick-slip friction (also known as dry friction) has been the major physical phenomenon of interest. Dry friction is the natural resistance between two contacting surfaces whose magnitude depends on the physical properties of the interacting surfaces. It is generally a combination of the coulomb and viscous frictions. The difficulty introduced into the control problem stems from the discontinuity of dry friction at zero velocity, [1, 22]. It is present in all machines involving parts with relative motion, [1], and the evolution of a model has been studied with details in the cited survey paper. It is realized that friction compensation has to be utilized efficiently to solve the tracking and regulation problems in motors and in any applications involving some sort of motor or mechanical assembly. The most common analysis tools used in the literature involve describing functions, algebraic analysis, phase plane analysis and simulations. Most of the developed controllers are unfortunately system specific and friction model dependent. Obviously modeling and control are the two essential components of nonlinearity compensation in general. Of the most commonly preferred techniques are, stiff position control, integral control with dead-band, direct force feedback, impulsive control, coulomb friction feed-forward and position dependent friction feed-forward. In the survey paper by Armstrong, Dupont and De Wit in 1994, it is quoted that there are about 2240 specific papers on dry friction modeling, control of systems with coulomb friction and compensation techniques. If we consider the extended study since then, this number grows to an even more significant level. Most of those techniques, at least those which exhibit nice performances, are nonlinear in nature. Generally a dithering is introduced to the system to modify its behavior, a variable structure control law is used, or the linear control is modified ("non-linearized") around the discontinuity. In any case, the control objective has been to eliminate the effects of friction in the control problem. In [4], a dynamic model for friction that describes the relatively complicated Stribeck effects as well as the hysteresis nature of the behavior due to frictional lag, spring-like characteristics during stiction and varying breakaway force depending on the rate of change of the applied force. The superiority of this model over many others is the ability to capture most of these characteristics at the same time and the good performance achieved at modeling the friction for low velocities especially when crossing zero velocity. The proposed friction model is modular and can be readily used in any application that involves a friction by identifying a few plant dependent parameters. In [21], an adaptive variable structure control methodology is proposed for friction compensation. The friction model as given in [4] is utilized to synthesize a controller around it while the proposed switching controller is designed for a more general class of nonlinear systems of the so-called special strict feedback form which combines the VSC scheme with back-stepping procedure. In [22], the proposed controller is a modified PD controller for robust nonlinear stick-slip friction compensation. It is observed that a discontinuous component in the applied force yields a significant improvement in the closed loop performance and helps unify the reference as the sole equilibrium point of the system. The precision issues in positioning and tracking are issued for a simple one degree of freedom system and the stability and robustness issues are addressed. In 1992, K.D. Young encounters the discontinuous right hand side phenomenon stemming from friction in sliding base isolated structures under earthquakes, [25]. Observing that the reaching condition of the designed sliding mode control law may be dominated by the friction force discontinuity under certain circumstances, he proposed an on-off modulation of the discontinuous feedback law that takes advantage of the friction force itself. The desired sliding manifold and the one imposed by the inherent nature of the system happened to coincide during certain modes of operation in that particular application. In [14], two models for stick-slip friction are proposed, one of which (referred as "bristle model") aims to capture the physical nature of the friction phenomenon in its most detailed form and the other (referred as "reset integrator model") targets numerical efficiency for effective use in simulation models. These models exhibit similar precision in terms of performance to those of Dahl's [14] and Karnopp's [18] models with significant improvements over the classical description of the friction. In [5], Dupont and Dunlop provide a relatively simple model for friction which is capable of describing the characteristics of this non-smooth nonlinearity especially at relatively lower speeds. They backup their results on the effectiveness of their proposed single-state variable friction model with experimental data. Their studies were primarily motivated towards finding an underlying compensation model for systematic PD-controller parameter selection to ensure closed loop stability. Their results seem to be in sync with the long-believed fact that if the controller stiffness dominates the machinery stiffness, then stability can be achieved. In [6], the problems that arise due to discontinuity in the highest order derivative term and due to the load dependent friction variation are addressed from simulation and control design points of views with applications in robotics. Armstrong and Amin, [2], formally demonstrate in 1996 that a stable stick-slip limit cycle exists for any single degree of freedom mechanism with nonzero static friction and minimal Coulomb friction using a stabilizing PID control. They have formalized the stick-slip phenomenon, solved the equations of motion involving discontinuities in the RHS for systems with Coulomb, static and/or viscous frictions. The most significant result, that when Coulomb and static frictions co-exists and are modeled as suggested in the paper, seems to justify the needs for nonlinear modifications to the PID control. Friedland and Park, [13], suggested a simple, reduced order observer to estimate the constant gain of the discontinuity and incorporate that in the control feedback loop to compensate for the unknown friction. The asymptotic convergence property of the observer law assured the estimation error go to zero with improvements in reference tracking in control. The observer structure makes it hard to generalize to more complicated friction models of higher order. The suggested observer structure is utilized in [23] to estimate friction, velocity and acceleration so as to employ an observer based friction compensation in tracking problem of servomechanisms, particularly those with non-negligible actuator dynamics. A novel approach can be found in [7] where Drakunov *et.al.* examine tools to overcome the coulomb friction (or stiction) in the tracking problem of a rod-less pneumatic servo-actuator. This approach is generalized via classifying nonlinearities in a structural form such that the nonlinear system of interest may, then, be examined systematically in the context of a more general class of systems involving signum functions operating on manifolds consisting of input(s), output(s), and states.

1.3 Nonlinear Systems with Discontinuous RHS

Nonlinear systems theory which deals with the behavior and properties of systems that can be represented as $\dot{x} = f(x(t), t)$ with the initial condition information $x(t_o) = x(t_o)$ imposes certain key constraints on the right hand side function $f(x(t), t)$ for the existence and uniqueness of a solution and the continuous dependence of solutions on the initial condition. In other words, for the described system be solvable over a finite time interval $[t_o, t_f]$, a continuous function $x : [t_o, t_f] \rightarrow \mathbb{R}^n$ satisfying the associated differential equation has to be found. In this case, if $f(x(t), t)$ is continuous both in the state variable x and time t, then the solution $x(t)$ will be a function that is continuously differentiable. If, however, $f(x(t), t)$ is continuous in $x(t)$ but only piecewise continuous in t then $x(t)$ will be piecewise continuously differentiable. Yet, these conditions are associated with the ***existence*** of a solution which is considered rather less restrictive than the ***uniqueness*** condition. The most standard (& well recognized) way of proving the local existence and uniqueness condition is called the *Lipschitz condition* which is stated in the form of a theorem as follows:

Theorem (Local Existence and Uniqueness), [19]
Given a right hand side function $f(x(t),t)$ that is piece-wise continuous in time, if $\forall x, y \in B = \{x \in \mathbb{R}^n \mid ||x - x_o|| \leq r\}$, $\forall t \in [t_o, t_f]$

$$||f(x(t),t) - f(y(t),t)|| \leq L||x - y||$$

Then, Lipschitz condition is said to be satisfied and it is concluded that $\exists \delta > 0$ such that

$$\dot{x} = f(x(t),t) \qquad x(t_o) = x_o$$

has a unique solution over $[t_o, t_o + \delta]$.

Although, many classical nonlinearities will satisfy this condition, it should be noted that any $f(x(t),t)$ containing a discontinuity in the state variable x would not satisfy the Lipschitz condition. So, the readily available theory of nonlinear systems can not be directly applied to the kind of nonlinear systems which involve discontinuities in their state space representations - at least not in their most conventional forms.

In 1964, Filippov has published a paper on systems with discontinuous right hand sides, [8], where he has examined the qualitative behavior of systems with piecewise continuous right hand sides (in the state variables and/or time). It was rather straight forward to show that the solutions of such systems consist of patches of continuous regions in each of which the right hand side of the differential equation remains continuous. Filippov, then, questioned the sort of potential difficulties one encounters in analysis (and consequently design) should the state trajectories of the solution hits a line (or surface) of discontinuity and within a finite time interval remains on this surface switching back and forth infinitely often.

Conventional methods -disappointingly but without any surprise- fail to yield the theory to analyze either of the existence of solutions, continuity, uniqueness or continuous dependence of solutions on the initial condition. In the same paper, [8], Filippov provides the conditions to analyze the qualitative properties (as mentioned above) for such systems. For the case, when a discontinuity surface is reached and the system remains on that surface for a finite time interval, the continuity of the solution is addressed by suggesting that the discontinuity on the right hand side be replaced with its continuous approximation which uniformly converges to the same discontinuity at the limit. The significance of this study is that the difficulties of dealing with systems involving non-smooth discontinuities are officially pointed out and the simplest form of the phenomenon which we refer in this study as "stiction in an open region" is observed.

1.4 Variable Structure Control

Variable structure control (VSC), after being inclusively studied by Russian researchers in 1950s has become very popular among many worldwide. What makes VSC so popular is its easy to analyze, easy to design, easy to implement nature. It is being used in almost all aspects of control, from electric motor drives, to automotive applications (such as throttle control, anti-lock brake systems control, fuel injection control), from space and aircraft flight control to industrial furnace control, power electronics, chemical processes, e.t.c.

Modeling involves approximations which result in mismatches between the actual plant and the developed model upon which the controller is based. The resulting uncertainty in the plant description needs to be handled in the control phases. There are various different control techniques that tend to compensate for parameter uncertainties and modeling errors and aim to come up with a design that stabilizes the closed loop system avoiding significant performance degradation. For example, adaptive schemes estimate the variations in the plant parameters, uncertainties and even the frequency response characteristics on/off-line and incorporate this information into the controller to achieve robustness. The pay in this approach is the increased complexity of the controller/identifier which becomes an important issue especially at the implementation phases and the underlying assumptions that need to be satisfied for parameter convergence (such as certainty equivalence).

Variable structure control theory has proven that, by means of deterministic fixed nonlinear control laws, robustness can be achieved without any sort of on-line identification of system parameters and thus global parameter convergence is not a necessary condition for the controller to exhibit satisfactory performance. Sliding mode control utilizes a special behavior of VSC known as the sliding mode regime and can exhibit perfect robustness and systems invariance for all underlying uncertainty -not in the average sense that is. The idea is simple: control law switches in a discontinuous manner on one or more manifolds to confine system trajectories on the intersection of user defined manifolds and once on it force them to stay there. One other advantage of VSC is that the manifold is reached in finite time and not in the asymptotic sense. Ideal infinite frequency switching causes the trajectories to remain on the manifold, but actually what happens is that system trajectories indeed cross to either side of the manifold as the control switches so that in the vicinity of the sliding manifold, state trajectories are always directed towards the designed manifold. The system is said to be in sliding mode when it reaches the manifold and stays on it thereafter. An equivalent continuous feedback control law can be derived for that system that -if initially left on the manifold- would have kept the system trajectories on that manifold. Equivalent control is known as the low frequency content of the infinitely high frequency switchings in sliding mode, [24].

Practical applications often require the discontinuous terms in the design be replaced with continuous nonlinear terms to avoid chattering in bounded bandwidth applications. Systems with uniformly convergent smooth approximations replacing the discontinuities are capable of replicating the nominal discontinuous system motions arbitrarily closely. All in all, VSC is very well suited for deterministic systems with matched uncertainties, can be easily implemented and widely used in many applications.

2 Generalized Stiction Phenomenon

2.1 Manifolds in $\mathbb{R}^n$

Definition 1. [3] A subset S in $\mathbb{R}^n$ is called a manifold of dimension p if every point x of S has a neighborhood in S which is homeomorphic to an open set in $\mathbb{R}^p$. A map $f : S \to \tilde{S}$ is called a homeomorphism if it is bijective and if f^{-1} is continuous as well as f itself.

For example, the ellipsoid described by $(x/\alpha)^2 + (y/\beta)^2 + (z/\gamma)^2 = 1$ is an $n-1$ dimensional manifold in $\mathbb{R}^n$, for $n \geq 3$. It should be noted that, by definition, a manifold can not lose dimension at any point over $S \subset \mathbb{R}^n$. For finite dimensional closed loop systems of the form

$$\dot{x} = f(x,t) \tag{1}$$

with $x \in \mathbb{R}^n$, stable integral manifolds are surfaces contained in $\mathbb{R}^n$ where the system trajectories evolve upon. The significant difference between the stable integral manifolds described in systems that have $f(x,t)$ satisfying the well known Lipschitz condition and those with discontinuous right hand sides is that the region of attraction absorbs the state trajectories onto the manifold within finite time interval with discontinuities whereas, asymptotically stable integral manifolds attract the states towards the manifold that are left in its vicinity but those states never actually reach the surface.

2.2 System Induced Manifolds

Consider the $n-$dimensional state space $\boldsymbol{x} = [x_1, \cdots, x_{n-1}, x_n]^T$, where $\boldsymbol{x} \in \mathbb{R}^n$. Assume the existence of k hyper-surfaces in this space, specified by,

$$\boldsymbol{s}(\boldsymbol{x},t) = \begin{bmatrix} s_1(\boldsymbol{x},t) \\ s_2(\boldsymbol{x},t) \\ \vdots \\ s_k(\boldsymbol{x},t) \end{bmatrix} = 0$$

where $\boldsymbol{s} : \mathbb{R}^n \to \mathbb{R}^k$. Consider now the following class of systems with right hand-side discontinuities on the k surfaces,

$$\dot{\boldsymbol{x}} = \boldsymbol{f}(\boldsymbol{x},t) + \boldsymbol{\mu} \cdot \mathrm{Sign}(\boldsymbol{s}(\boldsymbol{x},t)) + \boldsymbol{h}(\boldsymbol{x},t)\boldsymbol{u} \tag{2}$$

where $\boldsymbol{\mu} = [\boldsymbol{\mu}_1 \ \cdots \ \boldsymbol{\mu}_k] \in \mathbb{R}^{n\times k}$, $\boldsymbol{h}(\boldsymbol{x},t) \in \mathbb{R}^{n\times p}$, $\boldsymbol{f}(\boldsymbol{x},t) \in \mathbb{R}^n$ and f_i and h_{ij} are smoothly differentiable functions (in $\mathbb{C}^n$) and $\boldsymbol{u} \in \mathbb{R}^p$ is the control input. The signum operator is defined to operate on every entry of its argument (if it is an array or a matrix) individually, i.e.,

$$\mathrm{Sign}\left(\begin{bmatrix} s_1 \\ s_2 \\ \vdots \\ s_p \end{bmatrix}\right) = \begin{bmatrix} \mathrm{Sign}(s_1) \\ \mathrm{Sign}(s_2) \\ \vdots \\ \mathrm{Sign}(s_p) \end{bmatrix}$$

The system as given in Equation 2 appears in the form of an nth-order system with an input that has $k+p$ components k of which have already been specified in the form of sliding mode control with μ and $\sigma = \{\boldsymbol{x} \in \mathbb{R}^n \ : \ \boldsymbol{s}(\boldsymbol{x},t) = 0\}$ being the gain and the manifold. Note that, this is not exactly the case as these manifolds have not been designed and the associated gains have not been selected by the designer. Instead, they have been induced by the system itself. However, this sort of analogy allows us to analyze the system within the context of sliding mode theory using the mathematical tools that have been developed in this area.

Consider one of the candidate surfaces that might induce a stiction manifold, namely s_j for $j = 1, \cdots, k$. Under the assumption of the existence of sliding mode, i.e. when

$$\dot{s}_j \cdot s_j < 0; \quad j = 1, \cdots, p; \quad j \in \mathbb{N} \tag{3}$$

the system starts to slide on the manifold described by,

$$\sigma_j = \{[x_1 \ \cdots \ x_{n-1} \ x_n] \in \mathbb{R}^n \mid s_j(\boldsymbol{x},t) = 0\} \tag{4}$$

Note that the condition (3) defines an open region $\mathcal{A}_j$ in the state space. This region can be found by analyzing the derivative of s_j for $j = 1, \cdots, p$ where $j \in \mathbb{N}$,

$$\dot{s}_j = \frac{d}{dt} s_j \left(x_1, \cdots, x_{n-1}, x_n\right) = q_j \left(\dot{x}_1, \cdots, \dot{x}_{n-1}, \dot{x}_n\right) \tag{5}$$

where $q_j \in \mathbb{R}$ when confined to the trajectories described by (2), i.e., when the equality in (2) is used to replace the derivatives of the states in (5), becomes a function of the states $x_1, \cdots, x_n$, the control input $\boldsymbol{u}(t)$ and the combination of the discontinuities given on the right hand side of the system description.

$$\begin{aligned} \dot{s}_j = \tilde{q}_j \left(f_1(\boldsymbol{x}), \cdots, f_n(\boldsymbol{x}), \mathrm{Sign}(s_1(\boldsymbol{x})), \cdots, \mathrm{Sign}(s_k(\boldsymbol{x})), u_1, \cdots, u_p, \boldsymbol{x}\right) + \cdots \\ \cdots + g_j\left(\boldsymbol{x}\right) \mathrm{Sign}(s_j(\boldsymbol{x})) \end{aligned} \tag{6}$$

where $g_j : \mathbb{R}^n \to \mathbb{R}$ is the gain multiplying the jth discontinuous component, and $\tilde{q}_j : \mathbb{R}^n \to \mathbb{R}$. Then, $(\dot{s}_j \cdot s_j)$ becomes negative if,

$$-g_j\left(\boldsymbol{x},t\right) > |\tilde{q}_j(\cdot)| \tag{7}$$

$\forall \boldsymbol{x} \in \mathbb{R}^n$. The open region $\mathcal{A}_j$ is then described by,

$$\mathcal{A}_j = \{[x_1 \ \cdots \ x_{n-1} \ x_n] \in \mathbb{R}^n \mid -g_j\left(\cdot\right) > |\tilde{q}_j\left(\cdot\right)|\} \tag{8}$$

2.3 Stiction in an Open Region

Recall that this analysis is prior to the controller design. When the above condition (7) is satisfied for some j, the system trajectories of (2) will get stuck at $s_j = 0$ which is not a designed manifold. This phenomenon is induced due to the inherent right hand side discontinuities existing in the original system. Hence, an open stiction region can now be described in the state space as follows,

$$\mathcal{R}_s = (\sigma_1 \cap \mathcal{A}_1) \cup (\sigma_2 \cap \mathcal{A}_2) \cup \cdots \cup (\sigma_k \cap \mathcal{A}_k) = \bigcup_{j=1}^{k} (\sigma_j \cap \mathcal{A}_j) \tag{9}$$

Note that, $\mathcal{A}_j$ could differ from an empty set to the entire state space, but in general describes an open region which is a subset of $x \in \mathbb{R}^n$. It is also affected by the magnitude of the control input being generated.

2.4 Discussion

It is possible that, by making $\boldsymbol{u}$ large enough, one can violate the inequality in (8) for some $|\boldsymbol{u}| = \gamma$ which is magnitude-wise large enough for all stiction inducing "j"s if $\boldsymbol{\mu}$ lies in the image space of the (finitely generated subspace by the) smooth vector field $\boldsymbol{h}(\boldsymbol{x}, t)$. Then a control input can be chosen such that all $\mathcal{A}_j$'s will shrink into empty sets. This is guaranteed only if the $\boldsymbol{h}(\boldsymbol{x}, t)\boldsymbol{u}(t)$ term can be made arbitrarily large for all possible trajectories. The controllability condition, i.e. $\boldsymbol{h}(\boldsymbol{x}, t) \neq 0 \ \forall \boldsymbol{x} \in \mathbb{R}^n$, and $t > 0$ is thus one of the necessary conditions. Moreover, $\boldsymbol{u}(t)$ should never become 0. As a matter of fact $|\boldsymbol{u}| > \zeta > 0$ may be necessary to ensure $\boldsymbol{u}$'s domination in the inequality (8). If the control signal magnitude is picked larger than the supremum of such lower bounds for each component of $\boldsymbol{u}$, $\mathcal{A}_j$ can be forced into an empty set. The significance of this condition is that the stiction effects of the induced manifolds can *not* be eliminated entirely by a continuous function (Assuming $\mathcal{R}_s$ is non-empty) if the desired trajectory requires the input change sign in the course of tracking (which is the case in the context of general tracking problem). The other indication is that, using $|\boldsymbol{u}| = \gamma$ for some large enough γ, $\mathcal{R}_s$ can be shrunk into an empty set if $\boldsymbol{\mu}$ lies in the domain generated by $\boldsymbol{h}(\boldsymbol{x}, t)$ which in return implies that the generalized stiction problem can be entirely eliminated granted the discontinuities lie in the same subspace as the control which can be made discontinuous. However, it is generally the case that $\boldsymbol{u}$ might not be the actual input to the system to be controlled. An actuator may alter/filter that input before it affects the evolution of its trajectories. Discontinuities can not be totally eliminated with a continuous input, however, their effects may be minimized by careful design.

Up to this point only the statement of this phenomenon, which we call "generalized stiction" is given. In the sequel we will suggest a sliding mode controller design approach that will guarantee avoidance of generalized stiction $\mathcal{R}_s \backslash (\mathcal{R}_s \cap \mathcal{R}_c) \subset \mathbb{R}^n$ in tracking problem for a class of systems where $\mathcal{R}_c$ is the controlled manifold.

3 Systems with Discontinuous Right Hand Side Terms

Consider now the following class of SISO systems in their companion forms with right hand-side discontinuities on the p surfaces,

$$
\begin{aligned}
x^{(n)} &= f(x) + \boldsymbol{\mu} \cdot \mathrm{Sign}\,(s(x)) + h(x)\, v \qquad (10)\\
\dot{v} &= u, \qquad y = x
\end{aligned}
$$

where $\boldsymbol{\mu} = [\mu_1 \ \cdots \ \mu_p] \in \mathbb{R}^{1 \times p}$, $h(\cdot)$, $f(\cdot) : \mathbb{R}^n \to \mathbb{R}$ are smoothly differentiable functions (in $\mathbb{C}^n$) and moreover $h(\cdot) \neq 0$ for any $x = [x, \dot{x}, \cdots, x^{(n-1)}]^T$ (controllability condition over the entire state space), $u, v \in \mathbb{R}$, u is the control input.

4 VS Control Design

Consider the standard tracking problem where the controller needs to generate a control input u such that the output $y = x$ tracks the reference signal x_d. Define the tracking error as $\epsilon = x - x_d$.

The feedback linearizing control input v_d is,

$$v_d = \frac{1}{h(\cdot)} \left\{ \left[x_d^{(n)} - \kappa_1 \epsilon^{(n-1)} - \cdots - \kappa_{n-1}\dot{\epsilon} - \kappa_n \epsilon \right] - f(\cdot) - \mu \cdot \mathrm{Sign}(s(\cdot)) \right\} \quad (11)$$

and results in the following tracking error dynamics,

$$\epsilon^{(n)} + \kappa_1 \epsilon^{(n-1)} + \cdots + \kappa_{n-1}\dot{\epsilon} + \kappa_n \epsilon = 0 \quad (12)$$

where $\kappa_1, \kappa_2, \cdots, \kappa_n$ are design parameters. (12) can be made asymptotically stable by appropriate choice of a set of values so that $\epsilon \to 0$ at the desired rate. If a sliding manifold

$$S = v - v_d \quad (13)$$

is defined and the corresponding control input $\dot{v}$ is picked as,

$$u = \dot{v} = -\alpha \cdot \mathrm{Sign}(S), \quad \text{s.t.} \quad \alpha > (|\dot{v}_d| + \varepsilon) \quad (14)$$

then $\dot{S} \cdot S < 0$ is ensured. Therefore the system will start to slide on the manifold $S = 0$ after a finite amount of time for any $\varepsilon > 0$, resulting $v \to v_d$, which in return will yield the desired error dynamics and $x \to x_d$ at the desired rate. However, note that $\dot{v}_d$ is unbounded around $S = 0$ (which may be the case if the signal to be tracked lies on the hyper-surface described by the inherent discontinuity or requires crossing the mentioned hyper-plane) because of the discontinuous element in equation (11).

4.1 Control with Approximated Signum Function

If v_d is picked as,

$$v_d = \frac{1}{h(\cdot)} \left\{ \left[x_d^{(n)} - \kappa_1 \epsilon^{(n-1)} - \cdots - \kappa_{n-1}\dot{\epsilon} - \kappa_n \epsilon \right] - f(\cdot) - \mu \cdot \mathrm{Sign}_k(s(\cdot)) \right\} \quad (15)$$

where $\mathrm{Sign}_k(x) = \frac{2}{\pi}\arctan(kx)$ and $\Phi_k(x) = \mathrm{Sign}(x) - \mathrm{Sign}_k(x)$ are the continuous approximation for the discontinuous term and the approximation error respectively, tracking error dynamics becomes,

$$\epsilon^{(n)} + \kappa_1 \epsilon^{(n-1)} + \cdots + \kappa_{n-1}\dot{\epsilon} + \kappa_n \epsilon + \mu \Phi_k(s(\cdot)) = 0 \quad (16)$$

where $\mu \Phi_k(s(\cdot))$ can be made arbitrarily small out of a region $|s(\cdot)| > \gamma$. Recall that $s(\cdot)$ is not the designed manifold, but the stiction manifold to be avoided. Therefore, should the controller perform well, $|s(\cdot)| > \gamma$ should be satisfied except for a short (finite) duration of time. In either case,

$$|\Phi_k(s(\cdot))| \leq 1 \qquad \forall s \in \mathbb{R} \quad (17)$$

$$|\Phi_k(s(\cdot))| \leq \varepsilon \qquad |s| \geq \gamma,\ s \in \mathbb{R} \quad (18)$$

The error dynamics of (16) may be stabilized by appropriate choice of $\kappa_1, \cdots, \kappa_n$ where $\mu\Phi_k(s(\cdot))$ can be treated as an absolutely bounded disturbance to the system. The equilibrium point of (16) for constant $\mu\Phi_k(s(\cdot))$ becomes $\epsilon^{(n)} = \epsilon^{(n-1)} = \cdots = \dot{\epsilon} = 0$ and $\epsilon = -\mu\Phi_k(s(\cdot))/\kappa_n$ which can be made arbitrarily small by picking k large enough for $|s(\cdot)| \geq \gamma$. Note that, $|\dot{v}_d|$ can, now, be determined from (15) as everything becomes smoothly differentiable on the right hand side of the equality (15).

4.2 Multi-Layer Sliding Mode Controller Design

If there are uncertainties in $f(\cdot)$ and/or μ, and only some nominal values $\bar{f}(\cdot)$ and $\bar{\mu}$ are known instead of the exact values, and if the errors $\Delta f(\cdot) = |f(\cdot) - \bar{f}(\cdot)|$ and $\Delta\mu = |\mu - \bar{\mu}|$ are bounded, then multi layer sliding manifolds can be designed to compensate for the uncertainties. Modify 11 as

$$v_d = \frac{1}{h(\cdot)}\left\{\left[x_d^{(n)} - \kappa_1\epsilon^{(n-1)} - \cdots - \kappa_{n-1}\dot{\epsilon} - \kappa_n\epsilon\right] - \bar{f}(\cdot) - \bar{\mu}\cdot \mathrm{Sign}_k(s(\cdot)) + w\right\} \tag{19}$$

where w is a fictitious input and $\bar{f}(\cdot)$ & $\bar{\mu}$ are some nominal values for the model parameters. Then (16) becomes,

$$\epsilon^{(n)} + \kappa_1\epsilon^{(n-1)} + \cdots + \kappa_{n-1}\dot{\epsilon} + \kappa_n\epsilon + [\Delta f(\cdot) + \mu\Phi_k(s(\cdot)) \pm \Delta\mu \mathrm{Sign}_k(s(\cdot)) - w] = 0 \tag{20}$$

Let $\mathcal{S} = \epsilon^{(n-1)} + c_1\epsilon^{(n-2)} + \cdots + c_{n-2}\dot{\epsilon} + c_{n-1}\epsilon$ such that $\mathcal{S} = 0$ will exhibit stable dynamics with negative real poles. Then pick,

$$w = (\kappa_1 - c_1)\epsilon^{(n-1)} + \cdots + (\kappa_{n-1} - c_{n-1})\dot{\epsilon} + \kappa_n\epsilon - \beta \cdot \mathrm{Sign}_k(s) \tag{21}$$

which will ensure $(\dot{\mathcal{S}} \cdot \mathcal{S}) < 0$ for all $|\mathcal{S}| < \gamma$, provided that k is picked large enough and

$$\beta > |\Delta f(\cdot) + \mu\Phi_k(s(\cdot))| + |\Delta\mu \cdot \mathrm{Sign}_k(s(\cdot))| + \varepsilon \tag{22}$$

Note that the described control law will guarantee that the system trajectories will be directed towards the subspace described by $|s| \leq \gamma$ on the n-dimensional state space. The magnitude of γ can be manipulated by the designer, but can not be explicitly made zero.

4.3 Observer Design for Systems with Discontinuities

The calculation of $S(\cdot)$ and $\mathcal{S}(\cdot)$ require the existence of information $\epsilon = x - x_d$, $\dot{\epsilon} = \dot{x} - \dot{x}_d$, $\cdots$, $\epsilon^{(n-1)} = x^{(n-1)} - x_d^{(n-1)}$. The desired trajectory x_d and its derivatives $\dot{x}_d, \cdots, x_d^{(n-1)}$ are readily available, as they are being determined directly by the signal to be tracked. However, in practice some of the system states can not be directly measured. Those states need to be estimated using robust observers so that they can be used in the control input calculation.

Assume that only x is available as a measurement for the system (10) which is in companion form, rewrite system dynamics as:

$$\begin{aligned}
\dot{x}_1 &= x_2 \\
\dot{x}_2 &= x_3 \\
\vdots \;&\; \vdots \\
\dot{x}_n &= f\left(x_{n-1}, \cdots, x_2, x_1\right) + \mu \mathrm{Sign}\left(s\left(x_{n-1}, \cdots, x_2, x_1\right)\right) + h\left(x_{n-1}, \cdots, x_2, x_1\right) u
\end{aligned} \tag{23}$$

where $x_1 = x$ and $x_j = \frac{d^{j-1}x}{dt^{j-1}}$, $1 < j \leq n$; $j \in \mathbb{N}$. The conventional sliding mode observer structure will be in the form:

$$\begin{aligned}
\dot{\hat{x}}_1 &= \hat{x}_2 + k_1 \cdot \mathrm{Sign}(\tilde{x}_1) \\
\dot{\hat{x}}_2 &= \hat{x}_3 + k_2 \cdot \mathrm{Sign}(\tilde{x}_1) \\
\vdots \;&\; \vdots \qquad\quad \vdots \\
\dot{\hat{x}}_n &= \bar{f}(\cdot) + \bar{\mu} \cdot \mathrm{Sign}\left(\hat{s}\left(\cdot\right)\right) + h\left(\cdot\right) u + k_n \cdot \mathrm{Sign}(\tilde{x}_1)
\end{aligned} \tag{24}$$

where $\hat{x}_j$, for $j = 1, \cdots, n$ is the observer estimate of the state x_j and $\tilde{x}_j$ is the observation error, i.e., when $\tilde{x}_j = x_j - \hat{x}_j$. The error dynamics of the observer, then, becomes:

$$\begin{aligned}
\dot{\tilde{x}}_1 &= \tilde{x}_2 - k_1 \cdot \mathrm{Sign}(\tilde{x}_1) \\
\dot{\tilde{x}}_2 &= \tilde{x}_3 - k_2 \cdot \mathrm{Sign}(\tilde{x}_1) \\
\vdots \;&\; \vdots \qquad\quad \vdots \\
\dot{\tilde{x}}_n &= \underbrace{\Delta f(\cdot) + \bar{\mu} \cdot \left(\mathrm{Sign}\left(s\left(\cdot\right)\right) - \mathrm{Sign}\left(\hat{s}\left(\cdot\right)\right)\right) \mp \Delta\mu \cdot \mathrm{Sign}\left(s\left(\cdot\right)\right)}_{\Delta\eta} - k_n \cdot \mathrm{Sign}(\tilde{x}_1)
\end{aligned} \tag{25}$$

The uncertainty $\Delta f(\cdot) + \bar{\mu} \cdot \left(\mathrm{Sign}\left(s\left(\cdot\right)\right) - \mathrm{Sign}\left(\hat{s}\left(\cdot\right)\right)\right) \mp \Delta\mu \cdot \mathrm{Sign}\left(s\left(\cdot\right)\right)$ is combined into a single variable $\Delta\eta$. Assume that the dynamic uncertainty $\Delta\eta$ is explicitly bounded. The effect of $\Delta\eta$ can be compensated by using the information of the bound on this single lumped uncertainty.

When sliding mode occurs (within the observer structure), i.e. $\tilde{x}_1 = 0$ and $\dot{\tilde{x}}_1 = 0$, the equivalent control associated with the switchings becomes,

$$\left[\mathrm{Sign}(\tilde{x}_1)\right]_{eq} = \frac{1}{k_1}\tilde{x}_2 \tag{26}$$

The necessary condition for $\dot{\tilde{x}}_1 \cdot \tilde{x}_1 < 0$ is that $k_1 > |\tilde{x}_2|$. The reduced order $[(n-1)^{\text{th}}$ order$]$ system dynamics is determined from

$$\frac{d}{dt}\begin{bmatrix} \tilde{x}_2 \\ \tilde{x}_3 \\ \vdots \\ \tilde{x}_{n-1} \\ \tilde{x}_n \end{bmatrix} = \begin{bmatrix} -\frac{k_2}{k_1} & 1\,0 \cdots 0 \\ -\frac{k_3}{k_1} & 0\,1 \cdots 0 \\ \vdots & \vdots\,\vdots \ddots \vdots \\ -\frac{k_{n-1}}{k_1} & 0\,0 \cdots 1 \\ -\frac{k_n}{k_1} & 0\,0 \cdots 0 \end{bmatrix} \begin{bmatrix} \tilde{x}_2 \\ \tilde{x}_3 \\ \vdots \\ \tilde{x}_{n-1} \\ \tilde{x}_n \end{bmatrix} + \begin{bmatrix} 0 \\ 0 \\ \vdots \\ 0 \\ 1 \end{bmatrix} \Delta\eta \tag{27}$$

whose zero disturbance dynamics are governed by the roots of the polynomial,

$$s^{n-1} + \frac{k_2}{k_1}s^{n-2} + \frac{k_3}{k_1}s^{n-3} + \cdots + \frac{k_n}{k_1} = 0 \tag{28}$$

Assuming that $\Delta\eta$ is a constant, at steady state, the errors corresponding to x_j can be derived as $\tilde{x}_1 = 0$, $\tilde{x}_j = \frac{k_{j-1}}{k_n}\Delta\eta$ for $1 < j \leq n$. Therefore, it is suggested that k_n is picked much larger than k_j for $j = 1, \cdots, n-1$. Assume that $k_2/k_1, k_3/k_1, \cdots, k_n/k_1$ are picked accordingly to yield a critically damped observer system,

$$(s+\lambda)^{n-1} = \sum_{i=0}^{n-1} \binom{n-1}{i} s^i \lambda^{n-1-i} \tag{29}$$

Then, if $|\lambda|$ is picked large enough (say $|\lambda| = 10(n-1)$), all assumptions may be satisfied. There are more recently developed sliding mode observer design techniques which are relatively more complicated. See [15] for equivalent control theory based sliding mode observer design.

5 Example: Position Control of a System with Friction & Inherent Stiff-Position Feedback

The system involves a plant which is driven by an actuator with faster dynamics. The plant has inherent coulomb-viscous friction and stiff position feedback which are the two sources of stiction in the state space.

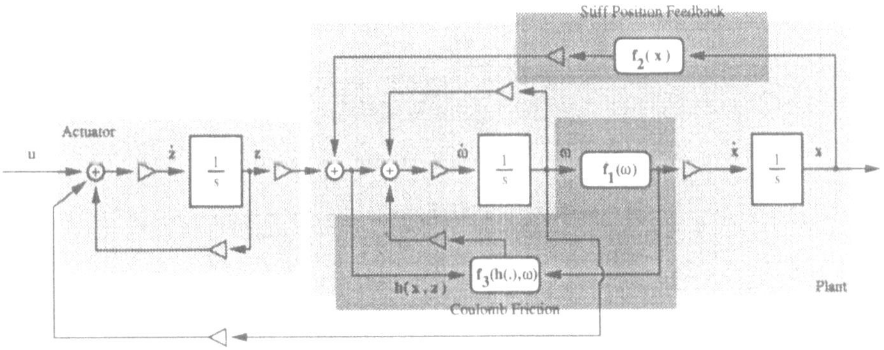

Fig. 1. The controlled system in Section 5.

5.1 The Plant Model

Consider the plant as displayed on Figure (1). The state space representation is given by

$$\dot{x} = \frac{1}{K_g} f_1(\omega) \tag{30}$$
$$\dot{\omega} = \frac{1}{J}\left(h(x,z) - \frac{C}{K_g}\omega - \frac{1}{K_g} f_3(h(x,z),\omega)\right)$$
$$\dot{z} = \frac{1}{L}\left(-Rz - K_t\omega + u\right)$$

where

$$f_1(\omega) = \text{deadzone}(\omega, \pm\delta, 1) \tag{31}$$
$$h(x,z) = K_t z - \frac{1}{K_g}\left(f_2(x) + K\cdot\left(\frac{180}{\pi}x - \theta_o\right)\right) \tag{32}$$
$$f_2(x) = \gamma \cdot \text{sat}\left(\frac{1}{\alpha}\left(\frac{180}{\pi}x - \theta_o\right)\right) \tag{33}$$
$$f_3(h(x,z),\omega) = \beta \cdot \text{sat}\left(\frac{1}{\delta} f_1(\omega)\right) + \beta \cdot \text{sat}(K_g h(x,z) f_4(f_1(\omega))) \tag{34}$$
$$f_4(f_1(\omega)) = 1 - (\text{relaywdzn}(f_1(\omega), \pm\delta, 1))^2 \tag{35}$$

where x is the position, ω is the angular velocity, and z is the auxiliary state variable that describes the dynamics of the first order actuator. The referred nonlinearities are defined as follows:

$$V = \text{deadzn}(U, \delta_1, \delta_2, m) = mU - \text{sat}(U, D_1, D_2, m)$$
$$V = \text{sat}(U, D_1, D_2, m) = \frac{1}{2}\{D_1 + D_2 - |mU - D_2| + |mU - D_1|\}$$
$$V = \text{sat}(U) = \text{sat}(U, -1, +1, 1) \tag{36}$$
$$V = \text{relaywdz}(U, \delta_1, \delta_2, D) = \frac{D}{2}\left(\text{Sign}(U - \delta_1) + \text{Sign}(U - \delta_2)\right)$$

There are physical limits on the position variable $x(t)$ which can not be less than 7° or more than 85°. The reference signals are also generated accordingly.

Table 1. Parameter Values for Example 5.

$L = 0.003$	$R = 2.6$	$K_t = 0.02$	$K_g = 14.2$
$J = 2 \times 10^{-6}$	$C = 2.2 \times 10^{-6}$	$K = 0.15$	$\theta_o = 10^\circ$
$\delta = 0.01$	$\gamma = 0.35$	$\alpha = 0.1$	$\beta = 0.85$

5.2 An Approximate Model for Control Design

Although friction has been modeled in details so as to include the Stribeck effects as well as stick-slip behavior, it is concluded that a simpler model suffices to describe the motion of the system with good precision while easing controller and observer design problems. Consider,

$$\dot{x} = a_{12}\omega \tag{37}$$
$$\dot{\omega} = a_{21}(x - x_o) + a_{22}\omega - \kappa\text{Sign}(x - x_o) - \mu\text{Sign}(\omega) + a_{23}z$$
$$\dot{z} = a_{32}\omega + a_{33}z + bu$$

where

$$a_{12} = \frac{1}{K_g}, \quad a_{21} = -\frac{K}{K_g J}, \quad a_{22} = -\frac{C}{K_g J}, \quad a_{23} = \frac{K_t}{J}, \quad a_{32} = -\frac{K_t}{L}$$

$$a_{33} = -\frac{R}{L}, \quad b = \frac{1}{L}, \quad x_o = \frac{\pi \theta_o}{180}, \quad \kappa = \frac{\gamma}{K_g J}, \quad \mu = \frac{\beta}{K_g J} \tag{38}$$

all of which can be calculated from the parameter values listed on Table 1. The unpowered system (when $u = 0$) converges the stable equilibrium point

$$\begin{bmatrix} x_{eq} \\ \omega_{eq} \\ z_{eq} \end{bmatrix} = \begin{bmatrix} x_o^* \\ 0 \\ 0 \end{bmatrix} \tag{39}$$

where $x_o^* = \{x \in \mathbb{R} : |x - x_o| \leq \zeta\}$ in the sliding mode sense starting from any initial conditions due to the existence of the discontinuous terms on the right hand side of the state space representation in (37). The coupling on the third equation is weak since $\frac{a_{33}}{a_{23}} > 127$, therefore, it is reasonable to assume $\dot{z} = 0$ (partially also due to the fast dynamics associated with the third equation, as $a_{33} = -860$) and concentrate on the dynamics of the first two equations, i.e. the slower subsystem. Based on the assumption $\dot{z} \approx 0$, from singular perturbation theory, the z term in the second equation can now be replaced with $z = -\frac{a_{32}}{a_{33}}\omega$.

There are two candidate stiction manifolds, namely $S_1 = \omega$ and $S_2 = (x - x_o)$. Since $\mu > \kappa$, for small enough $|x - x_o|$ the inequality $\dot{\omega}\omega < 0$ will be enforced and the manifold described by $\omega = 0$ will be reached in finite time. Note that the stiction region defined by $S_1 = 0$ and $\dot{S}_1 = 0$ is contained in the stiction region described by $S_2 = 0$ and $\dot{S}_2 = 0$. Also recall that this analysis is for the unpowered system only.

5.3 Controller Design

Let the position tracking error be defined as $\epsilon_x = x - x_r$ where x_r is the reference to be followed. From (37), one obtains,

$$\ddot{x} = a_{12}\left(a_{21}(x - x_o) + \frac{a_{22}}{a_{12}}\dot{x} - \kappa \text{Sign}(x - x_o) - \mu \text{Sign}\left(\frac{\dot{x}}{a_{12}}\right) + a_{23}z\right) \tag{40}$$

The feedback linearizing z_{fl} which yields the error dynamics:

$$(\ddot{x} - \ddot{x}_r) + \xi_1(\dot{x} - \dot{x}_r) + \xi_2(x - x_r) = 0 \tag{41}$$

can be derived (through straight-forward algebraic manipulations) as follows:

$$\begin{aligned} z_{fl} = \frac{1}{a_{12}a_{23}} &[\ddot{x}_r + \xi_1\dot{x}_r + \xi_2 x_r + a_{12}a_{21}x_o - (a_{12}a_{21} + \xi_2)x - (a_{22} + \xi_1)\dot{x} + \cdots \\ &\cdots + a_{12}\kappa \text{Sign}(x - x_o) + a_{12}\mu \text{Sign}(\dot{x})]. \end{aligned} \tag{42}$$

Design a sliding surface, $S_z = z - z_{fl}$ and consider its time derivative:

$$\dot{S}_z = \frac{a_{32}}{a_{12}}\dot{x} + a_{33}z + bu - \dot{z}_{fl} \tag{43}$$

Note that $\dot{S}_z S_z < 0$ may be achieved with the control input $u = -M\text{Sign}(S_z)$ if M dominates all the terms in the right hand side of (43) in magnitude. To achieve this, the signum functions need to be replaced with their continuous approximations. If there are uncertainties in some of the plant parameters, an auxiliary control input w can be injected in (42). Define,

$$\bar{z}_{fl} = z_{fl} + \frac{w}{a_{12}a_{23}} \tag{44}$$

Using $\bar{z}_{fl}$ yields the error dynamics:

$$\ddot{\epsilon}_x + \xi_1\dot{\epsilon}_x + \xi_2\epsilon_x + \Phi(\cdot) - w = 0 \tag{45}$$

where $\Phi(\cdot)$ is the lumped uncertainty originating from the bounded uncertainties in the plant parameters, which itself is also bounded. Define another layer of sliding manifold:

$$S_w = \dot{\epsilon}_x + C_1\epsilon_x \tag{46}$$

for some $C_1 > 0$ and let

$$w = (\xi_1 - C_1)\dot{\epsilon}_x + \xi_2\epsilon_x - \tilde{M}\text{Sign}(S_w) \tag{47}$$

where $\tilde{M} > |\Phi(\cdot)|$. This auxiliary control term injection provides robustness against bounded uncertainties since

$$\dot{S}_w = \ddot{\epsilon}_x + C_1\dot{\epsilon}_x = -\Phi(\cdot) - \tilde{M}\text{Sign}(S_w). \tag{48}$$

5.4 Observer Design

Consider the following observer structure,

$$\dot{\hat{x}} = a_{12}\hat{\omega} + \ell_1 w_x \tag{49}$$
$$\dot{\hat{\omega}} = a_{21}(\hat{x} - \bar{x}_o) + a_{22}\hat{\omega} - \bar{\kappa}\text{Sign}(\hat{x} - \bar{x}_o) - \bar{\mu}\text{Sign}(\hat{\omega}) + a_{23}\hat{z} + \ell_2 w_\omega \tag{50}$$
$$\dot{\hat{z}} = a_{32}\hat{\omega} + a_{33}\hat{z} + bu + \ell_3 w_z \tag{51}$$

where "bar" denotes the nominal value of the associated variable. This observer structure yields the following error dynamics:

$$
\begin{aligned}
\dot{e}_x &= a_{12}e_\omega - \ell_1 w_x \qquad (52)\\
\dot{e}_\omega &= a_{21}e_x + a_{21}\tilde{x}_o + a_{22}e_\omega + a_{23}e_z + \cdots\\
&\quad \cdots - [\kappa\text{Sign}(x - x_o) - \bar{\kappa}\text{Sign}(\hat{x} - \bar{x}_o)] - [\mu\text{Sign}(\omega) - \bar{\mu}\text{Sign}(\hat{\omega})] - \ell_2 w_\omega\\
\dot{e}_z &= a_{32}e_\omega + a_{33}e_z - \ell_3 w_z
\end{aligned}
$$

where e_x, e_ω and e_z denote the observation error of the state variable used in the subscript. If w_x is picked as $w_x = \text{Sign}(e_x)$, and $\ell_1 > 0$ selected such that $\ell_1 > |a_{12}e_\omega|$, then $\dot{e}_x e_x < 0$ can be enforced, resulting in sliding mode on the manifold described by $e_x = 0$. The corresponding equivalent control can, then, be calculated by letting $\dot{e}_x = e_x = 0$ as,

$$[\text{Sign}(e_x)]_{eq} = \frac{a_{12}e_\omega}{\ell_1} \tag{53}$$

Let $w_\omega = \text{Sign}([\text{Sign}(e_x)]_{eq})$. Then, pick ℓ_2 such that

$$\ell_2 > a_{22}|\tilde{x}_o| + 2|\bar{\kappa}| + |\tilde{\kappa}| + 2|\bar{\mu}| + |\tilde{\mu}| + |a_{23}e_z| \tag{54}$$

where "tilde" denotes the supremum of the difference between the nominal and the actual value of the corresponding variable. This choice ensures $\dot{e}_\omega e_\omega < 0$ while resulting in finite time convergence to $e_\omega = 0$. Finally, pick

$$w_z = [\text{Sign}(e_x)]_{eq}, \qquad \text{with } \ell_3 = \frac{a_{32}}{a_{12}}\ell_1. \tag{55}$$

This choice of w_z, if the equivalent control can be calculated effectively by low pass filtering the signum function, results in

$$\dot{e}_z = a_{32}e_\omega + a_{33}e_z - \frac{a_{32}}{a_{12}}\ell_1[\text{Sign}(e_x)]_{eq} = a_{32}e_\omega + a_{33}e_z - a_{32}e_\omega = a_{33}e_z \tag{56}$$

which is stable and fast as $a_{33} = -860$. In this scheme e_x and e_ω go to zero in the sliding mode sense whereas e_z converges to zero quickly but asymptotically.

5.5 Implementation

The plant is continuous. Assume that the controller can be implemented in continuous time as well. Let the reference signal and its derivatives be known. The only measured output is the position x. The other two states are obtained from the observer. The reference signal is picked to be a sinusoidal that crosses both of the inherent discontinuity surfaces. More specifically,

$$r(t) = 10 + 3\sin(2\pi t) \tag{57}$$

in degrees. The observer and controller parameters used in the simulation are as follows: $\ell_1 = 25$, $\ell_2 = 30000$, $\ell_3 = \frac{a_{32}}{a_{12}}\ell_1 = -2880$, $\xi_1 = 600$, $\xi_2 = 90000$, $C_1 = 300$, $\tilde{M} = 20000$, $M = 13.5$.

The observer developed in Section 5.4 generates the full-state feedback information based on the approximated model. In the simulation the plant is the sophisticated nonlinear model presented in Section 5.1. On Figure (2-a) the performance of the position tracking controller is displayed. The original system is relatively hard to control as there is not much damping in the system except for the frictional terms. Because of this reason the angular velocity trajectory, subjected to discontinuous control, chatters around its reference. The control input as displayed on Figure (2-b) is purely discontinuous and bounded. As a result, the corresponding $z(t)$ is bounded and this bound is less than that of the ideally required feedback linearizing $z_{fl}(t)$. However, the magnitude of z_{fl} has robustness implications as well as control, therefore, although not closely followed by the actual $z(t)$, the tracking control objective is met. Only a slight deviation can be noticed between x and x_d at the intervals when both of the discontinuities collaborate to oppose the motion in the direction of the reference signal.

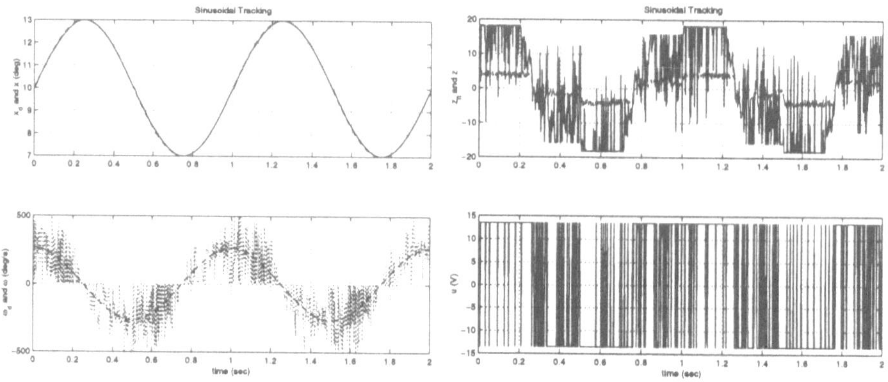

(a) (b)

Fig. 2. Continuous Plant, Continuous Controller, Reference Known: (a) The reference (dashed) and actual position and angular velocity signals, (b) The corresponding feedback linearizing $z_{fl}(t)$ (dashed), actual $z(t)$ and the generated control input u(t).

In the case that the controller/observer structure studied in the previous section can not be implemented in continuous time due to the need for utilizing a digital controller, some other major challenges are introduced into the control problem. The finite sampling rate T, no matter how small, introduces extra uncertainty in-between two consecutive samples. Moreover, the measurement is quantized and the observer has to be implemented in discrete time. Additionally, although the reference signal itself is known, it is likely to be generated by an independent higher level supervisory controller, thus, can assume arbitrary signal shapes. Observers need to be designed to estimate the derivative(s) of this reference signal so that we can use the designed VSC.

Position measurement is taken at every T seconds and a reference signal $x_r(t)$ is externally input to the system at the same frequency. A Kalman filter type estimator for the derivative term may be designed to estimate $\dot{x}_r$ and $\ddot{x}_r$ in discrete time. The final observer is implemented in the form:

$$\begin{bmatrix} \hat{x}_r[k+1] \\ \hat{\dot{x}}_r[k+1] \end{bmatrix} = \begin{bmatrix} A_{11} & A_{12} \\ A_{21} & A_{22} \end{bmatrix} \begin{bmatrix} \hat{x}_r[k] \\ \hat{\dot{x}}_r[k] \end{bmatrix} + \begin{bmatrix} B_1 \\ B_2 \end{bmatrix} x_r[k] \tag{58}$$

where $\hat{x}_r$ and $\hat{\dot{x}}_r$ represent the estimates of the reference (in degrees) and its derivative (in degrees/seconds) respectively. The other variable $x_r[k]$ (in degrees) is the actual reference signal to be tracked sampled at T intervals. This estimator introduces a phase lag in the estimate of the derivative signal which is not significant for relatively low frequency reference signals. The double derivative of the original signal can be estimated similarly by inputting the estimated derivative to an identical estimator, however, as will be justified next, double derivative information does not have a major contribution onto the overall control process for the choice of sliding manifolds and the controller can be simplified so as not to use this piece of information but rather by treating it as one component of a lumped uncertainty. The following observer parameters were used: $A_{11} = 0.7997$, $A_{12} = 0.0018$, $A_{21} = -10.0398$, $A_{22} = 0.9896$, $B_1 = 0.2003$, $B_2 = 10.0398$, $\hat{x}[0] = x[0]$, $\hat{\dot{x}}[0] = 0$.

The sampling frequency is fast ($f = 500$ Hz). By applying peak values of $u = \pm 13.5$ Volts in-between two sampling instants however, the switching control still can not approximate all low frequency signals (corresponding to the uncertainties within the system). Therefore, the observer needs to be modified and -if possible- the dependence of the controller on some state variables have to be dropped.

The dynamics associated with the signal $z(t)$ is fast and the coupling with $\omega(t)$ is very weak. Capitalizing on this fact and using singular perturbation methods, one can conclude $u \approx -\frac{a_{33}}{b} z$. Singular perturbation approximation is valid for continuous right hand sides (i.e. continuous control input case). It will be described, next, that for sampled data implementation, the discontinuous terms are replaced with nonlinearities as shown on Figure 3 which have a wide linear region. For all practical purposes, the above approximation is valid within this linear operating region. Once this assumption is made, the approximate feedback linearizing control input can be derived as

$$u_{fl} = -\frac{a_{33}}{b} z_{fl}. \tag{59}$$

This eliminates the sliding mode dynamics associated with the manifold S_z from the loop as u_{fl} can now be directly generated from (59). Consider,

$$\begin{aligned} u_{fl} = -\frac{a_{33}}{a_{12}a_{23}b} & [\ddot{x}_r + \xi_1 \dot{x}_r + \xi_2 x_r + a_{12}a_{21}x_o - (a_{12}a_{21} + \xi_2)x - (a_{22} + \xi_1)\dot{x} + \cdots \\ & \cdots + a_{12}\kappa \mathrm{Sign}(x - x_o) + a_{12}\mu \mathrm{Sign}(\dot{x}) + w]. \end{aligned} \tag{60}$$

The feedback linearizing control law enforces the very fast error dynamics whose rate of convergence is described by ξ_1 and ξ_2. Note that when u_{fl} manages to keep the error dynamics close to zero, the terms $\ddot{x}_r$ and $\xi_1(\dot{x}_r - \dot{x})$ become small compared to the discontinuous terms' gains when the tracking error performance goal is set to keep the error standard deviation within 0.1 degrees of the reference. If those signals were readily available, they would have been kept in the loop. However, their estimation introduces extra uncertainty which is not any easier to handle robustly in the control loop. The other terms x and x_d are not omitted due to their immediate availability despite little contribution in the determination of u_{fl}. The externally inputted auxiliary discontinuous input $w(t)$ should be capable of compensating for the additional uncertainties.

The superiority of the continuous time design comes from the infinite frequency switching capability which ensures approximation of unknown disturbances satisfying matching conditions with sufficiently large control authority and hence their rejection. Discrete time sliding mode control suffers chattering with purely discontinuous control. Normal is to calculate the equivalent control when on sliding mode and input that to the system which guarantees system trajectories to be on the sliding manifold at the next sampling instant. In-between two consecutive samples trajectories leave the surface but remain bounded. The equivalent control u_{eq} is obtained by letting $S[k+1] = 0$ for $S[k] = 0$. This method typically requires full state information availability. For this application it has been concluded, however, that $f = 500$ Hz sampling rate is fast enough, so the continuous time controller was discretized and then used for reference tracking. The signum function has been replaced with a saturation that has a smaller magnitude of discontinuity around zero. Saturation is utilized to avoid significant chattering and some discontinuity is kept in the loop to maintain some robustness (though reduced) with respect to uncertainties in the plant.

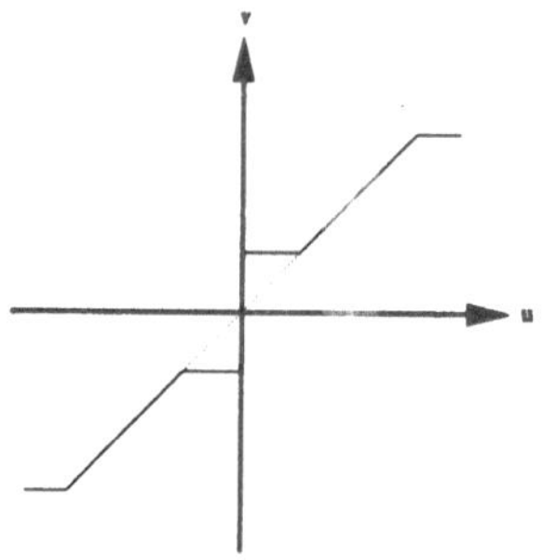

Fig. 3. The implemented nonlinearity for the sampled data system (instead of the signum function) that alleviates chattering to an extent.

In the simulations, the complex nonlinear system as given in (1) is used. To analyze the controller performance, assume that $\omega[k]$ and $x[k]$ are both available. On Figure 5 various simulation results are displayed based on this assumption. First of all, the sinusoidal tracking performance for three different references of distinct frequencies, all of which cross the discontinuity surfaces are displayed on Figure 5-a, and the generated control inputs to achieve the presented performances on Figure 5-b. Figure 5-c demonstrates robustness of the controller with respect to variations in the stiff position feedback center location. A 20% variation does not alter the closed loop system performance much. The discontinuous content of the control input on Figure 5-d manages to counter the introduced uncertainty effectively. Figures 5-e,f display the controller success at tracking various different reference signals. The apparent overshoot may be reduced/eliminated by low pass filtering the reference signal which spikes at the discontinuities of the reference signal. The plant has limited bandwidth and cannot respond too quickly.

The simulation results displayed on Figure 5 assume the availability of $\omega[k]$, which is not measured. In the continuous counterpart of the design, a robust sliding mode observer was proposed to estimate the state variables based on discontinuity surface devised on an equivalent control approach. It is observed that, discretization process affects the performance of the observer more severely for that matter. The equivalent control utilized in the observer was obtained by low pass filtering its argument. When the observer is discretized at T intervals, its bandwidth is largely reduced. Then, the equivalent control has to be obtained by means of a low pass filter with slower dynamics. Over filtering results in the loss of significant information embedded in the actual equivalent control signal regarding the error in the angular velocity estimation. As a result, the estimation of $\omega[k]$ involves estimation error. It is acknowledged that a better observer structure may be designed within the context of sampled data systems which can improve the performance of closed loop system. Here the proposed continuous time observer is implemented with bounded discrete integrators and the signum functions are replaced with saturations as shown on Figure 3. The achieved performance is displayed on Figure 4. Although tracking error is increased in comparison to the simulation results revealed on Figure 5-a, its standard deviation remains within 0.1 degrees. Without any measure of derivative information, the closed loop system chatters significantly around the reference.

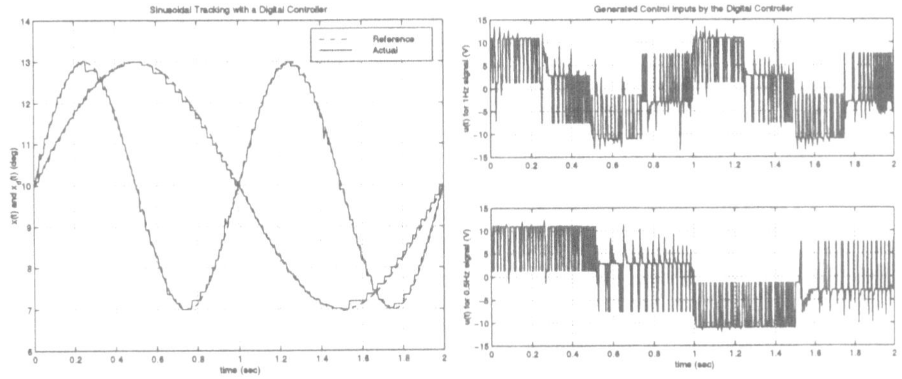

(a) (b)

Fig. 4. (a) The reference and the actual position signals for two different sinusoidals, (b) The corresponding control inputs generated by the digital controller.

6 Conclusions

Most nonlinear system analysis and control design techniques assume that the motions of a given system can be described in smooth vector fields satisfying Lipschitzian conditions. This assumption is often violated in systems involving non-smooth terms in their state space representations. Stick-slip friction behavior, for example, can not be modeled properly without the use of a signum operator around the zero velocity manifold (which violates Lipschitz continuity). Such plants describe a class of systems that typically induce "stiction" and "limit cycles" when subjected to conventional feedback control laws as the non-smooth terms on the state space representation occasionally dominate the trajectory evolution. It is observed that such anomalies introduced into control are primarily caused by the discontinuous and/or non-smooth terms in system's state space representation: Some stiction manifolds may be induced by the discontinuity existing in the original system state space description. Hence, some states may get stuck at some open region in $\mathbb{R}^n$.

The proposed variable structure control law involves the formulation of a feedback linearizing control law and design of a manifold to attain that desired dynamics by means of sliding mode control. The uncertainties in the system parameters are handled by means of an auxiliary discontinuous control injection into the feedback linearizing control term. Another layer of sliding manifold and a switching law on that reduced order manifold is defined so as to compensate for the bounded lumped uncertainty stemming from the difference between the actual and nominal plant parameters. This injection procedure may be carried out finitely many times until the last manifold depends on the error only. This scheme requires the discontinuous terms in the manifolds be replaced with uniformly convergent continuous approximations or their corresponding equivalent values to avoid having infinite derivative at the output. A controller/observer design procedure has been proposed for a class of systems in their companion forms. It should be noted, however, that feedback linearizing terms can be sought for a more generalized class of systems by means of reversible transformations that put the system in a normal form (or Fliess's generalized controller canonical form if the input-output relative degree of the system equals to the order of the system) where similar design procedures can be carried out and then transformed back into the original coordinated. Differential algebraic methods using Gröbner Bases for the polynomial equations with lexicographic term ordering yields a transformation method to obtain the associated normal form. Systems with discontinuous terms have to be replaced with their continuous approximations which can be represented as solutions to polynomial differential equations. This procedure may increase the order of the system significantly making the solution procedure tougher depending on the number of discontinuous terms. Moreover, the attained system is to exhibit different time-scale dynamics which need to be further investigated from redundancy and convergence stand points. The symbolic mathematical software program MAPLE© has Gröbner bases toolbox where normalization procedure can be effectively carried out.

In summary, this study focuses on the analyses of stiction phenomenon from a different perspective and its handling within the framework of variable structure controllers for continuous time systems. Observer couplings and controller linearity become crucial to avoid significant chattering in discrete time systems especially if the system bandwidth is close to the sampling frequency. Investigation of the proposed controller/observer design techniques in discrete time needs further attention.

References

1. B.H. Armstrong, "A survey of models, analysis tools and compensation methods for the control of machines with friction", *Automatica*, 30(9), 1994, pp. 1083-1138.
2. B.H. Armstrong and B. Amin, "PID control in the presence of static friction: a comparison of algebraic and describing function analysis", Automatica, vol. 32, no. 5, 1996, pp. 679-692.

3. M. Artin, *Algebra*, Prentice Hall, New Jersey, NY, 1991.
4. C. Canudas de Wit, H. Olson, K.J. Astrom and P. Lischinsky, "A New Model for Control of Systems with Friction", *IEEE Transactions on Automatic Control*, vol. 40, No. 3, March 1995.
5. P.E. Dupont and E.P. Dunlap, "Friction modeling and PD compensation at very low velocities", Journal of Dynamic Systems, Measurement and Control, vol. 117, no. 1, 1995, pp. 8-14.
6. P.E. Dupont, "The effect of friction on the forward dynamics problem", The Int. Jour. of Robotics Research, vol. 12, no. 2, 1993, pp. 164-179.
7. S. Drakunov, D. Hanchin, W. C. Su, Ü. Özgüner, "Nonlinear control of a rod-less pneumatic servo-actuator or sliding modes vs coulomb friction", *Automatica*, Vol 33, No 7, 1997, pp. 1401-1406.
8. A.F. Filippov, "Differential equations with discontinuous right hand side", *American Mathematical Society Translations*, vol 42, 1964, pp. 199-231.
9. M. Fliess, "Generalized controller canonical forms for linear and nonlinear dynamics", *Trans. Aut. Control*, AC-35(9), 1990, pp. 994-1001.
10. M. Fliess, H. Sira-Ramirez, "A module theoretic approach to sliding mode control in linear systems", Proc. of the 32nd CDC, San Antonio, TX, 1993, pp. 2465-2470.
11. M. Fliess, M. Hazewinkel, Eds., *Algebraic and Geometric Methods in Nonlinear Control Theory*, Dordrecht, The Netherlands: Riedel, 1986.
12. H. Fortell, "A generalized normal form and its applications to sliding mode control", Proc. of the 34th CDC, New Orleans, LA, 1995, pp. 13-18.
13. B. Friedland and Y.J. Park, "On adaptive friction compensation", IEEE Trans. on Automatic Control, vol. 37, no. 10, 1992, pp. 1609-1611.
14. D.A. Haessig and B. Friedland, "On the modeling and simulation of friction", Journal of Dynamic Systems, Measurement and Control, vol. 113, no. 3, 1992, pp. 354-362.
15. İ. Haskara, Ü. Özgüner, V.I. Utkin, "On variable structure observers", Proc. VSS '96, Tokyo, Japan, 1996, pp. 193-198.
16. C. Hatipoğlu, "Variable Structure Control of Continuous Time Systems Involving Non-smooth Nonlinearities ", Dissertation, The Ohio State University, Department of Electrical Engineering, 1998.
17. A. Isidori, *Nonlinear Control Systems*, Springer-Verlag, second edition, 1989.
18. B.E. Karnopp, "Computer simulation of slip stick friction in mechanical dynamic systems", ASME Jour. of Dynamic Systems, Measurement and Control, vol.107, no. 1, 1985, pp. 100-103.
19. H.K. Khalil, *Nonlinear Systems*, MacMillan Publishing Company, 1992.
20. H. Sira-Ramirez, "On the dynamical sliding mode control of nonlinear systems", *Int. Journal of Control*, 57(5), 1993, pp. 1039-1061.
21. S. Sankaranarayanan and F. Khorrami, "Adaptive Variable Structure Control and Applications to Frequency Compensation", *preprint*.
22. S.C. Southward, C.J. Radcliffe, C.R. MacCluer, "Robust nonlinear stick-slip friction compensation", *ASME JDMSC*, Vol 113, 1991, pp. 639-645.
23. S. Tafazoli, C.W. de Silva and P.D. Lawrence, "Tracking control of an electrohydraulic manipulator in the presence of friction", *IEEE Trans. on Control Systems Technology*, Vol 6, No 3, pp. 401-411, May 1998.
24. V.I. Utkin, *Sliding Modes in Control Optimization*, Springer-Verlag, 1992.
25. K.D. Young "Discontinuous control of sliding base isolated structures under earthquakes", Proc. of IMACS/SICE Int. Symp. on Robotics, Mechatronics and Manufacturing Systems, Kobe Japan, 1992, pp.375-380.

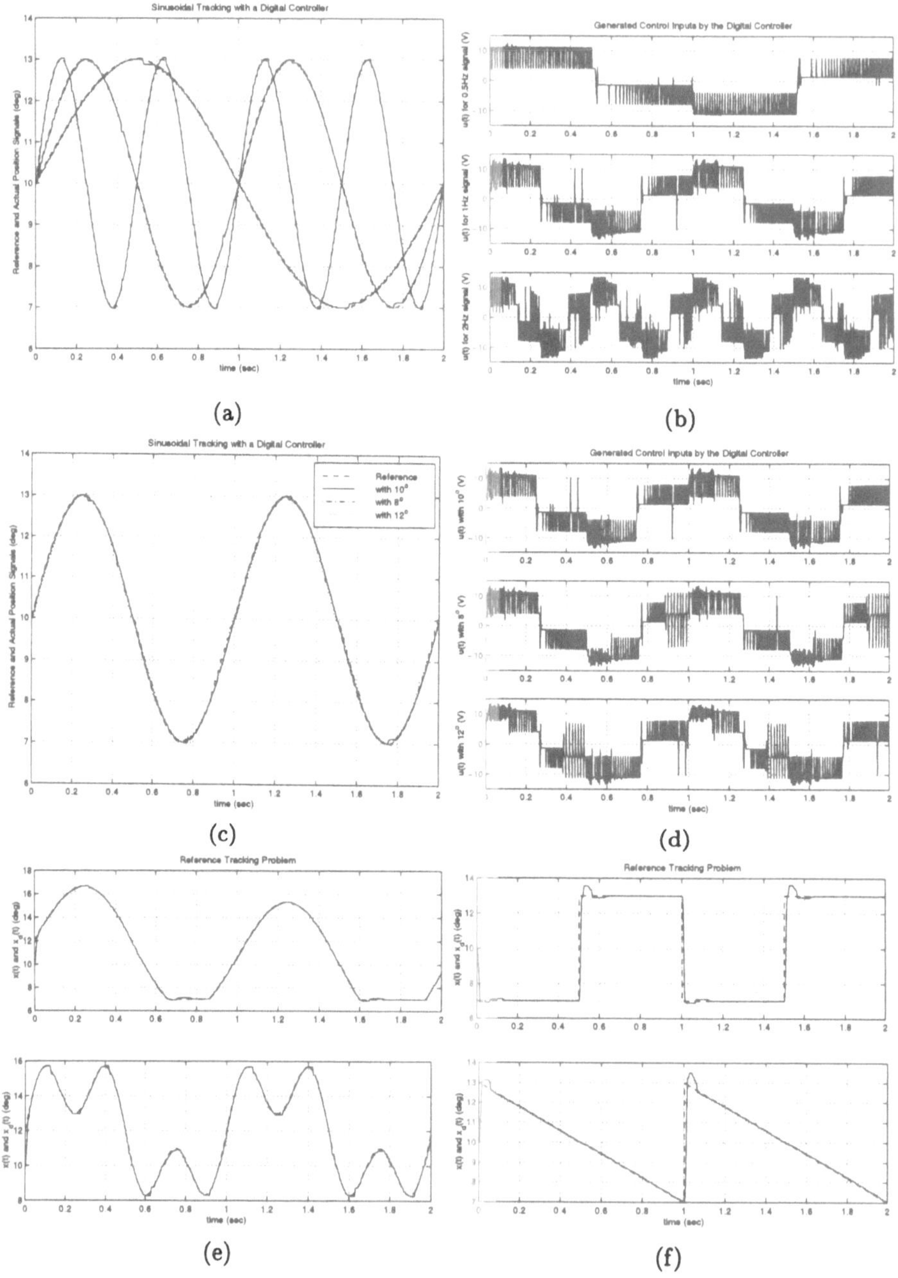

Fig. 5. (a) The reference and the actual position signals for three different sinusoidals, (b) The corresponding control inputs generated by the digital controller to achieve the tracking performance in (a), (c) Tracking performance for 20% variation in θ_o position, (d) The corresponding control inputs for (c), (e) and (f) Tracking performances for different reference signals.

Equivalent Value Filters in Disturbance Estimation and State Observation

İbrahim Haskara and Ümit Özgüner

The Ohio State University,
Department of Electrical Engineering,
2015 Neil Ave. 205 Dreese Labs,
Columbus, OH 43210

Abstract. In this paper, the implementation of the equivalent value operators by first order, high bandwidth, low pass filters is considered. An ultimate boundedness proof for the estimation/observation errors when the filters are used sequentially in the equivalent control based disturbance estimator and state observer designs is provided. A rule of thumb for the selection of filter time constants is also given. Initial results on this selection suggest the creation of a multiple time scale behavior since each of the filter time constants is to be as small as possible.

1 Introduction

Variable Structure Control (VSC) is one of the well-known robust control design approaches ([6], [7]). A standard VSC controller is designed in two stages. First, an appropriate lower order sliding manifold whose dimensionality is the same as that of the control vector is selected. The motion constrained on this manifold is to be prespecified. Second, a suitable control law is determined to steer the trajectories onto the manifold in finite time and to keep the subsequent motion on it. The resulting motion is called sliding mode.

The equivalent control methodology ([6], [7]) has been introduced originally as a regularization technique to analyze the sliding motion for systems affine in the control. It has been shown that the traditional discontinuous control switches in sliding mode so as to imitate the equivalent control in the average which can be broadly defined as the continuous control which would lead to the invariance conditions for the sliding motion. According to the equivalent control methodology, the slow component of the discontinuous control is actually the equivalent control and this component can be extracted by passing the discontinuous control through a low pass filter whose time constant is sufficiently large to filter out the high frequency switching terms, yet sufficiently small not to eliminate any slow component [7].

The equivalent control idea has also been used extensively in the design of disturbance estimators and state observers ([2], [3], [4], [5], [7]). In these designs, the overall problem is solved by a step-by-step procedure where the equivalent value operators provide the coupling between the two steps and first order, high bandwidth, low pass filters are used to implement the resulting sequential equivalent value operators based on the approximability property of the equivalent control by low pass filtering [7]. In this study, our main objective is to analyze the low pass filtering more closely in the disturbance estimator and the state observer designs as well as providing a possible selection guide for the filter time constants. To this end, we first present ideal illustrative designs for both the equivalent control based disturbance estimation and state observation problems as developed in the references cited herein and then perform an ultimate boundedness analysis by tracing the propagation of possible low pass filtering approximation errors through the sequential derivation. Utilizing the proof on the approximability of the equivalent control by a low pass filter ([7]) at each step, we demonstrate a relation between the estimation accuracy and the selection of filter time constants as well as the sampling time.

2 Equivalent Control Based Disturbance Estimation

Consider the following scalar, nonlinear system

$$\dot{x} = f(x) + g(x)u + d(x, u, t) \tag{1}$$

where $x \in \mathbf{R}$ is the state, $u \in \mathbf{R}$ is the control, $f(x)$ and $g(x)$ are sufficiently smooth, known functions over the domain of interest and $d(x, u, t)$ is an unknown disturbance function which lumps all possible uncertainties of the system. The disturbance function is assumed to be bounded, i.e;

$$|d(x, u, t)| \leq \rho(t) \tag{2}$$

where $\rho(t)$ is known and bounded for all $t \geq 0$. The problem considered is to estimate the disturbance and possibly its several time derivatives using the available state x. As in the sliding mode observer design, the nominal system is rewritten with an additional auxiliary input term:

$$\dot{\hat{x}} = f(x) + g(x)u + L\ sgn\ (x - \hat{x}) \tag{3}$$

where $L \in \mathbf{R}$ is the free design parameter to be chosen. Note that, $f(x) + g(x)u$ represents the nominal part which is available since the state and the control are known. Subtracting Eq. 3 from Eq. 1, the following error dynamic is produced.

$$\dot{e}_x = d(x, u, t) - L\ sgn\ e_x \tag{4}$$

where $e_x = x - \hat{x}$. Selecting $L = \rho(t) + \eta$ with $\eta > 0$, e_x converges and stays at zero ideally after a finite time. Furthermore, the selection of $\hat{x}(0) = x(0)$ eliminates the reaching phase. Using the equivalent control methodology, although the additional auxiliary input term is completely discontinuous, its equivalent value is equal to the disturbance on this manifold, i.e;

$$[\ L\ sgn\ e_x\ ]_{eq} = d(t) \tag{5}$$

Now, assume that the disturbance is also differentiable and its derivative is bounded by a known function, i.e;

$$\dot{d}(t) = D(t) \tag{6}$$

where $|D(t)| \leq \bar{\rho}(t)$. Selecting

$$\dot{\hat{d}}(t) = K \; sgn \; (d - \hat{d}) \tag{7}$$

$$\dot{\hat{d}}(t) = K \; sgn \; (\; [\; L \; sgn \; e_x \;]_{eq} - \hat{d} \;) \tag{8}$$

as the disturbance derivative estimator where $K = \bar{\rho}(t) + \kappa$ with $\kappa > 0$, the estimation error $d(t) - \hat{d}(t)$ converges to zero in finite time and after that one gets

$$[\; K \; sgn \; (\; [\; L \; sgn \; e_x \;]_{eq} - \hat{d}) \;]_{eq} = D(t) \tag{9}$$

Note that the design is sequential and it can ideally be continued to estimate higher order derivatives of the disturbance as long as the disturbance is continuously differentiable to some order. However, there are basically two problems which need to be addressed: First, a possible discrete time implementation of the signum functions may prevent the existence of an ideal sliding motion. Second, the equivalent value operators cannot be implemented ideally. Since these operators are used to pass information from one step to the next one, an implementation error of the current step can affect not only the accuracy of this step but also that of all the subsequent ones.

To address the possible implementation problems, suppose that $|\hat{e}_x - e_x| \leq \Delta$ where $\hat{e}_x$ denotes the available information on e_x and $\Delta \geq 0$ is the measurement error bound. If the disturbance estimator (Eq. 3) is implemented with a finite switching frequency using $\hat{e}_x$ instead of e_x the error variable satisfies $|e_x(t)| \leq \mathcal{O}(T) + \Delta$ after a finite time where T is the sampling time (note that, the effect of a discrete time implementation has been expressed as a sampling time order deviation). Theorem 1 states the main result.

Theorem 1. *Consider the following set of equations*

$$\begin{aligned} \dot{s} \quad &= h_1(t) - L_1 \; sgn \; s \\ \tau_1 \dot{v}_1 + v_1 &= L_1 \; sgn \; s \\ \tau_2 \dot{v}_2 + v_2 &= L_2 \; sgn \; [\; v_1 - \hat{h}_1(t) \;] \\ &\vdots \\ \tau_r \dot{v}_r + v_r &= L_r \; sgn \; [\; v_{r-1} - \hat{h}_{r-1}(t) \;] \end{aligned} \tag{10}$$

where $h_1(t)$ is r^{th} order differentiable, i.e; $|h_i(t)| \leq H_i$ where $\dot{h}_i(t) = h_{i+1}(t)$ for $i = 1, \cdots, r-1$ and

$$\begin{aligned} \dot{\hat{h}}_1(t) \quad &= L_2 \; sgn \; [\; v_1 - \hat{h}_1(t) \;] \\ \dot{\hat{h}}_2(t) \quad &= L_3 \; sgn \; [\; v_2 - \hat{h}_2(t) \;] \\ &\vdots \\ \dot{\hat{h}}_{r-1}(t) &= L_r \; sgn \; [\; v_{r-1} - \hat{h}_{r-1}(t) \;] \end{aligned} \tag{11}$$

Suppose that Eq. 10-11 are implemented with a sampling time of T and $L_i > H_i$ for $i = 1, \cdots, r$ using $\hat{s}$ instead of s where $|s - \hat{s}| \leq \Delta$. Then,

$$\lim_{\substack{T,\Delta/\tau_1\tau_2\tau_3\cdots\tau_i \to 0 \\ \tau_k/\tau_{k+1}\tau_{k+2}\cdots\tau_i \to 0 \quad k=1,\cdots,i-1 \\ \tau_i \to 0}} v_i(t) = h_i(t) \qquad \text{for } t \gg 0 \tag{12}$$

for $i = 1, \cdots, r$.

Lemma 2. *Let $\tau_{r-k} = \epsilon^{2^k}$ for $k = 0, 1, \cdots, r-1$ and select ϵ such that*

$$\epsilon^{2^r} > T, \Delta \tag{13}$$

Then

$$\lim_{\epsilon \to 0} v_i(t) = h_i(t) \quad \text{for } t \gg 0 \tag{14}$$

for $i = 1, \cdots, r$.

Proof. Consider

$$\dot{v}_1(t) = -\frac{1}{\tau_1} v_1(t) + \frac{1}{\tau_1}[h_1(t) - \dot{s}(t)] \tag{15}$$

Since $L_1 > H_1$, there exist a finite time t_0 such that $|s| \leq \mathcal{O}(T) + \Delta$ for $t \geq t_0$ provided that T is sufficiently small. As in the verification proof of the extraction of the equivalent value of a discontinuous signal by low pass filtering ([7]), integration by part yields

$$v_1(t) = v_1(t_0)e^{-(t-t_0)/\tau_1} + \frac{1}{\tau_1}\int_{t_0}^{t} e^{-(t-\gamma)/\tau_1}[h_1(\gamma) - \dot{s}(\gamma)]d\gamma \tag{16}$$

$$|v_1(t) - h_1(t)| \leq |v_1(t_0) - h_1(t_0)|\; e^{-(t-t_0)/\tau_1} + \tau_1 H_2 + \frac{3}{\tau_1}[\mathcal{O}(T) + \Delta] \tag{17}$$

for all $t \geq t_0$. Using the same idea, first consider the following error equation

$$\frac{d}{dt}[h_1 - \hat{h}_1] = h_2 - L_2\; sgn\,(\; h_1 - \hat{h}_1 + v_1 - h_1\;) \tag{18}$$

Note that $|v_1(t) - h_1(t)|$ is bounded according to Eq. 17 and the bound decreases exponentially to a constant with a convergence rate determined by the filter time constant. Although the convergence initially might be very fast one can select a sufficiently large L_2 so that time dependence effect of the bound is overcome at least after a finite time. Therefore there always exists a $t_1 > t_0$ such that

$$|h_1(t) - \hat{h}_1(t)| \leq \mathcal{O}(T) + \delta_1(t) \tag{19}$$

for all $t \geq t_1$ where $\delta_1(t)$ denotes the bound for $|v_1(t) - h_1(t)|$. Using the previous result, one gets

$$|v_2(t) - h_2(t)| \leq |v_2(t_1) - h_2(t_1)|e^{-(t-t_1)/\tau_2} + \frac{3}{\tau_2}|v_1(t_0) - h_1(t_0)|e^{-(t_1-t_0)/\tau_1}$$
$$+\tau_2 H_3 + 3H_2\frac{\tau_1}{\tau_2} + \frac{9}{\tau_1\tau_2}[\mathcal{O}(T) + \Delta] + \frac{3}{\tau_2}\mathcal{O}(T) \tag{20}$$

for all $t \geq t_1$. Proceeding in this manner, one can easily find an error bound expression for the error between the ith filter output and $h_i(t)$ for $t \geq t_{i-1}$ where t_k is the time instant after which the following condition is satisfied

$$|h_k(t) - \hat{h}_k(t)| \leq \mathcal{O}(T) + \delta_k(t) \tag{21}$$

and $\delta_k(t)$ denotes the ultimate error bound for $|v_k(t) - h_k(t)|$. Although the ultimate error bound expressions seem complicated, they have a certain pattern which can be summarized as follows:

$$\begin{aligned}|v_i(t) - h_i(t)| \leq\ & |v_i(t_{i-1}) - h_i(t_{i-1})|\, e^{-(t-t_{i-1})/\tau_i} \\ & +\mathcal{O}(\tau_i) + \mathcal{O}(\tau_{i-1}/\tau_i) + \cdots + \mathcal{O}(\tau_k/\tau_{k+1}\tau_{k+2} \\ & \cdots\tau_{i-1}\tau_i) + \cdots + \mathcal{O}(\tau_1/\tau_2\tau_3\cdots\tau_i) \\ & +\mathcal{O}(T/\tau_1\tau_2\cdots\tau_i) + \mathcal{O}(\Delta/\tau_1\tau_2\cdots\tau_i)\end{aligned} \tag{22}$$

for all $t \geq t_{i-1}$ and $i = 1, \cdots, r$ (note that, only the dominant terms are written in Eq. 22). For $t < t_{i-1}$, $v_i(t)$ will not necessarily be an accurate estimate for $h_i(t)$. However, since the filter input is either $\pm L_i$, $v_i(t)$ and $|v_i(t) - h_i(t)|$ are both bounded. Furthermore, each t_{i-1} can be made arbitrarily small increasing $L_1, L_2, \cdots, L_i$.

Note that, all the residual terms vanish while the conditions of Theorem 1 are satisfied. With the proposed filter time constant selection

$$\begin{aligned}& T/\tau_1\tau_2\cdots\tau_r < \epsilon \\ & \tau_k/\tau_{k+1}\tau_{k+2}\cdots\tau_{r-1}\tau_r = \epsilon\end{aligned} \tag{23}$$

for $k = 1, \cdots, r-1$. Therefore, all the residual terms of the ultimate error bound for each $|v_i(t) - h_i(t)|$ are at most in the order of $\epsilon^{2^{r-i}}$ for $i = 1, \cdots, r$ and they all converge to zero as ϵ goes to zero.

This selection is independent of the disturbance properties, however the actual values of the error bounds obviously depend on them. When the disturbance estimator equations are implemented with a finite switching frequency, the filter time constants will be bounded from below according to Eq. 13 which shows the tradeoff. ϵ should be as small as possible to make the disturbance estimation errors smaller, however without violating the condition of Eq. 13. In this case, the low pass filters cause time lag and the filter outputs might contain some unfiltered terms in addition to the related equivalent signals. The time lag and the maximum steady state error bounds reduce as the filter time constants approach to their ideal values and shrink to zero in the ideal situation. This recovers the performance of the ideal analysis. Note that, the filter time constant selection of Lemma 2 is based on balancing all the terms in the ultimate error bound expressions and it will not necessarily be the best selection since the whole analysis is performed by the order of magnitude arguments.

3 A Sliding Mode Observer Design for Nonlinear Systems in Strict Feedback Form

Consider the following system in strict feedback form

$$\begin{aligned} \dot{x}_1 &= x_2 + f_1(x_1) \\ \dot{x}_2 &= x_3 + f_2(x_1, x_2) \\ &\vdots \\ \dot{x}_{n-1} &= x_n + f_{n-1}(x_1, x_2, \cdots, x_{n-1}) \\ \dot{x}_n &= u + f_n(x_1, x_2, \cdots, x_n) \end{aligned} \tag{24}$$

where x_i's and u are scalars, f_i's are nonlinear functions known in form and they are Lipschitz in their arguments over the domain of interest. Although, this system is clearly a very special class of nonlinear systems it has been selected just to analyze the equivalent value filtering in the equivalent control based sliding mode observer design. The problem considered is to estimate the state vector from x_1.

3.1 Observer Design

The design idea is exactly the same as the linear case [2], [3], [4], [5], [7]. The overall design problem is transformed into independent smaller order stabilization problems which share information with each other through the equivalent value filters.

Let

$$\dot{\hat{x}}_1 = \hat{x}_2 + f_1(\hat{x}_1) + L_1 \, sgn \, (x_1 - \hat{x}_1) \tag{25}$$

be the observer equation for the variable x_1 where L_1 is the free parameter to be chosen. Subtracting Eq. 25 from the first line of Eq. 24 the following error dynamic is obtained:

$$\dot{\tilde{x}}_1 = \tilde{x}_2 + f_1(x_1) - f_1(\hat{x}_1) - L_1 \, sgn \, \tilde{x}_1 \tag{26}$$

where $\tilde{x}_i = x_i - \hat{x}_i$. Let $V_1 = \tilde{x}_1^2$ be a Lyapunov function candidate for this scalar system. By a direct substitution and using $|f_1(x_1) - f_1(\hat{x}_1)| \leq \alpha_1 |\tilde{x}_1|$ one gets

$$\dot{V}_1 \leq 2|\tilde{x}_1| \, [\, |\tilde{x}_2| + \alpha_1 |\tilde{x}_1| - L_1 \,] \tag{27}$$

where α_1 is the Lipschitz constant for $f_1(\bullet)$. Choosing L_1 sufficiently large $\tilde{x}_1$ can be guaranteed to converge to zero in finite time. Since

$$\tilde{x}_1 = 0 \Rightarrow f_1(x_1) - f_1(\hat{x}_1) = 0 \tag{28}$$

the equivalent value of the discontinuous injection becomes equal to $\tilde{x}_2$ in sliding mode, i.e;

$$[L_1 \; sgn \; \tilde{x}_1]_{eq} = \tilde{x}_2 \tag{29}$$

At the next step, choosing the following observer equation for the second line of Eq. 24

$$\dot{\hat{x}}_2 = \hat{x}_3 + f_2(\hat{x}_1, \hat{x}_2) + L_2 \; sgn \; (x_2 - \hat{x}_2) \tag{30}$$

$$\dot{\hat{x}}_2 = \hat{x}_3 + f_2(\hat{x}_1, \hat{x}_2) + L_2 \; sgn \; (\; [L_1 \; sgn \; \tilde{x}_1]_{eq} \;) \tag{31}$$

one gets the following error dynamic:

$$\dot{\tilde{x}}_2 = \tilde{x}_3 + f_2(x_1, x_2) - f_2(\hat{x}_1, \hat{x}_2) - L_2 \; sgn \; \tilde{x}_2 \tag{32}$$

Since

$$|f_2(x_1, x_2) - f_2(\hat{x}_1, \hat{x}_2)| \leq \alpha_2 \; \|X_2 - \hat{X}_2\| \tag{33}$$

where $X_2 = (x_1, x_2)^T$, $\hat{X}_2 = (\hat{x}_1, \hat{x}_2)^T$ and α_2 is the related Lipschitz constant, $\tilde{x}_2$ can also be guaranteed to converge to zero in finite time selecting L_2 sufficiently large. Since

$$X_2 - \hat{X}_2 = \mathbf{0} \Rightarrow f_2(x_1, x_2) - f_2(\hat{x}_1, \hat{x}_2) = 0 \tag{34}$$

the equivalent value of the discontinuous injection $L_2 \; sgn \; (\; [L_1 \; sgn \; \tilde{x}_1]_{eq} \;)$ gives $\tilde{x}_3$ in sliding mode, i.e;

$$[L_2 \; sgn \; (\; [L_1 \; sgn \; \tilde{x}_1 \;]_{eq} \;) \;]_{eq} = \tilde{x}_3 \tag{35}$$

Continuing in this manner, at the kth step, the observer equation for the variable x_k is chosen as follows:

$$\dot{\hat{x}}_k = \hat{x}_{k+1} + f_k(\hat{x}_1, \cdots, \hat{x}_k) + \Gamma_k \tag{36}$$

where

$$\Gamma_k = L_k \; sgn \; [\; \Gamma_{k-1} \;]_{eq}$$

with $[\; \Gamma_0 \;]_{eq} = x_1 - \hat{x}_1$, $\tilde{x}_k$ is also steered to zero in finite time with a proper L_k, the equivalent value of the discontinuous input of that step gives information required for the next step and so on and so forth. Theorem 3 summarizes the result when the equivalent value operators are implemented by first order, high bandwidth, low pass filters.

Theorem 3. *Let*

$$\begin{aligned} \dot{\hat{x}}_1 &= \hat{x}_2 + f_1(\hat{x}_1) + \Gamma_1 \\ \dot{\hat{x}}_2 &= \hat{x}_3 + f_2(\hat{x}_1, \hat{x}_2) + \Gamma_2 \\ &\vdots \\ \dot{\hat{x}}_{n-1} &= \hat{x}_n + f_{n-1}(\hat{x}_1, \hat{x}_2, \cdots, \hat{x}_{n-1}) + \Gamma_{n-1} \\ \dot{\hat{x}}_n &= u + f_n(\hat{x}_1, \hat{x}_2, \cdots, \hat{x}_n) + \Gamma_n \end{aligned} \tag{37}$$

be the observer for the system given by Eq. 24 where

$$\Gamma_i = L_i \, sgn \, v_{i-1}$$

$$\tau_i \dot{v}_i + v_i = L_i \, sgn \, v_{i-1}$$

for $i = 1, \cdots, n$ *with* $v_0 = x_1 - \hat{x}_1$, τ_i*'s are the equivalent value filter time constants and* L_i*'s are sufficiently large positive constants. Suppose that the switching frequency is finite because of a discrete time implementation with a sampling time of* T. *Then*

$$\begin{aligned} &\lim_{T \to 0} \hat{x}_1(t) = x_1(t) \quad \text{and} \\ &\lim_{\substack{T/\tau_1\tau_2\tau_3\cdots\tau_{i-1} \to 0 \\ \tau_k/\tau_{k+1}\tau_{k+2}\cdots\tau_{i-1} \to 0 \;\; k=1,\cdots,i-2 \\ \tau_{i-1} \to 0}} \hat{x}_i(t) = x_i(t) \quad \text{for } t \gg 0 \end{aligned} \tag{38}$$

for $i = 2, \cdots, n$.

Proof. As in the disturbance estimation proof, we characterize the effects of the the equivalent value filtering in each estimation error variable in the form of ultimate error bounds which depend on the filter time constants. First, consider the selection of a sufficiently large L_1 while all the other parameters have already been fixed. The effect of the largeness of L_1 in the dynamics of Eq. 26 can be studied in a stretched time scale t/μ with the singular perturbation parameter $\mu = 1/L_1$. Since the sliding mode condition holds in the stretched time scale for $\mu = 0$, an ideal sliding motion will also be achieved in finite time with a sufficiently small μ (sufficiently large L_1) provided that the initial value of $\tilde{x}_1$ is bounded [7]. Furthermore, for a discrete time implementation, the sampling time T should be compatible with α_1 and L_1 so that the maximum possible bound of $|f_1(x_1) - f_1(\hat{x}_1)|$ due to the deviation of $\tilde{x}_1$ from zero during the intersampling time can be overcome by L_1. Selecting L_1 and T with the above considerations, one gets

$$|\tilde{x}_1(t)| \leq O(T) \tag{39}$$

after a finite time t_0 and the solution of

$$\dot{v}_1 = -\frac{1}{\tau_1} v_1 + \frac{1}{\tau_1} \left[\tilde{x}_2 + f_1(x_1) - f_1(\hat{x}_1) - \dot{\tilde{x}}_1 \right] \tag{40}$$

for $t \geq t_0$ by integration by parts yields

$$\begin{aligned} |v_1(t) - \tilde{x}_2(t)| \leq {} & |v_1(t_0) - \tilde{x}_2(t_0)| \, e^{-(t-t_0)/\tau_1} + \\ & \tau_1 \, |\dot{\tilde{x}}_2|_\infty + \alpha_1 \, |\tilde{x}_1|_\infty + \frac{3 \, |\tilde{x}_1|_\infty}{\tau_1} \end{aligned} \tag{41}$$

where $| \bullet |_\infty$ denotes the maximum value of its argument for $t \geq t_0$.

The ultimate boundedness of $\tilde{x}_2$ after a finite time will be proven at the second step and also notice that all $\tilde{x}_i$'s (and therefore $\dot{\tilde{x}}_i$'s) are bounded for a finite time interval since the estimator is simply a replica of the original system with additional bounded inputs $+L_i$ to each equation and the nonlinearities of the system are assumed to be Lipschitz. Therefore, all the terms at the RHS of Eq. 41 can be made arbitrarily small selecting T and τ_1 according to Theorem 3. Rewrite the error equation for x_2 as follows:

$$\dot{\tilde{x}}_2 = \tilde{x}_3 + f_2(\bullet) - f_2(\hat{\bullet}) - L_2\, sgn\,(\tilde{x}_2 + v_1 - \tilde{x}_2) \tag{42}$$

With a suitable L_2, there exists a $t_1 > t_0$ such that

$$\begin{aligned} |\tilde{x}_2(t)| &\le \mathcal{O}(T) + \delta_1(t) \\ |\tilde{x}_2(t)| &\le |v_1(t_0) - \tilde{x}_2(t_0)|\, e^{-(t-t_0)/\tau_1} + \mathcal{O}(\tau_1) + \mathcal{O}(T/\tau_1) \end{aligned} \tag{43}$$

for all $t \ge t_1$ where $\delta_1(t)$ is the ultimate bound for $|v_1(t) - \tilde{x}_2(t)|$ which is controllable by τ_1 and T. Solving

$$\dot{v}_2 = -\frac{1}{\tau_2}v_2 + \frac{1}{\tau_2}\left[\tilde{x}_3 + f_2(\bullet) - f_2(\hat{\bullet}) - \dot{\tilde{x}}_2\right] \tag{44}$$

by integration by parts for $t \ge t_1$ gives

$$\begin{aligned} |v_2(t) - \tilde{x}_3(t)| \le\ & |v_2(t_1) - \tilde{x}_3(t_1)|\, e^{-(t-t_1)/\tau_2} \\ & + \tau_2\, |\dot{\tilde{x}}_3|_\infty + \alpha_2\, |\tilde{x}_1|_\infty + \alpha_2\, |\tilde{x}_2|_\infty + \frac{3|\tilde{x}_2|_\infty}{\tau_2} \end{aligned} \tag{45}$$

where $|\bullet|_\infty$ denotes the maximum value of its argument for $t \ge t_1$ at this time and all the error terms can also be made arbitrarily small according to Theorem 3. Rewriting the error equation for x_3 as follows:

$$\dot{\tilde{x}}_3 = \tilde{x}_4 + f_3(\bullet) - f_3(\hat{\bullet}) - L_3\, sgn(\tilde{x}_3 + v_2 - \tilde{x}_3) \tag{46}$$

and using Eq. 45, one can also get

$$\begin{aligned} |\tilde{x}_3(t)| \le\ & |v_2(t_1) - \tilde{x}_3(t_1)|\, e^{-(t-t_1)/\tau_2} + \mathcal{O}(\tau_2) \\ & + \mathcal{O}(\tau_1/\tau_2) + \mathcal{O}(T/\tau_1\tau_2) \end{aligned} \tag{47}$$

for $t \ge t_2$. Continuing with this manner, all the error variables can be shown to be ultimately bounded and the final result is summarized as follows:

$$\begin{aligned} |\tilde{x}_1(t)| \le\ & \mathcal{O}(T) \\ |\tilde{x}_i(t)| \le\ & |v_{i-1}(t_{i-2}) - \tilde{x}_i(t_{i-2})|\, e^{-(t-t_{i-2})/\tau_{i-1}} \\ & + \mathcal{O}(\tau_{i-1}) + \cdots + \mathcal{O}(\tau_k/\tau_{k+1}\tau_{k+3}\cdots\tau_{i-1}) \\ & + \cdots + \mathcal{O}(T/\tau_1\cdots\tau_{i-1}) \end{aligned} \tag{48}$$

for all $t \ge t_{i-1}$ for $i = 2, \cdots, n$ and $k = 1, \cdots, i-2$. All the residual terms vanish as the conditions of Theorem 3 are satisfied. For $t < t_{i-1}$, the observer does not necessarily provide an accurate estimate for $x_i, \cdots, x_n$. However, as before, each t_{i-1} can be made arbitrarily small by properly adjusting $L_1, L_2, \cdots, L_i$.

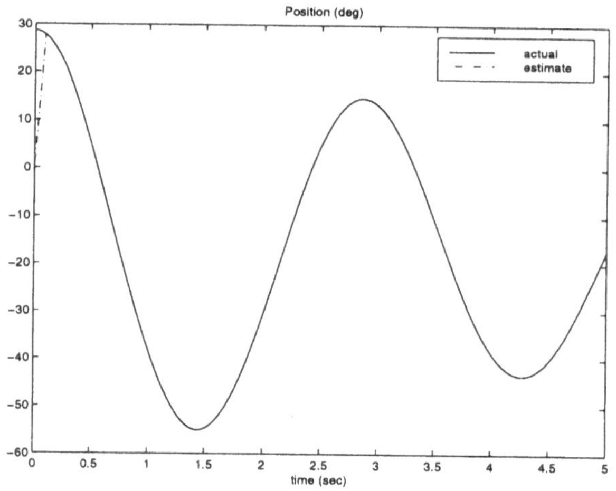

Fig. 1. Position

Lemma 4. *Let* $\tau_{n-k} = \epsilon^{2^{k-1}}$ *for* $k = 1, \cdots, n-1$ *where* ϵ *is such that*

$$\epsilon^{2^{n-1}} > T \tag{49}$$

Then

$$\lim_{\epsilon \to 0} \hat{x}_i(t) = x_i(t) \quad \text{for } t \gg 0 \tag{50}$$

for $i = 1, \cdots, n$.

As in the disturbance estimation, this selection provides all the residual terms to be at most in the order of ϵ and they all vanish as ϵ goes to zero. In a discrete time implementation, the filter time constants should be compatible with the sampling time. In this case, there might be steady state error and phase lag terms in the ultimate error bound expressions and the finite time convergence cannot be claimed any more. However, Theorem 3 guarantees that all the estimation errors can be confined to ultimately bounded boundary layers around zero after a finite time with a proper selection of the observer parameters and this behavior can be considered as the practical counterpart of the finite time convergence characteristic of the ideal observer. We also note that the results obtained are semiglobal since the observer gains should be in accordance with the initial conditions even though they can be arbitrarily large.

4 Simulation Example

In this section, the disturbance/state estimation problem is studied on a simple pendulum model (from [1]) whose motion equations are given by

$$\begin{aligned} \dot{x}_1 &= x_2 \\ \dot{x}_2 &= -\frac{g}{L} \sin x_1 - b x_2 + u - \frac{v}{L} \cos x_1 \end{aligned} \tag{51}$$

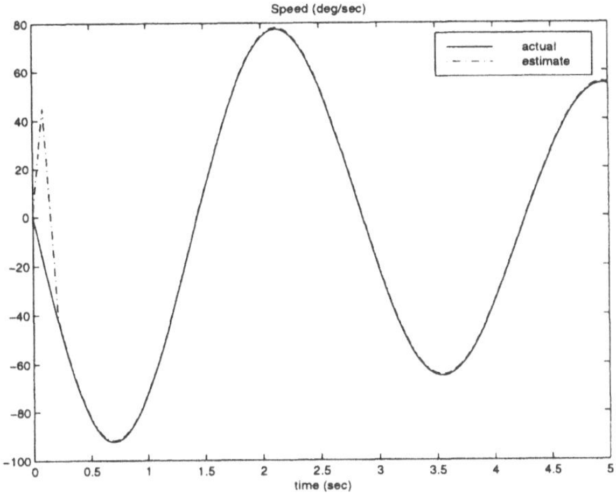

Fig. 2. Speed

where x_1, x_2 are the position and speed of the pendulum, respectively, v is the unknown horizontal acceleration of the pendulum's point of support, L is the pendulum length, b is the damping and g is the gravity constant. The problem considered is to estimate the disturbance $-(v/L)\cos x_1$ and its derivative without any information on the disturbance structure using only the position information. The state/disturbance estimator utilizes the following equations

$$\begin{aligned} \dot{\hat{x}}_1 &= \hat{x}_2 + L_1 \; sgn \; (\; x_1 - \hat{x}_1 \;) \\ \tau_1 \dot{v}_1 + v_1 &= L_1 \; sgn \; (\; x_1 - \hat{x}_1 \;) \\ \dot{\hat{x}}_2 &= -\frac{g}{L} \sin \hat{x}_1 - b\hat{x}_2 + u + L_2 \; sgn \; v_1 \\ \tau_2 \dot{v}_2 + v_2 &= L_2 \; sgn \; v_1 \\ \dot{z} &= L_3 \; sgn \; (\; v_2 - z \;) \\ \tau_3 \dot{v}_3 + v_3 &= L_3 \; sgn \; (\; v_2 - z \;) \end{aligned} \qquad (52)$$

where $\hat{x}_2$, v_2 and v_3 give estimates for the pendulum speed, the disturbance and its derivative, respectively. Figures 1-4 illustrate the simulation results for the case where the control is identically zero, $L = 2m$, $b = 0.25kgm^2/s$, $g = 9.81m/s^2$, $v = 3m/s^2$, $L_1 = 5$, $L_2 = 10$, $L_3 = 5$, $\tau_1 = 100\mu s$, $\tau_2 = 0.01s$, $\tau_3 = 0.03s$ and the simulation step size is selected as 1 μs. Note that, the disturbance and its derivative (Figure 3-4) are estimated with certain accuracy where the estimation errors are mainly due to the finite switching frequency and the non-ideal implementation of the equivalent value operators.

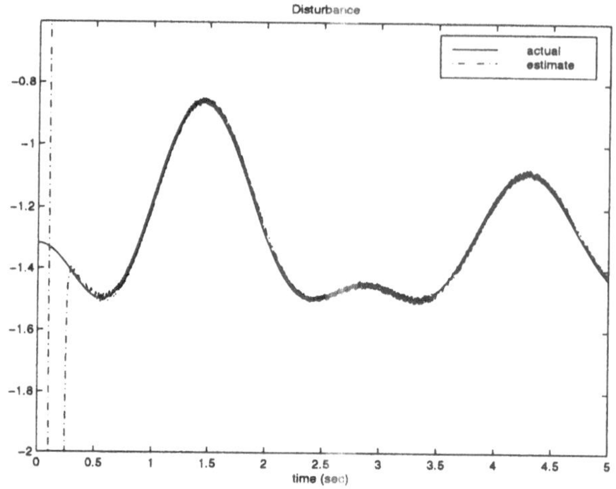

Fig. 3. Disturbance

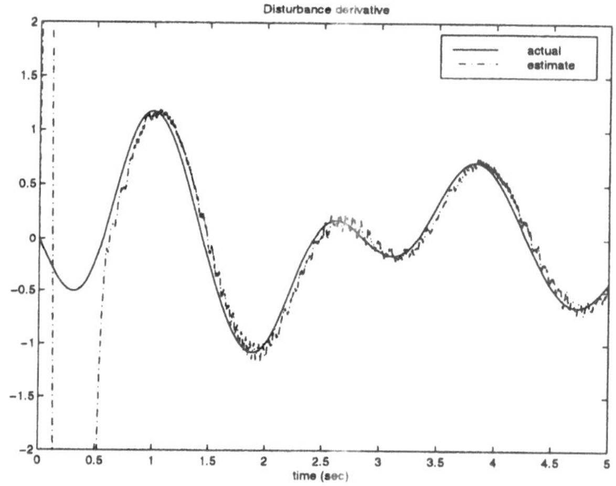

Fig. 4. Disturbance derivative

5 Conclusions

The implementation of the equivalent value operators by low pass filters has been analyzed. The filters have been used sequentially to transfer information between two consecutive steps for the disturbance estimation and the state observation problems. A guide for the selection of the filter time constants has also been provided. It has been shown that the estimation/observation accuracy strongly depends on the sampling time as well as the absolute and relative values of the filter time constants.

References

1. Corless, M.J. and G. Leitmann, "Continuous State Feedback Guaranteeing Uniform Ultimate Boundedness for Uncertain Dynamic Systems", *IEEE Trans. Automat. Contr.*, Vol. AC-26, No. 5, pp. 1139-1144, 1981.
2. Drakunov, S.V. and V.I. Utkin, "Sliding-Mode Observers. Tutorial", *Proc. of the 34th IEEE Conference on Decision and Control (CDC)*, New Orleans, LA, pp. 3376-3378, 1995.
3. Hashimoto, H., V.I. Utkin, J.X. Xu, H. Suzuki and H. Harashima, "VSS Observers for Linear Time Varying Systems", *IECON'90*, pp. 34-39, 1990.
4. Haskara, İ, Ü. Özgüner and V.I. Utkin, "On Variable Structure Observers", *Proc. of 1996 IEEE International Workshop on Variable Structure Systems*, Tokyo, Japan, pp. 193-198, 1996.
5. Haskara, İ, Ü. Özgüner and V.I. Utkin, "On Sliding Mode Observers via Equivalent Control Approach", *Int. J. Control*, Vol 71, No 6, pp. 1051-1067, 1998.
6. Utkin, V.I., "Variable Structure Systems with sliding modes", *IEEE Trans. Automat. Contr.*, Vol. AC-22, No. 2, pp. 212-222, 1977.
7. Utkin, V.I., *Sliding Modes in Control and Optimization*, Springer-Verlag, 1992.

Sliding Mode Control of a Car-Like Mobile Robot Using Single-Track Dynamic Model

A. Stotsky X. Hu

Control Engineering Laboratory
Department of Signals and Systems
SE - 412 96 Gothenburg, Sweden
e-mail: stotsky@s2.chalmers.se

Optimization and Systems Theory
Royal Institute of Technology
SE - 100 44 Stockholm, Sweden
e-mail: hu@math.kth.se

Abstract. An easy-to-implement global trajectory tracking algorithm in the class of variable structure systems is designed and analyzed within the single-track dynamical model framework. The algorithm uses measurements of the position, orientation and the yaw rate of the robot and is robust with respect to the parameter variations.

1 Introduction

Advanced industrial robot applications require robust trajectory tracking algorithms for mobile robots. Although this has been a well studied topic, see for example, [3], [5] and [11], many results are basically focused on a local stabilizing feedback design. Global trajectory tracking algorithms were proposed in [7] and [4], where the appropriate velocity control law and the steering angle commands were found. However, in those papers the controls are complex and not intuitive and the analysis is based on a kinematic model for mobile robot. It is well known that for high enough velocities the car motion can not be described accurately by kinematic models and one should use dynamical models to design accurate trajectory tracking algorithms. In this paper instead of using a traditional kinematic car model, we actually work with a dynamical model [1],[9], which is, of course, more realistic and includes orientation angle, yaw rate and a sideslip angle as state variables.

The main contribution of the paper is the following: we propose a straightforward and intuitive global trajectory tracking algorithm within the single-track dynamical model framework, using measurements of the position, orientation, and yaw rate of the vehicle under unknown vehicle parameters.

The idea for the algorithm design is very simple and uses the preview scheme. First of all we define the virtual point on the desired path ahead of the vehicle and design control algorithm in the class of variable structure systems so that the velocity of the vehicle is directed always (after short transient) towards the virtual point, which is moving on the curve. The second stage of the algorithm design is to devise a global update law for the virtual point. The key idea here, is to present the derivative of the distance between the vehicle and the virtual point in terms of the desired orientation of the robot velocity vector, velocity of robot and the velocity of the virtual point. Then via a proper choice of the velocity of vehicle and the velocity of the virtual point we drive the distance between the vehicle and the point to the arbitrarily small positive number, in order to guarantee the preview.

The paper is organized as follows. In the next Section we present dynamical model for a mobile robot. Simplified robot model is presented in Section 3. Section 4 is devoted to the description of the control aims and problem statement. We derive sliding mode observers for the sideslip angle in Section 5. In Section 6 we design our control algorithm and perform the stability analysis. Simulation results are presented in Section 7, and we finish with brief conclusions in Section 8.

2 Vehicle Model

An adequate mathematical model of the vehicle is probably the most important assumption about designing control systems. In [1], [9], the so called single-track, dynamical model can be found, which is based both on a description of the balanced forces acting on the vehicle in longitudinal and lateral directions, and on the torque conditions. Although the single track model has its limitations, and more complex models can be found, for example, in [10] and the references therein, in a low speed scenario like in our application, it should suffice.
Suppose that the vehicle drives on a circular path. The equations of the transient motion of the vehicle during the transient maneuver are formulated by expressing the absolute acceleration of the center of gravity of the vehicle in terms of the vehicle reference frame. The equations of motions for three degree of freedom in the horizontal plane are the following (see [12] page 304):

a) longitudinal motion

$$m(\dot{v}_x - v_y r) = f_x \tag{1}$$

b) lateral motion

$$m(\dot{v}_y + v_x r) = f_y \tag{2}$$

c) yaw motion

$$J\dot{r} = m_z \tag{3}$$

where

$$v_x = v\cos\beta \tag{4}$$

$$v_y = v\sin\beta \tag{5}$$

x is the longitudinal vehicle direction, y is the lateral vehicle direction, v_x and v_y are vehicle velocities in x and y directions, r is a yaw rate, β is a vehicle sideslip angle, m is a vehicle mass, J is momentum of inertia, v is the vehicle velocity, f_y and f_x are forces in x and y direction, m_z is a torque around z- axis (see Figure 1).

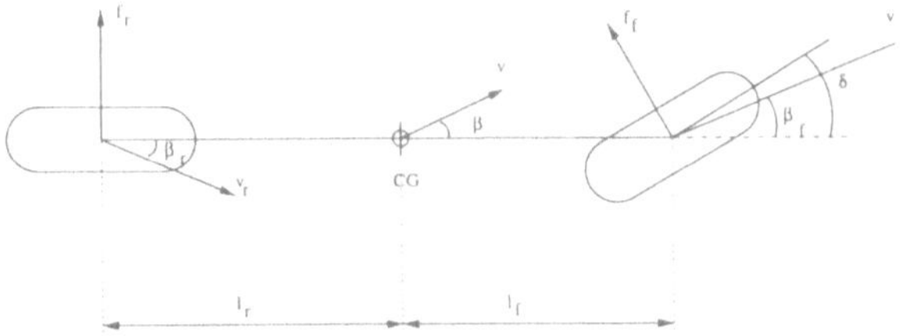

Fig. 1. Singletrack Model

Substituting (4),(5) in (1) and (2) we get

$$-(mv)(\dot{\beta}+r)\sin\beta + m\dot{v}\cos\beta = f_x \tag{6}$$

$$mvcos\beta(\dot{\beta}+r) + m\dot{v}\sin\beta = f_y \tag{7}$$

Equations (6) and (7) can be rewritten as follows

$$mv(\dot{\beta}+r) = -\sin\beta f_x + \cos\beta f_y \tag{8}$$

$$m\dot{v} = \cos\beta f_x + \sin\beta f_y \tag{9}$$

The side forces f_f and f_r are projected through the steering angle δ into chassis coordinates (x, y) as follows

$$f_x = -f_f \sin\delta \tag{10}$$

$$f_y = f_f \cos\delta + f_r \tag{11}$$

where f_f, f_r are side forces for front and rear wheels.

The torque around z axis is the following

$$m_z = f_f \cos\delta l_f - f_r l_r \tag{12}$$

where $l_f(l_r)$ is the distance from the center of gravity to the left (right) wheel along x axis.

The vehicle model is completed by specification the tire model. The front and rear side forces depend on the tire sideslip angles and can be written as follows

$$f_f = f_f(\beta_f) \tag{13}$$

$$f_r = f_r(\beta_r) \tag{14}$$

where β_f, β_r are tire sideslip angles for front and rear wheels. Tire sideslip angles can be found by equating the velocity components in longitudinal and lateral directions.

$$\beta_f = \delta - \arctan\frac{v\sin\beta + l_f r}{v\cos\beta} \tag{15}$$

$$\beta_r = -\arctan\frac{v\sin\beta - l_r r}{v\cos\beta} \tag{16}$$

3 Simplified Model of the Car-like Robot

By making the following assumptions we simplify the vehicle model.

Assumption 1 The sideslip angle is small. Then from (8), (9) and (3) we get

$$mv(\dot{\beta} + r) = -\beta f_x + f_y \tag{17}$$
$$m\dot{v} = f_x + \beta f_y \tag{18}$$
$$J\dot{r} = m_z \tag{19}$$

Substituting f_x from (18) in (17) and taking into account that $\beta^2 \ll 1$ we get

$$mv(\dot{\beta} + r) = f_y - \beta m\dot{v} \tag{20}$$

where $\dot{v}$ is treated as external input which we specify below. Notice, that similar model reduction was made in [2] when studying acceleration and braking effects for robustly decoupled car steering system.

Assumption 2 The steering angle δ is small. Then from (11) and (12) we have

$$f_y = f_f + f_r \tag{21}$$
$$m_z = f_f l_f - f_r l_r \tag{22}$$

Assumption 3. Nonlinear tire characteristic can be approximated by linear one, i.e.,

$$f_f = c_f \beta_f \tag{23}$$
$$f_r = c_r \beta_r \tag{24}$$

and

$$f_f = c_f(\delta - \beta - \frac{l_f r}{v}) \tag{25}$$
$$f_r = c_r(-\beta + \frac{l_r r}{v}) \tag{26}$$

where c_f and c_r are "cornering stiffnesses".

Then the model can be written as follows

$$\dot{x} = v cos(\psi + \beta) \tag{27}$$
$$\dot{y} = v sin(\psi + \beta) \tag{28}$$
$$a_0(\dot{\beta} + r) = \frac{a_{11}}{v}\beta + \frac{a_{12}}{v^2} r + \frac{\delta}{v} - a_0 \beta \frac{\dot{v}}{v} \tag{29}$$
$$\dot{\psi} = r \tag{30}$$
$$\dot{r} = a_{21}\beta + a_{22} r + b_{21}\delta \tag{31}$$

where

$$a_{11} = - \frac{c_r + c_f}{c_f} \tag{32}$$

$$a_{12} = \frac{c_r l_r - c_f l_f}{c_f} \tag{33}$$

$$a_{21} = \frac{c_r l_r - c_f l_f}{J} \tag{34}$$

$$a_{22} = - \frac{c_r l_r^2 + c_f l_f^2}{Jv} \tag{35}$$

$$b_{21} = \frac{c_f l_f}{J} \tag{36}$$

$$a_0 = \frac{m}{c_f} \tag{37}$$

4 Control Aims

Our problem is to find a steering angle $\delta(t)$ and velocity $v(t)$ so that the car follows a reference trajectory given by

$$x_r = p(s)$$
$$y_r = q(s)$$

where s is a parameter of the curve.

In other words, we require the following

$$\overline{\lim_{t \to \infty}} \rho(t) \leq d, \tag{38}$$

$$\psi(t) + \beta(t) - \psi_d(t) = 0, \quad \forall t \geq t_*, \quad t_* \geq 0 \tag{39}$$

where

$$\rho(t) = \sqrt{\Delta x^2 + \Delta y^2}$$
$$\Delta x = x - x_d(s_d), \quad \Delta y = y - y_d(s_d),$$
$$\psi_d = atan2(\Delta y, \Delta x),$$

where $d > 0$ is the "looking ahead" distance, s_d is a desired parameterization of the curve, which is defined below, $x_d(s_d)$ and $y_d(s_d)$ are the coordinates of the virtual point on the curve and $atan2(\cdot)$ is the same as $\arctan(\cdot)$ except that it ranges from $-\pi$ to π.

We measure the position of the robot in x, y coordinates, the orientation angle and the yaw rate. The sideslip angle is not measurable. Our next step is to design a observer for the sideslip angle.

5 Sliding Observers for the Side Slip Angle Estimation

The idea for sliding mode observers is to estimate the derivatives of $\dot{x}$ and $\dot{y}$.

Let us introduce the following sliding mode observer

$$\dot{\epsilon}_1 = \alpha_1(x - \epsilon_1) + \gamma_1 sign(x - \epsilon_1) \tag{40}$$

where $\alpha_1 > 0$ and $\gamma_1 > 0$ are chosen below. Subtracting (40) from (27) we get

$$\dot{x} - \dot{\epsilon}_1 = -\alpha_1(x - \epsilon) + v\cos(\psi + \beta) - \gamma_1 sign(x - \epsilon) \tag{41}$$

It is easy to show that sliding mode arises in the system if $\gamma_1 = v + \frac{\gamma_{01}}{2}$, $\gamma_{01} > 0$. Indeed, choosing a Lyapunov-like function candidate

$$V_1 = (x - \epsilon_1)^2 \tag{42}$$

we evaluate its derivative along the solutions of (41) and

$$\dot{V}_1 \leq -\gamma_{01}\sqrt{V_1} \tag{43}$$

Then for all $t \geq t_*$ where $t_* = \frac{2}{\gamma_{01}}\sqrt{V_1(0)}$, sliding mode appears in the system and in the sliding mode we have

$$\dot{x} = \gamma_1 sign(x - \epsilon_1) \tag{44}$$

The second filter is introduced to estimate $\dot{y}$

$$\dot{\epsilon}_2 = \alpha_2(y - \epsilon_2) + \gamma_2 sign(y - \epsilon_2) \tag{45}$$

where $\alpha_2 > 0$ and $\gamma_2 > 0$. If $\gamma_2 = v + \gamma_{02}$, $\gamma_{02} > 0$ then again a sliding mode appears in the system and

$$\dot{y} = \gamma_2 sign(y - \epsilon_2) \tag{46}$$

Thus for the sideslip angle we have the following estimate

$$\beta = atan2(\gamma_2 sign(y - \epsilon_2), \gamma_1 sign(x - \epsilon_1)) - \psi. \tag{47}$$

For implementation we should filter out high frequency components of the estimates and in this case

$$\beta = atan2(f_2, f_1) - \psi. \tag{48}$$

where f_1 and f_2 are outputs of the following filters:

$$\tau_1 \dot{f}_1 = -f_1 + \gamma_1 sign(x - \epsilon_1)$$
$$\tau_2 \dot{f}_2 = -f_2 + \gamma_2 sign(y - \epsilon_2)$$

where $\tau_i > 0$, $i = 1, 2$. Further we suppose that β is estimated by (48).

Our next step is to design the control action.

6 Control Algorithms and Stability Analysis

The key idea for the control algorithm design is to direct the velocity of vehicle to the virtual point on the curve.

Let us choose the control action as follows

$$\delta = -\gamma sign(\psi + \beta - \psi_d) + v\dot{\psi}_d \tag{49}$$

where $\psi_d = atan2((y - y_d(s_d)), (x - x_d(s_d))$, s_d we define below, $\gamma = v(\frac{\bar{a}_{11}}{v} \mid \beta \mid + \frac{\bar{a}_{12}}{v^2} \mid r \mid + \mid \beta \mid \bar{a}_0 \mid \frac{\dot{v}}{v} \mid) + \frac{\gamma_0}{2}$, where we denoted $\mid a_{11} \mid \leq \bar{a}_{11}$, $\mid a_{12} \mid \leq \bar{a}_{12}$, $\mid a_0 \mid \leq \bar{a}_0$. Taking the Lyapunov function candidate

$$V_2 = a_0(\psi + \beta - \psi_d)^2 \tag{50}$$

and evaluating its derivative along the solutions of the system (29), (49) we get

$$\dot{V}_2 \leq -\frac{\gamma_0}{\sqrt{\bar{a}_0}}\sqrt{V}_2 \tag{51}$$

The control aim (39) is reached, i.e., for all $t \geq t_*$, $t_* = \frac{2\sqrt{\bar{a}_0}}{\gamma_0}\sqrt{V_2(0)}$ sliding mode appears in the system and in the sliding mode the velocity of vehicle is directed to the virtual point on the curve. Now our aim is to find the update law for the virtual point and to prove the achievement of the aim (38).

Achievement of the aim (38)

First we present the expression for ρ in the following form

$$\dot{\rho} = -vcos(\psi_d - (\psi + \beta)) + v_p cos(\theta - \psi_d) \tag{52}$$

where v_p is the velocity of the point ahead, or the velocity of the "virtual vehicle", θ is the angle between x-axis and the velocity vector v_p, which is tangential to the curve at point $(x_d(s_d), y_d(s_d))$. Due to the achievement of the aim (39) in a finite time we can write (52) in the form:

$$\dot{\rho} = -v + v_p cos(\theta - \psi_d) \tag{53}$$

and

$$\dot{s}_d = \frac{v_p}{\sqrt{p'^2(s_d) + q'^2(s_d)}} \tag{54}$$

where $p'(s_d) = \frac{\partial p}{\partial s_d}$, $q'(s_d) = \frac{\partial q}{\partial s_d}$, $\theta = atan2(q', p')$. Now we have to select v and v_p so that to realize the aim (38). First we select the velocity of robot in the following way:

$$v = \gamma_p \rho + \lambda \tag{55}$$

where $\gamma_p > 0$ and λ is arbitrarily small positive number.

Now we are in position to choose v_p. We find convenient to choose the velocity v_p satisfying the following inequality:

$$v_p cos(\theta - \psi_d)(\rho - \varepsilon) \leq 0 \tag{56}$$

where ε is arbitrarily small positive number specified by the designer.
One of the choices to satisfy (56) is the following:

$$v_p = -\gamma_p cos(\theta - \psi_d)(\rho - \varepsilon) \tag{57}$$

and the desired parameterization is given by

$$\dot{s}_d = \frac{-\gamma_p cos(\theta - \psi_d)(\rho - \varepsilon)}{\sqrt{p'^2(s_d) + q'^2(s_d)}} \tag{58}$$

Choosing a positive definite function

$$Q = \frac{1}{2}(\rho - \varepsilon)^2 \tag{59}$$

we evaluate its derivative along the solutions of the system (53). Then taking into account (56) and (55) we get

$$\dot{Q} \leq -\frac{\gamma_p}{2}Q + \frac{\gamma_p}{2}\varepsilon^2 + \frac{1}{\gamma_p}\lambda^2 \tag{60}$$

and

$$Q(t) \leq Q(0)e^{-\frac{\gamma_p}{2}t} + \varepsilon^2 + \frac{2\lambda^2}{\gamma_p^2} \tag{61}$$

From (61) we get the transient bound for $\rho(t)$:

$$\rho(t) \leq \varepsilon + \sqrt{(\rho(0) - \varepsilon)^2 e^{-\frac{\gamma_p}{2}t} + 2\varepsilon^2 + \frac{4\lambda^2}{\gamma_p^2}} \tag{62}$$

and control aim (38) is reached with $d = \varepsilon + \sqrt{2}\varepsilon + \frac{2\lambda}{\gamma_p}$.

The upper bound for the velocity of vehicle is the following:

$$0 < \lambda \leq v(t) \leq \lambda + \gamma_p(\varepsilon + \sqrt{(\rho(0) - \varepsilon)^2 e^{-\frac{\gamma_p}{2}t} + 2\varepsilon^2 + \frac{2\lambda^2}{\gamma_p^2}}) \tag{63}$$

Notice, that there exists a lot of possibilities to choose v_p so that to realize the control aim (38).
Our last step is to prove the boundedness of the system states.

Boundedness of the System States.
We perform our stability analysis for the case where "cornering stiffnesses" are the same for front and rear wheels and the center of gravity is in the middle of the wheelbase, i.e., $c_f = c_r$ and $l_f = l_r$, $a_{21} = a_{12} = 0$.
In the sliding mode, according to the Fillipov's definition we get from (29) the expression for our control action:

$$\delta = (a_0\dot{v} - a_{11})\beta + a_0 v\dot{\psi}_d \tag{64}$$

Taking into account that $\beta + \psi - \psi_d = 0$ during the sliding motion we get

$$\delta = -(a_0\dot{v} - a_{11})\psi + (a_0\dot{v} - a_{11})\psi_d + a_0 v\dot{\psi}_d \tag{65}$$

Substituting (65) in (31) we get

$$\dot{\psi} = r \tag{66}$$

$$\begin{aligned}\dot{r} &= a_{22}r + b_{21}(-a_0\dot{v} + a_{11})\psi \\ &\quad + b_{21}(a_0\dot{v} - a_{11})\psi_d + b_{21}a_0 v\dot{\psi}_d\end{aligned} \tag{67}$$

It is easy to see that $b_{21}(a_0\dot{v} - a_{11})\psi_d + b_{21}a_0 v\dot{\psi}_d$ is bounded and to prove the boundedness of ψ and r we have to prove the stability of the system

$$\dot{\psi} = r \tag{68}$$

$$\dot{r} = -a_2 r + b_{21}(-a_0\dot{v} - a_1)\psi \tag{69}$$

where we denoted $a_1 = -a_{11}$, and $a_2 = -a_{22}$, a_1 and a_2 are positive. For the system (68), (69) we choose the following Lyapunov function candidate.

$$V = \frac{1}{2}a_1 b_{21}\psi^2 + \frac{1}{2}r^2 + c\psi r \tag{70}$$

where c is sufficiently small positive constant.
Evaluating the derivative of (70) along the solutions of (68), (69) we get after some calculations:

$$\dot{V} = -(a_2 - c)r^2 - (b_{21}a_0\dot{v} + a_2 c)\psi r - c(a_1 b_{21} + b_{21}a_0\dot{v})\psi^2 \tag{71}$$

for $\dot{V}$ to be negative definite we need:

$$(a_2 - c) > 0 \tag{72}$$

$$a_1 b_{21} + b_{21}a_0\dot{v} > 0 \tag{73}$$

$$(b_{21}a_0\dot{v} + a_2 c)^2 - 4(a_2 - c)c(a_1 b_{21} + b_{21}a_0\dot{v}) < 0 \tag{74}$$

It easy to see that as long as $\mid \dot{v} \mid$ and c are small the inequalities (72) - (74) are always true. Notice that $\mid \dot{v} \mid$ can be made arbitrarily small by choosing small enough γ_p.

7 Simulation Results

Let the initial position of the car be the following $x(0) = 2$,$y(0) = -1.5$ with the following initial condition on the orientation angle $\psi(0) = \frac{\pi}{2}$.
We design a control action such that a car should reach the desired curve, which a circle with radius R

$$\begin{aligned}x_r &= Rcos(s) \\ y_r &= Rsin(s);\end{aligned}$$

where s is a parameter of the curve, $R = 2$.

Simulation results for controller (49),(55) and (58) are presented in Figure 2. Sideslip angle β and its estimate (48) are presented in Figure 3. Finally, the velocity of the vehicle we plotted in Figure 4.

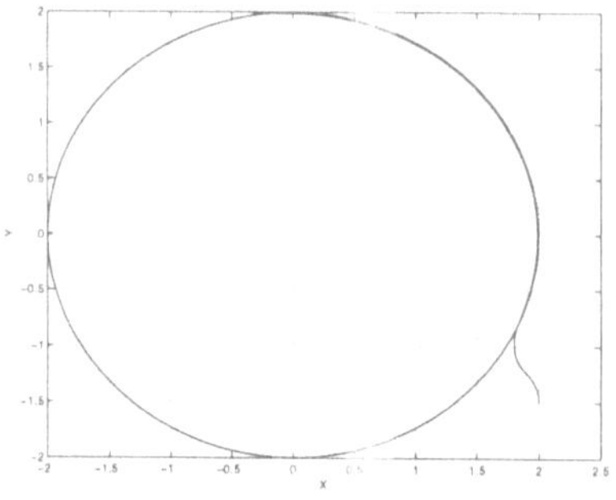

Fig. 2. Tracking

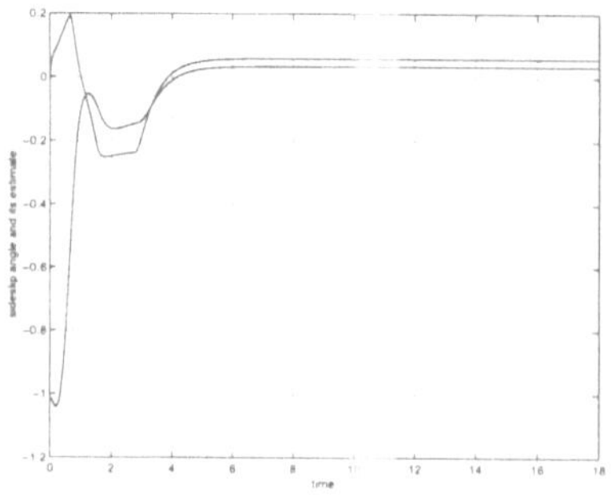

Fig. 3. Sideslip Angle and its Estimate

8 Conclusion

In this paper we proposed a new simple trajectory tracking algorithm which uses the idea of preview and sliding mode control so that the velocity vector

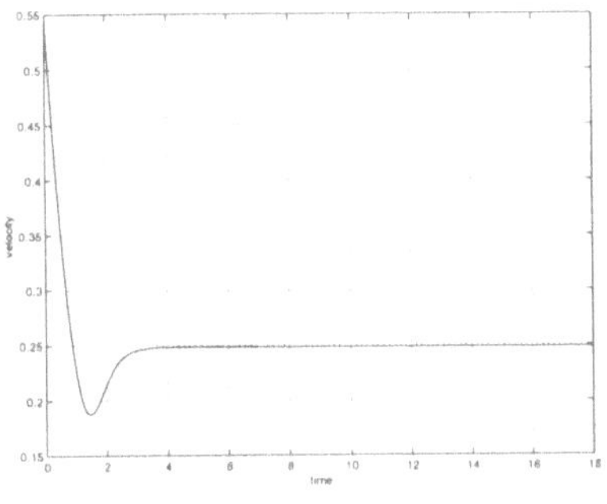

Fig. 4. Velocity

of the vehicle is directed always towards the virtual point, which is updated according to the position and orientation of the vehicle. The algorithm designed and analyzed within the single-track dynamical model framework and proved to be robust with respect to the parameter variations.

References

1. J. Ackermann: *Robust Control*, Springer-Verlag, London pp 371-375 1993,
2. J. Ackermann and X. Hu, Acceleration and braking effects on robustly decoupled car steering, Proc. of 3-rd ECC'95, Rome, Italy, Sept. 1995, pp.737-743 (see also DLR Internal Report No. 515-94-28).
3. R.W. Brockett: Asymptotic stability and feedback stabilization, in *Differential Geometric Control Theory* (Brockett,Millmann and Sussman, eds), pp.181-191, Boston, MA, USA, Birkhauser, 1983.
4. C. Canudas de Wit, Bruno Siciliano and G. Bastin: Theory of Robot Control, *Springer Verlag, 1996.*
5. Guldner J. and V. Utkin: Stabilization of nonholonomic mobile robots using Lyapunov functions for navigation and sliding mode control, *Proc. of the 33-rd CDC*, Lake Buena Vista, FL-December 1994, pp.2967-2972.
6. S.V. Gusev and I.A. Makarov: Stabilization of Program Motion of Transport Robot with Tracklaying Chassis, *Proceedings of LSU*, vol 1, issue 3, No15, 1989.
7. Jiang Z., Nijmeijer H., " Tracking Control of Mobile Robots: A case study in Backstepping", Automatica, vol.33, N7,pp. 1393-1399.
8. J-C. Latombe: *Robot Motion Planning*, Kluwer Academic Publishers, 1991.
9. E. Freund and R. Mayr: Nonlinear Path Control in Automated Vehicle Guidance, *IEEE Transactions on Robotics and Automation*, vol 13, No1, Feb 1997.
10. R. Majjad and U. Kiencke, Modular Design for the Computation of Vehicle Dynamic Behavior, Proc. of the 5th IEEE Med. Control Conference, July, 1997, Paphos, Cyprus.
11. R. Murray and S. Sastry: Nonholonomic motion planning: steering using sinusoids, *IEEE Transactions on Automatic Control*, vol. 38, No 5, pp. 700-716, 1993.
12. Wong J.Y. Theory of Ground Vehicles, John Wiley *and* Sons, Inc., 1993.

Park Vector Based Sliding Mode Control of UPS With Unbalanced and Nonlinear Load

Péter Korondi[1] * Hideki Hashimoto[2]

[1] Technical University of Budapest, Department of Automation
H-1111 Budapest XI.ker. Budafoki út 8, HUNGARY,
Tel:+36-1-463 2184 Fax: +36-1-463 3163 E-mail:korondi@elektro.get.bme.hu
[2] University of Tokyo, Institute of Industrial Science
7-22-1, Roppongi, Minato-ku Tokyo 106 JAPAN,
Fax:+81-3-3423 1484, E-mail: hashimoto@iis.u-tokyo.ac.jp

Abstract. The main contribution of this chapter is a new Park vector based variable structure control (VSC) method. In this chapter an inverter is taken to be a member of Variable Structure Multy Imput Multy Output System.
The design of a sliding mode controller consists of two main steps. Firstly, the design of the sliding surface, secondly, the design of the control law which holds the system trajectory on the sliding surface. Here a complex (Park vector based) sliding surface is proposed. The distance of the system state from the complex sliding surface measured by a complex vector. The inverter is switched in such a way that the system trajectory gets as close to the sliding surface as possible. In other words the complex distant vector should be decreased. This chapter focuses on the switching rule. Two switching strategies are compared. In the first approach, sliding mode exists only in the intersection of the switching surfaces. In the second approach a stable sliding mode may exist on any of the switching surfaces independently.
A modified definition of the Park vector is introduced to handle the effect of zero phase-sequence component caused by an asymmetrical load.
Experimental results of a 100 KVA inverter are presented.

1 Introduction

The principle of pulse-width modulation (PWM) plays a very important role in power electronics [1]. In the field of inverter technology, which produces sinusoidal voltage, a great number of "optimized PWM" techniques have been proposed in the literature. These types of PWM inverters have very good steady-state characteristics, but the voltage regulator response for sudden change in the load takes a few cycles, and nonlinear loads can cause high "load harmonics". This is not acceptable in Uninterruptable Power Supply (UPS) application for

* The authors wish to thank the Japanese Society of Promotion of Science and the Control Research Group of the Hungarian Academy of science for their financial support.

which instantaneous feedback is preferred [2],[3]. In most papers symmetrical load is assumed. There is an increasing need to supply not only one unit, but also the whole installation by UPS. The energy supply for the group of units of great importance (nuclear power plants, airports, computer and telecommunication centres etc.) is usually realized by three independent single phase UPS inverters, each having an independent control. This is because of the asymmetrical load of the phases due to the large number of possible single phase units connected to the network. A less expensive solution is to apply a single three phase inverter. In this case asymmetrical load causes problems because voltage in the three phases cannot be controlled independently. This chapter proposes a three phase controller which can be used under asymmetrical and nonlinear loading.

In the course of designing the controller, the inverter is taken as a member of Variable Structure Systems (VSS) [4]. The output filter is considered as the controlled plant with its control is realized by the inverter legs. The organization of the chapter is as follows: Section 2 describes the main circuits and basic equations of the UPS. Section 3 summarizes the theoretical background of sliding mode control of multi input multi output systems and it highlights the problem in case of the present application. Section 4 introduces the Park vector based sliding mode design, and describes the two main steps namely the design of sliding surface and the control law. Two switching strategies are compared. In the first approach, sliding mode exists only in the intersection of the switching surfaces. In the second approach a stable sliding mode may exist on any of the switching surfaces independently. Section 5 presents the experimental results. Section 6 draws the conclusions.

2 System description

2.1 Configuration

The main circuit and control system of the three phase inverter are illustrated in Fig.1. The figure does not show the battery and charge unit. The filter circuit consists of a special Δ/Y transformer and a capacitor unit connected in parallel to the secondary. For the sake of filtering, the leakage inductance on the primary is increased and the mutual main field inductance is decreased by an air gap in contrast to an ordinary transformer. The phase loads are connected between the star and the R-S-T terminals of the transformer, in general. Only the phase R is loaded in Fig.1.

The positive directions are shown in Fig. 1. This chapter is focused on control, so a simplified transformer model is applied for theoretical consideration. Resistances of transformer coils and semiconductor devices are ignored. For the calculations, the transformer ratio of the transformer is chosen in a such way that leakage inductances on the secondary are equal to zero [13]. The primary leakage inductance matrix ($\mathbf{L}_l \in R^{3x3}$) and the filter capacitor matrix ($\mathbf{C} \in R^{3x3}$) are diagonal. The mutual inductance matrix ($\mathbf{L}_m \in R^{3x3}$) is symmetrical.

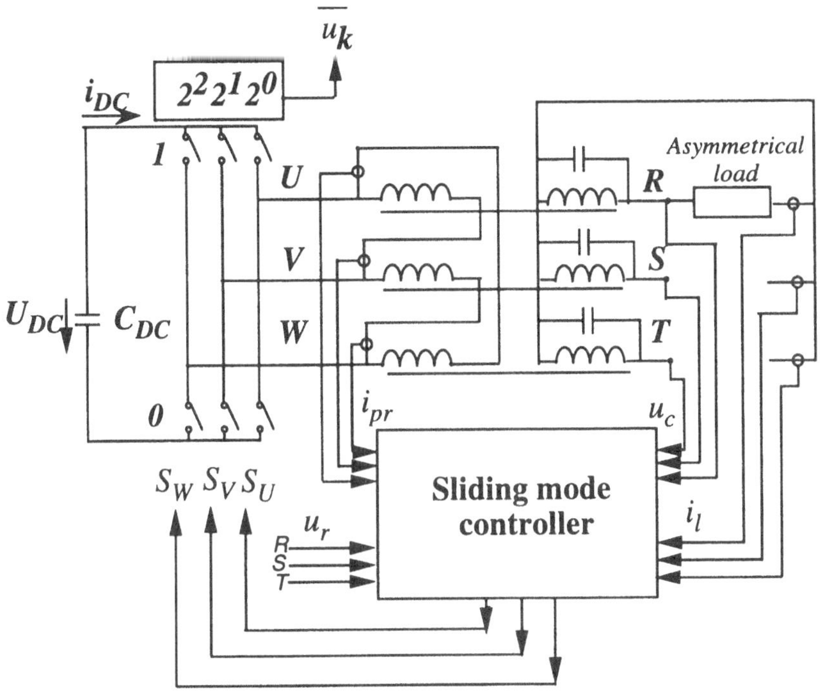

Fig. 1. Simplified diagram of the inverter

The control system senses three electrical variables (namely the filter circuit output voltage, $\mathbf{u_c} \in R^3$, the primary phase current, $\mathbf{i_p} \in R^3$ and the load current $\mathbf{i_l} \in R^3$), in each phase instantaneously. The sinusoidal reference signal,

$$\mathbf{u_r} = \left(U_r \sin(\omega t)\ U_r \sin(\omega t + \frac{2\pi}{3})\ U_r \sin(\omega t + \frac{4\pi}{3})\right)^T , \tag{1}$$

is stored in ROM memory. Let $\mathbf{u} \in R^3$ is a column vector of the potential of the inverter terminals U, V and W

$$\mathbf{u} = (u_U\ u_V\ v_W)^T . \tag{2}$$

The semiconductors of the inverter are controlled by a circuit which uses the reference signal and the three feed-back variables in each phase taking into account the switching frequency and current limitation requirements. The switches S_W, S_V and S_U are set to 1 if the upper transistor is switched on and 0 if the lower transistor is switched on in the corresponding inverter leg. Consequently,

$$\mathbf{u} = U_{DC}\,(S_U\ S_V\ S_W)^T . \tag{3}$$

The inverter has eight switching states, which are described by a decimal value of the three digit binary number, where the digits refer to the switching state of the corresponding inverter leg as shown in Fig. 1. For example, switching state

u_3 means that the upper transistors are switched on in the leg V and U and the lower one in the leg W since $3 = 011_2$.

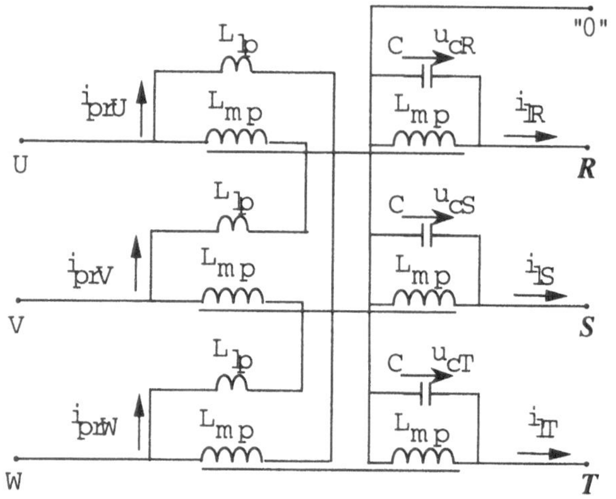

Fig. 2. Transformer and filter

2.2 Basic Equations

According to Fig. 2, the matrix differential equation for the filter circuit is as follows:

$$\mathbf{L_m T_{tr} u} = \mathbf{L_m L_l C} \frac{d^2 \mathbf{u_c}}{dt} + (\mathbf{L_m} + \mathbf{L_l})\mathbf{u_c} \ , \tag{4}$$

where

$$\mathbf{T_{tr}} = \begin{pmatrix} 1 & -1 & 0 \\ 0 & 1 & -1 \\ -1 & 0 & 1 \end{pmatrix} \tag{5}$$

is the transformer transformation matrix and the single phase equivalent leakage and mutual main field inductance are L_{lp} and L_{mp}, respectively. Let the following notations be introduced:

$$\mathbf{\Omega}^{-2} = \mathbf{L_m} \left(\mathbf{L_m} + \mathbf{L_l}\right)^{-1} \mathbf{L_l C}, \tag{6}$$

$$\mathbf{G} = \mathbf{L_m} \left(\mathbf{L_m} + \mathbf{L_l}\right)^{-1} \ . \tag{7}$$

The matrix $\mathbf{\Omega}$ can be considered as the eigen frequency matrix of the filter circuit. Using the above notation (4) can be rewritten.

$$\mathbf{G T_{tr} u} = \mathbf{\Omega}^{-2} \frac{d^2 \mathbf{u_c}}{dt} + \mathbf{u_c} \tag{8}$$

3 Sliding Mode Control Of A Multi-Input Multi-Output System

The following non-linear state-equations are considered.

$$\dot{\mathbf{x}} = \mathbf{f}(\mathbf{x}) + \mathbf{B}\mathbf{u}, \quad \mathbf{x} \in R^n, \quad \mathbf{u} \in R^m, \quad \mathbf{f} \in R^n , \tag{9}$$

where $\mathbf{x}$ is the state-variable vector; $\mathbf{f}(\mathbf{x})$ is a vector-vector function; $\mathbf{B} \in R^{nxm}$ is a positive or negative definite matrix; $\mathbf{u}$ is the control vector. The goal of the control is to drive the system into the origin in state space. The switching surfaces of the sliding mode, where control vector components have discontinuities, can be written in the following form

$$\sigma = \Lambda\mathbf{x}, \quad \sigma \in R^m \tag{10}$$

where Λ is an $m * n$ matrix. The condition for existence of sliding mode is

$$\sigma_i \dot{\sigma_i} < 0 , \tag{11}$$

where σ_i is the i-th element of vector $\boldsymbol{\sigma}$. The first derivative of vector σ with respect to the time is

$$\dot{\sigma} = \Lambda(\mathbf{f}(\mathbf{x}) + \mathbf{B}\mathbf{u}) . \tag{12}$$

If the matrix $\Lambda\mathbf{B}$ is not diagonal but its rank m, a $\mathbf{T}$ matrix can be introduced to decouple the system into m control loops

$$\sigma_\mathbf{t} = \mathbf{T}\sigma . \tag{13}$$

In other words $\mathbf{T}\Lambda\mathbf{B}$ should be diagonal in (14)

$$\dot{\sigma_\mathbf{t}} = \mathbf{T}\Lambda\mathbf{f}(\mathbf{x}) + \mathbf{T}\Lambda\mathbf{B}\mathbf{u} . \tag{14}$$

(It is remarked that matrix $\mathbf{T}$ can be time varying [5], in which case there is an additional term in (14)). The condition (11) must be true for each row of vector $\boldsymbol{\sigma}_\mathbf{t}$. That means that the sign of each row on the right hand side of (14) must be opposite to that of $\boldsymbol{\sigma}_\mathbf{t}$. If $\mathbf{T}$ exists then it can be chosen in a way so that the matrix $\mathbf{T}\Lambda\mathbf{B}$ is the identity matrix. Let a relay type control law be used

$$u_i = \text{sign}(\sigma_i) . \tag{15}$$

The condition (11) is held if the absolute value of each element the vector $\mathbf{u}$ is bigger than those of vector $\mathbf{T}\Lambda\mathbf{f}(\mathbf{x})$.

Applying the above method to a three phase inverter, the error of the output voltage

$$\mathbf{u_e} = \mathbf{u_r} - \mathbf{u_c} \tag{16}$$

is controlled. Let the state variables be chosen in the following way:

$$\mathbf{x} = (u_{eR} \; u_{eS} \; u_{eT} \; \dot{u}_{eR} \; \dot{u}_{eS} \; \dot{u}_{eT} \;)^T . \tag{17}$$

The system is described by

$$\mathbf{f}(\mathbf{x}) = \begin{pmatrix} \mathbf{Z} & \mathbf{I} \\ \Omega^2 & \mathbf{Z} \end{pmatrix} \mathbf{x} + \begin{pmatrix} \mathbf{Z_c} \\ (\Omega^2 - \omega^2 \mathbf{I_c})\mathbf{u_r} \end{pmatrix} , \tag{18}$$

$$\mathbf{B} = -\begin{pmatrix} \mathbf{Z} \\ \mathbf{C}^{-1}\mathbf{L_l}^2\mathbf{G} \end{pmatrix} \begin{pmatrix} 1 & -1 & 0 \\ 0 & 1 & -1 \\ -1 & 0 & 1 \end{pmatrix} , \tag{19}$$

where ω is the angular frequency of the sinusoidal reference signal, $\mathbf{Z} \in \mathbf{R}^{3x3}$ is zero matrix, $\mathbf{Z_c} \in R^3$ is zero column vector and

$$\mathbf{I_c} = (1 \;\; 1 \;\; 1)^T . \tag{20}$$

The discontinuous input is

$$\mathbf{u} = (S_u \;\; S_v \;\; S_w)^T U_{DC} . \tag{21}$$

The parameter matrix for the sliding surfaces has the follwing structure.

$$\Lambda = (\mathbf{I} \;\; \lambda \mathbf{I}) , \tag{22}$$

where $\mathbf{I} \in R^{3x3}$ is the identity matrix. Since $\det \mathbf{B} = 0$ the method of diagonalization can not be applied directly. From the point of view of the circuit, it is because of the floating potential of the star point of the load.

Hence, although the inverter has three legs, it is impossible to introduce three independent switching surfaces to control the three phase errors independently. Several papers have solved this problem by introducing two independent surfaces with a third, on which the system trajectory is automatically held. This kind of solution causes unnecessary switching which increase the stress of the semiconductors and decrease the efficiency.

4 A Park Vector Based Sliding Mode Control

4.1 Sliding surface design

To take the reduced degree of freedom into account a complex Park vector is introduced as a 3-phase to 2-phase transformation.

$$\bar{x} = \frac{2}{3}(x_R + \bar{a}x_S + \bar{a}^2 x_T) , \tag{23}$$

where x_R, x_S and x_T can be arbitrary time functions of the phase values of any three-phase signals and

$$\bar{a} = -\frac{1}{2} + \mathrm{j}\frac{\sqrt{3}}{2}, \quad \bar{a}^2 = -\frac{1}{2} - \mathrm{j}\frac{\sqrt{3}}{2} . \tag{24}$$

The Park transformation can be described by the following transformation matrix:

$$\mathbf{T}_{\mathbf{Park}} = \frac{2}{3}\begin{pmatrix} 1 & \frac{1}{2} & -\frac{1}{2} \\ 0 & \mathrm{j}\frac{\sqrt{3}}{2} & -\mathrm{j}\frac{\sqrt{3}}{2} \end{pmatrix} . \tag{25}$$

In the case of an unbalanced system, the zero phase-sequence component is calculated in the following way:

$$x_0 = \frac{1}{3}(x_R + x_S + x_T) . \tag{26}$$

The Park vector can be explain in a geometrical way. There are three symmetrical axes in the complex plane. The time functions of the three phase signals x_R, x_S and x_T are measured on the corresponding axes and they are added instantaneously (see in Fig. 3). The resulted vector is multiplied by $\frac{2}{3}$ to normalize it. Normalization means that the projection of the Park vector to the axes in any time instant equal to the corresponding instant phase value if $x_0 = 0$.

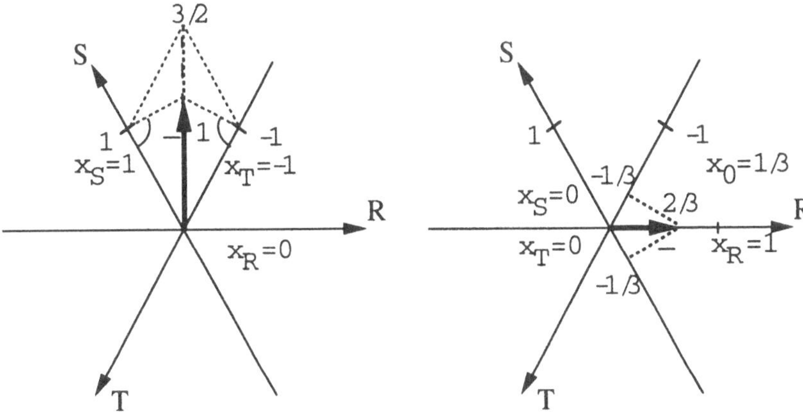

Fig. 3. Park vector in a geometrical approach

If $x_0 \neq 0$, the projection of the Park vector to the R, S and T axes are $(x_R - x_0)$, $(x_S - x_0)$ and $(x_T - x_0)$, respectively. Fig. 3 shows two cases. In the left hand side $x_R = 0$, $x_S = 1$ and $x_T = -1$, the system is balanced $x_0 = 0$. In the right hand side an unbalanced example is illustrated where $x_R = 1$, $x_S = 0$ and $x_T = 0$. Consequently, $x_0 = \frac{1}{3}$. According to the Park vector definition $\bar{x} = \frac{2}{3}x_R$ and its projection to the axes S and T are $-\frac{1}{3}$, since $x_S - x_0 = -\frac{1}{3}$ and $x_T - x_0 = -\frac{1}{3}$.

It is easy to prove that the end point of the Park vector of a symmetrical sinusoidal three-phase system moves alone a circular path with a constant speed. That is why some authors introduce a rotating coordinate frame where the reference vector is in a standstill. The so called synchronous rotating reference frame has several advantages but most engineers prefer the stationary reference frame which is used from now on.

The main advantage of the Park transformation for a system with symmetrical structure is that the two (real and imaginary) components are decoupled. The three phase values (x_R, x_S and x_T) can be arbitrary and asymmetrical, only the system matrixes should be symmetrical at any time instant [6] but it can be time varying. Using Park transformation, the all matrixes (including the non diagonal but symmetrical one) can be replaced by a scalar real value. In our case, the only problem is matrix $\mathbf{T_{tr}}$ in (8). The line-to-line voltage vector $\mathbf{v}$ and its Park equivalent $\bar{v}$ are introduced to avoid it

$$\mathbf{v} = \mathbf{T_{tr}}\mathbf{u} \ . \tag{27}$$

Expressing (8) in Park vector representation.

$$\frac{\mathrm{d}^2\bar{u}_c}{\mathrm{d}t} = \Omega_p^2 G_p \bar{v} - \Omega_p^2 G_p \bar{u}_c \tag{28}$$

where the subscription p denotes the Park equivalent of the corresponding matrixes. Ω_p^2 and G_p are scalar real values,

$$\Omega_p^{-2} = L_{mp}\left(L_{mp} + L_{lp}\right)^{-1} L_{lp} C_p, \tag{29}$$

$$G_p = L_{mp}\left(L_{mp} + L_{lp}\right)^{-1} . \tag{30}$$

The complex sliding surface

$$\bar{\sigma} = \bar{u}_e + \lambda\dot{\bar{u}}_e = 0 + 0\mathrm{j} \tag{31}$$

is introduced, where the control undergoes discontinuities. The λ is a time constant type parameter, it defines the transient behavior.

4.2 Control Law, Switching Strategies

The goal is the nullification of the complex vector $\bar{\sigma}$ which indicates the distance of actual system trajectory from the sliding surface. The inverter is switched such that the system trajectory gets as close to the sliding surface as possible. The inverter has eight switching states. The Park vectors of the eight switching states are shown in Fig 4, in the left hand side. It is trivial, if all upper or lower transistors are switched on, all line-to-line voltages are zero. In the other six cases one of the line-to-line voltages is zero and the other two have opposite sign. Because of the Δ/Y transformer connection the primary line-to-line voltage approximately equal to the secondary phase voltage that is why it is illustrated in the $R-S-T$ frame. For example, in case of $\bar{v}_3$

$$\begin{aligned} v_{UV} &= v_R = 0 \ ; \\ v_{VW} &= v_S = U_{CD} \ ; \\ v_{WU} &= v_T = -U_{CD} \ . \end{aligned} \tag{32}$$

From that eight switching state should be selected one which moves the system towards the sliding surface. The control law is a method for selection from the switching states of the inverter. For this reason, it should be derived, in which direction the vector $\bar{\sigma}$ is changing. Expressing the first time derivative of the vector $\bar{\sigma}$

$$\dot{\bar{\sigma}} = \dot{\bar{u}}_e - \lambda \ddot{\bar{u}}_e \ . \tag{33}$$

Since the reference signal is sinusoidal:

$$\ddot{\bar{u}}_r = -\omega^2 \bar{u}_r \ . \tag{34}$$

Substituding (28) and (34) into (33)

$$\dot{\bar{\sigma}} = \dot{\bar{u}}_e - \lambda \Omega_p^{-2} \bar{u}_e + \lambda \left[\left(\Omega_p^2 - \omega^2 \right) \bar{u}_r - \Omega_p^2 G_p \bar{v}_k \right] \ . \tag{35}$$

Since error in steady state is only a small percentage of the rotating reference vector, $\bar{u}_r$, a good approximation of $\dot{\bar{\sigma}}$ is the difference between the rotating reference and the corresponding switching state vector as shown in Fig. 4.

$$\dot{\bar{\sigma}} \ c_1 \bar{u}_r - c_2 \bar{v}_k \ , \tag{36}$$

where $c_1 = \lambda(\Omega_p^2 - \omega^2)$, $c_2 = -\lambda \Omega_p^2 G_p$. Fig. 4 shows the seven possible directions of $\dot{\bar{\sigma}}$ at an instant when the reference vector,$\bar{u}_r$, is pointed at P.

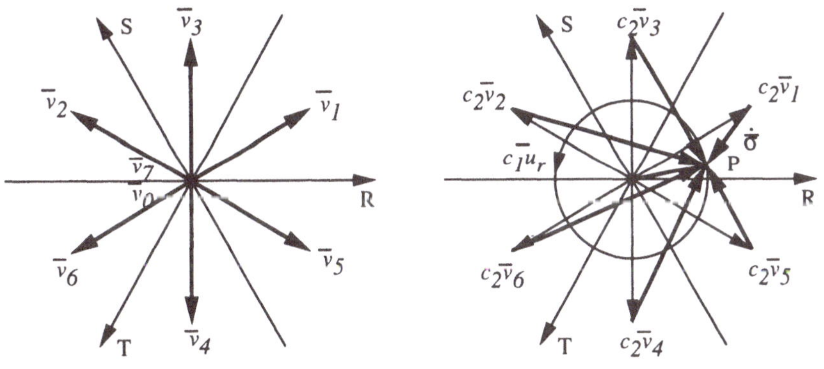

Fig. 4. Switching States and Direction of Vector Change

Usually none of them is directed exactly to the surface but more than one have a component in that direction. It is clear that the direction of $\dot{\bar{\sigma}}$ depends on the position of the reference signal, so the selection also depends on its actual state.

There are two main approaches of design of a control law for the sliding mode on the surface $\bar{\sigma} = 0 + 0\text{j}$. In the first approach, sliding mode exists only in the intersection of the switching surfaces. In this case, the condition for the existence of a sliding mode is that the vector $\bar{\sigma}$ and $\dot{\bar{\sigma}}$ have opposite components.

$$\mathrm{Re}(\bar{\sigma}\dot{\bar{\sigma}}^*) < 0 \ , \tag{37}$$

where $\dot{\bar{\sigma}}^*$ stands for the complex conjugation of $\dot{\bar{\sigma}}$. A realization of this approach is the triangular switching strategy below. In the second approach a stable sliding mode may exist on any of the two switching surfaces independently. In this case the condition for the existence of a sliding mode should hold for the two switching surfaces separately (see rhomboid switching strategy).

It is difficult to illustrate the complex sliding surface in the four dimensional state phase. The switching strategy is described by two figures. One shows the direction of the vector $\dot{\bar{\sigma}}$ (where are the system going) and the other is the plane of $\bar{\sigma}$ (where is the system).

Triangular Switching Strategy To adapt the most popular current vector control method [9] for sliding mode control of UPS, one period of the rotating vector,$\bar{u}_r$, is divided into six control intervals denotes by $T1$, $T2$, $T3$, $T4$, $T5$ and $T6$. Four switching states (i.e. only three directions of vector $\dot{\bar{\sigma}}$ since the two zero states have the same effects) are assigned to each control interval. In Fig. 5 the switching states of control interval $T2$ are shown. Since the switching states are in the vertexes of a triangle that includes $\bar{u}_r$ this switching strategy is referred to as "triangular" switching strategy. The complex $\bar{\sigma}$ plane is divided into six areas (a, b, c, d, e and f), as shown in Fig. 5. The six areas are joined in pairs in each control interval. In each pair of areas the orientation of one of the three possible vectors $\dot{\bar{\sigma}}$ is such that the condition (37) holds, as shown in Fig. 5. Assuming, that the vector $\bar{u}_r$ is pointed at P (control interval $T2$) and the distance vector $\bar{\sigma}$ is in the area a or b, the switching state $\bar{v}_1$ must be switched. If the distance vector $\bar{\sigma}$ is in c or d then state $\bar{v}_3$ must be switched. If the distance vector is in e or f then either state $\bar{v}_0$ or $\bar{v}_7$ must be switched. Regarding the above mentioned, the triangular switching strategy is summarized in Table 1.

Applying the triangular switching strategy some convergence problems emerge in the vicinity of the border of two control intervals discussed in [8] and [10]. In case of current vector control, the change of vector $\bar{\sigma}$ is exactly the difference between the rotating reference and the corresponding switching state vector. Because of the approximation (36) the convergence problem become more serious in case of UPS control. To avoid these problems a rhomboid switching strategy is introduced [12].

Rhomboid Switching Strategy In the second type of switching rule, the condition for the existence of a sliding mode should hold for two switching surfaces separately. Usually the two perpendicular components of the Park vector are controlled. Adapting this the switching surfaces are chosen as follows [5]

$$\mathrm{Im}(\bar{\sigma}) = 0 \qquad \mathrm{Re}(\bar{\sigma}) = 0 \tag{38}$$

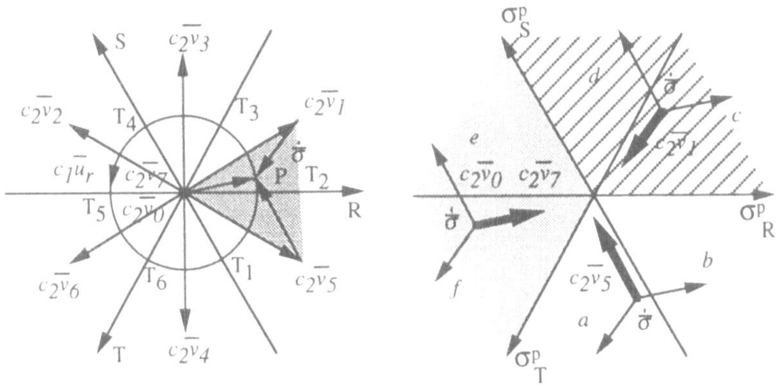

Fig. 5. Triangular switching strategy

Table 1. Triangular switching strategy

Control Interval	Areas of the Error Vector	Switched State	Control Interval	Areas of the Error Vector	Switched State
	f, a	$\bar{v}_5$		a, b	$\bar{v}_1$
$T1$	b, c	$\bar{v}_1$	$T2$	c, d	$\bar{v}_3$
	d, e	$\bar{v}_0, \bar{v}_7$		e, f	$\bar{v}_0, \bar{v}_7$
	b, c	$\bar{v}_3$		c, d	$\bar{v}_2$
$T3$	d, e	$\bar{v}_2$	$T4$	e, f	$\bar{v}_6$
	f, a	$\bar{v}_0, \bar{v}_7$		a, b	$\bar{v}_0, \bar{v}_7$
	d, e	$\bar{v}_6$		e, f	$\bar{v}_4$
$T5$	f, a	$\bar{v}_4$	$T6$	a, b	$\bar{v}_5$
	b, c	$\bar{v}_0, \bar{v}_7$		c, d	$\bar{v}_0, \bar{v}_7$

These switching surfaces are not suitable for the inverter structure since the Park vector cannot be measured directly. A simple switching strategy can be applied if only two of the three phase component (projection) of vector, $\bar{\sigma}$, are sensed and controlled simultaneously. (Remark: $x_R^p = x_R - x_0$, $x_S^p = x_S - x_0$ and $x_T^p = x_T - x_0$ are referred to as the phase component of the Park vector.) One period of the rotating vector, $\bar{u}_r$, is divided into six control intervals again. In each control interval the state of one leg is locked and only two legs are switched according to sign of the two sensed phase components. The control intervals and the six areas in the vector, $\bar{\sigma}$, plane are rotated to the previous case (a ' is used for the rhomboid strategy). Fig. 6 shows the case of the control interval $T2'$. The two sensed phase components are σ_R^p and σ_T^p, the locked switch is $S_U = 1$

and the two switched legs are S_V and S_W as shown in Fig. 6. For example, if vector $\bar{\sigma}$ is in the area b', c' or d' ($\sigma_T^p < 0$) and the switch S_W is switched to zero than the vector $\dot{\bar{\sigma}}$ has a component which is pointed to the switching line. After similar consideration the rhomboid switching strategy is summarized in the Table 2.

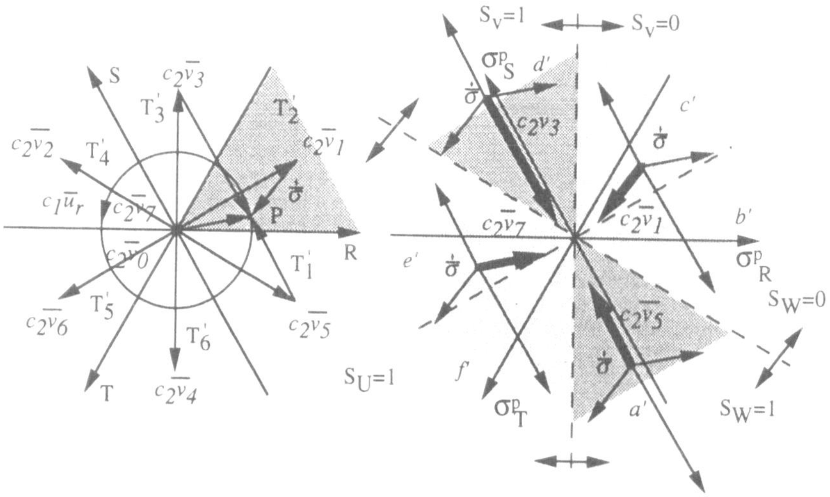

Fig. 6. Rhomboid switching strategy

4.3 Comparison of rhomboid and triangular switching strategies

The rhomboid and the triangular switching strategies are compared by simulation. A switching-time delay (T_{del}) and a hysteresis were taken into the account. The period between two consecutive switches are limited by T_{null}. Nominal symmetrical load was considered. The total harmonic distortion (THD), the number of switches (NS) during one period and the shortest period between two consecutive switches T_{min} are shown in Table 3.

Applying the triangular switching strategy a satisfactory reduction of distance could not always be achieved in the vicinity of the border of two control intervals. Fig. 7 shows the operation of triangular and rhomboid switching strategies in the critical areas. In case of triangular switching strategy, the trajectory of the vector, $\bar{\sigma}$, is chattering between the area e and d, the switching frequency should be limited artificially. The amplitude of the vector, $\bar{\sigma}$, is may not decreased since (36) gives only an approximation, the real direction of the vector $\dot{\bar{\sigma}}$ belonging to state $\bar{v}_{0,7}$ or $\bar{v}_1$ may have no component that is opposite to the vector, $\bar{\sigma}$. Consequently, the numbers of switching are asymmetrical in the three phases causing relatively big total harmonic distortion. In case of the rhomboid

Table 2. Rhomboid switching strategy

Control interval	Sensed phase components	Locked leg	Switching laws
$T1'$	$\bar{\sigma}_R^p$, $\bar{\sigma}_S^p$	$S_V = 0$	$S_U = 0.5 + 0.5\text{sign}\bar{\sigma}_R^p$ $S_W = 0.5 - 0.5\text{sign}\bar{\sigma}_S^p$
$T2'$	$\bar{\sigma}_R^p$, $\bar{\sigma}_T^p$	$S_U = 1$	$S_V = 0.5 - 0.5\text{sign}\bar{\sigma}_R^p$ $S_W = 0.5 + 0.5\text{sign}\bar{\sigma}_T^p$
$T3'$	$\bar{\sigma}_S^p$, $\bar{\sigma}_T^p$	$S_W = 0$	$S_U = 0.5 - 0.5\text{sign}\bar{\sigma}_T^p$ $S_V = 0.5 + 0.5\text{sign}\bar{\sigma}_S^p$
$T4'$	$\bar{\sigma}_R^p$, $\bar{\sigma}_S^p$	$S_V = 1$	$S_U = 0.5 + 0.5\text{sign}\bar{\sigma}_R^p$ $S_T = 0.5 - 0.5\text{sign}\bar{\sigma}_S^p$
$T5'$	$\bar{\sigma}_R^p$, $\bar{\sigma}_T^p$	$S_U = 0$	$S_V = 0.5 - 0.5\text{sign}\bar{\sigma}_R^p$ $S_W = 0.5 + 0.5\text{sign}\bar{\sigma}_T^p$
$T6'$	$\bar{\sigma}_S^p$, $\bar{\sigma}_T^p$	$S_W = 1$	$S_U = 0.5 - 0.5\text{sign}\bar{\sigma}_T^p$ $S_V = 0.5 + 0.5\text{sign}\bar{\sigma}_S^p$

Table 3. Comparison of rhomboid and triangular switching strategies

	Triangular[a]	Rhomboid[a]	Triangular[b]	Rhomboid[b]
THD_R	0.52%	0.203%	0.76%	0.229%
THD_S	0.46%	0.207%	0.71%	0.237%
THD_T	0.59%	0.205%	0.71%	0.210%
NS_U	64	34	36	32
NS_V	82	34	32	32
NS_W	58	34	34	32
T_{minU}	4 μs	116 μs	70 μs	85 μs
T_{minV}	4 μs	117 μs	70 μs	90 μs
T_{minW}	4 μs	115 μs	70 μs	79 μs

[a] $T_{del} = 4\mu s$ and $T_{null} = 4\mu s$
[b] $T_{del} = 7\mu s$ and $T_{null} = 70\mu s$

switching strategy the critical border is between the area e' and d'. The trajectory of the vector, $\bar{\sigma}$, gets close to the origin after a few switches. Artificial limitation of a period between two consecutive switches is not necessary.

Asymmetrical load The Park vector does not contain information about the zero phase-sequence component caused by the asymmetrical load. According to

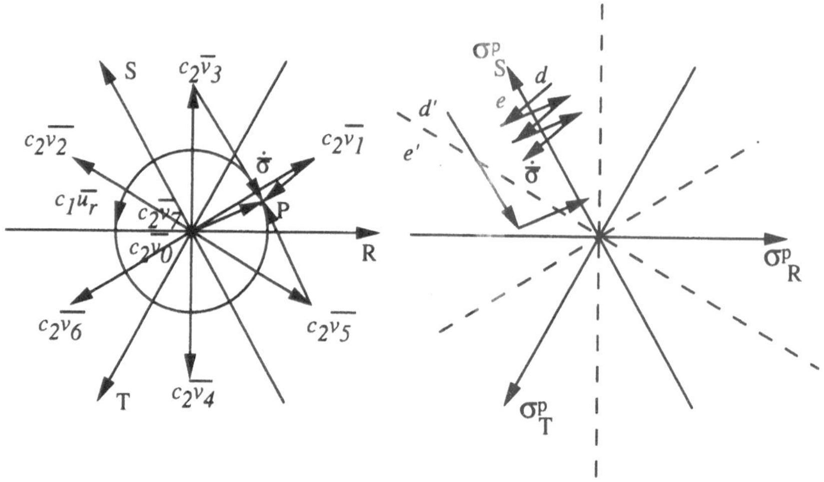

Fig. 7. Comparison of the operation of triangular and rhomboid switching strategies

the Park vector definition, the regular transformation

$$\begin{bmatrix} x_R^p \\ x_S^p \\ x_T^p \end{bmatrix} = \begin{pmatrix} \frac{2}{3} & -\frac{1}{3} & -\frac{1}{3} \\ -\frac{1}{3} & \frac{2}{3} & -\frac{1}{3} \\ -\frac{1}{3} & -\frac{1}{3} & \frac{2}{3} \end{pmatrix} \begin{bmatrix} x_R \\ x_S \\ x_T \end{bmatrix} , \tag{39}$$

which transforms the phase values into the Park vector and projects to the corresponding axes. To control the zero phase-sequence component, a K constant is introduced in the following way:

$$\begin{bmatrix} x_R^{p'} \\ x_S^{p'} \\ x_T^{p'} \end{bmatrix} = \begin{pmatrix} \frac{3-K}{3} & -\frac{K}{3} & -\frac{K}{3} \\ -\frac{K}{3} & \frac{3-K}{3} & -\frac{K}{3} \\ -\frac{K}{3} & -\frac{K}{3} & \frac{3-K}{3} \end{pmatrix} \begin{bmatrix} x_R \\ x_S \\ x_T \end{bmatrix} . \tag{40}$$

If $0 < K < 1$, it can be considered as negative feedback from the zero phase-sequence component. If the load is balanced then constant, K, has no effect on the control since

$$x_R + x_S + x_T = 0 . \tag{41}$$

5 Experimental Result

The parameters of the experimental system are given in the Table 4. Two mea-

Table 4. Nominal parameters

Nominal parameters			
Output Power	P_n	100	[kVA]
Output Voltage	U_c	220	[V]
Output Frequency	f	50	[Hz]
Switching Frequency	f_{sw}	2	[kHz]

surements are compared in Fig. 5 and Fig. 5. In both cases, the rhomboid switching strategy is applied and the load of phase R is changed from 0 to 80% of the nominal load. When $K = 0$, the two controlled phase voltages can perfectly follow the reference signals and the uncontrolled phase voltage can be sinusoidal as well but the uncontrolled phase voltage has phase angle and amplitude errors. At the beginning of each control interval, a transient phenomenon appears which increases the total harmonic distortion (D) (see in Fig. 5). Increasing the value of K, the total harmonic distortions are decreased in all phases but the asymmetry in the RMS values is increased. The optimum was found at $K = 0.7$ when the asymmetry in phase amplitudes was less than 2% and the total harmonic distortions were about 3% (see in Fig. 5)

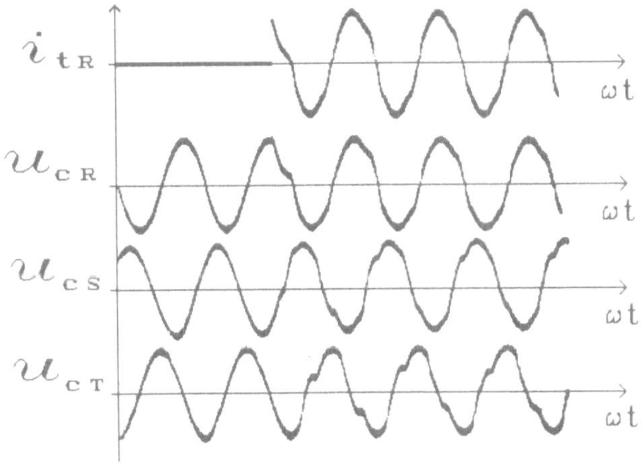

$K = 0$ $\quad I_{lR} = 120\text{A}$
$U_{cR} = 219\text{V}$ $U_{cS} = 221\text{V}$ $U_{cR} = 214\text{V}$
$D_R = 4.3\%$ $D_S = 6.3\%$ $D_T = 9.8\%$

Fig. 8. System response for 80% step change in load of phase R

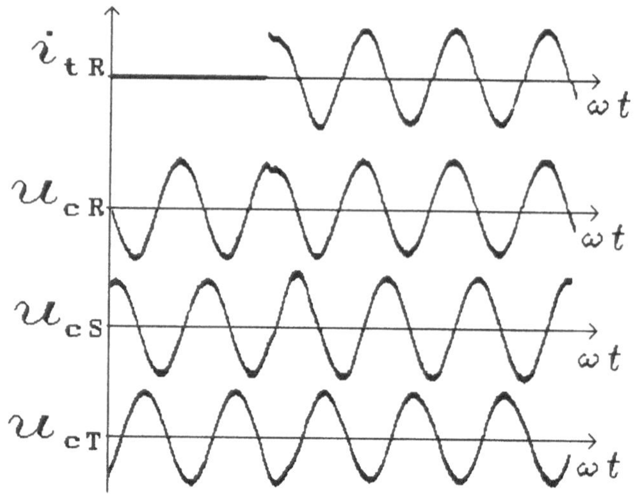

$K = 0.7$ $I_{lR} = 120A$
$U_{cR} = 219V$ $U_{cS} = 223V$ $U_{cR} = 216V$
$D_R = 2.1\%$ $D_S = 2.8\%$ $D_T = 3.1\%$

Fig. 9. System response for 80% step change in load of phase R

6 Conclusion

This chapter demonstrated that variable structure theory is a useful and practical tool by which to control inverters. The proposed controller structure performs well but it does not need a microcomputer, DSP or transputer. The effect of asymmetrical load can not be perfectly eliminated. However, there is an near-optimal solution where the asymmetry in phase amplitudes and the total harmonic distortions are acceptable.

References

1. Holtz J.(1992): Pulsewith Modulation - A Survey *IEEE Trans. on Industrial Electronics* vol. IE-39 no.5 pp. 410-420.
2. Kawamura,A.;Hoft,R.: Instantaneous Feedback Controlled PWM Inverter with Adaptive Hysteresis *IEEE Trans. on Industrial Application* Vol. IA-20, no, pp. 769-775.
3. Kawamura,A.;Yokoyama,T(1991): Comparison Five Control Methods for Digitally Feedback Controlled PWM Inverters *4th European Conference on Power Electronics*, Firenze Proc. Vol. 2. pp. 35-40.
4. Utkin V.I.(1993). *Variable Structure Control Optimalization*, Springer-Verlag.
5. Sabanovic,A.;Bilalovic,F.(1989):Sliding Mode Control of AC Drives *IEEE Trans. on Industrial Application,* Vol. IA-25, no. 1, pp. 70-75.

6. Jardan, K.R., Dewan,S.B. Slemon,G.R.(1969). "General Analysis of Three-Phase Inverters" em IEEE Trans. on Industry and General Applications. IGA-5, No.6,pp.672-679.
7. Korondi,P. Nagy,L. Nemeth,G.(1991) "Control of a three phase ups inverter with unbalanced and nonlinear load" *4th European Conference on Power Electronics Firenze*, vol. 3. pp. 3-180 - 3-184.
8. Korondi,P.(1993):"Comparison of two types of tolerance band controlled converters" *IEEE ISIE'93* Budapest, Proc. pp.128-133
9. Nabae A. Ogaswara S. Akagi H.(1985) "A novel control scheme of current controlled PWM inverter" *IEEE Conf. Record of Industry Applications Society* 1985. 473-477.
10. Nagy,I.(1993) "Control of bi-directional power conversion" *IEEE ISIE'93* Budapest Vol. pp.176-182
11. Nagy, I.(1994) Novel Adaptive Tolerance Band Based PWM for Field-Orientated Control of Induction Machines *IEEE Trans. on Industrial Electronics* vol. IE-41,
12. Korondi,P.; Yang,S-H. ;Hashimoto,H. ;Harashima,F.(1994):Pulse Modulated Variable Structure System Controller for Parallel Resonant Dual Converter *IEE Japan Ann. Meeting*
13. Retter,Gy.: *Matrix and space phasor theory of electrical machines*, Akademia Kiado, Budapest, 1987.

Dynamical Adaptive Sliding Mode Control of Observable Minimum-Phase Uncertain Nonlinear Systems

Miguel Ríos-Bolívar[1] and Alan S. I. Zinober[2]

[1] Departamento de Sistemas de Control, ULA, Mérida 5101, Venezuela
[2] Applied Mathematics Department, Sheffield University, Sheffield S10 2TN, UK

Abstract. We consider the application of a combined dynamical adaptive backstepping-sliding mode control (DAB–SMC) algorithm to nonlinear continuous chemical processes with both uncertainty and disturbances. The algorithm follows a systematic procedure for the design of dynamical adaptive SMC laws for the output regulation of observable minimum phase nonlinear systems.

1 Introduction

A simple and popular robust approach to the deterministic control of uncertain systems is the sliding mode control technique, which is based upon the special behaviour of variable structure systems in the so-called sliding regime. Sliding mode control is synthesized by means of high-frequency discontinuous regulation signals, which result in bang-bang control inputs with noticeable chattering in the controlled system responses. The chattering, associated with underlying discontinuous control policies, has traditionally been regarded as a serious drawback for the efficient regulation of certain types of continuous systems. This feature is particularly important in the regulation of either mechanical or chemical processes because discontinuities cannot be allowed in the actuator behaviour, and abrupt changes of the regulated variables are usually not tolerated.

On the other hand, dynamical SMC policies, based on state- and control input-dependent manifolds, provide outstanding features in comparison with traditional discontinuous feedback design schemes utilizing only state-dependent surfaces [13]. Both continuous control input signals and substantially smoothed chattering-free trajectories have been shown to be some of the several advantageous properties exhibited by the use of such input-dependent sliding manifolds. Dynamical feedback controllers have greatly improved the applicability of discontinuous feedback strategies leading to asymptotic stabilization and tracking in nonlinear systems [14]. Applications of this approach to mechanical and electromechanical systems have been reported by Sira-Ramírez *et al* [15]. Also, the feedback regulation of nonlinear chemical processes without uncertainty, via dynamical discontinuous compensators synthesized for systems in the Fliess generalized observability canonical form, has been addressed by Sira-Ramírez and Llanes-Santiago [16]. The control input signals thus obtained are substantially smoothed in comparison with their corresponding static alternatives.

Adaptive control is usually employed for the regulation of uncertain systems, when no information is available about the bounds of the unknown parameters. Recently, a new family of adaptive control algorithms, using the *backstepping* approach, has been developed [2, 5]. This control scheme allows the systematic design of adaptive controllers for triangular nonlinear systems containing unmatched parametric uncertainty. An alternative algorithm for the synthesis of dynamical adaptive backstepping (DAB) controllers has been proposed by Ríos-Bolívar *et al* [6]. This algorithm allows one to apply the backstepping approach to a higher class of uncertain nonlinear systems, satisfying observability and minimum phase conditions, which may be in either triangular or nontriangular canonical forms. Its application to the dynamic adaptive regulation of uncertain nonlinear chemical processes has been recently reported in [7].

In order to provide robustness in the presence of undesirable disturbances, a combined DAB-SMC design algorithm has been proposed by Ríos-Bolívar *et al* [8]. We consider here the application of this approach in the robust adaptive regulation of two nonlinear continuous chemical processes with uncertainty, and its validity is demonstrated via computer simulations.

2 Dynamical Adaptive Backstepping Sliding Mode Control

The DAB-SMC algorithm is based upon a combination of dynamical input-output linearization and the adaptive backstepping algorithm with tuning functions [5]. Its applicability to both triangular and nontriangular systems is guaranteed, but it requires that the controlled plant be *observable* and *minimum phase.* The observability condition is required to guarantee the existence of a local nonlinear mapping which transforms the plant into a convenient form of the error system. The need for the minimum phase property is to guarantee stability of the closed-loop system. At the final step of this algorithm, a sliding surface is defined in terms of the error variables and both an update law and a dynamical discontinuous feedback law are also synthesized [8].

Consider a single-input single-output nonlinear system with linearly parameterized uncertainty

$$\begin{aligned} \dot{x} &= f_0(x) + \Psi(x)\theta + \Big(g_0(x) + \varphi(x)\theta\Big)u \\ y &= h(x) \end{aligned} \tag{1}$$

where $x \in \Re^n$ is the state; $u, y \in \Re$ the input and output respectively; and $\theta = [\theta_1, \ldots, \theta_p]^T$ is a vector of unknown parameters. f_0, g_0 and the columns of the matrices $\Psi, \varphi \in \Re^{n\times p}$ are smooth vector fields in a neighbourhood R_0 of the origin $x = 0$ with $f_0(0) = 0$, $g_0(0) \neq 0$; and h is a smooth scalar function also defined in R_0.

In order to characterize the class of nonlinear systems for which this procedure is applicable, we set up a nonlinear mapping by considering the output $y(t)$

and its time derivatives as follows

$$\dot{y} = \frac{\partial h}{\partial x}\dot{x} = \frac{\partial h}{\partial x}\Big[f_0(x) + \Psi(x)\theta + (g_0(x) + \varphi(x)\theta)u\Big] \tag{2}$$

Due to the presence of the unknown parameter vector θ we rewrite (2) as

$$\begin{aligned}\dot{y} &= \mathcal{L}_h^1(x,\hat{\theta},u)\\ &= \frac{\partial h}{\partial x}\Big[f_0(x) + \Psi(x)\hat{\theta} + (g_0(x) + \varphi(x)\hat{\theta})u\Big] + \omega_1(\theta - \hat{\theta})\end{aligned} \tag{3}$$

where $\hat{\theta}$ is an estimate of θ, and ω_1 is the first regressor vector defined as

$$\omega_1 = \frac{\partial h}{\partial x}\Big(\Psi(x) + u\varphi(x)\Big) \tag{4}$$

In other words, (3) may be rewritten as

$$\dot{y} = \mathcal{L}_h^1(x,\hat{\theta},u) = \widehat{\mathcal{L}}_h^1(x,\hat{\theta},u) + \omega_1(\theta - \hat{\theta}) \tag{5}$$

with

$$\widehat{\mathcal{L}}_h^1(x,\hat{\theta},u) := \frac{\partial h}{\partial x}\Big[f_0(x) + \Psi(x)\hat{\theta} + (g_0(x) + \varphi(x)\hat{\theta})u\Big] \tag{6}$$

Note that $\widehat{\mathcal{L}}_h^1$ can be seem as a virtual output corresponding to the known part of the actual output time derivative, i.e.

$$\widehat{\mathcal{L}}_h^1(x,\hat{\theta},u) = \hat{\dot{y}}(x,\hat{\theta},u) \tag{7}$$

The time derivative of $\widehat{\mathcal{L}}_h^1$ is

$$\begin{aligned}\ddot{y} &= \frac{\partial\left(\widehat{\mathcal{L}}_h^1\right)}{\partial x}\dot{x} + \frac{\partial\left(\widehat{\mathcal{L}}_h^1\right)}{\partial\hat{\theta}}\dot{\hat{\theta}} + \frac{\partial\left(\widehat{\mathcal{L}}_h^1\right)}{\partial u}\dot{u}\\ &= \frac{\partial\left(\widehat{\mathcal{L}}_h^1\right)}{\partial x}\Big[f_0(x) + \Psi(x)\theta + (g_0(x) + \varphi(x)\theta)u\Big]\\ &\quad + \frac{\partial\left(\widehat{\mathcal{L}}_h^1\right)}{\partial\hat{\theta}}\dot{\hat{\theta}} + \frac{\partial\left(\widehat{\mathcal{L}}_h^1\right)}{\partial u}\dot{u}\end{aligned} \tag{8}$$

which can be rewritten as

$$\ddot{y} = \widehat{\mathcal{L}}_h^2(x,\hat{\theta},u,\dot{u}) + \omega_2(\theta - \hat{\theta}) + \frac{\partial\left(\widehat{\mathcal{L}}_h^1\right)}{\partial\hat{\theta}}(\dot{\hat{\theta}} - \tau_1) \tag{9}$$

with

$$\widehat{\mathcal{L}}_h^2 := \frac{\partial\left(\widehat{\mathcal{L}}_h^1\right)}{\partial x}\left[f_0(x) + \Psi(x)\hat{\theta} + (g_0(x) + \varphi(x)\hat{\theta})u\right] + \frac{\partial\left(\widehat{\mathcal{L}}_h^1\right)}{\partial\hat{\theta}}\tau_1 + \frac{\partial\left(\widehat{\mathcal{L}}_h^1\right)}{\partial u}\dot{u} \tag{10}$$

where we have replaced $\dot{\hat{\theta}}$ by the first tuning function τ_1 defined as

$$\tau_1 := \widehat{\mathcal{L}}_h^0\,\omega_1 \tag{11}$$

and the second regressor vector ω_2 is

$$\omega_2 = \frac{\partial\left(\widehat{\mathcal{L}}_h^1\right)}{\partial x}\Big(\Psi(x) + u\varphi(x)\Big) \tag{12}$$

Note that $\widehat{\mathcal{L}}_h^2$ can be seem as a recursive operator as shown below.

By proceeding successively in this manner, we obtain the j-th time derivative of the output

$$\hat{y}^{(j)} = \widehat{\mathcal{L}}_h^j(x, \hat{\theta}, u, \ldots, u^{(j-1)}) + \omega_j(\theta - \hat{\theta}) + \frac{\partial\left(\widehat{\mathcal{L}}_h^{j-1}\right)}{\partial\hat{\theta}}(\dot{\hat{\theta}} - \tau_{j-1}) \tag{13}$$

with

$$\widehat{\mathcal{L}}_h^j := \frac{\partial\left(\widehat{\mathcal{L}}_h^{j-1}\right)}{\partial x}\left[f_0(x) + \Psi(x)\hat{\theta} + (g_0(x) + \varphi(x)\hat{\theta})u\right] + \frac{\partial\left(\widehat{\mathcal{L}}_h^{j-1}\right)}{\partial\hat{\theta}}\tau_{j-1} + \sum_{k=0}^{j-2}\frac{\partial\left(\widehat{\mathcal{L}}_h^{j-1}\right)}{\partial u^{(k)}}u^{(k+1)} \tag{14}$$

$$\tau_{j-1} = \widehat{\mathcal{L}}_h^{j-2}\,\omega_{j-1} \tag{15}$$

and

$$\omega_j = \frac{\partial\left(\widehat{\mathcal{L}}_h^{j-1}\right)}{\partial x}\Big(\Psi(x) + u\varphi(x)\Big) \tag{16}$$

The expression (13) is valid if the relative degree is one. The general expression for systems with well-defined relative degree, i.e. $1 \leq \rho \leq n$, has the form

$$\hat{y}^{(j)} = \widehat{\mathcal{L}}_h^j(x, \hat{\theta}, u, \ldots, u^{(j-\rho)}) + \omega_j(\theta - \hat{\theta}) + \frac{\partial\left(\widehat{\mathcal{L}}_h^{j-1}\right)}{\partial\hat{\theta}}(\dot{\hat{\theta}} - \tau_{j-1}) \tag{17}$$

with

$$\widehat{\mathcal{L}}_h^j := \frac{\partial\left(\widehat{\mathcal{L}}_h^{j-1}\right)}{\partial x}\left[f_0(x) + \Psi(x)\hat{\theta} + (g_0(x) + \varphi(x)\hat{\theta})u\right] + \frac{\partial\left(\widehat{\mathcal{L}}_h^{j-1}\right)}{\partial\hat{\theta}}\tau_{j-1} + \sum_{k=0}^{j-\rho-1}\frac{\partial\left(\widehat{\mathcal{L}}_h^{j-1}\right)}{\partial u^{(k)}}u^{(k+1)} \tag{18}$$

In other words, the time derivatives of the output are obtained by the application of the following recursively defined operator

$$\begin{aligned}\widehat{\mathcal{L}}_h^0 &= h(x) \\ \widehat{\mathcal{L}}_h^j &:= \frac{\partial\left(\widehat{\mathcal{L}}_h^{j-1}\right)}{\partial x}\left[f_0(x) + \Psi(x)\hat{\theta} + (g_0(x) + \varphi(x)\hat{\theta})u\right] \\ &\quad + \frac{\partial\left(\widehat{\mathcal{L}}_h^{j-1}\right)}{\partial\hat{\theta}}\tau_{j-1} + \sum_{k=0}^{j-\rho-1}\frac{\partial\left(\widehat{\mathcal{L}}_h^{j-1}\right)}{\partial u^{(k)}}u^{(k+1)} \quad 1 \leq j \leq n\end{aligned} \tag{19}$$

which also characterizes the control dependent nonlinear mapping

$$z = \Xi(x, \hat{\theta}, u, \dots, u^{(n-\rho-1)}) = \begin{bmatrix} y \\ \hat{y}^{(1)} \\ \vdots \\ \hat{y}^{(n-1)} \end{bmatrix} = \begin{bmatrix} \widehat{\mathcal{L}}_h^0 \\ \widehat{\mathcal{L}}_h^1 \\ \vdots \\ \widehat{\mathcal{L}}_h^{n-1} \end{bmatrix} \tag{20}$$

Definition 1. System (1) is *observable* if the mapping (20) satisfies the rank condition

$$\text{rank}\frac{\partial \Xi(\cdot)}{\partial x} = n \tag{21}$$

in a subspace $R_1 \subset R_0 \subset \Re^n$.

Definition 2. System (1) is *minimum phase* in $R_1 \subset R_0 \subset \Re^n$ if the equilibrium of its zero-dynamics is asymptotically stable [1].

We consider here the adaptive control regulation of uncertain nonlinear systems of the form (1) which satisfy the observability and minimum-phase conditions. These two conditions replace the structural requirement, in the backstepping setting, that the system should be in a triangular form. This allows one to broaden the applicability of the backstepping approach to systems in nontriangular form.

2.1 The DAB-SMC algorithm

For observable minimum phase nonlinear systems of the form (1), the general problem of adaptively robustly tracking a bounded desired reference signal $y_r(t)$ with smooth and bounded derivatives can be solved through the DAB-SMC algorithm (see [8, 10]) summarized as follows:

Coordinate transformation

$$\begin{aligned} z_1 &:= y - y_r(t) = h^{(0)}(x) - y_r(t) \\ z_k &:= \hat{h}^{(k-1)}(\cdot) - y_r^{(k-1)}(t) + \alpha_{k-1}(\cdot), \quad 2 \le k \le n \end{aligned} \tag{22}$$

with

$$\begin{aligned} \hat{h}^{(k)} = {} & \frac{\partial \hat{h}^{(k-1)}}{\partial \hat{\theta}} \tau_k + \frac{\partial \hat{h}^{(k-1)}}{\partial x} \left[f_0 + \Psi \hat{\theta} + (g_0 + \varphi \hat{\theta}) v_1 \right] \\ & + \sum_{i=1}^{k-\rho-1} \frac{\partial \hat{h}^{(k-1)}}{\partial v_i} v_{i+1} + \frac{\partial \hat{h}^{(k-1)}}{\partial t} \end{aligned} \tag{23}$$

$$\omega_k = \left(\frac{\partial \hat{h}^{(k-1)}}{\partial x} + \frac{\partial \alpha_{k-1}}{\partial x} \right) \left(\Psi(x) + u \varphi(x) \right) \tag{24}$$

$$\begin{aligned} \alpha_k = z_{k-1} + {} & \left(\sum_{i=2}^{k-1} z_i \frac{\partial \hat{h}^{(i-1)}}{\partial \hat{\theta}} + \sum_{i=3}^{k-1} z_i \frac{\partial \alpha_{i-1}}{\partial \hat{\theta}} \right) \Gamma \omega_k^T \\ & + \sum_{i=1}^{k-\rho-1} \frac{\partial \alpha_{k-1}}{\partial v_i} v_{i+1} + \frac{\partial \alpha_{k-1}}{\partial \hat{\theta}} \tau_k + \frac{\partial \alpha_{k-1}}{\partial t} \\ & + \frac{\partial \alpha_{k-1}}{\partial x} \left[f_0 + \Psi \hat{\theta} + (g_0 + \varphi \hat{\theta}) v_1 \right] + c_k z_k \end{aligned} \tag{25}$$

$$\tau_k = \Gamma \sum_{i=1}^{k} \omega_k^T z_k \qquad 1 \le k \le n-1 \tag{26}$$

Sliding surface

Define the sliding surface

$$\sigma = k_1 z_2 + k_2 z_2 + \ldots + k_{n-1} z_{n-1} + z_n = 0 \tag{27}$$

with the design parameters k_i, $i = 1, \ldots, n-1$, chosen such that the polynomial

$$p(s) = k_1 + k_2 s + \ldots + k_{n-1} s^{n-2} + s^{n-1} \tag{28}$$

in the complex variable s is Hurwitz.

Parameter update law

$$\dot{\hat{\theta}} = \tau_n = \tau_{n-1} + \Gamma \sigma \left(\omega_n^T + \sum_{i=1}^{n-1} k_i \omega_i^T \right) \tag{29}$$

Dynamical adaptive SMC law

$$
\begin{aligned}
\dot{v}_1 &= v_2 \\
\dot{v}_2 &= v_3 \\
&\vdots \\
\dot{v}_{n-\rho} &= \frac{1}{\Delta}\Bigg[y_r^{(n)}(t) - \frac{\partial \hat{h}^{(n-1)}}{\partial t} - \frac{\partial \alpha_{n-1}}{\partial t} \\
&\quad - \left(\frac{\partial \hat{h}^{(n-1)}}{\partial \hat{\theta}} + \frac{\partial \alpha_{n-1}}{\partial \hat{\theta}} \right) \tau_n - \sum_{i=1}^{n-\rho-1} \left(\frac{\partial \hat{h}^{(n-1)}}{\partial v_i} + \frac{\partial \alpha_{n-1}}{\partial v_i} \right) v_{i+1} \\
&\quad - \left(\frac{\partial \hat{h}^{(n-1)}}{\partial x} + \frac{\partial \alpha_{n-1}}{\partial x} \right) \left(f_0 + \Psi \hat{\theta} + (g_0 + \varphi \hat{\theta}) v_1 \right) \\
&\quad - \sum_{i=2}^{n-1} \left(\frac{\partial \hat{h}^{(n-1)}}{\partial \hat{\theta}} + \frac{\partial \alpha_{n-1}}{\partial \hat{\theta}} \right) z_i \Gamma \left(\omega_n^T + \sum_{i=1}^{n-1} k_i \omega_i^T \right) \\
&\quad + \sum_{i=1}^{n-1} k_i \left(\frac{\partial \hat{h}^{(n-1)}}{\partial \hat{\theta}} + \frac{\partial \alpha_{n-1}}{\partial \hat{\theta}} \right) (\tau_n - \tau_i) \\
&\quad + \sum_{i=1}^{n-1} k_i \left(\sum_{j=2}^{i-1} z_j \frac{\partial \hat{h}^{(j-1)}}{\partial \hat{\theta}} + \sum_{j=3}^{i-1} z_j \frac{\partial \alpha_{j-1}}{\partial \hat{\theta}} \right) \\
&\quad - \sum_{i=1}^{n-1} k_i (-z_{i-1} - c_i z_i + z_{i+1}) - \kappa \left(\sigma + \beta \operatorname{sgn}(\sigma) \right) \Bigg]
\end{aligned} \tag{30}
$$

with

$$
\begin{aligned}
v_1 &= u \\
\Delta &= \left(\frac{\partial \hat{h}^{(n-1)}}{\partial v_{n-\rho}} + \frac{\partial \alpha_{n-1}}{\partial v_{n-\rho}} \right)
\end{aligned} \tag{31}
$$

where the c_i's are constant design parameters and $\Gamma = \Gamma^T > 0$ is the adaptation gain matrix. The control u is obtained implicitly as the solution of the nonlinear time-varying differential equation (30). The above described adaptive control allows one to formulate the following theorem.

Theorem 3. *The closed-loop system consisting of the plant (1), the dynamical adaptive SMC law (30) and the update law (29), has a locally stable equilibrium point at* $(z, \hat{\theta} - \theta) = (0, 0)$ *and* $\lim_{t \to \infty} z(t) = 0$, *which means that asymptotic tracking is achieved, i.e.*

$$
\lim_{t \to \infty} [y(t) - y_r(t)] = 0 \tag{32}
$$

Moreover, a sliding mode is generated on the sliding surface (27).

Proof. The proof follows the systematic control design algorithm as follows

Step 1. Define the output tracking error as follows:

$$z_1 := y - y_r(t) = h(x) - y_r(t) \tag{33}$$

whose time derivative is given by

$$\dot{z}_1 = h^{(1)}(x,\theta) - \dot{y}_r(t) = \frac{\partial h}{\partial x}\Big[f_0 + \Psi\theta + (g_0 + \varphi\theta)u\Big] - \dot{y}_r(t) \tag{34}$$

If the relative degree ρ with respect to u is greater than one,

$$\frac{\partial h}{\partial x}\Big(g_0(x) + \varphi(x)\theta\Big) = 0 \tag{35}$$

For the sake of generality, it is assumed that the relative degree ρ is greater than one. Nevertheless, this algorithm is also applicable to systems with $\rho = 1$. By adding to and subtracting from the actual value of the parameters θ their estimated values $\hat{\theta}$, (34) can be rewritten as

$$\dot{z}_1 = \hat{h}^{(1)}(x,\hat{\theta}) - \dot{y}_r(t) + \omega_1(\theta - \hat{\theta}) \tag{36}$$

with

$$\hat{h}^{(1)}(x,\hat{\theta}) = \frac{\partial h}{\partial x}\Big(f_0(x) + \Psi(x)\hat{\theta}\Big) \tag{37}$$

$$\omega_1 = \frac{\partial h}{\partial x}\Psi(x) \tag{38}$$

Consider the quadratic Lyapunov function

$$V_1 = \frac{1}{2}z_1^2 + \frac{1}{2}(\theta - \hat{\theta})^T\Gamma^{-1}(\theta - \hat{\theta}) \tag{39}$$

where $\Gamma = \Gamma^T > 0$ is a matrix of adaptation gains. The time derivative of V_1 is

$$\dot{V}_1 = z_1\Big(\hat{h}^{(1)}(x,\hat{\theta}) - \dot{y}_r(t)\Big) + (\theta - \hat{\theta})^T\Gamma^{-1}(-\dot{\hat{\theta}} + \Gamma\omega_1^T z_1) \tag{40}$$

One can achieve $\dot{V}_1 = -c_1 z_1^2$ with c_1 a positive scalar design constant, by choosing the tuning function

$$\dot{\hat{\theta}} = \tau_1 = \Gamma\omega_1^T z_1 \tag{41}$$

if the relation

$$\hat{h}^{(1)}(x,\hat{\theta}) - \dot{y}_r(t) = -c_1 z_1 \tag{42}$$

is satisfied. The expression (42) represents a desired algebraic relation for which effective stabilization of the output tracking error would be possible in combination with the estimation update law (41). However, since (42) is not valid from the outset and τ_1 is not considered as an update law but rather as the first tuning function, the deviation is taken as the second error variable, i.e.

$$z_2 := \hat{h}^{(1)}(x,\hat{\theta}) - \dot{y}_r(t) + \alpha_1 \tag{43}$$

with

$$\alpha_1 = c_1 z_1 \tag{44}$$

The closed-loop form is

$$\dot{z}_1 = -c_1 z_1 + z_2 + \omega_1(\theta - \hat{\theta}) \tag{45}$$

and

$$\dot{V}_1 = -c_1 z_1^2 + z_1 z_2 + (\theta - \hat{\theta})^T \Gamma^{-1}(-\dot{\hat{\theta}} + \tau_1). \tag{46}$$

By induction one obtains the following j-th generic step which characterizes the first steps prior to the explicit appearance of the control input in the transformed dynamical system.

Step j $(2 \leq j \leq \rho - 1)$

$$\begin{aligned}\dot{z}_j = {} & \hat{h}^{(j)}(x, \hat{\theta}, t) - y_r^{(j)}(t) + \frac{\partial \alpha_{j-1}}{\partial x}(f_0 + \Psi\hat{\theta}) + \frac{\partial \alpha_{j-1}}{\partial \hat{\theta}}\tau_j + \frac{\partial \alpha_{j-1}}{\partial t} \\ & + \omega_j(\theta - \hat{\theta}) + \left(\frac{\partial \hat{h}^{(j-1)}}{\partial \hat{\theta}} + \frac{\partial \alpha_{j-1}}{\partial \hat{\theta}}\right)(\dot{\hat{\theta}} - \tau_j)\end{aligned} \tag{47}$$

with

$$\hat{h}^{(j)}(x, \hat{\theta}, t) = \frac{\partial \hat{h}^{(j-1)}}{\partial x}(f_0 + \Psi\hat{\theta}) + \frac{\partial \hat{h}^{(j-1)}}{\partial \hat{\theta}}\tau_j + \frac{\partial \hat{h}^{(j-1)}}{\partial t} \tag{48}$$

$$\omega_j = \left(\frac{\partial \hat{h}^{(j-1)}}{\partial x} + \frac{\partial \alpha_{j-1}}{\partial x}\right)\Psi(x) \tag{49}$$

and τ_j the corresponding tuning function defined at this step. By augmenting the Lyapunov function

$$V_j = V_{j-1} + \frac{1}{2}z_j^2 = \frac{1}{2}\sum_{i=1}^{j} z_i^2 + \frac{1}{2}(\theta - \hat{\theta})^T \Gamma^{-1}(\theta - \hat{\theta}) \tag{50}$$

its time derivative is

$$\begin{aligned}\dot{V}_j = {} & -\sum_{i=1}^{j-1} c_i z_i^2 + (\theta - \hat{\theta})^T \Gamma^{-1}(-\dot{\hat{\theta}} + \tau_{j-1} + \Gamma\omega_j^T z_j) \\ & + z_j\left(\frac{\partial \hat{h}^{(j-1)}}{\partial \hat{\theta}} + \frac{\partial \alpha_{j-1}}{\partial \hat{\theta}}\right)(\dot{\hat{\theta}} - \tau_j) \\ & + \left(\sum_{i=2}^{j-1} z_i \frac{\partial \hat{h}^{(i-1)}}{\partial \hat{\theta}} + \sum_{i=3}^{j-1} z_i \frac{\partial \alpha_{i-1}}{\partial \hat{\theta}}\right)(\dot{\hat{\theta}} - \tau_{j-1}) \\ & + z_j\left[z_{j-1} + \hat{h}^{(j)}(x, \hat{\theta}, t) - y_r^{(j)}(t) + \frac{\partial \alpha_{j-1}}{\partial t}\right. \\ & \left. + \frac{\partial \alpha_{j-1}}{\partial \hat{\theta}}\tau_j + \frac{\partial \alpha_{j-1}}{\partial x}(f_0 + \Psi\hat{\theta})\right]\end{aligned} \tag{51}$$

The parameter estimate error $(\theta - \hat{\theta})$ can be eliminated from $\dot{V}_j$ by choosing the update law

$$\dot{\hat{\theta}} = \tau_j = \tau_{j-1} + \Gamma \omega_j^T z_j \tag{52}$$

However, τ_j will instead be used as a new tuning function. Thus, noting that

$$\dot{\hat{\theta}} - \tau_{j-1} = \dot{\hat{\theta}} - \tau_j + \tau_j - \tau_{j-1} = \dot{\hat{\theta}} - \tau_j + \Gamma \omega_j^T z_j, \tag{53}$$

one can rewrite $\dot{V}_j$ as

$$\begin{aligned}
\dot{V}_j = &- \sum_{i=1}^{j-1} c_i z_i^2 + (\theta - \hat{\theta})^T \Gamma^{-1} (-\dot{\hat{\theta}} + \tau_j) \\
&+ \left(\sum_{i=2}^{j} z_i \frac{\partial \hat{h}^{(i-1)}}{\partial \hat{\theta}} + \sum_{i=3}^{j} z_i \frac{\partial \alpha_{i-1}}{\partial \hat{\theta}} \right) (\dot{\hat{\theta}} - \tau_j) \\
&+ z_j \left[\left(\sum_{i=2}^{j-1} z_i \frac{\partial \hat{h}^{(i-1)}}{\partial \hat{\theta}} + \sum_{i=3}^{j-1} z_i \frac{\partial \alpha_{i-1}}{\partial \hat{\theta}} \right) \Gamma \omega_j^T + \hat{h}^{(j)}(x, \hat{\theta}, t) - y_r^{(j)}(t) \right. \\
&\left. + \frac{\partial \alpha_{j-1}}{\partial x} (f_0 + \Psi \hat{\theta}) + \frac{\partial \alpha_{j-1}}{\partial \hat{\theta}} \tau_j + \frac{\partial \alpha_{j-1}}{\partial t} + z_{j-1} \right]
\end{aligned} \tag{54}$$

One can achieve $\dot{V}_j = -\sum_{i=1}^{j} c_i z_i^2$, with the c_i's being positive scalar design constants, if τ_j is the update law and the relation

$$\begin{aligned}
&\left(\sum_{i=2}^{j-1} z_i \frac{\partial \hat{h}^{(i-1)}}{\partial \hat{\theta}} + \sum_{i=3}^{j-1} z_i \frac{\partial \alpha_{i-1}}{\partial \hat{\theta}} \right) \Gamma \omega_j^T + \hat{h}^{(j)}(x, \hat{\theta}, t) - y_r^{(j)}(t) \\
&+ \frac{\partial \alpha_{j-1}}{\partial x} (f_0 + \Psi \hat{\theta}) + \frac{\partial \alpha_{j-1}}{\partial \hat{\theta}} \tau_j + \frac{\partial \alpha_{j-1}}{\partial t} + z_{j-1} = -c_j z_j
\end{aligned} \tag{55}$$

is satisfied. Since (55) is not valid from the outset, its deviation is taken as the $(j+1)$-th error variable

$$z_{j+1} := \hat{h}^{(j)}(x, \hat{\theta}, t) - y_r^{(j)}(t) + \alpha_j(x, \hat{\theta}, t) \tag{56}$$

with

$$\begin{aligned}
\alpha_j = z_{j-1} &+ \left(\sum_{i=2}^{j-1} z_i \frac{\partial \hat{h}^{(i-1)}}{\partial \hat{\theta}} + \sum_{i=3}^{j-1} z_i \frac{\partial \alpha_{i-1}}{\partial \hat{\theta}} \right) \Gamma \omega_j^T + \frac{\partial \alpha_{j-1}}{\partial \hat{\theta}} \tau_j \\
&+ \frac{\partial \alpha_{j-1}}{\partial x} (f_0(x) + \Psi(x) \hat{\theta}) + \frac{\partial \alpha_{j-1}}{\partial t} + c_j z_j
\end{aligned} \tag{57}$$

obtaining the closed-loop form for $\dot{z}_j$ as

$$\begin{aligned}
\dot{z}_j = &-z_{j-1} - c_j z_j + z_{j+1} + \omega_j (\theta - \hat{\theta}) + \left(\frac{\partial \hat{h}^{(j-1)}}{\partial \hat{\theta}} + \frac{\partial \alpha_{j-1}}{\partial \hat{\theta}} \right) (\dot{\hat{\theta}} - \tau_j) \\
&- \left(\sum_{i=2}^{j-1} z_i \frac{\partial \hat{h}^{(i-1)}}{\partial \hat{\theta}} + \sum_{i=3}^{j-1} z_i \frac{\partial \alpha_{i-1}}{\partial \hat{\theta}} \right) \Gamma \omega_j^T
\end{aligned} \tag{58}$$

and

$$\dot{V}_j = -\sum_{i=1}^{j} c_i z_i^2 + z_j z_{j+1} + (\theta - \hat{\theta})^T \Gamma^{-1}(-\dot{\hat{\theta}} + \tau_j)$$
$$+ \left(\sum_{i=2}^{j} z_i \frac{\partial \hat{h}^{(i-1)}}{\partial \hat{\theta}} + \sum_{i=3}^{j} z_i \frac{\partial \alpha_{i-1}}{\partial \hat{\theta}} \right) (\dot{\hat{\theta}} - \tau_j). \tag{59}$$

Now the steps containing the control input and its derivatives are summarized in the following generic step.

Step k ($\rho \leq k \leq n-1$)

$$\dot{z}_k = \hat{h}^{(k)}(x, \hat{\theta}, u, \ldots, u^{(k-\rho)}, t) - y_r^{(k)}(t) + \frac{\partial \alpha_{k-1}}{\partial t} + \frac{\partial \alpha_{k-1}}{\partial \hat{\theta}} \tau_k$$
$$+ \frac{\partial \alpha_{k-1}}{\partial x} \left[f_0 + \Psi \hat{\theta} + (g_0 + \varphi \hat{\theta}) u \right] + \sum_{i=1}^{k-\rho} \frac{\partial \alpha_{k-1}}{\partial u^{(i-1)}} u^{(i)} + \omega_k (\theta - \hat{\theta})$$
$$+ \left(\frac{\partial \hat{h}^{(k-1)}}{\partial \hat{\theta}} + \frac{\partial \alpha_{k-1}}{\partial \hat{\theta}} \right) (\dot{\hat{\theta}} - \tau_k) \tag{60}$$

with

$$\hat{h}^{(k)}(x, \hat{\theta}, u, \ldots, u^{(k-\rho)}, t) = \frac{\partial \hat{h}^{(k-1)}}{\partial \hat{\theta}} \tau_k + \frac{\partial \hat{h}^{(k-1)}}{\partial x} \left[f_0 + \Psi \hat{\theta} + (g_0 + \varphi \hat{\theta}) u \right]$$
$$+ \sum_{i=1}^{k-\rho} \frac{\partial \hat{h}^{(k-1)}}{\partial u^{(i-1)}} u^{(i)} + \frac{\partial \hat{h}^{(k-1)}}{\partial t} \tag{61}$$

$$\omega_k = \left(\frac{\partial \hat{h}^{(k-1)}}{\partial x} + \frac{\partial \alpha_{k-1}}{\partial x} \right) (\Psi + \varphi u) \tag{62}$$

and τ_k the tuning function defined at this step. By augmenting the Lyapunov function

$$V_k = V_{k-1} + \frac{1}{2} z_k^2 = \frac{1}{2} \sum_{i=1}^{k} z_i^2 + \frac{1}{2} (\theta - \hat{\theta})^T \Gamma^{-1} (\theta - \hat{\theta}) \tag{63}$$

and its time derivative is

$$\dot{V}_k = -\sum_{i=1}^{k-1} c_i z_i^2 + (\theta - \hat{\theta})^T \Gamma^{-1} \left(-\dot{\hat{\theta}} + \tau_{k-1} + \Gamma \omega_k^T z_k \right)$$
$$+ z_k \left(\frac{\partial \hat{h}^{(k-1)}}{\partial \hat{\theta}} + \frac{\partial \alpha_{k-1}}{\partial \hat{\theta}} \right) (\dot{\hat{\theta}} - \tau_k)$$
$$+ \left(\sum_{i=2}^{k-1} z_i \frac{\partial \hat{h}^{(i-1)}}{\partial \hat{\theta}} + \sum_{i=3}^{k-1} z_i \frac{\partial \alpha_{i-1}}{\partial \hat{\theta}} \right) (\dot{\hat{\theta}} - \tau_{k-1})$$

$$+ z_k \Bigg[z_{k-1} + \hat{h}^{(k)} - y_r^{(k)} + \frac{\partial \alpha_{k-1}}{\partial \hat{\theta}} \tau_k + \sum_{i=1}^{k-\rho} \frac{\partial \alpha_{k-1}}{\partial u^{(i-1)}} u^{(i)} + \frac{\partial \alpha_{k-1}}{\partial t} + \frac{\partial \alpha_{k-1}}{\partial x} \Big[f_0 + \Psi \hat{\theta} + (g_0 + \varphi \hat{\theta}) u \Big] \Bigg] \tag{64}$$

The parameter estimate error $(\theta - \hat{\theta})$ can be eliminated from $\dot{V}_k$ by choosing the update law

$$\dot{\hat{\theta}} = \tau_k = \tau_{k-1} + \Gamma \omega_k^T z_k. \tag{65}$$

However, τ_k will instead be used as a new tuning function. Thus, noting that

$$\dot{\hat{\theta}} - \tau_{k-1} = \dot{\hat{\theta}} - \tau_k + \tau_k - \tau_{k-1} = \dot{\hat{\theta}} - \tau_k + \Gamma \omega_k^T z_k, \tag{66}$$

$\dot{V}_k$ can be rewritten as

$$\begin{aligned} \dot{V}_k = & - \sum_{i=1}^{k-1} c_i z_i^2 + (\theta - \hat{\theta})^T \Gamma^{-1} (-\dot{\hat{\theta}} + \tau_k) \\ & + \left(\sum_{i=2}^{k} z_i \frac{\partial \hat{h}^{(i-1)}}{\partial \hat{\theta}} + \sum_{i=3}^{k} z_i \frac{\partial \alpha_{i-1}}{\partial \hat{\theta}} \right) (\dot{\hat{\theta}} - \tau_k) \\ & + z_k \Bigg[\left(\sum_{i=2}^{k-1} z_i \frac{\partial \hat{h}^{(i-1)}}{\partial \hat{\theta}} + \sum_{i=3}^{k-1} z_i \frac{\partial \alpha_{i-1}}{\partial \hat{\theta}} \right) \Gamma \omega_k^T + \hat{h}^{(k)} - y_r^{(k)} \\ & + \frac{\partial \alpha_{k-1}}{\partial \hat{\theta}} \tau_k + \frac{\partial \alpha_{k-1}}{\partial x} \Big[f_0 + \Psi \hat{\theta} + (g_0 + \varphi \hat{\theta}) u \Big] \\ & + \sum_{i=1}^{k-\rho} \frac{\partial \alpha_{k-1}}{\partial u^{(i-1)}} u^{(i)} + \frac{\partial \alpha_{k-1}}{\partial t} + z_{k-1} \Bigg] \end{aligned} \tag{67}$$

One can achieve $\dot{V}_k = -\sum_{i=1}^{k} c_i z_i^2$, with the c_i's being positive scalar design constants, if τ_k were the update law and the relation

$$\begin{aligned} z_{k-1} + & \left(\sum_{i=2}^{k-1} z_i \frac{\partial \hat{h}^{(i-1)}}{\partial \hat{\theta}} + \sum_{i=3}^{k-1} z_i \frac{\partial \alpha_{i-1}}{\partial \hat{\theta}} \right) \Gamma \omega_k^T + \hat{h}^{(k)} - y_r^{(k)} + \sum_{i=1}^{k-\rho} \frac{\partial \alpha_{k-1}}{\partial u^{(i-1)}} u^{(i)} \\ & + \frac{\partial \alpha_{k-1}}{\partial \hat{\theta}} \tau_k + \frac{\partial \alpha_{k-1}}{\partial x} \Big[f_0 + \Psi \hat{\theta} + (g_0 + \varphi \hat{\theta}) u \Big] + \frac{\partial \alpha_{k-1}}{\partial t} = -c_k z_k \end{aligned} \tag{68}$$

were satisfied. However, since (68) is not valid from the outset, its deviation is taken as the $(k+1)$-th error variable

$$z_{k+1} := \hat{h}^{(k)}(x, \hat{\theta}, u, \ldots, u^{(k-\rho)}, t) - y_r^{(k)} + \alpha_k(x, \hat{\theta}, u, \ldots, u^{(k-\rho)}, t) \tag{69}$$

with

$$\begin{aligned} \alpha_k = z_{k-1} + & \left(\sum_{i=2}^{k-1} z_i \frac{\partial \hat{h}^{(i-1)}}{\partial \hat{\theta}} + \sum_{i=3}^{k-1} z_i \frac{\partial \alpha_{i-1}}{\partial \hat{\theta}} \right) \Gamma \omega_k^T + \sum_{i=1}^{k-\rho} \frac{\partial \alpha_{k-1}}{\partial u^{(i-1)}} u^{(i)} \\ & + \frac{\partial \alpha_{k-1}}{\partial x} \Big[f_0 + \Psi \hat{\theta} + (g_0 + \varphi \hat{\theta}) u \Big] + \frac{\partial \alpha_{k-1}}{\partial \hat{\theta}} \tau_k + \frac{\partial \alpha_{k-1}}{\partial t} + c_k z_k . \end{aligned} \tag{70}$$

We obtain the closed-loop form

$$\dot z_k = -z_{k-1} - c_k z_k + z_{k+1} + \omega_k(\theta - \hat\theta)^T + \left(\frac{\partial \hat h^{(k-1)}}{\partial \hat\theta} + \frac{\partial \alpha_{k-1}}{\partial \hat\theta}\right)(\dot{\hat\theta} - \tau_k)$$
$$- \left(\sum_{i=2}^{k-1} z_i \frac{\partial \hat h^{(i-1)}}{\partial \hat\theta} + \sum_{i=3}^{k-1} z_i \frac{\partial \alpha_{i-1}}{\partial \hat\theta}\right) \Gamma \omega_k^T \tag{71}$$

and

$$\dot V_k = -\sum_{i=1}^{k} c_i z_i^2 + z_k z_{k+1} + (\theta - \hat\theta)^T \Gamma^{-1}(-\dot{\hat\theta} + \tau_k)$$
$$+ \left(\sum_{i=2}^{k} z_i \frac{\partial \hat h^{(i-1)}}{\partial \hat\theta} + \sum_{i=3}^{k} z_i \frac{\partial \alpha_{i-1}}{\partial \hat\theta}\right)(\dot{\hat\theta} - \tau_k) \tag{72}$$

Step n. At this step we obtain the update law and the *dynamical adaptive sliding mode* tracking controller. After the $k = (n-1)$-th step of the DAB algorithm, the transformed system is

$$\dot z_1 = -c_1 z_1 + z_2 + \omega_1(\theta - \hat\theta)$$
$$\dot z_2 = -z_1 - c_2 z_2 + z_3 + \omega_2(\theta - \hat\theta) + \frac{\partial \hat h^{(1)}}{\partial \hat\theta}(\dot{\hat\theta} - \tau_2)$$
$$\vdots$$
$$\dot z_k = -z_{k-1} - c_k z_k + z_{k+1} + \omega_k(\theta - \hat\theta) + \left(\frac{\partial \hat h^{(k-1)}}{\partial \hat\theta} + \frac{\partial \alpha_{k-1}}{\partial \hat\theta}\right)(\dot{\hat\theta} - \tau_k)$$
$$- \left(\sum_{i=2}^{k-1} z_i \frac{\partial \hat h^{(i-1)}}{\partial \hat\theta} + \sum_{i=3}^{k-1} z_i \frac{\partial \alpha_{i-1}}{\partial \hat\theta}\right) \Gamma \omega_k^T \tag{73}$$
$$\vdots$$
$$\dot z_n = \hat h^{(n)}(x, \hat\theta, u, \ldots, u^{(n-\rho)}, t) - y_r^{(n)}(t) + \alpha_n(x, \hat\theta, u, \ldots, u^{(n-\rho-1)}, t)$$
$$+\omega_n(\theta - \hat\theta) + \left(\frac{\partial \hat h^{(n-1)}}{\partial \hat\theta} + \frac{\partial \alpha_{n-1}}{\partial \hat\theta}\right)(\dot{\hat\theta} - \tau_n)$$

$$\tau_{n-1} = \Gamma \sum_{i=1}^{n-1} \omega_i^T z_i$$

and the time derivative of V_{n-1} is

$$\dot V_{n-1} = -\sum_{i=1}^{n-1} c_i z_i^2 + z_{n-1} z_n + \left(\sum_{i=2}^{n-1} z_i \frac{\partial \hat h^{(i-1)}}{\partial \hat\theta} + \sum_{i=3}^{n-1} z_i \frac{\partial \alpha_{i-1}}{\partial \hat\theta}\right)\left(\dot{\hat\theta} - \tau_{n-1}\right)$$
$$+(\theta - \hat\theta)^T \Gamma^{-1}\left(-\dot{\hat\theta} + \tau_{n-1}\right) \tag{74}$$

We now define the sliding surface

$$\sigma = k_1 z_1 + k_2 z_2 + \ldots + k_{n-1} z_{n-1} + z_n = 0 \tag{75}$$

with the positive design parameters k_i, $i = 1, \ldots, n-1$, chosen in such a manner that the polynomial

$$p(s) = k_1 + k_2 s + \ldots + k_{n-1} s^{n-2} + s^{n-1} \tag{76}$$

in the complex variable s is Hurwitz. Extending the Lyapunov function as

$$V_n = V_{n-1} + \frac{1}{2}\sigma^2 = \frac{1}{2}\sum_{i=1}^{n-1} z_i^2 + \frac{1}{2}\sigma^2 + \frac{1}{2}(\theta - \hat{\theta})^T \Gamma^{-1}(\theta - \hat{\theta}) \tag{77}$$

the time derivative of V_n is

$$\begin{aligned}
\dot{V}_n = &-\sum_{i=1}^{n-1} c_i z_i^2 + z_{n-1} z_n + \left(\sum_{i=2}^{n-1} z_i \frac{\partial \hat{h}^{(i-1)}}{\partial \hat{\theta}} + \sum_{i=3}^{n-1} z_i \frac{\partial \alpha_{i-1}}{\partial \hat{\theta}}\right)\left(\dot{\hat{\theta}} - \tau_{n-1}\right) \\
&+\sigma\Bigg[\hat{h}^{(n)} - y_r^{(n)}(t) + \alpha_n + \left(\frac{\partial \hat{h}^{(n-1)}}{\partial \hat{\theta}} + \frac{\partial \alpha_{n-1}}{\partial \hat{\theta}}\right)\left(\dot{\hat{\theta}} - \tau_n\right) \\
&+\sum_{i=1}^{n-1} k_i \left(-z_{i-1} - c_i z_i + z_{i+1} - \left(\sum_{j=2}^{i-1} z_j \frac{\partial \hat{h}^{(j-1)}}{\partial \hat{\theta}} + \sum_{j=3}^{i-1} z_j \frac{\partial \alpha_{j-1}}{\partial \hat{\theta}}\right)\Gamma \omega_i^T\right) \\
&-\sum_{i=1}^{n-1} k_i \left(\frac{\partial \hat{h}^{(i-1)}}{\partial \hat{\theta}} + \frac{\partial \alpha_{i-1}}{\partial \hat{\theta}}\right)\left(\dot{\hat{\theta}} - \tau_i\right)\Bigg] \\
&+(\theta - \hat{\theta})^T \Gamma^{-1}\left(-\dot{\hat{\theta}} + \tau_{n-1} + \Gamma\sigma\left(\omega_n^T + \sum_{i=1}^{n-1} k_i \omega_i^T\right)\right)
\end{aligned} \tag{78}$$

We can eliminate $(\theta - \hat{\theta})$ from $\dot{V}_n$ by choosing the update law

$$\dot{\hat{\theta}} = \tau_n = \tau_{n-1} + \Gamma\sigma\left(\omega_n^T + \sum_{i=1}^{n-1} k_i \omega_i^T\right). \tag{79}$$

Now, noting that

$$\dot{\hat{\theta}} - \tau_{n-1} = \tau_n - \tau_{n-1} = \Gamma\sigma\left(\omega_n^T + \sum_{i=1}^{n-1} k_i \omega_i^T\right) \tag{80}$$

$\dot{V}_n$ can be rewritten as

$$\begin{aligned}
\dot{V}_n = &-\sum_{i=1}^{n-1} c_i z_i^2 + z_{n-1} z_n \\
&+\sigma\Bigg[+\left(\sum_{i=2}^{n-1} z_i \frac{\partial \hat{h}^{(i-1)}}{\partial \hat{\theta}} + \sum_{i=3}^{n-1} z_i \frac{\partial \alpha_{i-1}}{\partial \hat{\theta}}\right)\Gamma\left(\omega_n^T + \sum_{i=1}^{n-1} k_i \omega_i^T\right)
\end{aligned}$$

$$+\sum_{i=1}^{n-1} k_i \left(-z_{i-1} - c_i z_i + z_{i+1} - \left(\sum_{j=2}^{i-1} z_j \frac{\partial \hat{h}^{(j-1)}}{\partial \hat{\theta}} + \sum_{j=3}^{i-1} z_j \frac{\partial \alpha_{j-1}}{\partial \hat{\theta}} \right) \Gamma \omega_i^T \right)$$

$$\left. - \sum_{i=1}^{n-1} k_i \left(\frac{\partial \hat{h}^{(i-1)}}{\partial \hat{\theta}} + \frac{\partial \alpha_{i-1}}{\partial \hat{\theta}} \right) (\tau_n - \tau_i) + \hat{h}^{(n)} - y_r^{(n)}(t) + \alpha_n \right] \quad (81)$$

Finally, to achieve

$$\dot{V}_n = -\sum_{i=1}^{n-1} c_i z_i^2 + z_{n-1} z_n - \kappa \sigma^2 - \kappa \beta \, |\sigma \quad (82)$$

the bracketed term multiplying σ should be $-\kappa(\sigma + \beta \operatorname{sgn}(\sigma))$, where κ and β are positive design parameters and sgn is the *signum* function. So

$$\hat{h}^{(n)} - y_r^{(n)}(t) + \alpha_n + \left(\sum_{i=2}^{n-1} z_i \frac{\partial \hat{h}^{(i-1)}}{\partial \hat{\theta}} + \sum_{i=3}^{n-1} z_i \frac{\partial \alpha_{i-1}}{\partial \hat{\theta}} \right) \Gamma \left(\omega_n^T + \sum_{i=1}^{n-1} k_i \omega_i^T \right)$$

$$+\sum_{i=1}^{n-1} k_i \left(-z_{i-1} - c_i z_i + z_{i+1} - \left(\sum_{j=2}^{i-1} z_j \frac{\partial \hat{h}^{(j-1)}}{\partial \hat{\theta}} + \sum_{j=3}^{i-1} z_j \frac{\partial \alpha_{j-1}}{\partial \hat{\theta}} \right) \Gamma \omega_i^T \right)$$

$$- \sum_{i=1}^{n-1} k_i \left(\frac{\partial \hat{h}^{(i-1)}}{\partial \hat{\theta}} + \frac{\partial \alpha_{i-1}}{\partial \hat{\theta}} \right) (\tau_n - \tau_i) = -\kappa \Big(\sigma + \beta \operatorname{sgn}(\sigma) \Big) \quad (83)$$

The update law (79) together with the dynamical discontinuous adaptive feedback law (83) achieve a sliding mode on the sliding surface (75). Note that (82) can be rewritten as

$$\dot{V}_n = -z^T Q z - \kappa \beta |\sigma| \quad (84)$$

where Q is a symmetric matrix with the following form

$$Q = \begin{bmatrix} c_1 + \kappa k_1^2 & \dots & \kappa k_1 k_{n-1} & \kappa k_1 \\ \kappa k_2 k_1 & \dots & \kappa k_2 k_{n-1} & \kappa k_2 \\ \vdots & \ddots & \vdots & \vdots \\ \kappa k_{n-1} k_1 & \dots & c_{n-1} + \kappa k_{n-1}^2 & -\frac{1}{2} + \kappa k_{n-1} \\ \kappa k_1 & \dots & -\frac{1}{2} + \kappa k_{n-1} & \kappa \end{bmatrix}$$

Noting that the determinants of the principal minors of Q are all positive, a sufficient condition to guarantee that Q is positive definite is

$$|Q| = \left[-\frac{1}{4} + \kappa (c_{n-1} + k_{n-1}) \right] \prod_{i=1}^{n-2} c_i - \frac{1}{4} \kappa \sum_{i=1}^{n-2} (c_1 \dots c_{i-1} k_i^2 c_{i+1} \dots c_{n-2}) > 0. \quad (85)$$

Note that the discontinuous feedback control law (83) can be rewritten in the form of the dynamical adaptive sliding mode control law (30) by replacing the control input u and its derivatives $\dot{u}, \ddot{u}, \dots$ by the state variables $v_1, v_2, v_3, \dots$ respectively and solving for $\dot{v}_{n-\rho}$. Therefore, stability is guaranteed and asymptotic output tracking is achieved. Moreover, since the condition $\sigma \dot{\sigma} \leq 0$ holds, a sliding mode is generated on the sliding surface $\sigma = 0$.

An important advantage arises from the dynamical adaptive sliding mode control: ***the output tracking error function $z_1(t)$ asymptotically approaches zero with substantially reduced chattering*** [9, 10, 11]. A symbolic toolbox, which implements the above DAB-SMC algorithm, has been developed [11].

Better performance of the closed-loop system is achieved when the parameter estimates converge to the actual unknown parameters. This is guaranteed if the rank condition

$$\operatorname{rank}\big[F(x,u) = \Psi(x) + \varphi(x)u\big]_{(x,u)=(X,U)} = p \tag{86}$$

is satisfied for the desired equilibrium point [10].

3 Illustrative Examples

We consider in this section the application of the above DAB-SMC algorithm to the output regulation of two nonlinear chemical processes with uncertain parameters. In both examples, the relative degree is one and, thus, the resulting adaptive controllers are dynamical, due to the presence of time derivatives of the control input.

3.1 Example 1: A gravity-flow tank/pipeline

Consider the following system taken from Karjala and Himmelblau [3] which includes an elementary static model for an "equal percentage valve"

$$\begin{aligned} \dot{x}_1 &= \frac{A_p g}{L} x_2 - \frac{K_f}{\rho A_p^2} x_1^2 \\ \dot{x}_2 &= \frac{1}{A_t}\left(F_{Cmax}\alpha^{-(1-u)} - x_1\right) \\ y &= x_2 \end{aligned} \tag{87}$$

with

- x_1: volumetric flow rate of liquid leaving the tank
- x_2: height of the liquid in the tank
- F_{Cmax}: maximum value of the volumetric rate of fluid entering the tank
- g: gravitational acceleration constant
- L: length of the pipe
- K_f: friction factor
- ρ: density of the liquid
- A_p: cross sectional area of the pipe
- A_t: cross sectional area of the tank
- α: rangeability parameter of the valve
- u the valve position (control input) taking values in the closed interval [0,1].

For a constant value $U \in [0,1]$ of the control input u, the system has an equilibrium point given by

$$X_1 = F_{Cmax}\alpha^{-(1-U)} \quad ; \quad X_2 = \frac{LK_f}{A_p^3 g\rho} X_1^2$$

The operating region for system (87) is given by points strictly located in the first quadrant of $\Re^2$, i.e. the region defined by

$$\chi = \{x = (x_1, x_2)^T \in \Re^2, \quad \text{s.t.} \quad x_1 > 0 \text{ and } x_2 > 0\}$$

It has been shown by Sira-Ramírez and Delgado [12] that the system (87) is minimum phase for the system output $y = x_2$. Therefore, the DAB-SMC algorithm can be applied to synthesize a dynamical adaptive compensator for its regulation.

We consider the control input term via the following auxiliary control input $v = F_{Cmax}\alpha^{-(1-u)}$. Assuming that the friction factor K_f is constant but unknown,

$$\begin{aligned} \dot{x}_1 &= \frac{A_p g}{L} x_2 - \theta \frac{x_1^2}{\rho A_p^2} \\ \dot{x}_2 &= \frac{1}{A_t}(v - x_1) \\ y &= x_2 \end{aligned} \tag{88}$$

where $\theta = K_f$.

STEP 1. Defining the new error coordinate z_1 as $z_1 = y - X_2 = x_2 - X_2$, its time derivative is

$$\dot{z}_1 = \frac{1}{A_t}(v - x_1) \tag{89}$$

We can stabilize (89) with respect to the quadratic Lyapunov function $V_1 = \frac{1}{2}z_1^2$ whose time derivative is

$$\dot{V}_1 = z_1\left[\frac{1}{A_t}(v - x_1)\right] \tag{90}$$

We would achieve $\dot{V}_1 = -c_1 z_1^2$ with

$$\frac{1}{A_t}(v - x_1) = -c_1 z_1 \tag{91}$$

but, since this desired behaviour cannot be obtained arbitrarily, we define the control input-dependent error variable z_2

$$z_2 = \frac{1}{A_t}(v - x_1) + c_1 z_1 \tag{92}$$

Thus, the closed-loop form of $\dot{z}_1$ is

$$\dot{z}_1 = -c_1 z_1 + z_2 \tag{93}$$

and the time derivative of V_1 yields

$$\dot{V}_1 = -c_1 z_1^2 + z_1 z_2 \tag{94}$$

STEP 2. The time derivative of z_2 is

$$\dot{z}_2 = \frac{1}{A_t}\left[\dot{v} - \left(\frac{A_p g}{L}x_2 - \theta\frac{x_1^2}{\rho A_p^2}\right)\right] + c_1\left[\frac{1}{A_t}(v - x_1)\right] \tag{95}$$

By adding and subtracting the estimated value $\hat{\theta}$ we can rewrite $\dot{z}_2$ as

$$\dot{z}_2 = \frac{1}{A_t}\left[\dot{v} - \left(\frac{A_p g}{L}x_2 - \hat{\theta}\frac{x_1^2}{\rho A_p^2}\right)\right] + c_1\left[\frac{1}{A_t}(v - x_1)\right] + (\theta - \hat{\theta})\frac{x_1^2}{A_t \rho A_p^2} \tag{96}$$

We consider now the augmented Lyapunov function

$$V_2 = V_1 + \frac{1}{2}\sigma^2 + \frac{1}{2\gamma}(\theta - \hat{\theta})^2 = \frac{1}{2}z_1^2 + \frac{1}{2}\sigma^2 + \frac{1}{2\gamma}(\theta - \hat{\theta})^2 \tag{97}$$

with the sliding surface

$$\sigma = k_1 z_1 + z_2 = 0 \tag{98}$$

The time derivative of V_2 is

$$\begin{aligned}\dot{V}_2 = {} & -c_1 z_1^2 + z_1 z_2 + \frac{(\theta - \hat{\theta})}{\gamma}\left(-\dot{\hat{\theta}} + \gamma\frac{\sigma}{A_t}\frac{x_1^2}{\rho A_p^2}\right) \\ & + \sigma\left[\frac{1}{A_t}\left[\dot{v} - \left(\frac{A_p g}{L}x_2 - \hat{\theta}\frac{x_1^2}{\rho A_p^2}\right)\right] + (k_1 + c_1)\left[\frac{v - x_1}{A_t}\right]\right]\end{aligned} \tag{99}$$

To eliminate the estimate error from $\dot{V}_2$ we choose the update law

$$\dot{\hat{\theta}} = \tau_2 = \gamma\frac{\sigma}{A_t}\frac{x_1^2}{\rho A_p^2} \tag{100}$$

Thus, the transformed control function v can be obtained as the solution of the nonlinear time-varying differential equation

$$\dot{v} = A_t\left[\frac{1}{A_t}\left(\frac{A_p g}{L}x_2 - \hat{\theta}\frac{x_1^2}{\rho A_p^2}\right) - (k_1 + c_1)\left(\frac{v - x_1}{A_t}\right) - \kappa(\sigma + \beta\ \mathrm{sign}(\sigma))\right] \tag{101}$$

and then the actual control input is readily obtained as

$$u = 1 + \frac{1}{\log\alpha}\log\left(\frac{v}{F_{Cmax}}\right) \tag{102}$$

Thus, from the stability condition (85), we obtain the relation

$$\kappa(c_1 + k_1) > \frac{1}{4} \tag{103}$$

which should be satisfied in order to guarantee asymptotic output regulation, i.e. convergence of $y = x_2$ to the desired value X_2.

We have used in computer simulations the system parameters

$$g = 9.81\ m/s^2, \quad L = 914\ m, \quad \rho = 998\ Kg/m^2, \quad A_p = 0.653\ m^2$$

$$A_t = 10.5\ m^2, \quad \alpha = 10, \quad F_{Cmax} = 2\ m^3/s$$

The unknown parameter was set to be $K_f = 4.41\ N\ s^2/m^3$. The required equilibrium point for $y = x_2$ was chosen to be $X_2 = 5\ m$, with $X_1 = 1.8360\ m^3/s$. This corresponds to a steady state value of the control input $U = 0.9627$. The design parameters were chosen to be

$$c_1 = 0.5, \quad c_2 = 0.8, \quad \gamma = 15, \quad \kappa = 1, \quad \beta = 0.5, \quad k_1 = 2$$

In order to verify the robustness of this SMC law, a random input signal η, bounded in the closed interval $[-0.1, 0.1]$, has been incorporated additively in the first equation of system (87), i.e.

$$\begin{aligned} \dot{x}_1 &= \frac{A_p g}{L} x_2 - \theta \frac{x_1^2}{\rho A_p^2} + \eta \\ \dot{x}_2 &= \frac{1}{A_t}(v - x_1) \end{aligned} \tag{104}$$

Figure 1 shows the robustly controlled closed-loop response of the gravity-flow tank/pipe system with good stabilization features; no overshoot and a settling time of less than 150 seconds.

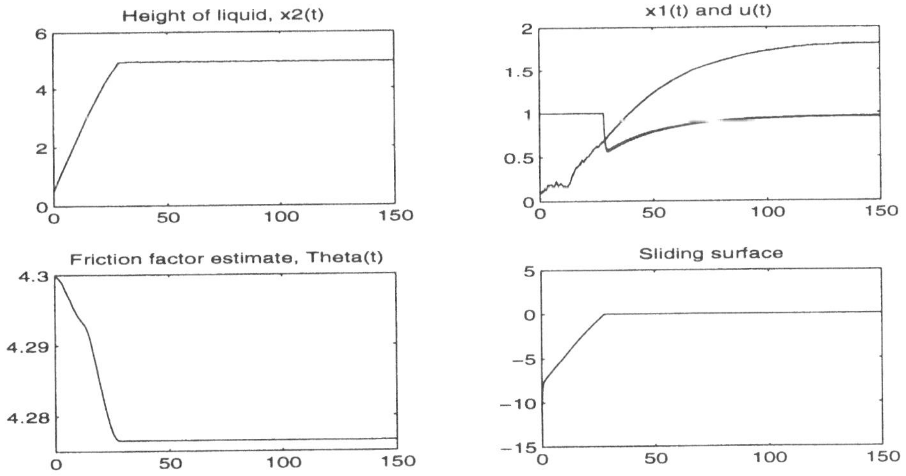

Fig. 1. Controlled responses of the gravity-tank/pipe system.

3.2 Example 2: Continuously Stirred Tank Reactor

Consider the following dynamic model taken from Kravaris and Palanki [4] of a Continuously Stirred Tank Reactor (CSTR) in which an isothermal liquid-phase multicomponent chemical reaction takes place

$$\begin{aligned}\dot{x}_1 &= 1 - (1 + D_{a1})x_1 + D_{a2}x_2^2 \\ \dot{x}_2 &= D_{a1}x_1 - x_2 - (D_{a2} + D_{a3})x_2^2 + u \\ \dot{x}_3 &= D_{a3}x_2^2 - x_3 \\ y &= x_3\end{aligned} \tag{105}$$

with

- x_1: normalized concentration C_A/C_{AF} of a species A
- x_2: normalized concentration C_B/C_{AF} of a species B
- x_3: normalized concentration C_C/C_{AF} of a species C
- C_{AF}: the feed concentration of the species A $(mol \cdot m^{-1})$
- u: the ratio of the per-unit volumetric molar feed rate of species B, denoted by N_{BF}, and the feed concentration C_{AF}, i.e. $u = \frac{N_{BF}}{FC_{AF}}$
- F: volumetric feed rate $(m^3 s^{-1})$
- $D_{a1} = \frac{k_1 V}{F}$ constant parameter
- $D_{a2} = \frac{k_2 V C_{AF}}{F}$ constant parameter
- $D_{a3} = \frac{k_3 V C_{AF}}{F}$ constant parameter
- V: the volume of the reactor (m^3)
- k_1, k_2, k_3 first order rate constants (s^{-1}).

The system has a constant stable equilibrium point $X = (X_1, X_2, X_3)^T$, for every constant volumetric feed rate of value $u = U$, which is located in a minimum phase region of the system [12].

The operating region of the system is the strict orthant in $\Re^3$, where all concentrations are positive, i.e.

$$\chi = \{x \in \Re^3, \quad \text{s.t.} \quad x_i > 0 \quad \text{for } i = 1, 2, 3\}$$

We assume that the constant parameters D_{a1}, D_{a2} and D_{a3} are all unknown. Thus, system (105) can be rewritten as

$$\begin{aligned}\dot{x}_1 &= 1 - x_1 + \varphi_1^T(x_1, x_2)\theta \\ \dot{x}_2 &= -x_2 + u + \varphi_2^T(x_1, x_2)\theta \\ \dot{x}_3 &= -x_3 + \varphi_3^T(x_2)\theta \\ y &= x_3\end{aligned} \tag{106}$$

with $\theta = [\theta_1 \ \theta_2 \ \theta_3]^T = [D_{a1} \ D_{a_2} \ D_{a3}]^T$ the unknown parameter vector and

$$\varphi_1^T = [-x_1 \ \ x_2^2 \ \ 0]; \quad \varphi_2^T = [x_1 \ \ -x_2^2 \ \ -x_2^2]; \quad \varphi_3^T = [0 \ \ 0 \ \ x_2^2]$$

Applying the DAB-SMC algorithm, we synthesize a dynamical adaptive SMC compensator for the regulation of system (106). This compensator is characterized by the following expressions:

Coordinate transformation

$$\begin{aligned} z_1 &= y - X_3 = x_3 - X_3 \\ z_2 &= -x_3 + \varphi_3^T(x_2)\hat{\theta} + c_1 z_1 \\ z_3 &= \alpha(x, \hat{\theta}) + \frac{\partial \varphi_3^T}{\partial x_2}\hat{\theta} u \end{aligned} \tag{107}$$

Sliding surface

$$\sigma = k_1 z_1 + k_2 z_2 + z_3 = 0$$

Parameter update law

$$\begin{aligned} \dot{\hat{\theta}} &= \tau_3 = \tau_2 + \Gamma\sigma(k_1\varphi_3 + k_2\omega_2 + \omega_3) \\ &= \Gamma\Big[z_1\varphi_3 + z_2\omega_2 + \sigma(k_1\varphi_3 + k_2\omega_2 + \omega_3)\Big] \end{aligned} \tag{108}$$

with

$$\begin{aligned} \omega_2^T &= (c_1 - 1)\varphi_3^T(x_2) + \frac{\partial \varphi_3^T}{\partial x_2}\hat{\theta}\varphi_2^T(x_1, x_2) \\ \omega_3^T &= \frac{\partial \alpha}{\partial x_1}\varphi_1^T + \Big(\frac{\partial \alpha}{\partial x_2} + \frac{\partial^2 \varphi_3^T}{\partial x_2^2}\hat{\theta} u\Big)\varphi_2^T + \frac{\partial \alpha}{\partial x_3}\varphi_3^T(x_2) \\ \alpha(x, \hat{\theta}) &= z_1 - (c_1 - 1)x_3 - \frac{\partial \varphi_3^T}{\partial x_2}\hat{\theta} x_2 + \omega_2^T\hat{\theta} + \varphi_3^T\Gamma(z_1\varphi_3 + z_2\omega_2) + c_2 z_2 \end{aligned} \tag{109}$$

Dynamical adaptive SMC law

$$\begin{aligned} \dot{u} = \frac{1}{\frac{\partial \varphi_3^T}{\partial x_2}\hat{\theta}}\Bigg[&- (k_2 + z_2)\varphi_3^T(\tau_3 - \tau_2) - k_1(-c_1 z_1 + z_2) \\ &- \omega_3^T\hat{\theta} - k_2(-z_1 - c_2 z_2 + z_3) - \frac{\partial \alpha}{\partial x_1}(1 - x_1) \\ &- \Big(\frac{\partial \alpha}{\partial x_2} + \frac{\partial^2 \varphi_3^T}{\partial x_2^2}\hat{\theta} u\Big)(-x_2 + u) + \frac{\partial \alpha}{\partial x_3}x_3 \\ &- \Big(\frac{\partial \alpha}{\partial \hat{\theta}} + u\frac{\partial \varphi_3^T}{\partial x_2}\Big)\tau_3 - \kappa(\sigma + \beta \mathrm{sign}(\sigma))\Bigg] \end{aligned} \tag{110}$$

where $\Gamma = \Gamma^T > 0$ is a diagonal matrix containing the adaptation parameter gains.

Finally, by satisfying the stability condition (85)

$$|Q| = \Big[-\frac{1}{4} + \kappa(c_2 + k_2)\Big]c_1 - \frac{1}{4}\kappa k_1^2 c_2 > 0$$

the output $y = x_3$ asymptotically converges to the desired value X_3.

Computer simulations were performed using the above DAB-SMC law for the robust output regulation of a CSTR with the "unknown" parameters

$$D_{a1} = 3.0, \quad D_{a2} = 0.5, \quad D_{a3} = 1.0$$

The desired equilibrium, corresponding to a constant value of u given by $U = 1$, is

$$X_1 = 0.3467, \quad X_2 = 0.8796, \quad X_3 = 0.7753$$

whilst the design parameters were selected to be

$$c_1 = 2, \quad c_2 = 1, \quad c_3 = 2, \quad \Gamma = 2I_3, \quad \kappa = 2, \quad \beta = 1, \quad k_1 = 1, \quad k_2 = 2$$

Figure 2 shows the good performance of the controlled CSTR in the absence of disturbances.

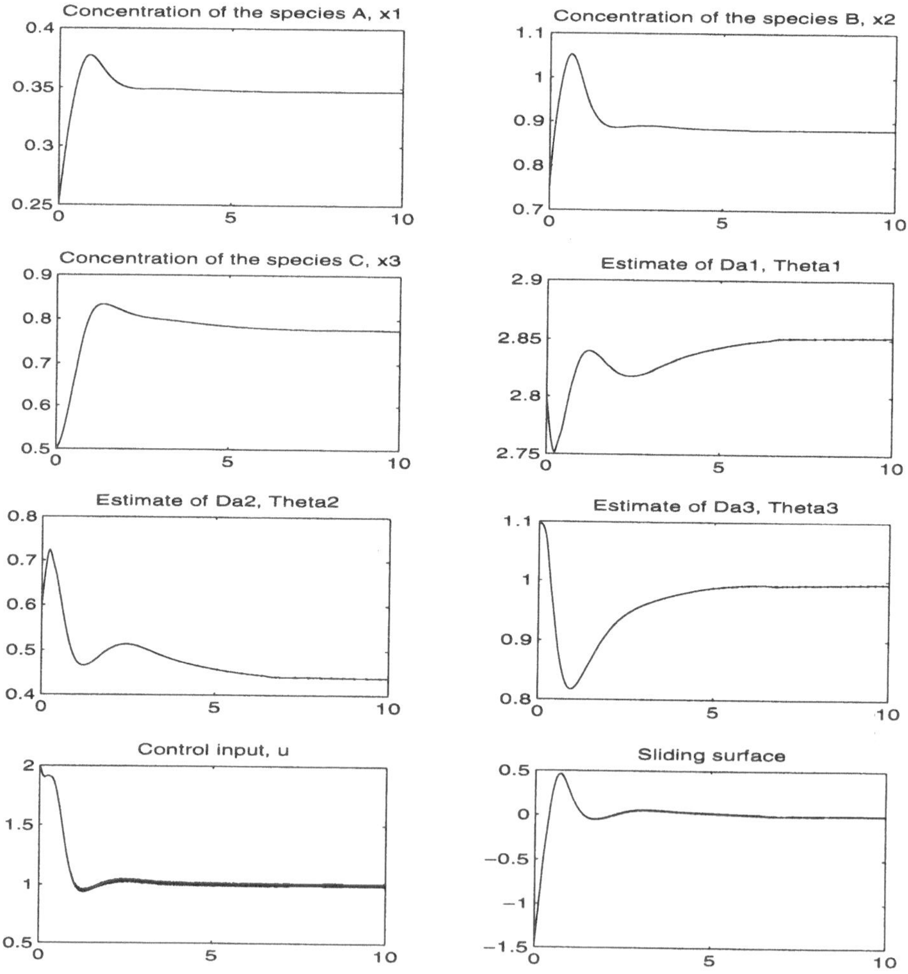

Fig. 2. Controlled responses of the Isothermal CSTR.

In order to verify the robustness of this DAB-SMC law, a random input signal η, bounded in the closed interval $[-1, 1]$, was incorporated additively in the first equation of system (105), i.e.

$$\begin{aligned} \dot{x}_1 &= 1 - x_1 + \varphi_1^T(x_1, x_2)\theta + \eta \\ \dot{x}_2 &= -x_2 + u + \varphi_2^T(x_1, x_2)\theta \\ \dot{x}_3 &= -x_3 + \varphi_3^T(x_2)\theta \end{aligned} \tag{111}$$

Figure 3 depicts the asymptotic behaviour of the robustly controlled CSTR output response whilst the other state variables remain bounded.

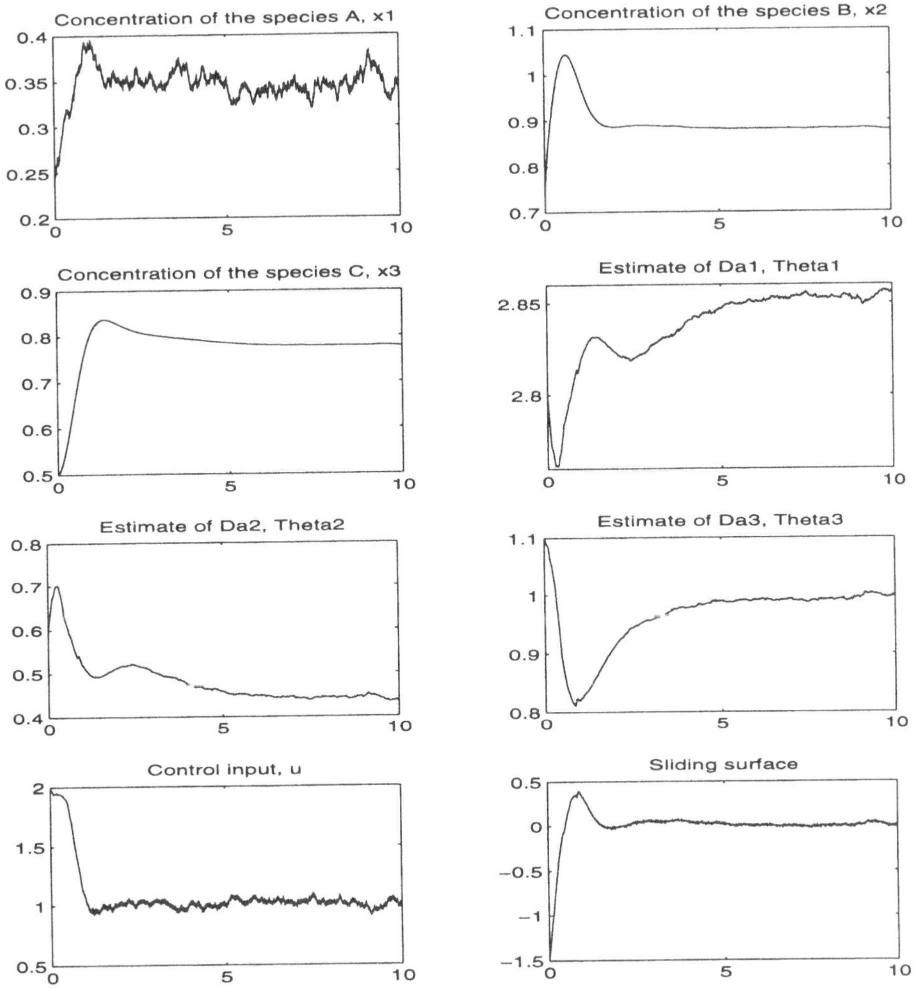

Fig. 3. Controlled responses of the Isothermal CSTR with disturbances

4 Conclusions

The application of the combined DAB-SMC algorithm to two nonlinear continuous chemical processes has been presented. Both mathematical models satisfy the minimum phase and observability conditions which are stringent conditions for the applicability of this control design approach. Furthermore, from computer simulations, the adaptively controlled responses of these two chemical processes exhibit both robustness, in the presence of bounded disturbances, and good transient performance with substantially reduced chattering in the controlled trajectories. These features motivate the applicability of the DAB-SMC algorithm for the robust output regulation of this kind of dynamical system.

References

1. A. Isidori, *Nonlinear Control Systems.* 2nd ed., Springer-Verlag, Berlin, 1989.
2. I. Kanellakopoulos, P.V. Kokotović and A. S. Morse, "Systematic Design of Adaptive Controllers for Feedback Linearizable Systems", *IEEE Trans. on Aut. Cont.*, Vol. 36, pp. 1241–1253, 1991.
3. T. Karjala and D. Himmelblau, *AIChE Journal*, Vol. 42, pp. 2225–2239, 1996.
4. C. Kravaris and S. Palanki, *AIChE Journal*, Vol. 34, pp. 1119–1127, 1988.
5. M. Krstić, I. Kanellakopoulos and P.V. Kokotović, *Nonlinear and Adaptive Control Design*, New York: John Wiley & Sons, 1995.
6. M. Ríos-Bolívar, H. Sira-Ramírez and A. S. I. Zinober, "Output Tracking Control via Adaptive Input-Output Linearization: A Backstepping Approach", *Proceedings of the 34th IEEE Cnference on Decision and Control*, New Orleans, USA, December 1995, Vol. 2, pp. 1579–1584.
7. M. Ríos-Bolívar and A. S. I. Zinober, "Dynamic Adaptive Regulation of Nonlinear Continuous Processes", *Proc. 6th IEEE Mediterranean Conference on Control and Systems*, Alghero, Italy, 1998.
8. M. Ríos-Bolívar, A. S. I. Zinober and H. Sira-Ramírez, "Dynamical Adaptive Sliding Mode Output Tracking Control of a Class of Nonlinear Systems", *International Journal of Robust and Nonlinear Control*, Vol. 7, pp. 387–405, 1997.
9. M. Ríos-Bolívar, A. S. I. Zinober and H. Sira-Ramírez,"Dynamical Sliding Mode Control via Adaptive Input-Output Linearization: A Backstepping Approach", in: *Robust Control via Variable Structure and Lyapunov Techniques* (F. Garofalo and L. Glielmo, Eds.), Springer-Verlag. pp. 15–35, 1996.
10. M. Ríos-Bolívar, "Adaptive Backstepping and Sliding Mode Control of Uncertain Nonlinear Systems", Ph.D. Thesis, The University of Sheffield, 1997.
11. M. Ríos-Bolívar and A. S. I. Zinober, "A Symbolic Computation Toolbox for the Design of Dynamical Adaptive Nonlinear Controllers", *Appl. Math. and Comp. Sci.*, Vol. 8, No. 1, pp. 73–88, 1998.
12. H. Sira-Ramírez and M. Delgado, "Passivity Based Regulation of Nonlinear Continuous Processes", to be published in *Advances in Control.*
13. H. Sira-Ramírez, "Asymptotic Output Stabilization for Nonlinear Systems via Dynamical Variable Control", *Dynamics and Control*, Vol. 2, pp. 45–58, 1992.

14. H. Sira-Ramírez, "A Dynamical Variable Structure Control Strategy in Asymptotic Output Tracking Problems", *IEEE Trans. on Automatic Control*, Vol. 38, pp. 615–620, 1993.
15. H. Sira-Ramírez, S. Ahmad, and M. Zribi, "Dynamical Feedback Control of Robotic Manipulators with Joint Flexibility", *IEEE Trans. on Systems, Man and Cybernetics*, Vol. 22, pp. 736–747, 1992.
16. H. Sira-Ramírez and O. Llanes-Santiago, "Dynamical Feedback Strategies in the Regulation of Nonlinear Chemical Processes", *IEEE Trans. on Control Systems Technology*, Vol. 2, pp. 11-21, 1994.

Symbolic Computing Tools for Nonsmooth Dynamics and Control

C. Teolis
Techno-Sciences, Incorporated, 10001 Derekwood Lane, Suite 204
Lanham, MD 20706

H. G. Kwatny
Department of Mechanical Engineering & Mechanics, Drexel University
Philadelphia, PA 19104

M. Mattice
Advanced Drives and Weapon Stabilization Lab, U. S. Army ARDEC
Picatinny Arsenal, NJ 07806

Abstract In this paper we describe a set of symbolic computing tools for variable structure control system design. The software implements all aspects of a design approach for input-output linearizable systems. It is part of a comprehensive symbolic computing environment for nonlinear and adaptive control system design that has been under continuous development for several years. Current work is focused on plants with nondifferentiable nonlinearities. Some preliminary results are reported.

1 Introduction

The purpose of this paper is twofold. First, to describe a set of symbolic computing tools developed to assist in the design and implementation of variable structure control systems. The tools enable the efficient design of sliding surfaces and reaching controllers including the inclusion of 'smoothing' and 'moderating' functions and the assembly of C-source code for simulation and real time implementation. These functions extend the capabilities of the symbolic modeling and control design software described in [1] and elsewhere.

The second purpose is to introduce a new backstepping methodology for systems with uncertain nondifferentiable nonlinearities. The key innovations in our approach are (1) that the states are grouped depending on where an uncertainty enters the system and the robustification is attempted only where the uncertainty is identified, and (2) that the control designed at each step is a variable structure control.

The variable structure methods we have implemented are developed in [2-4]. These references deal with variable structure control system design for smooth affine systems that are feedback linearizable in the input-output sense. Such systems are of the form:

$$\begin{aligned} \dot{x} &= f(x) + G(x)u \\ y &= h(x) \end{aligned} \tag{1}$$

where f, G, h are sufficiently smooth and satisfy certain feedback linearizability conditions [5]. All of the basic functions needed for design including reduction to regular form (as described in [1]) and computation of the zero dynamics (as previously reported in [6]), as well as functions for designing sliding surfaces and switching controllers, have been integrated into a convenient *Mathematica* package[1]. Ongoing work is focused on extending these techniques to plants containing hard nonlinearities such as dead zone, backlash, hysteresis and coulomb friction. To do this has required extending *Mathematica's* facilities for working with nondifferentiable nonlinear functions. Our examples describe applications to friction compensation.

In Section 2 we summarize the methods and computations that we have implemented. We include some preliminary remarks concerning nonsmooth plant dynamics and we briefly discuss chattering reduction techniques. Section 3 describes and illustrates the symbolic computing tools. A very simple system with nonsmooth friction is used for illustrative purposes. The effects of control smoothing and moderation are illustrated. Section 4 describes ongoing work involving plants with nondifferentiable nonlinearities. A simple example is given which demonstrates the problems that can occur when applying methods designed for smooth system to those with nonsmooth nonlinearities by simply approximating the nonsmooth nonlinearities by smooth functions. A backstepping approach to the variable structure control design is shown to solve the problem in this simple case. In Section 5, the symbolic computing tools are used to design a friction compensating slewing controller for the US Army Apache Helicopter 30-mm chain gun. Simulation results are given that show the control robustness to parameter variation. Some concluding remarks are given in Section 6.

2 Variable Structure Control Design

There are 2 basic steps to designing a variable structure control. The first is the design of the sliding control or equivalently the sliding surface. The second is the design of the reaching or switching control. The system is typically reduced to normal, or regular, form before the design begins. Also, in order to avoid exciting higher order unmodeled dynamics, 'smoothing' and 'moderating' functions are used to reduce chattering. In this section, these methods are summarized. In addition, a simple example is given with nonsmooth friction and local asymptotic stability of the variable structure control is proven.

2.1 Normal Form

Denote the kth Lie (directional) derivative of the scalar function $\phi(x)$ with respect to the vector field $f(x)$ by $L_f^k(\phi)$. Now, by successive differentiation of the outputs y in (1) we arrive at the following definitions for the list of integers r_i, the column vector $\alpha(x)$ and the matrix $\rho(x)$:

---

[1] More information can be found at the website: www.technosci.com.

$$r_i := \inf\{k \mid L_{g_j}(L_f^{k-1}(h_i)) \neq 0 \textit{ for at least one } j\}$$

$$\alpha_i(x) := L_f^{r_i}(h_i),\ i = 1,..,m$$

$$\rho_{ij}(x) := L_{g_j}(L_f^{r_i-1}(h_i)),\ i, j = 1,..m$$

Also define the partial state transform $x \rightarrow z \in R^r$, $r = r_1 + \ldots + r_m \leq n$ as

$$z := \begin{bmatrix} z_1 \\ z_2 \\ \vdots \\ z_m \end{bmatrix},\ z_i \in \mathrm{R}^{r_i}, i = 1,\ldots,m \tag{2a}$$

where

$$z_i^k(x) = L_f^{k-1}(h_i),\ k = 1,..,r_i \textit{ and } i = 1,..,m \tag{2b}$$

It is a straightforward calculation to verify that the variables z defined by (2) satisfy the relation

$$\dot{z} = Az + E[\alpha(x) + \rho(x)u]$$

$$y = Cz$$

where the only nonzero rows of E are the m rows $r_1, r_1+r_2,..,r$ and these form the identity I_m, the only nonzero columns of C are the columns $1, r_1+1, r_1+r_2+1,..,r-r_m+1$ and these form the identity I_m, and

$$A = diag(A_1,\ldots,A_m), A_i = \begin{bmatrix} 0 & I_{r_i-1} \\ 0 & 0 \end{bmatrix} \in \mathrm{R}^{r_i \times r_i}$$

The variables z are referred to as the linearizable coordinates. The remaining part of the transform can be defined by arbitrarily choosing additional independent coordinates. The condition $\det\{\rho(x)\} \neq 0$ insures the existence of a local (around x_0) change of coordinates $x \rightarrow (\xi, z)$, $\xi \in R^{n-r}$, $z \in R^r$ such that

$$\dot{\xi} = F(\xi, z) \tag{3a}$$

$$\dot{z} = Az + E[\alpha(x(\xi, z)) + \rho(x(\xi, z))] \tag{3b}$$

$$y = Cz \tag{3c}$$

Equation (3) is frequently referred to as the *local normal form* of (1). It is common to refer to (3a) as the *internal dynamics* and (3b) as the *linearizable dynamics*. If z is set to zero in (3a) then we have a local representation of the zero dynamics.

Equation (3) is the point of departure for the variable structure design as described in [2]. It constitutes a *regular form* in the sense of [7].

2.2 Sliding

The reduction to this normal form is commonly associated as the first step in the process of feedback linearization. Here instead of feedback linearization, we construct a variable structure control law with switching surface of the form, $s(x)=Kz(x)$. We can prove that during sliding, the equivalent control is $u_{eq} = Kz$, so that we achieve feedback linearized behavior in the sliding phase (see, [2-4, 9]).

2.3 Reaching

The second step in VS control system design is the specification of the control functions $u_i^{\pm}$ such that the manifold s(x)=0 contains a stable submanifold which insures that sliding occurs. There are many ways of approaching the reaching design problem, Utkin [10]. We consider only one. Consider the positive definite quadratic form in s

$$V(x) = s^T Qs$$

A sliding mode exists on a submanifold of s(x)=0 which lies in a region of the state space on which the time rate of change V is negative. Upon differentiation we obtain

$$\frac{d}{dt}V = 2\dot{s}^T Qs = 2[KAz + \alpha]^T QKz + 2u^T \rho^T QKz$$

If the controls are bounded, $|u|_i \le \overline{U}_i > 0$ $(0 > U_{\min,i} \le u_i \le U_{\max,i} > 0)$ then obviously, to minimize the time rate of change of V, we should choose

$$u_i = U_{\min,i}\,\text{step}(s_i^*) + U_{\max,i}\,\text{step}(-s_i^*)\,,\ i = 1,\dots,m,$$

$$s^*(x) = \rho^T(x)QKz(x)$$

Notice that if $U_{\min,i} = -U_{\max,i}$, the control reduces to

$$u_i = -U_{\max,i}\,\text{sign}(s_i^*)$$

In this case it follows that $\dot{V}$ is negative provided

$$\left|U_{\max}{}^T \rho^T QKz\right| > \left|[KAz + \alpha]^T QKz\right| \tag{4}$$

A useful sufficient condition is that

$$\left|(\rho(x)U_{\max})_i\right| > \left|(KAz(x) + \alpha(x))_i\right| \tag{5}$$

Condition (4) or (5) may be used to insure that the control bounds are of sufficient magnitude to guarantee sliding and to provide adequate reaching dynamics. This rather simple approach to reaching design is satisfactory when a "bang-bang" control is acceptable.

2.4 Chattering Reduction

The state trajectories of ideal sliding motions are continuous functions of time contained entirely within the sliding manifold. These trajectories correspond to the equivalent control $u_{eq}(t)$. However, the actual control signal, $u(t)$ – definable only for nonideal trajectories – is discontinuous as a consequence of the switching mechanism that generates it. Persistent switching or 'chattering' is undesirable in some applications. Several techniques have been proposed to reduce or eliminate chattering. These include: 'regularization' of the switch by replacing it with a continuous approximation; 'extension' of the dynamics by using additional integrators to separate an applied discontinuous pseudo-control from the actual plant inputs; and 'moderation' of the reaching control magnitude as errors become small.

Switch regularization entails replacing the ideal switching function, $\mathrm{sign}(s(x))$, with a continuous function such as

$$\mathrm{sat}\left(\frac{1}{\varepsilon}s(x)\right) \text{ or } \frac{s(x)}{\varepsilon+|s(x)|} \text{ or } \tanh\left(\frac{s(x)}{\varepsilon}\right)$$

This intuitive approach is employed by Young & Kwatny [11] and Slotine and Sastry [12, 13] and there are probably historical precedents. Regularization induces a boundary layer around the switching manifold whose size is $O(\varepsilon)$. The justification for this approach for linear systems is provided by the results in [14]. Some of those results have been extended to single input–single output systems nonlinear systems by Marino [9]. Switch regularization for nonlinear systems has been extensively discussed by Slotine and coworkers, e.g. [12, 13]. With nonlinear systems there are subtleties and regularization can result in an unstable system.

Dynamic extension is another effective approach to control input smoothing, Emelyanov et al [15]. A sliding mode is said to be of p-th order relative to an output y if the time derivatives $\dot{y}, \ddot{y}, \ldots, y^{(p-1)}$ are continuous in t but y^{p} is not. The following observation is a straightforward consequence of the regular form theorem: Suppose (1) is input-output linearizable with respect to the output $y = h(x)$ with vector relative degree $(r_1,\ldots,r_m)$. Then the sliding mode corresponding to the variable structure control law is of order $p=\min(r_1,\ldots,r_m)$ relative to the output y. We may modify the relative degree by augmenting the system with input dynamics as described. Hence, we can directly control the smoothness of the output vector y.

Control moderation involves design of the reaching control functions $u_i(x)$ such that $|u_i(x)| \to$ small as $|e(x)| \to 0$. For example,

$$u_i(x) = |e(x)|\,\mathrm{sign}(s_i(x))$$

Control moderation was used by Young and Kwatny [11] and the significance of this approach for chattering reduction in the presence of parasitic dynamics was discussed by Kwatny and Siu [16].

2.4.1 Example 1: Simple Friction

The following is a simple rotor with friction and input torque:

$$\dot{x}_1 = x_2$$

$$\dot{x}_2 = -\phi_{fr}(x_2) + u$$

Suppose the input torque u is bounded, say, $u \in [-U, U]$. We can easily show that the controller $u = -U\,\mathrm{sgn}(cx_1 + x_2)$, $c > 0$ and U sufficiently large stabilizes the origin for all piecewise smooth friction functions with a discontinuity at the origin such that $\phi_{fr}(0)$ is bounded.

Consider a VS controller with

$$u(x) = \begin{cases} u^+(x) & s(x) > 0 \\ u^-(x) & s(x) < 0 \end{cases}, \quad s(x) = cx_1 + x_2,\ c > 0$$

Imposing the sliding condition $s(x) \equiv 0$ leads to

$$\dot{x}_1 = -cx_1,\ u_{eq} = -cx_2 + \phi_{fr}(x_2)$$

Now, we need to design the reaching control. Choose $V(x) = s^2(x)$ and compute

$$\dot{V} = 2(cx_1 + x_2)(cx_2 - \phi_{fr}(x_2) + u)$$

If u is bounded, say, $u \in [-U, U]$, choose $u = -U\,\mathrm{sgn}(cx_1 + x_2)$. Then

$$\dot{V} = 2\,\mathrm{abs}(cx_1 + x_2)\{\mathrm{sgn}(cx_1 + x_2)(cx_2 - \phi_{fr}(x_2)) - U\}$$

Certainly, $\dot{V} < 0$ if $\mathrm{abs}(cx_2 - \phi_{fr}(x_2)) < U$. It follows that so long as $U > \sup \mathrm{abs}(\phi_{fr}(0))$ there is a neighborhood of the origin N such that each trajectory beginning in N converges to the origin.

3 Computing Tools

We need to be able to reduce the system to normal form, compute an appropriate switching surface, assemble the switching control and insert smoothing and/or moderating functions as desired. Functions that we have implemented to do this are defined in Table 1 and Table 2.

3.1 Sliding Surface Computations

There are several methods for determining the sliding surface, $s(x) = Kz(x)$, once the system has been reduced to normal form. We have included a function `SlidingSurface` that implements two alternatives depending on the arguments provided. The function may be called via

```
{rho,s}=SlidingSurface[f,g,h,x,lam]
```

or

```
s=SlidingSurface[rho,vro,z,lam]
```

Function Name	Operation
VectorRelativeOrder	computes the relative degree vector
DecouplingMatrix	computes the decoupling matrix
IOLinearize	computes the linearizing control
NormalCoordinates	computes the partial state transformation,
LocalZeroDynamics	computes the local form of the zero dynamics
StructureAlgorithm	computes the parameters of an inverse system
DynamicExtension	applies dynamic extension as a remedy for singular decoupling matrix

Table 1. Nonlinear systems: Geometric Control

Function Name	Operation
SlidingSurface	generates the sliding (switching) surface for feedback linearizable nonlinear systems
SwitchingControl	computes the switching functions – allows the inclusion of smoothing and moderating functions
SmoothingFunctions	an option for SwitchingControl that introduces specified smoothing functions
ModeratingFunctions	an option for SwitchingControl that introduces specified moderating functions

Table 2. Nonlinear systems: Variable Structure Control

In the first case the data provided is the nonlinear system definition f, g, h, x and an m-vector lam which contains a list of desired exponential decay rates, one for each channel. The function returns the decoupling matrix rho and the switching surfaces s as functions of the state x. The matrix K is obtained by solving the appropriate Ricatti equation.

The second use of the function assumes that the input-output linearization has already been performed so that the decoupling matrix rho, the vector relative degree and the normal coordinate (partial) transformation $z(x)$ are known. In this case the dimension of each of the m switching surfaces is known so that it is possible to specify a complete set of eigenvalues for each surface. Thus, lam is a list of m-sublists containing the specified eigenvalues. Only the switching surfaces are returned. In this case K is obtained via pole placement.

3.2 Switching Control

The function `SwitchingControl[rho,s,bounds,Q,opts]` returns the variable structure control, where rho is the decoupling matrix, s is the vector of switching surfaces, 'bounds' is a list of controller bounds each in the form {lower bound, upper bound}, Q is an mxm positive definite matrix (a design parameter), and 'opts' are options that allow the inclusion of smoothing and/or moderating functions in the control.

Smoothing functions are specified by a rule of the form

```
SmoothingFunctions[x_]->{function1[x],...,functionm[x]}
```

Where m is the number of controls. Moderating functions are similarly specified by a rule

```
ModeratingFunctions->{function1[z],...,functionm[z]}
```

The smoothing function option replaces any pure switch sign by a smooth switch function as specified. The moderating function option multiplies the switch by the specified function. We give an example below.

3.2.1 Example 1 Continued.

We will apply some of the above computations to Example 1. For illustrative purposes the friction function is taken to be

$$\phi_{fr} = \operatorname{sign} \omega .$$

```
{rho2, s2} = SlidingSurface[f, g, h, {theta, omega}, {2}]
Computing Decoupling Matrix
Computing linearizing/decoupling control
{{{1}}, {8.03066 omega + 16.1844 theta}}
```

Now, we compute the switching control using various combinations of smoothing and moderating functions. The particular functions chosen for this example are shown below in Figure 1. Results can change significantly when other functions are used or when the parameters of the functions are varied.

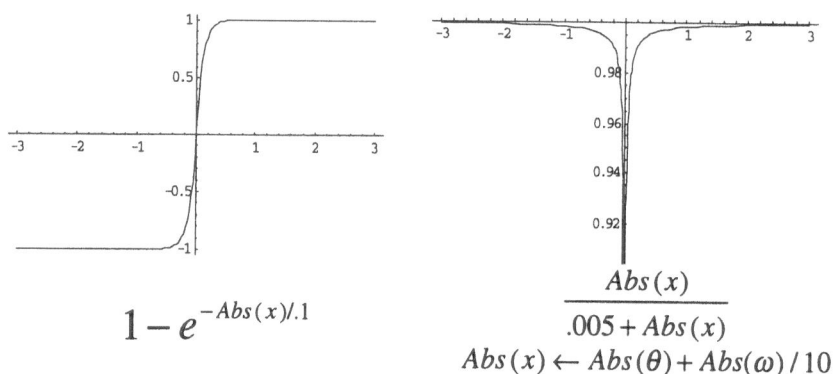

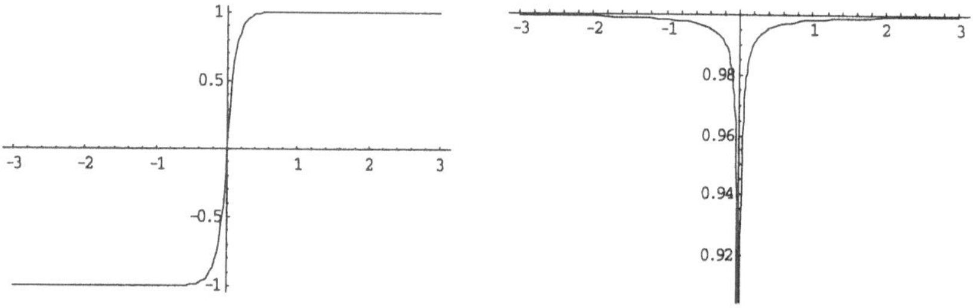

$$1 - e^{-Abs(x)/.1}$$

$$\frac{Abs(x)}{.005 + Abs(x)}$$

$$Abs(x) \leftarrow Abs(\theta) + Abs(\omega)/10$$

Smoothing Function **Moderating Function**

Figure 1. Smoothing and moderating functions used in the example.

We specify the control bounds as ± 5 and $Q = 1$. The following computation yields the four controls.

```
In[29]:= vsc1 = SwitchingControl[rho2, s2, ctrlbnds, Q]
         vsc2 = SwitchingControl[rho2, s2,
           ctrlbnds, Q, SmoothingFunctions[x_] -> {(1 - Exp[-Abs[x/ .1]])}]
         vsc3 = SwitchingControl[rho2, s2, ctrlbnds, Q, ModeratingFunctions ->
            {(Abs[theta] + Abs[omega] / 10) / (.002 + Abs[theta] + Abs[omega] / 10)}]
         vsc4 = SwitchingControl[rho2, s2, ctrlbnds, Q, ModeratingFunctions ->
            {(Abs[theta] + Abs[omega] / 10) / (.002 + Abs[theta] + Abs[omega] / 10)},
          SmoothingFunctions[x_] -> {(1 - Exp[-Abs[x/ .1]])}]

Out[29]= {5 Sign[-8.03066 omega - 16.1844 theta]}

Out[30]= {5 Sign[-8.03066 omega - 16.1844 theta] -
           5 E^(-10. Abs[-8.03066 omega - 16.1844 theta]) Sign[-8.03066 omega - 16.1844 theta]}

Out[31]= {(Abs[omega] Sign[-8.03066 omega - 16.1844 theta]) / (2 (0.002 + Abs[omega]/10 + Abs[theta])) +
           (5 Abs[theta] Sign[-8.03066 omega - 16.1844 theta]) / (0.002 + Abs[omega]/10 + Abs[theta])}

Out[32]= {(Abs[omega] Sign[-8.03066 omega - 16.1844 theta]) / (2 (0.002 + Abs[omega]/10 + Abs[theta])) -
           (E^(-10. Abs[-8.03066 omega - 16.1844 theta]) Abs[omega] Sign[-8.03066 omega - 16.1844 theta]) / (2 (0.002 + Abs[omega]/10 + Abs[theta])) +
           (5 Abs[theta] Sign[-8.03066 omega - 16.1844 theta]) / (0.002 + Abs[omega]/10 + Abs[theta]) -
           (5 E^(-10. Abs[-8.03066 omega - 16.1844 theta]) Abs[theta] Sign[-8.03066 omega - 16.1844 theta]) / (0.002 + Abs[omega]/10 + Abs[theta])}
```

Notice that the controllers do not depend on the specific parameters of the friction function. Figure 2 compares the closed loop performance of the first three controllers.

4 Nonsmooth Plants

Many important systems contain so-called 'hard' or 'nonsmooth' nonlinearities such as dead zone, backlash, hysteresis and coulomb friction. These nonlinearities can have a profound influence on the performance of a control system. While there exist standard models for these frequently neglected (often considered parasitic) effects, the parameters associated with them are almost always highly uncertain. Approaches to control system design that directly address hard nonlinearities must account for that uncertainty. Several alternatives have been suggested including a variety of adaptive [17, 18] and variable structure control methods.

Both adaptive and variable structure control designs are simple and effective if the system is input-output feedback linearizable and minimum phase [1, 2, 4, 19, 20]. When this is the case, the first step in design is to reduce the system the regular form described above. The basic reduction process applies to affine systems that are sufficiently smooth so that functions can be differentiated an appropriate number of times.

In this case we are interested in a more general class of models than given by (1):

$$\dot{x} = f(x) + G(x)\varphi(u)$$
$$y = h(x) \tag{1a}$$

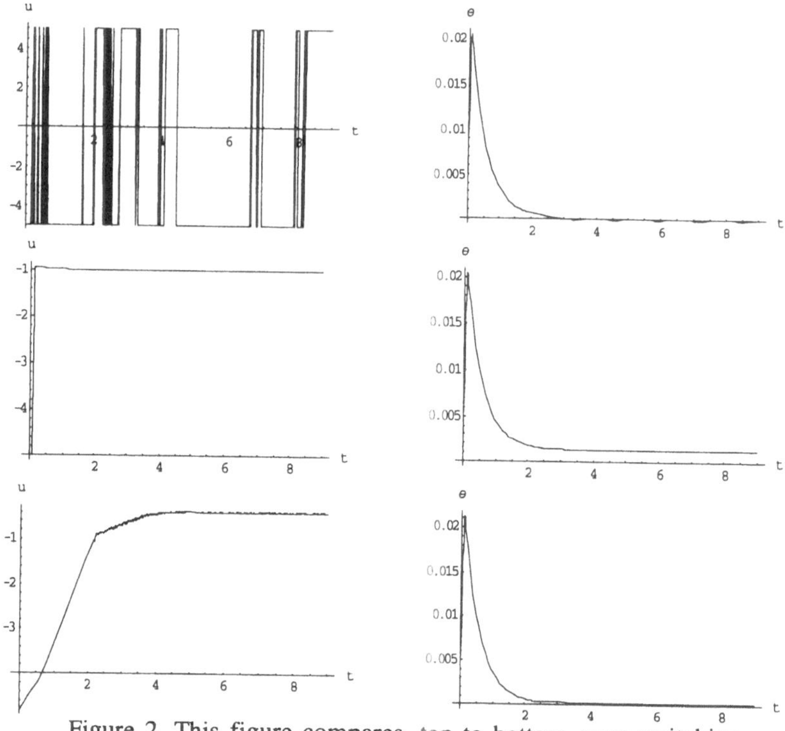

Figure 2. This figure compares, top to bottom, pure switching control, switching with smoothing and switching with moderating. From left to right: control, u, and position, θ. Chattering is virtually eliminated with either smoothing or moderating. However, smoothing leaves significantly larger steady state error because the effective gain is now bounded.

where $\varphi : R^m \to R^m$ is an invertible mapping, and f, G, h are only piecewise smooth functions.

Our primary interest has been applications to various pointing control systems associated with relatively small (Apache helicopter) to very large (Abrams tank) weapons. In these cases friction is a very significant issue and, depending on the drive system, backlash may also be important.

4.1 Controller Design with Nonsmooth Plants

One approach to dealing with nonsmooth nonlinearities is to approximate the nonsmooth function by a smooth one. In particular, we might consider replacing a piecewise smooth function $f(x)$ by a smooth ε-approximation $\hat{f}(x,\varepsilon)$ such that

$\lim_{\varepsilon \to 0} \hat{f}(x,\varepsilon) \to f(x)$. Then the design proceeds using the approximate system with ε sufficiently small. It is important to realize that there is no *a priori* assurance that the resulting control system when applied to the original nonsmooth plant will produce closed loop behavior close to that designed for the approximate smooth plant. There are many examples in which any smooth approximation to nonsmooth nonlinear dynamics produces qualitatively different behavior.

As a matter of fact, a naïve application of the above approach for designing variable structure controllers, i.e., reduction to normal form, smooth ε-approximation of the nonsmooth friction, then variable structure control design (sliding and reaching control), will almost certainly fail. We will give a simple explanation below. As an alternative, we will use a backstepping approach, introduced in [21] for adaptive control design and adapted for recursive Lyapunov design in [22]. Now, let us consider the following simple example which highlights the essential issues.

4.1.1 Example 2: Sandwiched Friction

Suppose we reduce the system

$$\dot{x}_1 = x_2$$
$$\dot{x}_2 = -\phi_{fr}(x_2) + x_3$$
$$\dot{x}_3 = u$$

to normal form. Let us write the friction model in the form of a nominal plus an uncertain part: $\phi_{fr}(x_2) = \phi_{fr0}(x_2) + \delta\phi_{fr}(x_2)$, where $\phi_{fr0}(x_2)$ is smooth. For example, $\phi_{fr0}(x_2) = \tanh(x_2 / \varepsilon), \varepsilon > 0$ and $\delta\phi_{fr}(x_2) = \text{sign}(x_2) - \tanh(x_2 / \varepsilon)$.

Then we have the coordinate transform

$$z_1 = x_1$$
$$z_2 = x_2$$
$$z_3 = -\phi_{fr}(x_2) + x_3$$

which yields the transformed system

$$\dot{z}_1 = z_2$$
$$\dot{z}_2 = z_3$$
$$\dot{z}_3 = -\phi'_{fr0}(z_2) + \delta\phi'_{fr}(z_2) + u$$

Thus, any error in the friction function produces an uncertainty that depends on the derivative $\delta\phi'_{fr}(z_2)$. Obviously, if the friction function is nondifferentiable, this will produce an unbounded (although matched) uncertainty. The variable structure control, which has bounded control authority, cannot be made robust to this type of unbounded uncertainty. See Figure 5 for simulation results.

Let us instead base the normal form reduction on the smooth nominal system. Then we have the coordinate transform

$$z_1 = x_1$$
$$z_2 = x_2$$
$$z_3 = -\phi_{fr0}(x_2) + x_3$$

which yields the transformed system

$$\dot{z}_1 = z_2$$
$$\dot{z}_2 = z_3 + \delta\phi_{fr}(z_2)$$
$$\dot{z}_3 = -\phi'_{fr0}(z_2) + u$$

Now we have a bounded, although not matched, uncertainty. It is precisely because the uncertainty is unmatched that we use a backstepping approach. Before proceeding with this example we describe the backstepping process.

4.2 The VS Backstep Procedure

We give a brief description of the backstepping procedure we propose for SISO VS control system design in the presence of uncertain nonsmooth nonlinearities. Technical details and stability proofs will be given elsewhere. The key innovations in our approach for nonsmooth plants are (1) that the states are grouped depending on where an uncertainty enters the system and the robustification is attempted only where the uncertainty is identified, and (2) that the control designed at each step is a variable structure control.

Consider a SISO nonlinear system in the (multi-state back-stepping) form:

$$\begin{aligned} x_i^{(n_i)} &= x_{i+1} + \Delta_i(x,t), \quad i = 1,\ldots,p-1 \\ x_p^{(n_p)} &= \alpha(x) + \rho(x)u + \Delta_p(x,t) \\ y &= x_1 \end{aligned} \tag{6}$$

We assume that the (possibly nonsmooth) uncertainties $\Delta_i(x,t)$ are bounded by smooth, non-negative functions $\varepsilon_i(x)$, i.e.,

$$0 \le |\Delta_i(x,t)| \le \varepsilon_i(x)$$

Such a model might arise by reduction of a smooth nominal system to regular from and applying the transformation to the uncertain system.

At each of p-1 stages we design a 'pseudo-control' v_i. The kth control is obtained by designing a stabilizing smoothed VS controller for a system in the form

$$y_i^{(n_i)} = y_{i+1}, \quad i = 1,\ldots,k-2$$
$$y_{k-1}^{(n_{k-1})} = x_k,$$
$$x_k^{(n_k)} = v_k$$
$$y_k = x_k - v_{k-1}$$

To design the control v_k we first reduce the system to normal form by successive differentiation:

$$y_k^{(n_k)} = v_k - L_f^{n_k}(x_k - v_{k-1})$$

Thus, we identify the evolution equation in the new coordinate y_k that will replace x_k. Notice that the zero dynamics of this system are

$$y_i^{(n_i)} = y_{i+1}, \quad i = 1,\ldots,k-2$$
$$y_{k-1}^{(n_{k-1})} = v_{k-1}$$

Now, we design a VS stabilizing controller, $v_k(y_k,\ldots,y_k^{(n_k)})$ such that $y_k(t) \to 0$ as $t \to \infty$. For each $k < p$ we smooth the controller so that the process can be continued. Working in this way through the p stages, and redefining the states ($x \to y$) at each stage we arrive at the final set of dynamical equations. Notice the triangular structure.

$$\begin{aligned} y_i^{(n_i)} &= y_{i+1} + v_i(y_i,\ldots y_i^{(n_i)}) \quad i = 1,\ldots,p-1 \\ y_p^{(n_p)} &= \alpha + \rho u(y_p,\ldots y_p^{(n_p)}) \end{aligned} \tag{7}$$

This structure, upon which a stability analysis is based, is illustrated in Figure 3. The basic idea is roughly as follows. A VS controller is designed for system p, (7), via methods described above. The system is stable if and only if the zero dynamics,

$$y_i^{(n_i)} = y_{i+1} + v_i(y_i,\ldots y_i^{(n_i)}) \quad i = 1,\ldots,p-1, \tag{8}$$

are stable. But, v_{p-1} is itself a (smoothed) VS control so that (8) is stable if its zero dynamics:

$$y_i^{(n_i)} = y_{i+1} + v_i(y_i,\ldots y_i^{(n_i)}) \quad i = 1,\ldots,p-2$$

are stable. The argument proceeds in this way. There are subtleties because of the smoothing. And we must also establish the robust stability properties.

4.2.1 Example 2 continued

Since the example system is already in multi-state back stepping form (6) no transformation is necessary. We break the system into two parts, treating x_3 as a temporary control and ignoring the uncertainty:

Step 1 Design a *smoothed* VS control, $v(x_1, x_2)$, for:

$$\dot{x}_1 = x_2$$
$$\dot{x}_2 = -\phi_{fr}(x_2) + v$$
$$y = x_1$$

Then, we design a VS control for the composite nominal system with modified output equation.

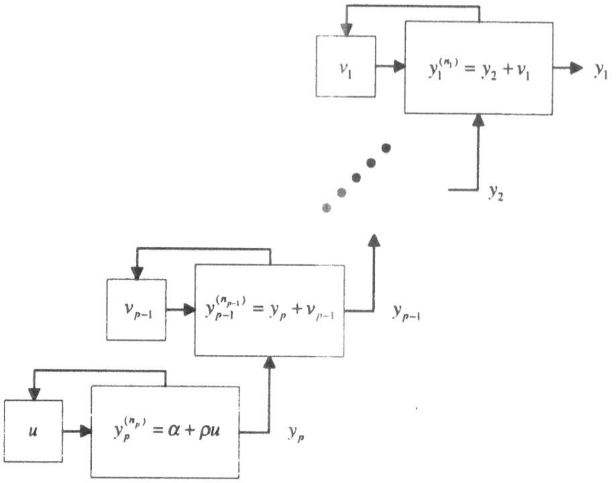

Figure 3. The triangular structure of the closed loop dynamics achieved with the multistate backstep control design.

Step 2 Design a VS control, u for

$$\dot{x}_1 = x_2$$
$$\dot{x}_2 = -\phi_{fr}(x_2) + x_3$$
$$\dot{x}_3 = u$$
$$y = x_3 - v(x_1, x_2)$$

Now, we will implement these calculatons. The Mathematica code is shown below where $[x_1, x_2, x_3] \rightarrow [\text{theta}, \text{omega}, \text{uu}]$. Using the previously described tools we have for Step 1:

```
f1 = {omega, -Tanh[omega / .02]};
g1 = {0, 1};
h1 = {theta};
{rho1, s1} = SlidingSurface[f1, g1, h1, {theta, omega}, {2}]
ctrlbnds = {{-5, 5}};
Q = {{1}};
vsc0 = SwitchingControl[rho1, s1, ctrlbnds, Q,
  SmoothingFunctions[x_] -> {Tanh[x/ .01]}]

Out[4]= {-5 Tanh[100. omega + 423.607 theta]}
```

and Step 2:

```
f = {omega, -Tanh[omega / .02] + uu, 0};
g = {0, 0, 1};
h = {uu - vsc0[[1]]};
{rho2, s2} = SlidingSurface[f, g, h, {theta, omega, uu}, {20}]
ctrlbnds = {{-5, 5}};
Q = {{1}};
vsc1 = SwitchingControl[rho2, s2, ctrlbnds, Q]

Out[19]= {5 Sign[-uu - 5 Tanh[100. omega + 423.607 theta]]}
```

Simulation results obtained with this controller are illustrated by the trajectory in Figure 4. For comparison purposes, Figure 5 illustrates the failure of the non-backstepping controller to eliminate the position output error – as anticipated.

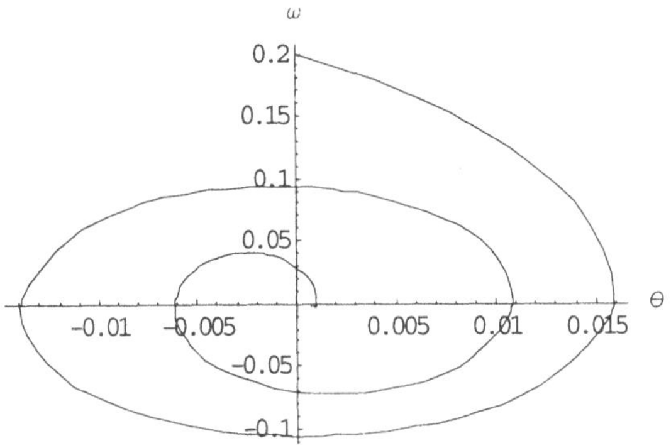

Figure 4. The projection of a state trajectory on the $\omega - \theta$ plane illustrates asymptotic convergence.

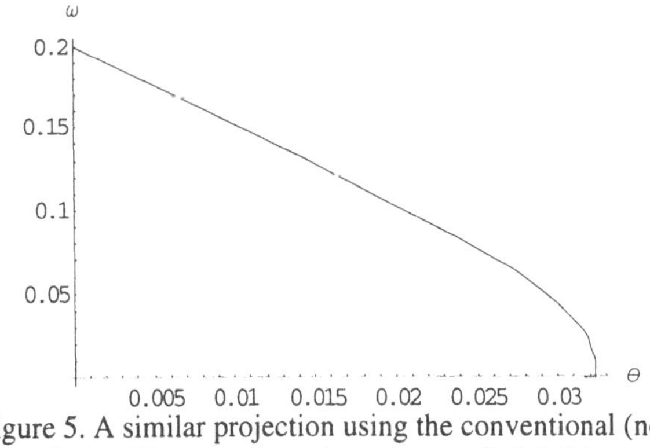

Figure 5. A similar projection using the conventional (non-backstep) design illustrates how the trajectory "sticks" because of the large matched uncertainty.

5 Apache Helicopter 30-mm Chain Gun

As a test bed problem for the variable structure control design tools, we used the US Army Apache Helicopter 30-mm chain gun. The goal of the control design is to increase the pointing and tracking performance of the gun by including friction compensation.

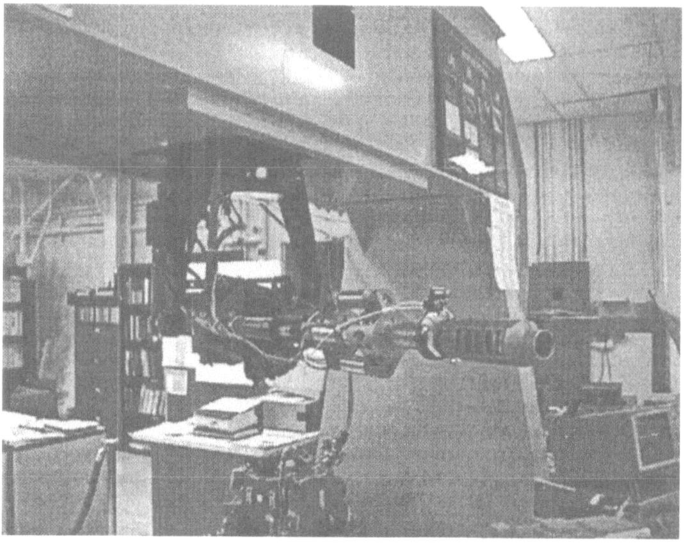

Figure 6. Apache 30mm Chain Gun Test Bed ADAWS Lab, Picatinny Arsenal

The control will be tested at the Apache 30mm Chain Gun Test Bed ADAWS Lab, Picatinny Arsenal Figure 6. The test bed gun is driven by a direct drive electric motor that is simply modeled as an input torque. The friction in the motor is dominant, so this is a simple friction control problem, not sandwiched friction problem.

5.1 Dynamic Model of Apache Gun

A four-degree of freedom (DOF) model of the apache gun system was developed. A schematic of the multibody-flex model appears in Figure 7. The model consists of a rigid turret with flexible forks that connect to a rigid gun which has a flexible barrel attached to it. A rigid blast suppressor is attached to the muzzle end of the barrel. A two channel bending actuator, developed by TSi to increase pointing accuracy, is mounted to the flexible barrel. The actuator can deliver two pairs of torques to produce muzzle angular deflection in both azimuth and elevation. The bending actuator will not be used for this study.

The gun system model was developed using the Mathematica package *ProPac* [23]. Figure 7 shows the three bodies into which the gun system is broken for modeling. Each body has a local reference frame that is located at the inboard joint.

The reference frames and their associated degrees of freedom are also shown in Figure 7.

The model was developed for designing and testing slewing controls. In order to keep the model dimension to a minimum; motions not related to slewing or slewing disturbances are not modeled. For example, the elevation of the gun is assumed a fixed value, thus eliminating a potential degree of freedom. In addition, any bending of the forks such as that which might be caused by a firing disturbance is not modeled.

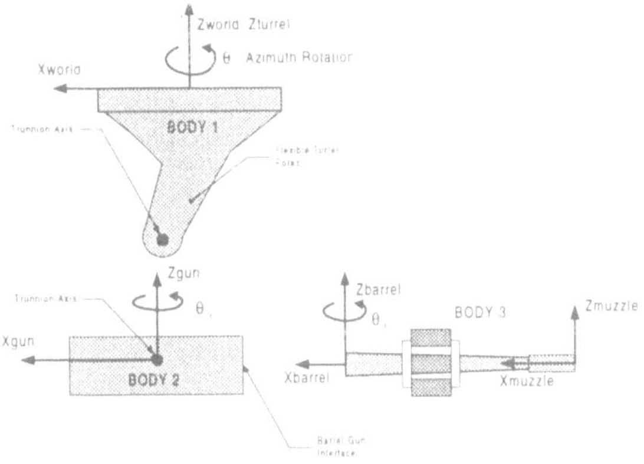

Figure 7. Body coordinate reference frames and degrees of freedom.

As depicted in Figure 7, the turret has one degree of freedom, θ_1, about the azimuth axis. The flexibility of the forks is modeled by a torsional spring with one degree of freedom, θ_2. The majority of the vibration energy in the system comes from the approximately 12 Hz flexure mode of the forks in the azimuth direction. The barrel-gun interface is allowed one degree of freedom, θ_3. This joint allows for the unintentional motion at the gun-barrel interface due to clearances in the fitting. The barrel-gun interface joint was added to the model during the model validation process to improve the matching of the transfer functions.

The flexible barrel is modeled using a reduced order FEM modal model and includes the first lateral bending modes in the azimuth plane. The reasoning behind including only the first mode is that for a cantilevered beam, 75% of the energy is in the first mode. The complete non linear system model of the apache gun contains 4 degrees of freedom which are $[\theta_1, \theta_2, \theta_3, x_1,]^T$, where x_1 is the barrel modal coordinate.

For the slewing control model, there is one control input to the system: a torque, $\tau_{tur,}$ generated by the turret motor and applied to the turret about the Z_{tur} axis. There are two outputs measurements from the system: turret azimuth, θ_{tur}, and muzzle azimuth acceleration, a_{maz}, both muzzle variables are measured with respect to the world coordinate frame.

5.1.1 Modal Analysis

Under the assumption that the fork deflections are small, i.e., θ_2 is small, then the non-linear model can be linearized about θ_1 to generate a representative linear state space model. The state space equations take the following form

$$\dot{x} = Ax + Bu$$

$$y = Cx + Du$$

where $u = [\tau_{tur}]$ is the control input vector, $y = [\theta_{tur} \quad a_{maz}]^T$ is the system measurement (output) vector, and $x = [q \quad \dot{q}]^T$ is the system state vector, and $q = [\theta_1 \quad \theta_2 \quad \theta_3 \quad x_1]^T$ is the coordinate vector.

Mode Number	Frequency [Hz]	Mode Description
1	0	Rigid body rotation about the turret azimuth axis
2	12.6	Fork azimuth flexure. Motion at gun/barrel interface in phase with gun motion. Barrel flexure negligible.
3	54.3	System moves as three bodies with the first and last moving out of phase with the second. In addition, the last body, the barrel, flexes in phase with the turret and gun/barrel interface motion.
4	322.0	Primarily barrel flexure in azimuth plane. Interface motion and barrel flexure out of phase. Other motion negligible.

Table 3: Description of mode shapes and frequencies of chain gun model

An eigenvalue decomposition of the *A* matrix gives the linearized approximation of the system modal frequencies. The system mode frequencies and mode shapes are described in Table 3. The lowest frequency mode is a rigid body mode corresponding to turret azimuth angular displacement. The next lowest frequency mode is predicted at 12.6 Hz and corresponds to fork azimuth flexure. This mode is characterized by angular displacements of the rigid turret and the gun/barrel assembly that are 180° out of phase with each other. The predicted frequency of this mode agrees with observations made of the production gun system frequency, which was determined to be around 10 Hz[2]. This is probably the most problematic mode in the system, since the firing rate of the gun is approximately 10 Hz.

[2] The slight increase in frequency of this mode may be due to differences in the testbed system from the real system.

The remaining system modes are associated with barrel flexure. The third mode excites the turret and barrel to move out of phase with the gun. In addition, the barrel flexes in phase with the turret and gun/barrel interface motion. The fourth mode is azimuth barrel flexure mode. In this mode, the flexure of the barrel is out of phase with motion of the barrel/gun interface.

5.2 Friction Modeling

The goal of this control design is to increase the pointing and tracking performance of the apache gun testbed at ARDEC by including friction compensation. Experimentalists have observed several characteristic properties of friction. These properties can be broken into two categories: static and dynamic. The static characteristics of friction, including the stiction force, the kinetic force, the viscous force, and the Stribeck effect, are functions of steady state velocity. The dynamic phenomena include pre-sliding displacement, varying breakaway force, and frictional lag. Many empirical friction models have been developed which attempt to capture specific parts of observed friction behavior.

With all the models available one must decide which friction model should be used in the friction compensating control. It is unclear whether complicated friction models improve control performance. One problem is the difficulty in obtaining *good* parameter estimates. In experiments we have performed to obtain parameter estimates, we found it particularly difficult to estimate the dynamic friction parameters [24]. The problem is complicated by the fact that the parameters may vary considerably based on such factors as temperature, lubricant condition, and material wear [25, 26]. Moreover, the various friction models in the literature represent many empirical features; however, realistically the friction present in any physical system may be different from that described in the model. We have chosen to use a simple static friction model.

Some sort of control is needed which is robust to the inaccuracy in parameter measurement, variation of parameters, and model inaccuracy. In the simulation results below, we demonstrate that the variable structure control achieves the first two. Testing is scheduled to determine the control performance on the test bed gun system.

5.2.1 Friction Experiments

The tests performed to determine static friction parameter estimates are described below. It is assumed that friction enters only at joint 1 and is a function of the angular velocity of the turret.

The steady state friction parameters including static, Coulomb, viscous, and Stribeck friction terms (F_S, F_C, F_V, v_{str}) were estimated. In the estimation problem, nonlinear constrained optimization methods from the Matlab Optimization Toolbox [27] were applied. The objective function used was the error between the observed and predicted value of friction. Being non-convex with respect to some of the optimization parameters, the objective function may have local minima. It is thus important to start from a reasonable initial guess for the parameters. This is not too difficult for the steady state parameters.

To estimate the steady state friction parameters a friction versus velocity map is constructed. To construct the friction versus velocity map, several constant velocity experiments were run with reference velocities ranging from -0.03 radians/second to +0.05 radians/second. A closed-loop PI velocity control law was implemented for the tests. The velocity for feedback control was estimated from a low pass filtered derivative of the motor encoder output. Average steady state velocity and friction force[4] were computed from the time histories of each experiment to produce the data points 'o' in Figure 8. The zero velocity data point was obtained by computing the average of the break away force from experiments where the driving force was a linearly increasing input force.

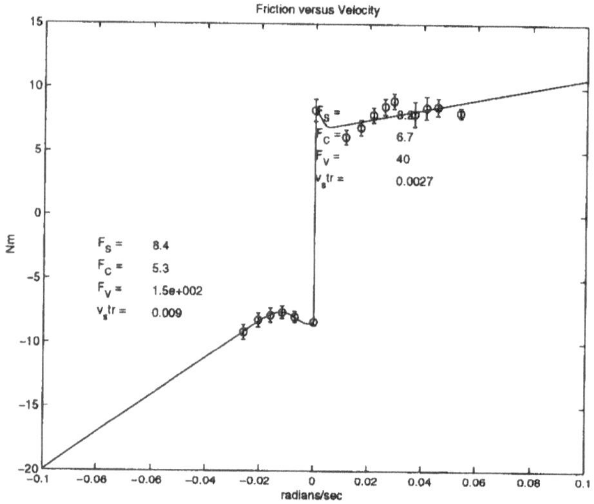

Figure 8. Friction versus Velocity

The parameter estimates were obtained by minimizing over $\hat{F}_S, \hat{F}_C, \hat{F}_V, \hat{v}_{str}$ the cost function

$$\sum_{i=1}^{n} \left\{ F_{SS}(v_i) - \hat{F}_{SS}(v_i) \right\}^2$$

where $F_{SS}(v_i)$ are the friction data obtained during constant velocity motion and

$$\hat{F}_{SS}(v_i) = \begin{cases} \left[\hat{F}_C + (\hat{F}_S - \hat{F}_C)\exp\left(-(v_i / \hat{v}_{str})^2\right)\right]\mathrm{sgn}(v_i) + \hat{F}_V v_i & v_i \neq 0 \\ \min(\hat{F}_S, F) & v_i = 0 \end{cases}$$

[3] The measured velocity data is from the motor encoder. In these low constant velocity motions the system is assumed rigid.

[4] The friction force is approximately equal to the negative of the input torque in constant velocity motion.

The parameters were all constrained to be greater than or equal to zero, with the additional constraints that $0.00001 \le \hat{v}_{str} \le 0.01$, and $\hat{F}_S \ge \hat{F}_C$. The optimization was done using the Matlab constrained optimization function Constr with tolerances of 1.0e-5 and the maximum number of iteration set to 1000.

The steady state friction parameter estimates are given in Table 4. The estimates for static, F_S, and coulomb, F_C, friction parameters were obtained by averaging the values for positive (clockwise) and negative (counterclockwise) motion. The viscous, F_V, and Stribeck, v_{str}, friction parameters were taken directly from the positive motion data since it seems that estimates from the negative motion data may be incorrect due to insufficient measurements at high enough velocity.

5.2.2 Simulation

Simulations of the nonlinear apache gun model were run using a simple PID control with anti-windup to demonstrate the effects of friction on simple control strategies. The Simulink block diagram of the closed loop system is displayed in Figure 9.

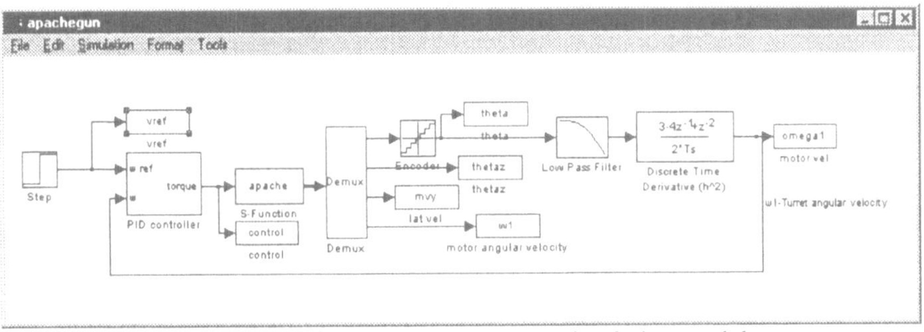

Figure 9. Simulink Apache gun simulation model

Parameter	Value
Static friction, F_S	8.3 N m
Coulomb friction, F_C	6.0 N m
Viscous friction, F_V	40 N m s/rad
Stribeck velocity, v_{str}	0.003 rad/s

Table 4: Friction Parameters

In addition to friction, there are other nonlinearities in the system that must be accounted for if one hopes to obtain realistic simulation results. For example, care must be taken to model the nonlinearities imposed by actuator saturation as well as measurement quantization. Accounting for these factors can alter significantly the simulation results. The PID controller output torque is saturation limited. The saturation values are ± 162.5 Nm (32.5 Nm/V*5 V). In addition, a quantizer with quantization interval $2\pi/2^{20}$ is included to model the encoder output. A low pass filter is added before the derivative block to smooth the encoder output. Care must be taken when adding a filter to a feedback loop. The phase delay of the filter could cause the closed loop system to be come unstable.

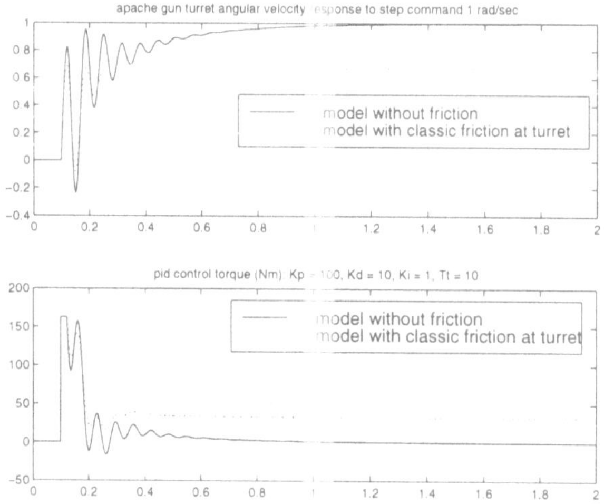

Figure 10. Step response with PID control with and without friction

In the simulations, friction is modeled using the classical model with the values obtained during the friction experiments described above, see Table 4. The friction enters the system at the first joint and is a function of the turret angular velocity, ω_1.

For comparison purposes, simulations were run both with and without friction. A step velocity command of 1 rad/sec at 0.1 sec was input to the PID controller. The simulation results are shown in Figure 10. It is clear from the simulation results that accounting for friction in the gun model simulation has significantly deteriorated the PID controlled tracking performance. The system with friction has a large steady state error from the commanded velocity.

5.3 Variable Structure Control

For control design purposes we used a rigid two-body model of the apache gun. The flexibility in the barrel and the movement at the gun barrel interface were ignored. The friction enters the system at the first joint and is a function of the turret angular velocity ω_1, as described in above.

Once the control design model was generated, the control was simple to obtain. The Mathematica/*ProPac* code for the control design is shown in Figure 11. A simple VSC control was generated. In addition, controls with smoothing using the function $\tanh(x)$, and moderation using the function $\dfrac{|\omega_1|/10+|\omega_2|/10}{0.02+|\omega_1|/10+|\omega_2|/10}$, and both moderation and smoothing were generated.

```
■ VSC Controls

Outputs = Chop[{w1}]
{rho2, s2} = SlidingSurface[F, G, Outputs, States, {2}]

{{{19.502/(41.4593 + 19.1083 Cos[theta2]^2 + 1.02759 Cos[theta2] Sin[theta2] + 22.9619 Sin[theta2]^2)}},
 {w1/(-2 + √5)}}

(*  ctrlbnds = {{-162.5,162.5}}; *) (* 5 volts: 32.5 Nm/V *)
ctrlbnds = {{-325, 325}}; (* 10 volts: 32.5 Nm/V *)
Q = {{1}};

vsc1 = Chop[SwitchingControl[rho2, s2, ctrlbnds, Q]]
vsc2 = SwitchingControl[rho2, s2, ctrlbnds, Q, SmoothingFunctions[x_] -> {Tanh[x]}]
vsc3 = Chop[SwitchingControl[rho2, s2, ctrlbnds, Q,
   ModeratingFunctions -> {(Abs[theta1] + Abs[theta2] + Abs[w1] / 10 + Abs[w2] / 10) /
      (.002 + (Abs[theta1] + Abs[theta2] + Abs[w1] / 10 + Abs[w2] / 10))}]]
vsc4 = SwitchingControl[rho2, s2, ctrlbnds, Q,
  ModeratingFunctions -> {(Abs[theta1] + Abs[theta2] + Abs[w1] / 10 + Abs[w2] / 10) /
     (.002 + (Abs[theta1] + Abs[theta2] + Abs[w1] / 10 + Abs[w2] / 10))},
SmoothingFunctions[x_] -> {Tanh[x/ .1]}];
```

Figure 11. Mathematica code for VSC design

Simulations were run from Simulink testing control of the 3-body apache model against each of the controls. The goal of each of the controls is to drive the angular velocity ω_1 to zero. The results are shown below. It is clear from the simulation that for this command, the VSC with smoothing performs best, i.e., induces the least vibration in the system.

6 Conclusions

In this paper we have described a set of symbolic computing tools that enable efficient design and implementation of variable structure control systems. The functionality includes reduction to regular form, computation of zero dynamics, design of sliding modes, assembly of the switching controller, the addition of smoothing and moderating functions and assembly of C-code for real-time implementation. The toolbox in its present form applies to smooth, (partially) feedback linearizable systems.

In developing the toolbox for variable structure control, we have had to tackle several fundamental issues involved in symbolic computing with nonsmooth functions in general, i.e., whether control or modeling related. Ongoing work is focused on extending the design method as well as the software to plants with nondifferentiable nonlinearities other than friction. We have provided a simple example above that illustrates some of the issues involved in the control design and explains why we have adopted a backstepping approach to plants of this type. Our results to date suggest that this formulation can be very effective for systems involving uncertain nonlinear friction. The backstepping approach to handling robust control is not new, however in our approach we put a new twist on it. The key innovations in our approach for nonsmooth plants are (1) that the states are grouped depending on where an uncertainty enters the system and the robustification is attempted only where the uncertainty is identified, and (2) that the control designed at each step is a smoothed variable structure control.

An additional example was given which demonstrates the power of the computing tools to easily tackle real industrial control problems with simple nonsmooth uncertainties. The variable structure control designed for the apache gun system is scheduled to be tested using the real-time C-code implementation.

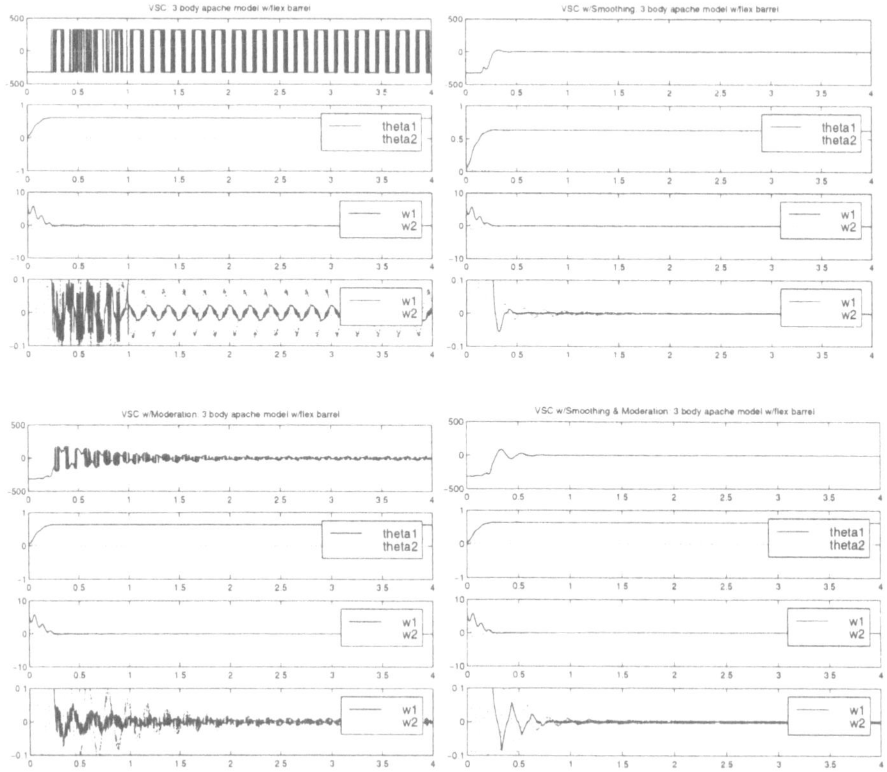

Figure 12. Apache VSC simulations

In order to address the issue of robustness to parameter variation, simulations were performed over a large range of friction parameters. Of course, with less friction the control performed better, i.e., less induced vibration in the system. In Figure 13, simulation results are shown for the case when the friction parameters are increased from their nominal values, see Table 4, to approximately double the friction, $F_v = 45$ Nm, $F_C = 12$ Nm, $F_S = 17$ Nms / rad, $v_S = 0.003$ rad / sec and then to approximately eight times the friction, $F_v = 45$ Nm, $F_C = 48$ Nm, $F_S = 64$ Nms / rad, $v_S = 0.003$ rad / s.

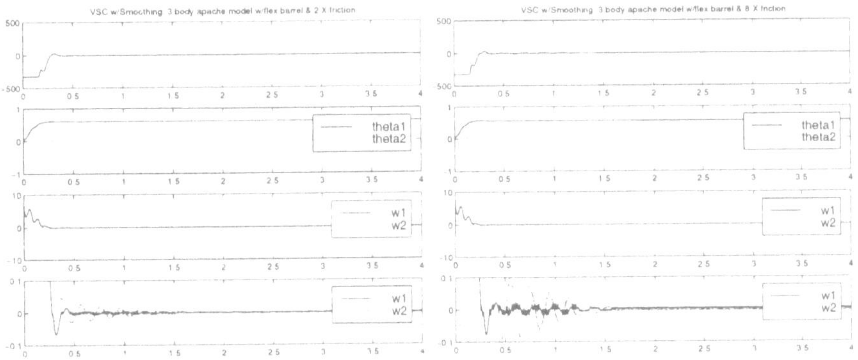

Figure 13. Robustness to varying friction parameters.

Acknowledgement: Supported, in part, by U. S. Army ARDEC, Contract No.: DAAE30-96-C-0063

7 References

1. Blankenship, G.L., *et al.*, *Integrated tools for Modeling and Design of Controlled Nonlinear Systems.* IEEE Control Systems, 1995. **15**(2): p. 65-79.
2. Kwatny, H.G. and H. Kim, *Variable Structure Regulation of Partially Linearizable Dynamics.* Systems & Control Letters, 1990. **15**: p. 67–80.
3. Kwatny, H.G., *Variable Structure Control of AC Drives*, in *Variable Structure Control for Robotics and Aerospace Applications*, K.D. Young, Editor. 1993, Elsevier: Amsterdam.
4. Kwatny, H.G. and J. Berg, *Variable Structure Regulation of Power Plant Drum Level*, in *Systems and Control Theory for Power Systems*, J. Chow, R.J. Thomas, and P.V. Kokotovic, Editors. 1995, Springer–Verlag: New York. p. 205-234.
5. Isidori, A., *Nonlinear Control Systems.* 1989, NY: Springer-Verlag.
6. Kwatny, H.G. and G.L. Blankenship. *Symbolic Tools for Variable Structure Control System Design: The Zero Dynamics.* in *IFAC Symposium on Robust Control via Variable Structure and Lyapunov Techniques.* 1994. Benevento, Italy.
7. Luk'yanov, A.G. and V.I. Utkin, *Methods of Reducing Equations of Dynamic Systems to Regular Form.* Avtomatica i Telemechanika, 1981(4): p. 5–13.
8. Isidori, A., *Nonlinear Control Systems.* 3 ed. 1995, London: Springer-Verlag.
9. Marino, R., *High Gain Feedback Non–Linear Control Systems.* International Joural of Control, 1985. **42**(6): p. 1369–1385.
10. Utkin, V.I., *Sliding Modes and Their Application.* 1974 (in Russian) 1978 (in English), Moscow: MIR.
11. Young, K.D. and H.G. Kwatny, *Variable Structure Servomechanism and its Application to Overspeed Protection Control.* Automatica, 1982. **18**(4): p. 385-400.

12. Slotine, J.J. and S.S. Sastry, *Tracking Control of Non–Linear Systems Using Sliding Surfaces, With Application to Robot Manipulators.* International Journal of Control, 1983. **38**(2): p. 465–492.
13. Slotine, J.J.E., *Sliding Controller Design for Non–Linear Control Systems.* International Journal of Control, 1984. **40**(2): p. 421–434.
14. Young, K.D., P.V. Kokotovic, and V.I. Utkin, *Singular Perturbation Analysis of High Gain Feedback Systems.* IEEE Transactions on Automatic Control, 1977. **AC–22**(6): p. 931–938.
15. Emelyanov, S.V., S.K. Korovin, and L.V. Levantovsky, *A Drift Algorithm in Control of Uncertain Processes.* Problems of Control and Information Theory, 1986. **15**(6): p. 425–438.
16. Kwatny, H.G. and T.L. Siu. *Chattering in Variable Structure Feedback Systems.* in *10th IFAC World Congress.* 1987. Munich.
17. Friedland, B., *Advanced Control System Design.* 1996, Englewood Cliffs: Prentice hall.
18. Tao, G. and P.V. Kokotovic, *Adaptive Control of Systems with Actuator and Sensor Nonlinearities.* 1996, New York: John Wiley and Sons, Inc.
19. Bennett, W.H., *et al.*, *Nonlinear and Adaptive control of Flexible Space Structures.* Transactions ASME, Journal of Dynamic Systems, Measurement and Control, 1993. **115**(1): p. 86–94.
20. Bennett, W.H., H.G. Kwatny, and M.J. Baek, *Nonlinear Dynamics and Control of Articulated Flexible Spacecraft: Application to SSF/MRMS.* AIAA Journal on Guidance, Control and Dynamics, 1994. **17**(1): p. 38–47.
21. Kanellakapoulos, I., P.V. Kokotovic, and A.S. Morse, *Systematic design of Adaptive Controllers for Feedback Linearizable Systems.* IEEE Transactions on Automatic Control, 1991. **AC–36**(11): p. 1241–1253.
22. Freeman, R.A. and P.V. Kokotovic, *Design of 'Softer' Robust Nonlinear Control Laws.* Automatica, 1993. **29**(6): p. 1425-1437.
23. Kwatny, H.G. and C. LaVigna, *TSi Dynamics User's Guide*, . 1994, Techno–Sciences, Inc.: Lanham, MD.
24. Teolis, C., *Contract Summary Report: Adaptive Control of Systems with Friction and Backlash*, . 1997, Techno-Sciences, Inc.: Lanham.
25. Armstrong-Helouvry, B., *Control of Machines with Friction.* 1991, Boston: Kluwer Academic Publishers.
26. Armstrong-Helouvry, B., P. Dupont, and C.C.d. Wit, *A survey of models, analysis tools and compensation methods for the control of machines with friction.* Automatica, 1994. **30**(7): p. 1083-1138.
27. Branch, M.A. and A. Grace, *Optimization Toolbox.* 1996, Natick, MA: The Mathworks, Inc.

Sliding Mode Control with Gain Scheduled Hyperplane for LPV Plant

Kenzo Nonami and Selim Sivrioglu

Department of Electronics and Mechanical Engineering,
Chiba University, 1-33 Yayoi-cho,Inage-ku,Chiba 263-8522, Japan

Abstract. This study presents sliding mode hyperplane design for a class of linear parameter-varying (LPV) plants, the state-space matrices of which are an affine function of time-varying physical parameters. The proposed hyperplane, involving a linear matrix inequality(LMI) approach, has continuous dynamics due to scheduling parameters and provides stability and robustness against parametric uncertainties. We have designed a time-varying hyperplane for a rotor-magnetic bearing system with a gyroscopic effect, which can be considered an LPV plant due to parameter dependence on rotational speed. The obtained hyperplane is continuously scheduled with respect to rotational speed. We successfully carried out experiments using a commercially available turbomolecular pump system and results were reasonable and good.

1 Introduction

A variable structure system or sliding mode control uses a discontinuous control structure to control nonlinear systems [1]-[3]. It is a powerful control method with increasing application in many areas of control engineering. Gain scheduling is a special type of nonlinear feedback used in a variety of control [4]. In conventional gain scheduling approach, a nonlinear control system is linearized at several operating conditions and for each operating condition a linear controller is designed [5]. Implementation of the resulting family of linear controllers as a single controller whose parameters are changed by monitoring the scheduling variables are called gain scheduling of controller. One advantage of gain scheduling is its potential to incorporate linear robust control methodologies into nonlinear control design. In conventional gain scheduling techniques, the scheduling variable should capture the plant nonlinearities and should vary slowly [6]. The slow variation condition of the scheduling variable restricts the conventional gain scheduling applications in practice.

Recent advances in robust control theory have offered a new theoretical framework and systematic gain scheduling for LPV plants. The gain scheduled control proposed in recent studies has some significant differences compared with the conventional gain scheduled control. Using LPV plant definition, the gain-scheduled H_∞ control design is extensively presented in [7] and [8] in terms of linear matrix inequalities(LMI's), the solution of which are within the scope of efficient convex optimization techniques. Here, our aims are to extend this approach to designing a sliding mode hyperplane and applying it to a practical system. All definitions and theorems given for LPV plants are also valid in our study and will not be repeated here.

2 Background for LPV Plant

A linear parameter varying plant can be described as model of linear time-varying plants or nonlinear plants which are linearized using a vector of time-varying parameter such as $\theta(t)$. The state-space representation of LPV plants is described as

$$\begin{aligned} \dot{x} &= A[\theta(t)]x + B[\theta(t)]u \\ y &= C[\theta(t)]x + D[\theta(t)]u \end{aligned} \tag{1}$$

where x, y and u denote the state vector, the measured output vector and the control input vector, respectively. $\theta(t)$ is a vector of time varying plant parameters and all plant matrices are affine functions of the $\theta(t)$. For a frozen value of θ, such as θ_τ, the transfer function of the LPV system becomes a linear-time-invariant(LTI) system, such as

$$G(s) = C(\theta_\tau)[sI - A(\theta_\tau)]^{-1}B(\theta_\tau) + D(\theta_\tau) \tag{2}$$

In practice, θ can be the time varying physical parameters, such as velocity, damping, stiffness and etc., and be given between known extremal values such as

$$\theta_i(t) \in [\underline{\theta}_i, \overline{\theta}_i] \tag{3}$$

If measurements of the $\theta(t)$ are available in real time during control operation, it is possible to design a controller which has the same parameter dependence as an LPV plant. The controller form is

$$\begin{aligned} \dot{x}_K &= A_K[\theta(t)]x + B_K[\theta(t)]y \\ u &= C_K[\theta(t)]x + D_K[\theta(t)]y \end{aligned} \tag{4}$$

This controller has continuous adjustment with the parameter measurements against the variations in the plant dynamics and maintains stability and good performance.

3 Linear Matrix Inequalities(LMI's)

Consider state-space representations of the plant and the controller given by

$$\begin{bmatrix} \dot{x} \\ q \\ y \end{bmatrix} = \begin{bmatrix} A & B_1 & B_2 \\ C_1 & D_{11} & D_{12} \\ C_2 & D_{21} & 0 \end{bmatrix} \begin{bmatrix} x \\ w \\ u \end{bmatrix} \tag{5}$$

$$\begin{bmatrix} \dot{x}_c \\ u \end{bmatrix} = \begin{bmatrix} A_c & B_c \\ C_c & D_c \end{bmatrix} \begin{bmatrix} x_c \\ y \end{bmatrix} \tag{6}$$

where $x \in \Re^n$ and $x_c \in \Re^k$ are the plant and the controller state vectors, respectively. q and y denote the controlled output and the measured output vectors. u is the control input and w is the disturbance input vector. Combining the two systems, the closed-loop system can be obtained by

$$\begin{bmatrix} \dot{x}_{cl} \\ q \end{bmatrix} = \begin{bmatrix} A_{cl} & B_{cl} \\ C_{cl} & D_{cl} \end{bmatrix} \begin{bmatrix} x_{cl} \\ w \end{bmatrix} \tag{7}$$

Closed loop matrices $A_{cl}, B_{cl}, C_{cl}, D_{cl}$ are:

$$\begin{bmatrix} A_{cl} & B_{cl} \\ C_{cl} & D_{cl} \end{bmatrix} = \begin{bmatrix} A_0 + \overline{B}\Omega\overline{C} & B_0 + \overline{B}\Omega\overline{D}_{21} \\ C_0 + \overline{D}_{12}\Omega\overline{C} & D_{11} + \overline{D}_{12}\Omega\overline{D}_{21} \end{bmatrix} \tag{8}$$

where

$$\begin{bmatrix} A_0 & B_0 & \overline{B} \\ C_0 & D_{11} & \overline{D}_{12} \\ \overline{C} & \overline{D}_{21} & \Omega \end{bmatrix} = \left[\begin{array}{cc|c|cc} A & 0 & B_1 & 0 & B_2 \\ 0 & 0 & 0 & I_k & 0 \\ \hline C_1 & 0 & D_{11} & 0 & D_{12} \\ \hline 0 & I_k & 0 & A_c & B_c \\ C_2 & 0 & D_{21} & C_c & D_c \end{array}\right] \tag{9}$$

Note that controller matrices are collected into a single matrix Ω. The Lyapunov function $V(x) = x^T P x$, $P > 0$ establishes global asymptotic stability for the closed-loop system (7). The $\mathcal{L}_2$-induced norm from w to q for LTI systems is bounded as

$$\|q\|_2 < \gamma\|w\|_2 \qquad \gamma > 0 \tag{10}$$

Finally, there exists a positive definite Lyapunov function $V(x) = x^T P x$, $P > 0$ that satisfies

$$\frac{d}{dt}V(x) + q^T q - \gamma^2 w^T w < 0 \tag{11}$$

The validity of the inequality (11) is proved in [9]. The H_∞ suboptimal control problem is equivalent to the existence of a solution to the following inequality for $X_{cl} > 0$

$$\begin{bmatrix} A_{cl}^T X_{cl} + X_{cl} A_{cl} & X_{cl} B_{cl} & C_{cl}^T \\ B_{cl}^T X_{cl} & -\gamma I & D_{cl}^T \\ C_{cl} & D_{cl} & -\gamma I \end{bmatrix} < 0 \tag{12}$$

Solution of the LMI (12) requires to find two symmetric matrices R and S such that

$$N_R^T \begin{bmatrix} AR + RA^T & RC_1^T & B_1 \\ C_1 R & -\gamma I & D_{11} \\ B_1^T & D_{11}^T & -\gamma I \end{bmatrix} N_R < 0 \tag{13}$$

$$N_S^T \begin{bmatrix} A^T S + SA & SB_1 & C_1^T \\ B_1^T S & -\gamma I & D_{11}^T \\ C_1 & D_{11} & -\gamma I \end{bmatrix} N_S < 0 \tag{14}$$

$$\begin{bmatrix} R & I \\ I & S \end{bmatrix} \geq 0 \tag{15}$$

where N_R and N_S denote bases of the null spaces of (B_2^T, D_{12}^T) and (C_2, D_{12}), respectively.

The above H_∞ control problem is valid only for LTI system and can also be extended for LPV systems. Consider a state-space representation of LPV plant in general form

$$\begin{bmatrix} \dot{x} \\ q \\ y \end{bmatrix} = \begin{bmatrix} A(\theta) & B_1(\theta) & B_2 \\ C_1(\theta) & D_{11}(\theta) & D_{12} \\ C_2 & D_{21} & 0 \end{bmatrix} \begin{bmatrix} x \\ w \\ u \end{bmatrix} \tag{16}$$

where matrices $A(\cdot), B_1(\cdot), C_1(\cdot)$ and $D_{11}(\cdot)$ are fixed function of the θ. B_2, C_2, D_{12}, D_{21} matrices are independent of the parameter θ because of tractability reasons. Finally, the solution of H_∞ control problem for LPV plants has the same form of LTI plants as follows:

$$N_R^T \begin{bmatrix} A_i R + R A_i^T & R C_{1i}^T & B_{1i} \\ C_{1i} R & -\gamma I & D_{11i} \\ B_{1i}^T & D_{11i}^T & -\gamma I \end{bmatrix} N_R < 0 \tag{17}$$

$$N_S^T \begin{bmatrix} A_i^T S + S A_i & S B_{1i} & C_{1i}^T \\ B_{1i}^T S & -\gamma I & D_{11i}^T \\ C_{1i} & D_{11i} & -\gamma I \end{bmatrix} N_S < 0 \tag{18}$$

$$\begin{bmatrix} R & I \\ I & S \end{bmatrix} \geq 0 \tag{19}$$

where A_i, B_{1i}, C_{1i}, and D_{11i} denote the parameter values of $A(\theta)$, $B_1(\theta)$, $C_1(\theta)$ and $D_{11}(\theta)$ at the vertices $\theta = \theta_i$ of the parameter polytope.

The inequalities (17), (18), and (19) can be solved using the toolbox [10] which uses convex optimization algorithms. The construction of the controller matrix Ω from R and S matrices can be done by the same convex programs.

4 Frequency-Shaped Hyperplane Design

4.1 LTI Control Systems

The frequency-shaped based hyperplane design is proposed in reference [11] and this approach is extended using H_∞/μ control theory in [12]. If a plant has some unknown or neglected dynamics, such as truncation of high frequency modes in a flexible system, the frequency-shaped-based hyperplane for sliding mode maintains stability and good performance. Suppose that a transformed system is given by

$$\begin{bmatrix} \dot{x}_1 \\ \dot{x}_2 \end{bmatrix} = \begin{bmatrix} A_{11} & A_{12} \\ A_{21} & A_{22} \end{bmatrix} \begin{bmatrix} x_1 \\ x_2 \end{bmatrix} + \begin{bmatrix} 0 \\ B_2 \end{bmatrix} u + \begin{bmatrix} D \\ 0 \end{bmatrix} w \tag{20}$$

$$y = C x_1$$

where $x_1 \in \Re^n$, $x_2 \in \Re^m$, $u \in \Re^m$ and B_2 is assumed full rank. y and w represent the measured output vector and the disturbance input vector, respectively. Specifically, the control system is considered as a servo type and the error signal is defined by:

$$e = r - C x_1 \tag{21}$$

where r is the reference input signal. Now, we can offer a switching function for this control system as follows:

$$\Phi = S(x_1) + x_2 \tag{22}$$

where $S(x_1)$ is a linear operator of x_1. The offered switching function Φ will have some dynamics comparing with a conventional switching function in terms of the defined new variable z

$$\begin{aligned} \dot{z} &= Fz + Ge \\ S(x_1, z) &= -Hz - Le \end{aligned} \tag{23}$$

After combining (20) and (23), the extended system can be obtained by

$$\begin{bmatrix} \dot{z} \\ \dot{x}_1 \\ \dot{x}_2 \end{bmatrix} = \begin{bmatrix} F & -GC & 0 \\ 0 & A_{11} & A_{12} \\ 0 & A_{21} & A_{22} \end{bmatrix} \begin{bmatrix} z \\ x_1 \\ x_2 \end{bmatrix} + \begin{bmatrix} 0 \\ 0 \\ B_2 \end{bmatrix} u + \begin{bmatrix} G \\ 0 \\ 0 \end{bmatrix} r + \begin{bmatrix} 0 \\ D \\ 0 \end{bmatrix} w \tag{24}$$

$$\Phi = \begin{bmatrix} -H & LC \end{bmatrix} \begin{bmatrix} z \\ x_1 \end{bmatrix} - Lr + x_2 \tag{25}$$

Using the known equation for sliding mode $\Phi = \dot{\Phi} = 0$, the equivalent control input is

$$\begin{aligned} u_{eq} = -B_2^{-1}[&-HFz + (HGC + LCA_{11} + A_{21})x_1 \\ &+(LCA_{12} + A_{22})x_2 - (HG + L)r] \end{aligned} \tag{26}$$

From the (22), we have

$$x_2 = Hz - LCx_1 + Lr \tag{27}$$

Supposing that the sliding mode occurs on Φ=0, the equation of sliding mode is given by

$$\begin{bmatrix} \dot{x}_1 \\ \dot{z} \end{bmatrix} = \begin{bmatrix} A_{11} - A_{12}LC & A_{12}H \\ -GC & F \end{bmatrix} \begin{bmatrix} x_1 \\ z \end{bmatrix} + \begin{bmatrix} A_{12}L \\ G \end{bmatrix} r + \begin{bmatrix} D \\ 0 \end{bmatrix} w \tag{28}$$

If the system (A_{11}, A_{12}) is controllable, (24) and (25) are uniquely decided with (F, G, II, L). The sliding mode hyperplane as an H_∞ control for an LTI system is shown in Fig.1. In this figure, filters W_1 and W_2 are used for frequency shaping.

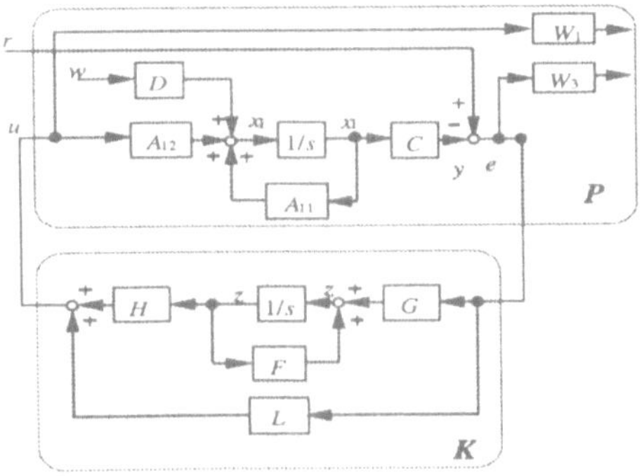

Fig.1 Sliding mode hyperplane as H_∞ control for LTI systems

4.2 LPV Control Systems

Frequency-shaped hyperplane design can also be extended to LPV plants. Although the LPV plant is described in a very general form, the control design for LPV plant requires some limitations, such as parameter independence of control inputs and measured outputs due to tractability reasons. Therefore the canonical form will be obtained directly from state-space equation of the considered plant by using a pre-filter. Let us consider the state space representation of the LPV plant:

$$\begin{aligned} \dot{x} &= A(\theta)x + Bu + D_1(\theta)w \\ y &= Cx \end{aligned} \tag{29}$$

Choosing a filter, such as

$$\begin{aligned} \dot{x}_w &= A_w x_w + B_w u \\ y_w &= C_w x_w + D_w u \end{aligned} \tag{30}$$

If we combine the state-space model of the plant with the above filter, we can obtain an augmented structure:

$$\begin{aligned} \begin{bmatrix} \dot{x} \\ \dot{x}_w \end{bmatrix} &= \begin{bmatrix} A(\theta) & BC_w \\ 0 & A_w \end{bmatrix} \begin{bmatrix} x \\ x_w \end{bmatrix} + \begin{bmatrix} BD_w \\ B_w \end{bmatrix} u + \begin{bmatrix} D_1(\theta) \\ 0 \end{bmatrix} w \\ y &= \begin{bmatrix} C & 0 \end{bmatrix} \begin{bmatrix} x \\ x_w \end{bmatrix} \end{aligned} \tag{31}$$

For D_w=0, (31) reduces the canonical form, as given in (20).

If measurements of the time-varying parameter θ are available in real time, then the switching function will have parameter dependence as follows:

$$\begin{aligned} \dot{z} &= F(\theta)z + G(\theta)e \\ S(x_1, z) &= -H(\theta)z - L(\theta)e \end{aligned} \tag{32}$$

Finally, in the general form, the extended system can be written as

$$\begin{aligned} \begin{bmatrix} \dot{z} \\ \dot{x}_1 \\ \dot{x}_2 \end{bmatrix} = \begin{bmatrix} F(\theta) & -G(\theta)C & 0 \\ 0 & A_{11}(\theta) & A_{12} \\ 0 & A_{21} & A_{22} \end{bmatrix} \begin{bmatrix} z \\ x_1 \\ x_2 \end{bmatrix} + \begin{bmatrix} 0 \\ 0 \\ B_2 \end{bmatrix} u \\ + \begin{bmatrix} G(\theta) \\ 0 \\ 0 \end{bmatrix} r + \begin{bmatrix} 0 \\ D(\theta) \\ 0 \end{bmatrix} w \end{aligned} \tag{33}$$

$$\Phi(\theta) = \begin{bmatrix} -H(\theta) & L(\theta)C \end{bmatrix} \begin{bmatrix} z \\ x_1 \end{bmatrix} - L(\theta)r + x_2 \tag{34}$$

The reduced-order system with scheduling parameter is shown in Fig.2 for LPV plants.

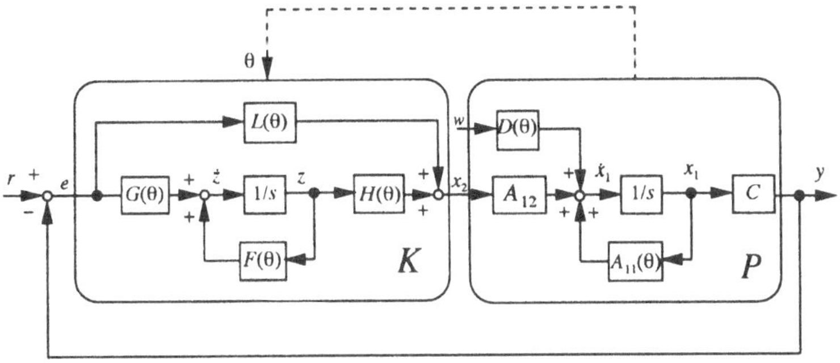

Fig.2 Reduced order system for LPV plants

5 VSS Controller Design For LPV Plants

We seek a Lyapunov function such that

$$V = \frac{1}{2}\Phi\Phi \tag{35}$$

The following condition should be satisfied for existence of sliding mode and stabilization of the closed-loop system:

$$\dot{V} = \Phi\dot{\Phi} < 0 \tag{36}$$

For affine parameter-dependent plants, the fixed Lyapunov function is replaced by a parameter-dependent Lyapunov function [13]. Affine parameter-dependent systems can be converted to polytopic parameter ones. In our study, the compensator matrices are obtained as polytopic ones. For simplicity, we have supposed that the compensator is strictly proper, $L(\theta) = 0$. The same design procedure is also valid for the proper case. For a single parameter case, such as θ_1, the affine parameter-dependent case is

$$H(\theta_1) = \tilde{H}_0 + \theta_1\tilde{H}_1, \qquad \theta_1(t) \in \left[\underline{\theta}_1\ \overline{\theta}_1\right] \tag{37}$$

where $\tilde{H}_0$ and $\tilde{H}_1$ are known fixed matrices. This can be converted to a polytopic parameter-dependent matrix as follows:

$$H(\alpha) = \alpha_1 H_1 + \alpha_2 H_2 \tag{38}$$

where H_1, H_2, α_1 and α_2 are obtained by

$$H_1 = H(\underline{\theta}_1), \quad H_2 = H(\overline{\theta}_1), \qquad \underbrace{\frac{\overline{\theta}_1 - \theta_1(t)}{\overline{\theta}_1 - \underline{\theta}_1}}_{\alpha_1} + \underbrace{\frac{\theta_1(t) - \underline{\theta}_1}{\overline{\theta}_1 - \underline{\theta}_1}}_{\alpha_2} = 1 \tag{39}$$

The derivative of the hyperplane can be obtained by

$$\begin{aligned}\Phi &= (\alpha_1 H_1 + \alpha_2 H_2)z + x_2 \\ \dot{\Phi} &= \frac{d\Phi(z, x_2, \alpha_1, \alpha_2)}{dt} \\ &= -(\alpha_1 H_1 + \alpha_2 H_2)\dot{z} + \dot{x}_2 - (\dot{\alpha}_1 H_1 + \dot{\alpha}_2 H_2)z \\ &= -\sum_{i=1}^{2}(\alpha_i H_i \dot{z} + \dot{\alpha}_i H_i z) + \dot{x}_2\end{aligned} \tag{40}$$

From (36),

$$\begin{aligned}\dot{V} &= \Phi\dot{\Phi} \\ &= \Phi(-\sum_{i=1}^{2}(\alpha_i H_i \dot{z} + \dot{\alpha}_i H_i z) + \dot{x}_2) \\ &= \Phi\{-\sum_{i=1}^{2}(\alpha_i H_i F_i + \dot{\alpha}_i H_i)z + [\sum_{i=1}^{2}(\alpha_i H_i G_i C) + A_{21}]x_1 + A_{22}x_2 + B_2 u\} \\ &= \Phi B_2(u - u_{eq})\end{aligned} \tag{41}$$

where the equivalent control input u_{eq} is obtained by using the condition $\Phi = \dot{\Phi} = 0$

$$u_{eq} = -B_2^{-1}[-\sum_{i=1}^{2}(\alpha_i H_i F_i + \dot{\alpha}_i H_i)z + [\sum_{i=1}^{2}(\alpha_i H_i G_i C) + A_{21}]x_1 + A_{22}x_2] \tag{42}$$

The condition of (36) becomes

$$\dot{V} = \Phi B_2(u - u_{eq}) < 0 \tag{43}$$

The control input u, which satisfies the condition of (43) can be chosen by

$$u = \begin{cases} u_{eq} - u_n & B_2\Phi > 0 \\ u_{eq} + u_n & B_2\Phi < 0 \end{cases} \tag{44}$$

Discontinuous control input u_n is given by

$$u_n = \epsilon \frac{B_2\Phi}{\|B_2\Phi\| + \eta} \tag{45}$$

where $\epsilon > 0$ and $\eta > 0$ are selected as 0.1 and 0.01, respectively. η is used for a smooth discontinuous control input. ϵ is specified with experimental experiences. For slow parameter variation, the derivative of parameter $\dot{\alpha}$ becomes very small and $\dot{\alpha}_i H_i z$ term can be neglected.

6 Design Example

Control system design for active magnetic bearings (AMBs) is an advanced topic for control engineers, because of their highly complex structures, precise design requirements, and increasing applications in industry. The advantages of AMBs, such as contactless and frictionless operation in normal running, without lubrication, make them virtually maintenance free, and attractive for various applications [14]. In many AMB systems in use today, proportional-integral-derivative (PID) controllers have been used because of certain practical advantages. However, it is not easy to satisfy the requirements for robust performance using PID control. Recently, robust control design approaches have been effectively studied for AMB system design [15]-[16]. The control problem presented here can be a good test for proposed gain-scheduled sliding mode control.

6.1 Control Object

The control object presented here is a commercially available turbomolecular pump system (Fig.3). Gain-scheduled H_∞ control design has been previously studied for this system [17]. We designed the controller only for radial directions. Axial direction is controlled by PID controllers. The plant has a lower side permanent magnetic bearing which has constant stiffness and damping effect and an upper side AMB which produces control inputs. Gyroscopic couples have considerable effects on the rotor and natural frequencies of the system vary with rotational frequency.

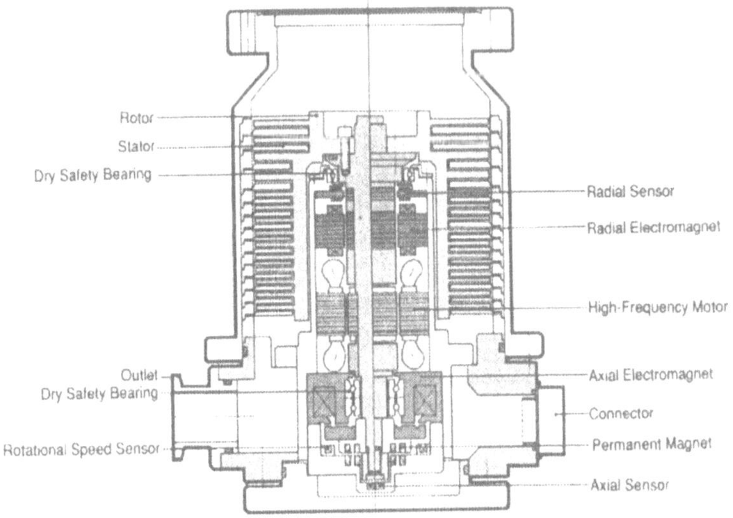

Fig.3 Cross-sectional view of turbomolecular pump

6.2 Modeling of AMB System with Gyroscopic Effect

The equations of motion of rotor-magnetic bearing system given in Fig.4 can be written by

$$\begin{aligned}
m\ddot{x}_g &= f_{x_u} + f_{x_b} + m_{un} l \omega_z^2 Cos\omega_z t \\
J_r\ddot{\beta} &= -J_a\omega_z\dot{\alpha} + L_u f_{x_u} - L_b f_{x_b} \\
m\ddot{y}_g &= f_{y_u} + f_{y_b} + m_{un} l \omega_z^2 Sin\omega_z t \\
J_r\ddot{\alpha} &= J_a\omega_z\dot{\beta} - L_u f_{y_u} + L_b f_{y_b}
\end{aligned} \tag{46}$$

Control forces produced by upper side AMB can be given by

$$\begin{aligned}
f_{xu} &= 2K_{du}x_g + 2L_u K_{du}\beta + 2K_{iu}i_{xu} \\
f_{yu} &= 2K_{du}y_g - 2L_u K_{du}\alpha + 2K_{iu}i_{yu}
\end{aligned} \tag{47}$$

and also lower side forces are

$$\begin{aligned}
f_{xb} &= -2C_b\dot{x}_g + 2C_b L_b\dot{\beta} - 2K_b x_g + 2K_b L_b \beta \\
f_{yb} &= -2C_b\dot{y}_g - 2C_b L_b\dot{\alpha} - 2K_b y_g - 2K_b L_b \alpha
\end{aligned} \tag{48}$$

Table 1 shows the parameters of AMB system. The above equations can be rewritten in a more compact form by

$$\tilde{M}\ddot{z}_r + (\tilde{G}_c + \tilde{G}_j)\dot{z}_r + \tilde{K}z_r = \tilde{F}u + \tilde{E}w \tag{49}$$

where $\tilde{M}, \tilde{G}_c, \tilde{G}_j$ and $\tilde{K}$ are the mass matrix, the damping matrix, the gyroscopic matrix, and the stiffness matrix, respectively. $\tilde{F}$ and $\tilde{E}$ indicate the control input and the disturbance input locations, respectively. The vector z_r represents the displacement variables of the rotor with respect to the center of gravity measured in the fixed coordinate as follow:

$$z_r = \begin{bmatrix} x_g \ \beta \ y_g \ \alpha \end{bmatrix}^T \tag{50}$$

The gyroscopic effects are characterized by a skew-symmetric matrix $\tilde{G}_j = -\tilde{G}_j^T$, which contains the rotor speed ω_z as a linear factor. The unbalance effects also vary with the square function of rotational speed ω_z. The gyroscopic matrix and linearized unbalance matrix using some approximation are

$$\tilde{G}_j(\omega_z) = \begin{bmatrix} 0 & 0 & 0 & 0 \\ 0 & 0 & 0 & J_a \\ 0 & 0 & 0 & 0 \\ 0 & -J_a & 0 & 0 \end{bmatrix} \omega_z \quad \tilde{E}(\omega_z) = \begin{bmatrix} m_{un} l \gamma_p & 0 \\ 0 & 0 \\ 0 & m_{un} l \gamma_p \\ 0 & 0 \end{bmatrix} \omega_z \tag{51}$$

where γ_p is a constant. Finally, the state-space model of this system can be obtained by

$$\begin{aligned}
\dot{x}_s &= A(\omega_z)x_s + Bu + D_1(\omega_z)w \\
y &= Cx_s
\end{aligned} \tag{52}$$

where x_s is the state vector, u is the control input vector, w is the disturbance input vector and y is the measured output vector. The system matrices are

$$A(\omega_z) = \begin{bmatrix} 0 & I \\ -\tilde{M}^{-1}\tilde{K} & -\tilde{M}^{-1}(\tilde{G}_c + \tilde{G}_j(\omega_z)) \end{bmatrix}$$
$$D_1(\omega_z) = \begin{bmatrix} 0 \\ \tilde{M}^{-1}\tilde{E}(\omega_z) \end{bmatrix}, B = \begin{bmatrix} 0 \\ \tilde{M}^{-1}\tilde{F} \end{bmatrix} \quad (53)$$
$$C = K_s \begin{bmatrix} 1 & l_u & 0 & 0 & 0 & 0 & 0 & 0 \\ 0 & 0 & 1 & -l_u & 0 & 0 & 0 & 0 \end{bmatrix}$$

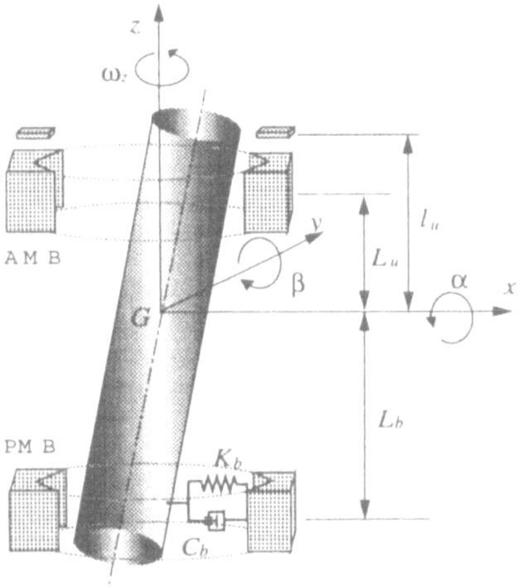

Fig.4 Model of the rotor-active bearing system

Table 1. Parameters of the AMB system

Parameter	Symbol	Value	Unit
Mass of the rotor	m	1.595	kg
Moment of inertia around radial axis	J_r	0.00383	kgm^2
Polar moment of inertia	J_a	0.00161	kgm^2
Distance of upper AMB to center of gravity	L_u	0.0128	m
Distance of lower PMB to center of gravity	l_b	0.0843	m
Distance of sensor to center of gravity	l_u	0.0314	m
Linearized force/current factor	K_{iu}	200	N/A
Linearized force/displacement factor	K_{du}	2.8×10^5	N/m
Stiffness coefficient of PMB	K_b	10^5	N/m
Damping coefficient of PMB	C_b	48	kg/s
Unbalance mass	m_{un}	0.6×10^{-3}	kg
Distance of unbalance mass from center	l	0.02	m
Sensor gain	K_s	10000	V/m

6.3 Augmentation of State Space by Using a Pre-Filter

We used a pre-filter to augment the state space of the plant to canonical form. The filter given in Fig.5 is selected, so as to give integral action to the sliding mode controller. Combining the filter and the plant, the augmented system is obtained as

$$\begin{aligned}\begin{bmatrix}\dot{x}_s\\ \dot{x}_w\end{bmatrix} &= \begin{bmatrix}A(\omega_z) & BC_w\\ 0 & A_w\end{bmatrix}\begin{bmatrix}x_s\\ x_w\end{bmatrix} + \begin{bmatrix}0\\ B_w\end{bmatrix}u + \begin{bmatrix}D_1(\omega_z)\\ 0\end{bmatrix}w\\ y &= \begin{bmatrix}C\ 0\end{bmatrix}\begin{bmatrix}x_s\\ x_w\end{bmatrix}\end{aligned} \tag{54}$$

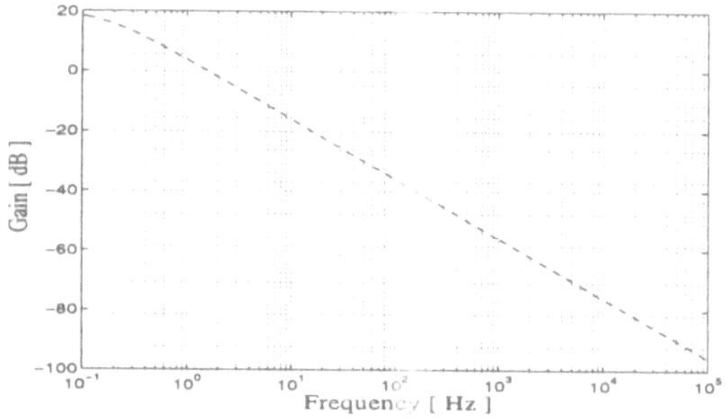

Fig.5 Bode plot of the pre-filter

6.4 Gain-Scheduled Compensator Design

The control system accelerates from 0 to 30000 rpm in 2 minutes and decelerates in breaking mode in the same amount of time. We define parameter dependence of the plant due to rotational speed as follows:

$$\omega_z \in [\underline{\omega}_z, \overline{\omega}_z] \quad or \quad \omega_z \in [0, 30000]\ \text{r/min} \tag{55}$$

In this parameter variation range, the plant can be defined by

$$\begin{aligned}G(\alpha_1,\alpha_2) = \begin{bmatrix}0 & \begin{bmatrix}0\ B_2\end{bmatrix}\\ C & 0\end{bmatrix} &+ \alpha_1\begin{bmatrix}A(\underline{\omega}_z) & \begin{bmatrix}D_1(\underline{\omega}_z)\ 0\end{bmatrix}\\ 0 & 0\end{bmatrix}\\ &+\alpha_2\begin{bmatrix}A(\overline{\omega}_z) & \begin{bmatrix}D_1(\overline{\omega}_z)\ 0\end{bmatrix}\\ 0 & 0\end{bmatrix}\end{aligned} \tag{56}$$

The parameters α_1 and α_2 are obtained by the following convex decomposition

$$\underbrace{\frac{\overline{\omega}_z - \omega_z(t)}{\overline{\omega}_z - \underline{\omega}_z}}_{\alpha_1} + \underbrace{\frac{\omega_z(t) - \underline{\omega}_z}{\overline{\omega}_z - \underline{\omega}_z}}_{\alpha_2} = 1 \quad \text{or} \quad \sum_{i=1}^{2}\alpha_i = 1, \qquad \alpha_i \geq 0 \tag{57}$$

We specified two frequency shaping filters for robust stability and sensitivity reduction. Our plant is open-loop unstable and has right- half-plane poles and zeros. A good way to select the corner frequency of the filters is to consider the unstable poles and zeros of the plant [18]. Unfortunately, this rule does not work completely for multi-input-multi-output (MIMO) systems. Therefore, the following way is chosen from experimental experience:

$$W_1(s) = \frac{k_1(s+q)}{(s+k_2p)}, \qquad W_2(s) = \frac{k_2p}{(s+k_2p)}$$

where p is the unstable pole of the plant and $q = 1$, $k_1 = 0.06$, $k_2 = 5$. For every direction, the same frequency-shaping filter is used. Finally, using the parameter dependent plant with the above filters, the gain-scheduled compensator is computed using LMI Control Toolbox in MATLAB [10]:

$$\begin{bmatrix} F(\omega_z) & G(\omega_z) \\ H(\omega_z) & L(\omega_z) \end{bmatrix} = \sum_{i=1}^{2} \alpha_i \begin{bmatrix} F_i & G_i \\ H_i & L_i \end{bmatrix} \tag{58}$$

The interpolated structure of this compensator has the form

$$\begin{bmatrix} F(\omega_z) & G(\omega_z) \\ H(\omega_z) & L(\omega_z) \end{bmatrix} = \frac{\overline{\omega}_z - \omega_z(t)}{\overline{\omega}_z - \underline{\omega}_z} \begin{bmatrix} F_1 & G_1 \\ H_1 & L_1 \end{bmatrix} + \frac{\omega_z(t) - \underline{\omega}_z}{\overline{\omega}_z - \underline{\omega}_z} \begin{bmatrix} F_2 & G_2 \\ H_2 & L_2 \end{bmatrix} \tag{59}$$

7 Simulations

The order of plant is eight. Using two weighting functions, the compensator is computed as twelfth order. The Bode plot of the compensator is given in Fig.6 as two-dimensional(2-D) and (3-D) plots. As can be seen in these figures, the compensator dynamics vary with rotational speed.

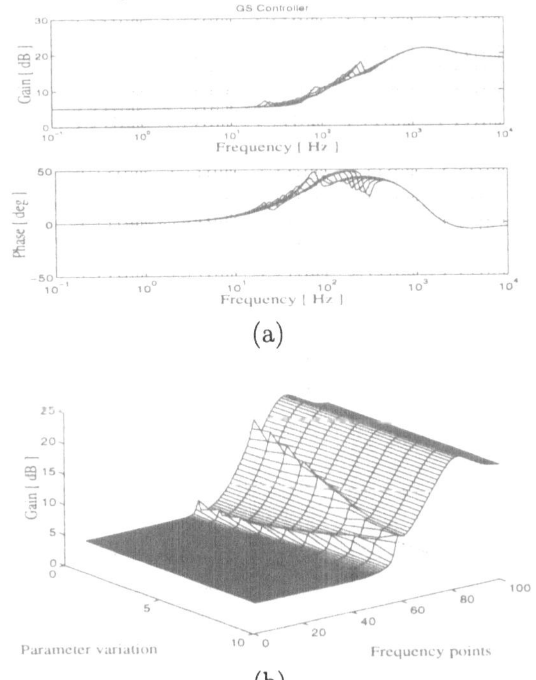

(a)

(b)

Fig.6 Bode plot of the gain scheduled compensator (a) 2D (b) 3D

8 Gain-Scheduling Implementation

In this approach, the designed gain scheduled compensator has a vertex property. Basically, the vertex controllers are placed at the lower and upper bound of the parameter space. The scheduling of controller is realized using vertex controllers with a convex combination of the parameter measurements. The controller is obtained using the VSS controller structure. The computation of controller output is realized by using the following steps.

$$\begin{aligned}
&z_{k+1} = [(1-\alpha_k)F_d^{(1)} + \alpha_k F_d^{(2)}]x_k + [(1-\alpha_k)G_d^{(1)} + \alpha_k G_d^{(2)}]y_k \\
&S_k = [(1-\alpha_k)H_d^{(1)} + \alpha_k H_d^{(2)}]x_k \\
&\Phi_k = B_{w_k}(S_k + x_{w_k}) \\
&u_{n_k} = \epsilon_s \frac{B_{w_k}\Phi_k}{\|B_{w_k}\Phi_k\| + \eta_s} \\
&u_{eq_k} = -B_{w_k}^{-1}[-[(1-\alpha_k)H_d^{(1)} + \alpha_k H_d^{(2)}]z_{k+1} + x_{w_{k+1}}] \\
&u_{t_k} = u_{eq_k} + u_{n_k} \\
&x_{w_{k+1}} = A_{w_k}x_{wk} + B_{w_k}u_{t_k} \\
&u_k = C_{w_k}x_{wk}
\end{aligned} \tag{60}$$

where $\alpha_k = \frac{\omega_{z_k}}{\overline{\omega}_z}$. $F_d^{(i)}, G_d^{(i)}$, and $H_d^{(i)}$, (i=1,2) are the discretized vertex compensator matrices.

9 Experimental Results

The configuration of the experimental setup is schematically shown in Fig.7. The actual turbomolecular pump is used for experiment. Two displacements measured by two position sensors in the x and y directions and rotational speed signal ω_z go to the digital signal processor(DSP) through A/D converters. Two control inputs are supplied to electromagnets through D/A converters and power amplifiers. The sampling time of compensator is 0.2 msec. The rotational speed signal ω_z is continuously measured with a pulse modulation of 0.2 msec between 0 to 2.5 V amplitude. The measured rotational speed signal values has a tolerance between $\mp$ 100 rpm.

Generally speaking, it is impossible for rotating machinery to remove unbalance effects completely. Here our aim is to maintain stability and robustness even if the rotor has some unbalance effect. For this reason, we tested the controller by mounting 0.6 g unbalance mass on the rotor. For four different rotational speeds, the orbits of the rotor central axis are shown in Fig.8. Except ω_z=5000 rpm, the obtained orbits are synchronized and reasonable. The critical speed of the rigid rotor occurred at 5000 rpm and for this reason the diameter of the orbit at this rotational speed was increased but closed-loop system was still stable. High-speed rotation of the rotor with gain-scheduled sliding mode controller is shown in Fig.9. These results also demonstrate very good synchronization. In the case of PID control, it was impossible to increase the rotational speed beyond 10000 rpm any more with the 0.6 g unbalance mass.

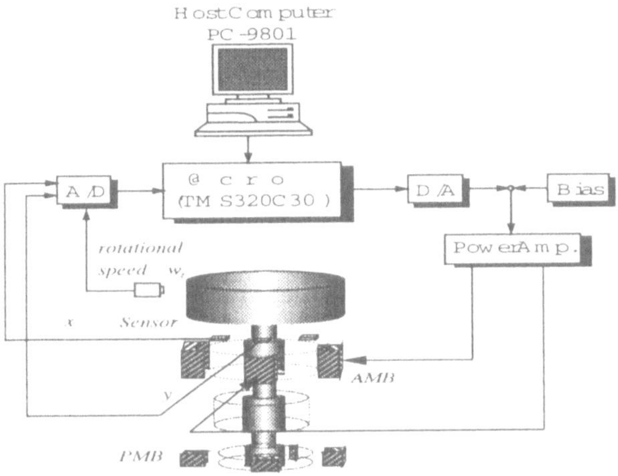

Fig.7 Configuration of the experimental setup

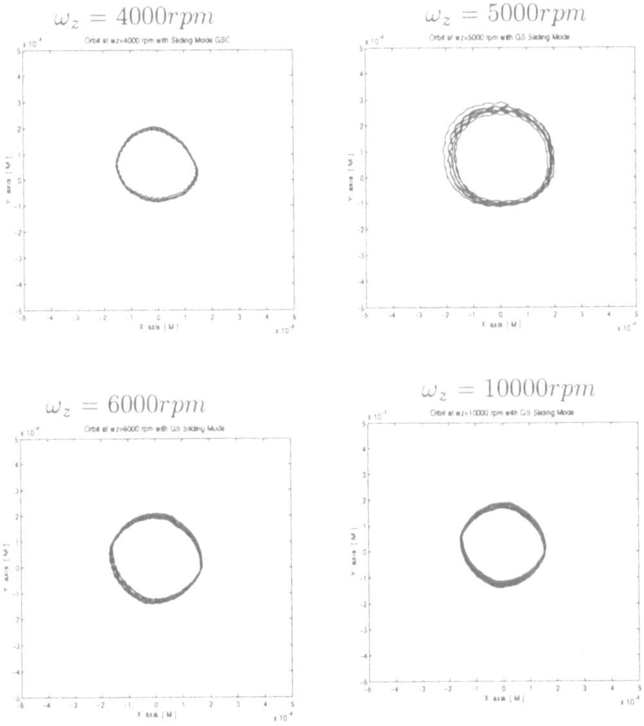

Fig. 8 Orbits with sliding mode control(Axes wide $50\mu m$)

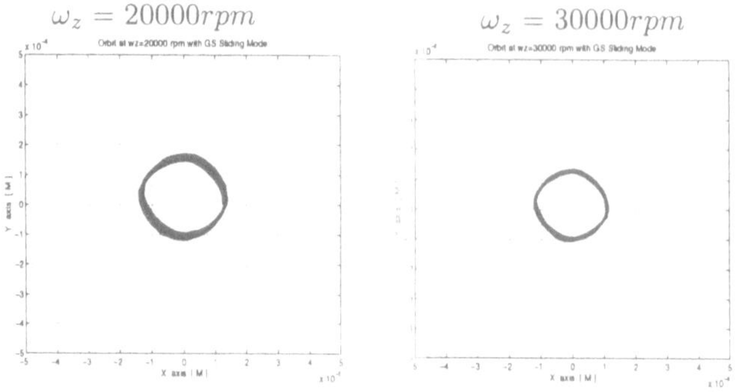

Fig.9 Orbits with sliding mode control(Axes wide $20\mu m$)

10 Conclusions

In this study, we have proposed a new type of hyperplane design based on gain-scheduled H_∞ control. The LPV plant eigenvalues move by means of linear parameters, and control of this class of system using conventional method is generally difficult. The gain-scheduled compensator has the same parameter dependence, and the eigenvalues of the compensator also vary with the parameters. As a result of the synchronized behavior of the plant and compensator, stability and robustness can be maintained. Due to parameter dependence of the LPV plant, the existence of sliding mode by a static hyperplane may not be guaranteed. The offered hyperplane has continuous dynamics with respect to time-varying parameters and the sliding mode occurs on this hyperplane in every case.

References

1. V.I., Utkin, "Sliding Modes in Control Optimization," *Springer-Verlag*, 1992.
2. K. Nonami, H. Tian, "Sliding Mode Control," *Corona Publication*, 1994(in Japanese).
3. K.D. Young, V.I., Utkin, U. Ozguner, "A Control Engineer's Guide to Sliding Mode Control" *Procs. IEEE International Workshop on VSS*, p.1-14, Tokyo, 1996.
4. K. J. Astrom and B. Wittenmark, "Adaptive Control," *Reading, MA: Addison-Weslay*, 1989.
5. H.K Khalil, "Nonlinear Systems," Second Edition, *Prentice Hall*, 1997.
6. J. Shamma and M. Athans, "Guaranteed properties of gain scheduled control of linear parameter-varying plants," *Automatica*, vol.27, no.3, pp.559-564, 1991.
7. P. Apkarian, P. Gahinet, "A Convex Characterization of Gain Scheduled H_∞ Controllers," *IEEE. Trans. on Automatic Control*, vol.40, pp.853-863, 1995.
8. P. Apkarian, P. Gahinet, G. Becker, " Self-scheduled H_∞ Control of Linear Parameter-varying Systems: a Design Example," *Automatica*, vol.31-9, pp.1251-1261, 1995.
9. S. Boyd, L. E.Ghaoui, E. Feron, V. Balakrishnan, "Linear Matrix Inequalities in System and Control Theory," *SIAM Publication*, 1994.

10. P. Gahinet, A. Nemirovski, A.J. Laub, M. Chilali, "LMI Control Toolbox, For Use with MATLAB," *Mathworks*, 1995.
11. D. Young, U. Ozguner, "Frequency Shaping Compensator Design for Sliding Mode Control," *Int. J. Control*, vol.57, No.5, pp.1005, 1993.
12. K. Nonami, H. Nishimura, H. Tian, "H_∞/μ Control-based Frequency-shaped Sliding Mode Control for Flexible Structures," *JSME International Journal*, Series C, vol.39, No.3, pp. 493-501, 1996.
13. P. Gahinet, P. Apkarian, M. Chilali, "Affine Parameter-Dependent Lyapunov Functions and Real Parametric Uncertainty," *IEEE. Trans. on Automatic Control*, vol.41, pp.436-442, 1996.
14. G. Schweitzer, H. Bleuler, A. Traxler, "Active Magnetic Bearings," *vdf Hochschuverlag AG an der ETH* Zurich, 1994.
15. K. Nonami, H. Ueyama and Y. Segawa, "H_∞ Control of Milling AMB Spindle," *JSME International Journal*, vol.39, no.3, pp.502-508, 1996.
16. K. Nonami, T. Ito, "μ Synthesis of Flexible Rotor Bearing Systems," *IEEE. Trans. on Control Systems Technology*, vol.4, No.5, pp.503-512, 1996.
17. S. Sivrioglu, K. Nonami, "LMI Approach to Gain Scheduled H_∞ Control Beyond PID Control for Gyroscopic Rotor-Magnetic Bearing System," *Proc. of the 35th CDC*, pp.3694-3699, 1996.
18. J.C. Doyle, B.A. Francis, A.R. Tannenbaum, "Feedback Control Theory," *Macmillan Publishing Company*, 1992.

A Study on Parameterized Output Feedback Sliding Mode Controller

Kenichiro Nonaka[1] and Katsuhisa Furuta[2]

[1] Musashi Institute of Technology,
1–28–1, Tamazutsumi Setagaya, Tokyo, 158–8557, Japan.
[2] Tokyo Institute of Technology,
2-12-1 Ookayama, Meguro-ku, Tokyo, 152–8552, Japan.

Abstract. In this study, a parameterization of output feedback dynamic sliding mode controller is proposed and the BIBO stability problem of generalized plant is studied. It is shown that by the proposed controller, the sliding mode is achieved in finite time, and thereafter, the ideal sliding mode controller stabilizes the generalized plant in the sense that the slightly modified BIBO stability is assured. Also, the model matching problem is formulated for this controller and the difference between linear dynamic controller and sliding mode controller is highlighted.

1 Introduction

Nowadays Sliding Mode Controller (SMC) is recognized to be a powerful tool of nonlinear robust controller [1][2][3]. One of the most distinguishing feature is that, if the plant's state is accessible, perfect rejection of matching disturbances is achieved. On the other hand, if the plant's state is not accessible and only plant output is available, some methods can be applied. In [5][6][7], observer is combined with the state feedback SMC, where SMC is implemented using the estimated state, and closed loop stability and is proved. In [8][9][11], under some assumption of plant, static output feedback SMC is formulated and matching disturbance rejection is shown. In [6][12][13][14], under some weaker assumption of plant, dynamic output feedback SMC is proposed and stronger disturbance rejection is asserted. These existing methods are in fact effective for each problem formulation, but the authors think that SMC should be applied to more general problems to clarify its advantages and disadvantages.

In this study, for a linear time invariant system with generalized plant[16], a new output feedback SMC is proposed and its disturbance rejection measure is shown. Our output feedback SMC is described by coprime factorization on $\mathbf{RH}_\infty$ and is parameterized by a Youla's free parameter. There is no restrictive assumptions on relative degree, unstable zeros, matching/unmatching disturbance. Also, the disturbance rejection measure is formulated as a model matching problem and may be minimized by a existing optimization methods like H_∞ or L_1 etc.

The outline of this article is as follows. The coprime factorization on $\mathbf{RH}_\infty$ is briefly reviewed and our PSMC (Parameterized Sliding Mode Controller) is proposed which is described by a Youla-parameterization on $\mathbf{RH}_\infty$. Then, it is shown that PSMC achieves that so called finite time reaching condition and ideal sliding mode assures slightly modified BIBO(Bounded-input, bounded-output) stability for generalized plant[16]. The difference between PSMC and linear controller is discussed. Also, PSMC is applied to three representative examples, plant with relative degree one, relative degree two, and plant with unstable zero.

For the convenience, the following notations are introduced: a linear time invariant system with zero initial condition,

$$\begin{cases} \dot{x}(t) = A\ x(t) + B\ u(t) & x(0) = 0 \\ y(t) = C\ x(t) + D\ u(t) \end{cases}$$

is described using both t(time variable) and s(Laplace operator) together as

$$y(t) = H(s)\ u(t), \quad H(s) := C\ (sI - A)^{-1}\ B + D$$

For a matrix $M(\in \mathbf{R}^{l\times m})$,$(i,j)$-th element is represented as $[M]_{(i,j)}$, and for a vector $v(\in \mathbf{R}^n)$, k-th element is $[v]_k$. Also, a diagonal matrix $G \in \mathbf{R}^{m\times m}$ whose diagonal elements are $g_i \in \mathbf{R}(i = 1,\cdots,m)$, is $G = \mathrm{diag}\,(g_1,\cdots,g_m)$. For $a \in \mathbf{R}$, sign function $\mathrm{sgn}(a)$ is defined as

$$\mathrm{sgn}(a) := \begin{cases} +1 & \text{if } a > 0 \\ 0 & \text{if } a = 0 \\ -1 & \text{if } a < 0 \end{cases}$$

and for $v \in \mathbf{R}^m$, vector valued function $\mathrm{sgn}(v)$ is

$$\mathrm{sgn}(v) := [\mathrm{sgn}([v]_1),\cdots,\mathrm{sgn}([v]_m)]^T$$

Similarly, for any scalar a and $\delta > 0$, saturation function $\mathrm{sat}(a,\delta)$ is defined as

$$\mathrm{sat}(a,\delta) := \begin{cases} sgn(a) & \text{if } |a| > \delta \\ \dfrac{a}{\delta} & \text{if } |a| \leq \delta \end{cases}$$

For $v \in \mathbf{R}^m$, $\delta_i > 0(i = 1,\cdots,m)$ and $\Delta := \mathrm{diag}\,(\delta_1,\cdots,\delta_m)$, the vector valued saturation function $\mathrm{sat}(v,\Delta)$ is defined as

$$\mathrm{sat}(v,\Delta) := \left[\mathrm{sat}([v]_1,\delta_1),\ \cdots,\ \mathrm{sat}([v]_m,\delta_m)\right]^T$$

Note that if $|[v]_i| \leq \delta_i\ (i = 1,\cdots,m)$,

$$\mathrm{sat}(v) = \Delta^{-1}v$$

2 The Model-matching Problem

In this section, parameterization on $\mathbf{RH}_\infty$ of generalized plant, its stabilizing controller and model-matching problem[16] are briefly reviewed. The generalized plant is represented as follows;

$$\begin{bmatrix} z(t) \\ y(t) \end{bmatrix} = \begin{bmatrix} G_{11}(s) & G_{12}(s) \\ G_{21}(s) & G_{22}(s) \end{bmatrix} \begin{bmatrix} w(t) \\ u(t) \end{bmatrix} \tag{1}$$

where $z \in \mathbf{R}^p$ is output to be controlled, $u \in \mathbf{R}^m$ is control input, $y \in \mathbf{R}^q$ is measured output $w \in \mathbf{R}^r$ is exogenous input. $G_{22}(s)$ is assumed to be strictly proper. A right coprime factorization of G_{22} on $\mathbf{RH}_\infty$ is as follows;

$$G_{22}(s) = N(s)M(s)^{-1} = \bar{M}(s)^{-1}\bar{N}(s) \tag{2}$$

where $N(s), M(s), \bar{N}(s), \bar{M}(s) \in \mathbf{RH}_\infty$, and without loss of generality, we assume that

$$\lim_{s\to\infty} M(s) = I_m, \quad \lim_{s\to\infty} \bar{M}(s) = I_q \tag{3}$$

There exist $Y(s), X(s), \bar{Y}(s), \bar{X}(s) \in \mathbf{RH}_\infty$, as a doubly coprime factorization of G_{22}, which satisfy

$$\begin{bmatrix} \bar{X}(s) & -\bar{Y}(s) \\ -\bar{N}(s) & \bar{M}(s) \end{bmatrix} \begin{bmatrix} M(s) & Y(s) \\ N(s) & X(s) \end{bmatrix} = \begin{bmatrix} M(s) & Y(s) \\ N(s) & X(s) \end{bmatrix} \begin{bmatrix} \bar{X}(s) & -\bar{Y}(s) \\ -\bar{N}(s) & \bar{M}(s) \end{bmatrix} = I_{m+q} \tag{4}$$

$$\lim_{s\to\infty} X(s) = I_q, \quad \lim_{s\to\infty} \bar{X}(s) = I_m \tag{5}$$

Note 1. A state space realizations of $Y(s), X(s), \bar{Y}(s), \bar{X}(s) \in \mathbf{RH}_\infty$ are given in [16].

All stabilizing controller $K(s)$ is parameterized by a Youla's free parameter $Q_L(s) \in \mathbf{RH}_\infty$ as follows;

$$K(s) = \left(\bar{X}(s) - Q_L(s)\bar{N}(s)\right)^{-1}\left(\bar{Y}(s) - Q_L(s)\bar{M}(s)\right) \tag{6}$$

$$= \left(Y(s) - M(s)Q_L(s)\right)\left(X(s) - N(s)Q_L(s)\right)^{-1} \tag{7}$$

$T_1(s), T_2(s), T_3(s) \in \mathbf{RH}_\infty$ [16] is defined as

$$\begin{aligned} T_1(s) &:= G_{11}(s) + G_{12}(s)M(s)\bar{Y}(s)G_{21}(s) \\ T_2(s) &:= G_{12}(s)M(s) \\ T_3(s) &:= \bar{M}(s)G_{21}(s) \end{aligned} \tag{8}$$

Note 2.

$$T_1(\infty) = D_{11}, \; T_2(\infty) = D_{12}, \; T_3(\infty) = D_{21} \tag{9}$$

By (1)(2)(8),

$$\bar{M}(s)y - \bar{N}(s)u = \bar{M}(s)\, G_{21}(s)w$$
$$= T_3(s)w \tag{10}$$

Applying (8)(10) to (1)(6), transfer matrix $G_{zw}^{Lin}(s)$ from w to z is described with model-matching representation:

$$G_{zw}^{Lin}(s) = T_1(s) - T_2(s)Q_L(s)T_3(s) \tag{11}$$

Thus, $G_{zw}^{Lin}(s)$ is parameterized by $Q_L(s) \in \mathbf{RH}_\infty$ with $T_1(s), T_2(s), T_3(s)$ in (8) [16].

3 Motivation: An Intuitive Derivation of PSMC

Before the proposed PSMC appears, some intuitive discussion may help to understand the property and merit of PSMC.

As reviewed in Section 2, all linear controllers have left coprime parameterization of $Q(s) \in \mathbf{RH}_\infty$ as

$$K(s) = \left(\bar{X}(s) - Q(s)\bar{N}(s)\right)^{-1}\left(\bar{Y}(s) - Q(s)\bar{M}(s)\right) \tag{12}$$

Note that $Q(s)$ is proper so that $K(s)$ is well-posed and proper.

For disturbance rejection, $Q(s) \in \mathbf{RH}_\infty$ should be adequately designed so that the transfer function from w to z, (11), satisfies some specification.

Consider a simple example designing K for

$$z = y = \frac{1}{s+1}(u + w) \tag{13}$$

A doubly coprime factorization is

$$\bar{M}(s) = 1,\ \bar{N}(s) = \frac{1}{s+1},\ \bar{X}(s) = 1,\ \bar{Y}(s) = 0 \tag{14}$$

These apparently satisfy (4), and the best solution to minimize

$$T_1 - T_2QT_3 = \frac{1}{s+1} - \frac{1}{(s+1)^2}Q(s)$$

is

$$Q = s + 1 \tag{15}$$

because $T_1 - T_2QT_3 = 0$ (disturbance is completely rejected). But since such $Q(s)$ is not proper, this cannot be achieved practically.

Next, examine an improper $Q(s)$. In this case, $K(s)$ is not proper, for $Q(s) = 2(s+1)$ makes $K(s)$ improper:

$$K(s) = 2(s+1)$$

Also, $\bar{X}(s) - Q(s)\bar{N}(s)$ is not necessary regular. For example, (14) satisfies (4), while improper $Q(s) = s + 1$ makes

$$\bar{X}(s) - Q(s)\bar{N}(s) = 0$$

, so (12) is not well-posed.

On the other hand, SMC claims that matching disturbance can be completely rejected in some system. In fact for (13), let the switching function σ be described as

$$\sigma = y$$

Differentiating σ with respect to t results in

$$\dot{\sigma} = -\sigma + u + w \tag{16}$$

Ordinally sliding mode control consists of linear part which cancels known variables (σ) and switching part which suppresses unkwon part(w). So it would be

$$u = \sigma - G\ \mathrm{sgn}(\sigma) \quad \text{with } G > \sup_t |w| \tag{17}$$

and this results in $z = y = 0$ in finite time, because

$$\dot{\sigma} = w - G\ \mathrm{sgn}(\sigma) \tag{18}$$

$z = y = 0$ means that the effect of disturbance w is completely eliminated which is just similar to the case of (15). In other words, the above sliding mode controller realizes improper controller with $Q = s + 1$ in (15).

These facts motivates us to parameterize the sliding mode controller by the solution of (4) and possibly improper $Q(s)$.

Now let us derive the PSMC. Output feedback linear control

$$u = K(s)y \tag{19}$$

with $K(s)$ in (6) is transformed to the following form:

$$0 = \left(\bar{X}(s)u(t) - \bar{Y}(s)y(t)\right) + Q(s)\left(\bar{M}(s)y(t) - \bar{N}(s)u(t)\right) \tag{20}$$

At this point, we do not assume that $Q \in \mathbf{RH}_\infty$.

It may be natural to select the right hand side of (20) to be a switching function for sliding mode control, because $\sigma = 0$ is usually the desired system. But in the meantime, σ should be proper. So we construct the sliding function as

$$\sigma = L(s)^{-1}\left(\bar{X}(s)u(t) - \bar{Y}(s)y(t)\right) + R(s)\left(\bar{M}(s)y(t) - \bar{N}(s)u(t)\right) \tag{21}$$

where $L(s)$ is a polynomial matrix such that $R(s) := L(s)^{-1}Q(s)$ is proper.

Though (21) just seems to be a transformation of (6), it is not trivial, since sliding mode control with switching function (21) is different from linear controller (19) in that the free parameter $Q(s)$ is not $\mathbf{RH}_\infty$. A typical example is

$$\det\left(\bar{X}(s) - Q(s)\bar{N}(s)\right) = 0$$

which is not accepted for (6).

To derive sliding mode control, multiplying (21) by $L(s)$ gives

$$L(s)\sigma = \left(\bar{X}(s)u(t) - \bar{Y}(s)y(t)\right) + Q(s)\left(\bar{M}(s)y(t) - \bar{N}(s)u(t)\right) \tag{22}$$

If the degree of $L(s)$ is 1, i.e. $L(s) = sI + L_0$, only $\dot{\sigma}$ appears on the left hand side of this equation, and is preferred for finite time reaching condition. So we restrict the degree of $L(s)$ as 1, and $Q(s)$ is improper at most degree 1 hereafter.

The second term of the right hand side is known to be a innovation, and is represented with w as shown in (10). Substitution of (10) gives

$$L(s)\sigma = \left(\bar{X}(s)u(t) - \bar{Y}(s)y(t)\right) + Q(s)T_3(s)w \tag{23}$$

It should be noted that $Q(s)\left(\bar{M}(s)y(t) - \bar{N}(s)u(t)\right)$ in (22) cannot be canceled by proper control u, because Q may be improper.

Then the sliding mode controller is naturally constructed as

$$u(t) = \bar{X}(s)^{-1}\left(L(0)\sigma(t) + \bar{Y}(s)y(t) - G\mathrm{sgn}(\sigma)\right)$$

or

$$u(t) = L(0)\sigma + \left(I - \bar{X}(s)\right)u(t) + \bar{Y}(s)y(t) - G\ \mathrm{sgn}(\sigma) \tag{24}$$

, this results in

$$\dot{\sigma} = Q(s)T_3(s)w - G\ \mathrm{sgn}(\sigma) \tag{25}$$

Since $Q(s)T_3(s)$ is improper at most degree 1, $Q(s)T_3(s)w$ is bounded, for each case that w is smooth and $Q(s)T_3(s)$ is improper, and w is bounded and $Q(s)T_3(s)$ is proper. So $\sigma = 0$ is achieved in finite time, if G is set to be larger than QT_3w.

For the factorization in (14) with (15) and $L(s) = s + 1$, the above sliding mode controller (24) and resulting dynamics (25) is described as

$$u = \sigma - G\ \mathrm{sgn}(\sigma) \tag{26}$$

$$\dot{\sigma} = w - G\ \mathrm{sgn}(\sigma) \tag{27}$$

These are equivalent to (17) and (18). G should be chosen so that it suppresses w.

The formal and more generalized representation of PSMC is given in the next section.

4 Parameterized Sliding Mode Controller

In this section, PSMC(Parameterized Sliding Mode Controller) is proposed.

Let $\sigma(t) = 0$ be a sliding mode, and so called switching function $\sigma(t)$ is given as

$$\sigma(t) = H(s)\left[\bar{X}(s)u(t) - \bar{Y}(s)y(t)\right] + R(s)\left[\bar{M}(s)y(t) - \bar{N}(s)u(t)\right] \tag{28}$$

where $R(s) \in \mathbf{RH}_\infty$ and $H(s)$ satisfies the following two restrictions

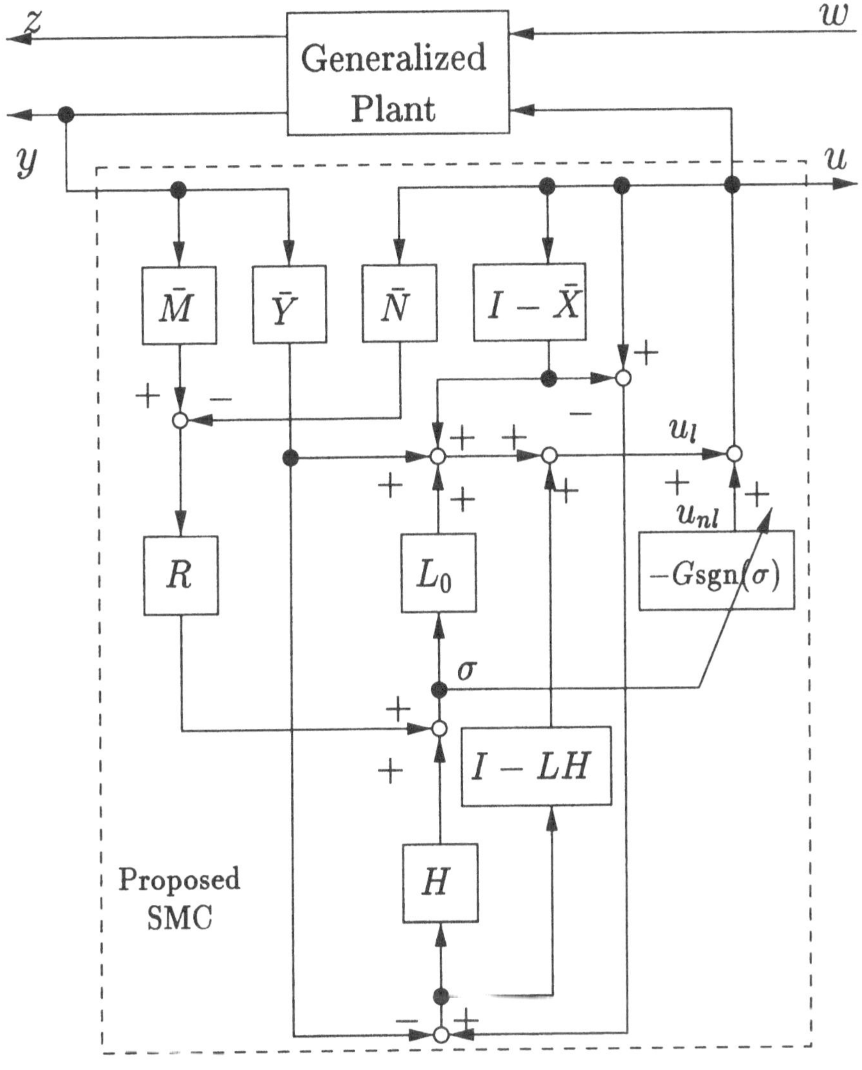

Fig. 1. Block diagram of parameterized SMC

1. Let a minimum realization of $H(s)$ be

$$H(s) =: \left[\begin{array}{c|c} A_H & B_H \\ \hline C_H & D_H \end{array}\right]$$

then,

$$C_H B_H = I_m, \quad D_H = 0 \tag{29}$$

2. $H(s)$ does not have unstable invariant zeros

Note 3. At the switching function (21) in section 3,

$$H(s) = L(s)^{-1} \tag{30}$$

Note 4. It should be noted that

$$L(s) := sI_m + L_0 \tag{31}$$

satisfies

$$(L(s)H(s))\,|_{s=\infty} = I_m \tag{32}$$

If L_0 does not have non-positive eigenvalues, then $(LH)^{-1} \in \mathbf{RH}_\infty$. The simplest example of $H(s)$ is $H(s) = L(s)^{-1}$.

Note 5. As mentioned in section 3, the controller which is derived by just solving $\sigma = 0$, is not equivalent to (6), since $Q(s) := H(s)^{-1}R(s)$ is not $\mathbf{RH}_\infty$, and possibly $\det\left(\bar{X}(s) - Q(s)\bar{N}(s)\right) = 0$.

Now an assumption on exogenous input $w(t)$ is given.

Assumption 1 *For a first order polynomial matrix (31), there exists a* $t_s (< \infty)$ *and a known* $\gamma_i > 0 (i = 1, \cdots, m)$ *which satisfy*

$$\gamma_i \geq \sup_{t \geq t_s} \left| [L(s)R(s)T_3(s)w(t)]_i \right| \tag{33}$$

Note 6. If $R(\infty)T_3(\infty) = R(\infty)D_{21} \neq 0$, then $L(s)R(s)T_3(s)$ is improper by 1, so $R(\infty)D_{21}\dot{w}(t)$ must be bounded.

Note 7. Above bounds on w is required to assure finite time reaching condition, and is usually assumed in SMC. [1][2][3]. If $\gamma_i (i = 1, \cdots, m)$ are unknown, same results can be derived using adaptive gains, for example [18]

Let the control input u be represented as a sum of a linear part u_l which is a linear combination of input and output, and nonlinear part $u_{nl} = -G\,\mathrm{sgn}(\sigma)$ which is a switching function with a sliding surface $\sigma = 0$;

$$\begin{aligned} u &:= u_l + u_{nl} && (34)\\ u_l &:= L_0\sigma + \left(I - LH\bar{X}\right)u + LH\bar{Y}y && (35)\\ &= L_0\sigma + \left(I - \bar{X}\right)u + \bar{Y}y + (I - LH)\left(\bar{X}u - \bar{Y}y\right) &&\\ u_{nl} &:= -G\,\mathrm{sgn}(\sigma) && (36) \end{aligned}$$

where $G (\in \mathbf{R}^{m \times m})$ is defined with

$$G := \mathrm{diag}\,(g_1, \cdots, g_m),\ g_i \geq \gamma_i + \epsilon_0\ (i = 1, \cdots, m) \tag{37}$$

(ϵ_0 is a positive constant.)

Fig.1 is a block diagram of this PSMC.

Note 8. Although (35)(36) use u, for (35), $I - LH\bar{X}$ is strictly proper by (5)(32), and, for (36), the transfer matrix $H(s)\bar{X}(s) - R(s)\bar{N}(s)$ from u to σ in (28) is strictly proper, above PSMC is implemented with proper system. Also, though it is output feedback controller, there is no assumption like relative degree or stability of zeros.

In general, discontinuous switching function sgn(·) causes chattering phenomena, saturation function sat(·) may replace it. If so,

$$u_{nl} := -G\,\mathrm{sat}(\sigma, \Delta) \tag{38}$$

is used instead of (36), where Δ is defined as $\Delta := \mathrm{diag}\,(\delta_1, \cdots, \delta_m)$, and $\delta_i (i = 1, \cdots, m)$ is adequate positive constant.

Note 9. If $|\sigma_i| \leq \delta_i$ for all $i = 1, \cdots, m$

$$u_{nl} = -G\,\Delta^{-1}\,\sigma \tag{39}$$

5 Proof of the BIBO Stability

In this section, the BIBO stability of closed loop system[16] is shown.

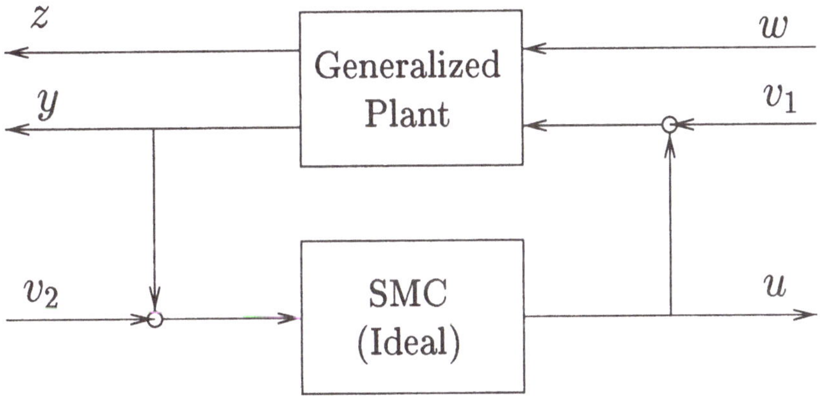

Fig. 2. Standard problem: generalized plant and ideal SMC

For system in Fig.2, if the nine transfer matrix from w,v_1,$v_2 \in \mathbf{L}_\infty$ to z,y,u is stable, the closed loop system is the BIBO stable [16]. We show that the PSMC satisfies the BIBO stability under some adequate assumptions. Let the generalized plant be represented as

$$z = G_{11}w + G_{12}(u + v_1) \tag{40}$$
$$y = G_{21}w + G_{22}(u + v_1) \tag{41}$$

The BIBO stability of sliding mode controlled system is divided into three cases, considering of (1) the boundness of $\hat{w}$ and v_2 which effects $R(\infty)$ and (2) sgn or sat is used.

5.1 In case sgn(·) and $R(s) \in \mathbf{RH}_2$ is used

Definition 10 SM BIBO Stability. For bounded v_1,v_2,w and sufficiently large G, system in Fig.2 is Sliding Mode BIBO Stable(SM BIBO Stable), if there exists a time $t_f (< \infty)$ such that sliding mode $\sigma(t) = 0$ is achieved for $t \geq t_f$, and z,y,u are bounded.

Now PSMC of previous section is reformulated for the set-up in Fig.2. Since accessible plant output is $y + v_2$, (28) is redefined as follows;

$$\sigma = H(s)\left[\bar{X}(s)u - \bar{Y}(s)(y + v_2)\right] + R(s)\left[\bar{M}(s)(y + v_2) - \bar{N}(s)u\right] \tag{42}$$

where $R(s) \in \mathbf{RH}_2$. Then by the assumption of $H(s)$,

$$Q_S(s) := H(s)^{-1}R(s) \in \mathbf{RH}_\infty \tag{43}$$

Similarly, u is redefined as

$$u := u_l + u_{nl} \tag{44}$$

$$u_l := L_0\sigma + \left(I - LH\bar{X}\right)u + LH\bar{Y}(y + v_2) \tag{45}$$

$$u_{nl} := -\bar{G}\,\mathrm{sgn}(\sigma) \tag{46}$$

where

$$\bar{G} := \mathrm{diag}\left(\bar{g}_1, \cdots, \bar{g}_m\right) \tag{47}$$

and for adequate positive constant ϵ_0, $\bar{g}_i$ $(i = 1, \cdots, m)$ is chosen so that

$$\bar{g}_i \geq \sup_{t \geq t_s} |[\hat{w}]_i| + \epsilon_0 \tag{48}$$

$$\hat{w} := LRT_3w + LR\bar{N}v_1 + LR\bar{M}v_2 \tag{49}$$

Note 11. Since $L(s)$ is a first order polynomial, $R(s) \in \mathbf{RH}_2$, $T_3(s), \bar{N}(s), \bar{M}(s) \in \mathbf{RH}_\infty$, $\hat{w}$ is bounded, so bounded $\bar{g}_i$ exists too.

Theorem 12. *Let $R(s) \in \mathbf{RH}_2$. For (47) which satisfies (48), the system (40) (41) (42) (44) (45) (46) is SM BIBO Stable.*

Proof. By (41),

$$\bar{M}(y + v_2) - \bar{N}u = T_3w + \bar{N}v_1 + \bar{M}v_2$$

and apply it for (42) results

$$\sigma = H\left(\bar{X}u - \bar{Y}(y + v_2)\right) + RT_3w + R\bar{N}v_1 + R\bar{M}v_2 \tag{50}$$

Multiplying $L(s)$ from left and applying (49) gives

$$L\sigma = LH\bar{X}u - LH\bar{Y}(y+v_2) + \hat{w} \tag{51}$$
$$= u - \left(I - LH\bar{X}\right)u - LH\bar{Y}(y+v_2) + \hat{w} \tag{52}$$

Fist, the transfer matrix from w, v_1, v_2 to u, y, z is derived. u is derived from (52) as

$$u = \left(I - LH\bar{X}\right)u + LH\bar{Y}(y+v_2) - \hat{w} + L\sigma \tag{53}$$

Let $\bar{L}(s)$ be a first order Hurwitz polynomial. Then, by deleting y, we have from (41) and (51) that

$$\begin{aligned} u = &\left(Y - MH^{-1}R\right)T_3 w + \left(Y - MH^{-1}R\right)\bar{N}v_1 \\ &+ \left(Y - MH^{-1}R\right)\bar{M}v_2 + M\left(\bar{L}H\right)^{-1}\bar{L}\sigma \end{aligned} \tag{54}$$

Substituting (54) to (41) gives

$$\begin{aligned} y = &\left(X - NH^{-1}R\right)T_3 w + \left(X - NH^{-1}R\right)\bar{N}v_1 \\ &+ N\left(\bar{Y} - H^{-1}R\bar{M}\right)v_2 + N\left(\bar{L}H\right)^{-1}\bar{L}\sigma \end{aligned} \tag{55}$$

Also by deleting u from (40) and (54), we have

$$\begin{aligned} z = &\left(T_1 - T_2H^{-1}RT_3\right)w + T_2\left(\bar{X} - H^{-1}R\bar{N}\right)v_1 \\ &+ T_2\left(\bar{Y} - H^{-1}R\bar{M}\right)v_2 + T_2\left(\bar{L}H\right)^{-1}\bar{L}\sigma \end{aligned} \tag{56}$$

By collecting (54)(55)(56), and using $Q_S(s)$ in (43),

$$\begin{bmatrix} z \\ y \\ u \end{bmatrix} = \begin{bmatrix} T_1 - T_2Q_ST_3 & T_2\left(\bar{X} - Q_S\bar{N}\right) \\ \left(X - NQ_S\right)T_3 & \left(X - NQ_S\right)\bar{N} \\ \left(Y - MQ_S\right)T_3 & \left(Y - MQ_S\right)\bar{N} \end{bmatrix}$$
$$\begin{matrix} T_2\left(\bar{Y} - Q_S\bar{M}\right) & T_2(\bar{L}H)^{-1} \\ N\left(Y - Q_S\bar{M}\right) & N\left(\bar{L}H\right)^{-1} \\ M\left(\bar{Y} - Q_S\bar{M}\right) & M(\bar{L}H)^{-1} \end{matrix} \Bigg] \begin{bmatrix} w \\ v_1 \\ v_2 \\ \bar{L}\sigma \end{bmatrix} \tag{57}$$

Since $(\bar{L}H)^{-1} \in \mathbf{RH}_\infty$ and $Q_S \in \mathbf{RH}_\infty$, the transfer matrix of the right in (57) is stable, so SM BIBO stability from w, v_1, v_2 to z, y, u is concluded, if it can be shown that $\bar{L}\sigma$ is bounded.

From (44) and (52), we have

$$\dot{\sigma} = -\bar{G}\,\mathrm{sgn}(\sigma) + \hat{w} \tag{58}$$

then, as shown in lemma below, $\sigma(t) \equiv 0$ and $\dot{\sigma}(t) \equiv 0$ is achieved in finite time.

Lemma 13. *Let*

$$t_f := \max_i(\left|[\sigma(t_s)]_i\right|/\epsilon_0) + t_s$$

Then $\sigma(t) = 0$ and $\dot{\sigma}(t) = 0$ for any $t \geq t_f$.

Proof. Let $\sigma_i := [\sigma]_i$and $\hat{w}_i := [\hat{w}]_i$, then (58) is rewritten as

$$\dot{\sigma}_i = -\bar{g}_i \,\mathrm{sgn}(\sigma_i) + \hat{w}_i$$

Let $V_i := |\sigma_i|$. If $\sigma_i > 0$ then for $t \geq t_s$, the inequality below is derived by (48)

$$\dot{V}_i = -\bar{g}_i \,\mathrm{sgn}(\sigma_i) + \hat{w}_i \leq -\epsilon_0$$

Also, if $\sigma_i < 0$ then

$$\dot{V}_i \leq -\epsilon_0 (t \geq t_s)$$

So in both cases, there exists a time $t(\leq V_i(t_s)/\epsilon_0 + t_s)$ such that $V_i(t) = 0$. Then by [4], it is proved that $V_i(t) = 0$ for

$$t \geq t_f := V_i(t_s)/\epsilon_0 + t_s$$

Therefore, $\sigma(t) = 0$ for any $t \geq t_f$, and $\dot{\sigma}(t) = 0$ $(t \geq t_f)$.

Since $L(s)$ is first order polynomial, it is proved that $L\sigma = 0$ since $\sigma(t) = 0$ and $\dot{\sigma}(t) = 0$. Also by assumption, $w, v_1, v_2, \in \mathbf{L}_\infty$, $z, y, u \in \mathbf{L}_\infty$ is concluded for (57). Therefore the BIBO stability from w, v_1, v_2 to z, y, u is proved

5.2 In case sat $(\cdot, \cdot)$ is used

In this section, $R(s) \in \mathbf{RH}_\infty$ is assumed. Since the saturation function sat$(\cdot)$ is used instead of sgn$(\cdot)$, $\sigma = 0$ is no longer achieved and only boundness of σ is shown. So in this section, the following stability is proved instead of SM BIBO Stability in 5.2.

Definition 14 SMR BIBO Stability. The system in Fig.2 is Sliding Mode Region BIBO Stable(SMR BIBO Stable), if for bounded v_1,v_2,w and sufficiently large G, there exists a $t_b(< \infty)$ such that for $t \geq t_b$, the state is restricted in the neighborhood of sliding surface $\sigma = 0$ and z,y,u are bounded.

Instead of (46), SMR BIBO Stability is shown for

$$u_{nl} := -\bar{G}\,\mathrm{sat}(\sigma, \Delta) \tag{59}$$

$$\Delta := \mathrm{diag}\left(\delta_1, \cdots, \delta_m\right), \left(\delta_i > 0,\ i = 1, \cdots, m\right) \tag{60}$$

Theorem 15. *If (47) satisfies (48), the system (40) (41) (42) (44) (45) (59) is SMR BIBO Stable.*

Proof. (57) is also valid as in the previous section.

The following lemma which indicates boundness of $L\sigma$ is shown.

Lemma 16. *Let*

$$t_b := \max_i ((|\sigma_i(t_s)| - \delta_i)/\epsilon_0) + t_s$$

For any $t \geq t_b$, $|\sigma_i| \leq \delta_i$ $(i = 1, \cdots, m)$ *is satisfied.*

Proof. Applying (44)(45)(59) to (52) results

$$\dot{\sigma} = -\bar{G}\ \mathrm{sat}(\sigma, \Delta) + \hat{w} \tag{61}$$

If $|\sigma_i| \geq \delta_i$, then

$$\dot{\sigma}_i = -\bar{g}_i\ \mathrm{sgn}(\sigma_i) + \hat{w}_i \tag{62}$$

so we get $|\sigma_i| \leq \delta_i$ for $t \geq t_b$ by (48) like lemma 13.

Especially or $t \geq t_b$, $|[\sigma(t)]_i| \leq \delta_i$ $(i = 1, \cdots, m)$, so

$$u_{nl} = -\bar{G}\Delta^{-1}\sigma$$

, we get

$$\dot{\sigma} = -\bar{G}\Delta^{-1}\sigma + \hat{w}$$

, i.e.

$$\sigma = \left(sI_m + \bar{G}\Delta^{-1}\right)^{-1}\hat{w} \tag{63}$$

Substituting (63) into (57), we get the transfer matrix from w,v_1,v_2 to z,y,u;

$$\begin{bmatrix} z \\ y \\ u \end{bmatrix} = \begin{bmatrix} T_1 - T_2 Q_B T_3 & T_2\left(\bar{X} - Q_B \bar{N}\right) & T_2\left(\bar{Y} - Q_B \bar{M}\right) \\ \left(X - NQ_B\right) T_3 & \left(X - NQ_B\right)\bar{N} & N\left(\bar{Y} - Q_B \bar{M}\right) \\ \left(Y - MQ_B\right) T_3 & \left(Y - MQ_B\right)\bar{N} & M\left(\bar{Y} - Q_B \bar{M}\right) \end{bmatrix} \begin{bmatrix} w \\ v_1 \\ v_2 \end{bmatrix} \tag{64}$$

$$Q_B := H^{-1}\left(sI_m + G\Delta^{-1}\right)^{-1}\left(G\Delta^{-1} - L_0\right) R \tag{65}$$

and since $G\Delta^{-1} = \mathrm{diag}\left(\bar{g}_1\delta_1^{-1}, \cdots, \bar{g}_m\delta_m^{-1}\right)$ and $\bar{g}_i\delta_i^{-1} > 0$ $(i = 1, \cdots, m)$ means $\left(sI_m + G\Delta^{-1}\right)^{-1} \in \mathbf{RH}_\infty$ and $R \in \mathbf{RH}_\infty$ results $Q_B(s) \in \mathbf{RH}_\infty$. So in (64), all transfer matrix from w,v_1,v_2 to z,y,u are stable, SMR BIBO Stability is proved.

5.3 In case sgn $(\cdot)$ and $R(s) \in \mathbf{RH}_\infty$ is used

In this section, sgn $(\cdot)$ is used together with $R(s) \in \mathbf{RH}_\infty$. In section 5.1, $R(s) \in \mathbf{RH}_2$ i.e. $R(0) = 0$ is assumed, but if $R(0) \neq 0$, LRT_3,$L R\bar{M}$ in (49) is improper by 1st order. So there exists $\bar{g}_i$ which satisfies (48), if $D_{21}\dot{w}$, $\dot{v}_2 \in \mathbf{L}_\infty$ in addition to $w, v_1, v_2 \in \mathbf{L}_\infty$. So, in the case of $R(s) \in \mathbf{RH}_\infty$, additional assumption on $D_{21}w$ and v_2 that they have first order derivatives is necessary.

Definition 17 SM BIBO Stability for C^1 OD. The system in **Fig.2** is Sliding Mode BIBO Stable for C^1 Output Disturbance (SM BIBO Stable for C^1 OD), if for bounded v_1,v_2,w, $D_{21}\dot{w}$,$\dot{v}_2$ and sufficiently large G, there exists a time $t_f(< \infty)$ such that sliding mode $\sigma(t) = 0$ is achieved for $t \geq t_f$, and z,y,u are bounded.

Theorem 18. *Let $R(s) \in \mathbf{RH}_\infty$. If (47) satisfies (48), the system (40) (41) (42) (44) (45) (46) is SM BIBO Stable for C^1 OD.*

Proof. As in section 5.1, (57) is also valid. Since $v_1, v_2, w, D_{21}\dot{w}, \dot{v}_2 \in \mathbf{L}_\infty$, $\hat{w} \in \mathbf{L}_\infty$ in (49). For $\bar{g}_i$ which satisfies (48), $\sigma = 0$, $\dot{\sigma} = 0$ $(t \geq t_f)$ is achieved as Theorem 1. This means $L\sigma = 0$ $(t \geq t_f)$. For (57), $T_1 - T_2Q_ST_3$, MQ_ST_3, and $T_2Q_S\bar{M}$, $MQ_S\bar{M}$j can be improper by 1st order, but since $w, v_1, v_2, D_{21}\dot{w}, \dot{v}_2, \in \mathbf{L}_\infty$,$L\sigma = 0$, $z, y, u \in \mathbf{L}_\infty$ is concluded. Now BIBO stability from w, v_1, v_2 to z, y, u is proved.

6 Model-matching Problem and Comparison between PSMC and Linear Controller

In this section, to assess the disturbance rejection measure, the transfer matrix G_{zw} from w to z is formulated as a model-matching form, and the difference between PSMC and linear controller is discussed.

Let G_{zw}^{Lin} and G_{zw}^{SMC} be that of linear controller and SMC respectively. The class of G_{zw}^{Lin} is derived from (11) as

$$\mathcal{G}_{zw}^{Lin} = \{T_1 - T_2Q_LT_3 \mid Q_L \in \mathbf{RH}_\infty\} \tag{66}$$

Also, the class of G_{zw}^{SMC} with sgn$(\cdot)$ is derived from (57) as

$$\mathcal{G}_{zw}^{SMC} = \left\{T_1 - T_2Q_ST_3 \mid Q_S = H^{-1}R\right\} \tag{67}$$

Especially, if $R(0) \neq 0$ and γ_i which satisfies (33) is known, then Q_S can be a transfer matrix with stable finite pole and improper by 1st order, we see that

$$\mathcal{G}_{zw}^{Lin} \subset \mathcal{G}_{zw}^{SMC} \tag{68}$$

Example 1. The following example[16] shows that (68) means PSMC decreases the relative degree of model-matching problem.

$$z = y = \frac{1}{s+1}(u + w) \tag{69}$$

For this plant,

$$G_{11} = G_{12} = G_{21} = G_{22} = \frac{1}{s+1} \tag{70}$$

A solution of (4) is

$$X = \bar{X} = 1,\ Y = \bar{Y} = 0,\ M = \bar{M} = 1,\ N = \bar{N} = \frac{1}{s+1}$$

Then by (8),

$$T_1(s) = T_2(s) = T_3(s) = \frac{1}{s+1}$$

As shown in [16], there does not exist $Q_L(s) \in \mathbf{RH}_\infty$ which achieves the infimum

$$\inf_{Q_L \in \mathbf{RH}_\infty} \left\|G_{zw}^{Lin}(s)\right\|_\infty = 0$$

But for PSMC with

$$H(s) = \frac{1}{s+1}, \ R(s) = 1, \ L(s) = s + 1$$

and $Q_S = H^{-1}R = s + 1$ gives

$$G_{zw}^{SMC}(s) := T_1(s) - T_2(s)Q_S(s)T_3(s) = 0$$

, and the infimum is exactly achieved.

The set of G_{zw}^{SMC} with saturation function sat $(\cdot,\cdot)$ is

$$\mathcal{G}_{zw}^{SMC} = \{T_1 - T_2 Q_B T_3 \mid Q_B \in \mathbf{RH}_\infty\} \tag{71}$$

from (64), so we have

$$\mathcal{G}_{zw}^{Lin} = \mathcal{G}_{zw}^{SMC} \tag{72}$$

It means, at least in steady state, the disturbance rejection performance is equivalent to that of linear controller. In this case, the difference is that PSMC globally stabilize the plant even though control input is partially saturated.

7 Simulation

7.1 Plant with relative degree 1

For (69) with $w = 0.1\cos(2\pi t)$ is considered.

$$X = M = 1, \ Y = 0, \ N = G_{22} = \frac{1}{s+1}$$

$$H = \frac{1}{s+1}, \ R = 1, \ L = s + 1$$

It should be noted that $G_{zw}^{SMC} = 0$. To compare, free parameter Q_L of linear controller is selected as

$$Q_L = \left.\frac{s+1}{\tau s + 1}\right|_{\tau=0.001} = \frac{1000(s+1)}{s+1000}$$

z and u are shown in **Fig.3** and **Fig.4** respectively. Although, the maximum value of u is equivalent, z of PMSC is smaller than that of linear controller.

7.2 Plant with relative degree 2

Plant with relative degree 2 is considered.

$$z = y = \frac{1}{(s+1)^2}(u+w), \;\; w = 0.1\cos(2\pi t)$$

A coprime factorization is selected as

$$X = 1, \; Y = 0, \; N = \frac{1}{(s+1)^2}, \;\; M = 1$$

SMC (28)(35)(36) is given as

$$H = \frac{1}{s+1}, \;\; R = \frac{1000(s+1)}{s+1000}, \;\; L = s+1$$

z and u are shown in **Fig.5** and **Fig.6** respectively. We can see that the plant is stabilized and disturbance is effectively rejected.

7.3 Plant with unstable zero

Generalized plant with two unstable poles and one unstable zero is considered.

$$\begin{bmatrix} z \\ y \end{bmatrix} = \begin{bmatrix} \frac{1}{(s-3)^2} & \frac{s-1}{(s-3)^2} \\ \frac{1}{(s-3)^2} & \frac{s-1}{(s-3)^2} \end{bmatrix} \begin{bmatrix} w \\ u \end{bmatrix}$$

The external disturbance w is set as

$$w(t) = 0.1 \text{ sgn}(sin(2\pi t))$$

but is assumed to be unknown except its upper bound $|w(t)| \leq 0.1$.

A doubly coprime factorization is selected as

$$X = 1, \; Y = -8, \; N = \frac{s-1}{(s+1)^2}, \;\; M = \frac{(s-3)^2}{(s+1)^2}$$

In this case,

$$T_1(s) = \frac{1}{(s+1)^2}, \qquad T_2(s) = \frac{s-1}{(s+1)^2}, \qquad T_3(s) = \frac{1}{(s+1)^2}$$

Then the model matching problem is represented by

$$\begin{aligned} G_{zw}^{SMC} &= T_1(s) - T_2(s)Q_S T_3 \\ &= \frac{1}{(s+1)^2} - \frac{s-1}{(s+1)^2}R(s)\frac{s+1}{(s+1)^2} \end{aligned}$$

where

$$Q_S =: H(s)^{-1}R(s), \text{ with } H(s) = \frac{1}{s+1}$$

$R(s) \in \mathbf{RH}_\infty$ is designed so that $||G_{zw}^{SMC}||_\infty$ is minimized. Solving such $R(s)$ is equivalent to design a H_∞ sub-optimal controller $u' = Ry'$ for the following generalized plant:

$$\begin{bmatrix} z \\ y' \end{bmatrix} = \begin{bmatrix} \frac{1}{(s+1)^2} & \frac{s-1}{(s+1)^2} \\ \frac{1}{(s+1)^2} & 0 \end{bmatrix} \begin{bmatrix} w \\ u' \end{bmatrix}$$

Computed $R(s)$ is

$$R(s) = \frac{3932.18s^2 + 15465.9s + 11533.7}{s^2 + 264.811s + 15466.9}$$

and, corresponding H_∞-norm is 0.25

For this $R(s)$, γ_i in (33) is computed. Since L_1-norm of $L(s)R(s)T_3(s)$ can be computed numerically as 20.6, so we get

$$\sup_{t\geq 0} |L(s)R(s)T_3(s)w(t)| \leq ||L(s)R(s)T_3(s)||_1 \cdot \sup_{t\geq 0} |w(t)| < 2.1$$

, then the switching gain G is chosen as

$$G = [2.1]$$

z and u are shown in **Fig.7** and **Fig.8** respectively. We can see that the plant is stabilized and disturbance is effectively rejected.

8 Conclusion

In this article, a parameterized sliding mode controller is proposed, and its BIBO stability is proved for a slightly modified condition. This PSMC is constructed for generalized plant, so there is no restrictive constraints of plant on relative degree, unstable zero, matching/unmatching disturbance. The difference between SMC and linear controller on disturbance rejection issue is clarified through model-matching problem. Since our PSMC is described as a coprime factorization and Youla's free parameter, we think that large class of existing SMC is also parameterized by this PSMC, then its disturbance rejection performance can be estimated as a model-matching problem.

References

1. R.A.DeCarlo, S.H.Zak, G.P.Matthews, "Variable structure control of nonlinear multivariable system: a tutorial", *Proc. of the IEEE.*, Vol. 76, 1988.
2. V.I.Utkin, "Sliding Modes in Control and Optimization", Springer-Verlag, 1992.
3. Nonami, Den, "Slidig Mode Control", Corona pub., 1994 (in Japanese).
4. D.Shevitz and B.Paden, "Lyapunov Stability Theory of Nonsmooth Systems", *IEEE Trans. Automat. Contr.*, vol. 39, no. 9, 1994.

5. A.G.Bondarev, S.A.Bondarev, N.E.Kostyleva, and V.I.Utkin, "Sliding modes in systems with asymptotic state observers", *Automation and Remote Control*, No.6, pp.679-684, 1985.
6. B.M.Diong and J.V.Medanic, "Dynamic output feedback variable structure control for system stabilization", *Int. J. Control*, Vol.56, No.3, 1992.
7. B.M.Diong and J.V.Medanic, "Robust implementation of a variable structure control scheme", *Proc. of the 29th Conf. on Decision and Control*, December, 1990
8. R.El-Khazali and R.A.DeCarlo, "Output Feedback Variable Structure Control Design Using Dynamic Compensation for Linear Systems", *Proc. of the American Control Conf.*, June, 1992.
9. S.H.Zak and S.Hui, "On Variable Structure Output Feedback Controllers for Uncertain Dynamic Systems", *IEEE Trans. on Automatic Control*, Vol.38, No.10, 1993.
10. C.M.Kwan, "On Variable Structure Output Feedback Controllers", IEEE Transactions on Automatic Control, Vol.41, No.11, November 1996.
11. C.Edwards and S.K.Spurgeon, "Sliding mode stabilization of uncertain systems using only output information", *Int. J. Control*, Vol.62, No.5, pp.1129-1144, 1995.
12. S.V.Yallapragada and B.S.Heck, "Optimal Control Design for Variable Structure Systems with Fixed Order Compensators", *Proc. of the American Control Conference*, June, 1992.
13. M.A.Zohdy and M.S.Fadali and J.Liu, "Variable Structure Dynamic Output Feedback", *Proc. of the American Control Conference*, June, 1995.
14. K.S.Yeung, C.Cheng and C.Kwan, "A Unifying Design of Sliding Mode and Classical Controllers", *IEEE Trans. on Auto. Control*, Vol.38, No.9, 1993.
15. M. Vidyasagar, "Control System Synthesis", MIT Press, Boston, 1985.
16. B.A.Francis, "A Course in H_∞ Control Theory", Springer-Verlag, 1986.
17. T.Mita, "H_∞ control", Shokodo, 1994.
18. K. Nonaka, M. Yamakita, K. Furuta: "A Study on Model Reference Variable Structure Control with Adaptive Sliding Surface", 35th IEEE Conference on Decision and Control, Kobe, December, 1996.
19. W.J.Wang and Y.T.Fan, "New Output Feedback Design in Variable Structure Systems", Journal of Guidance, Control, and Dynamics, Vol.17, No.2, 1994.
20. S.K.Bag, S.K.Spurgeon, and C.Edwards, "Output feedback sliding mode design for linear uncertain systems", IEE Proc. Control Theory Appl., Vol. 144, No.3, May 1997.
21. C.M.Kwan, "On Variable Structure Output Feedback Controllers", *IEEE Trans. on Automatic Control*, Vol.41, No.11, 1996.

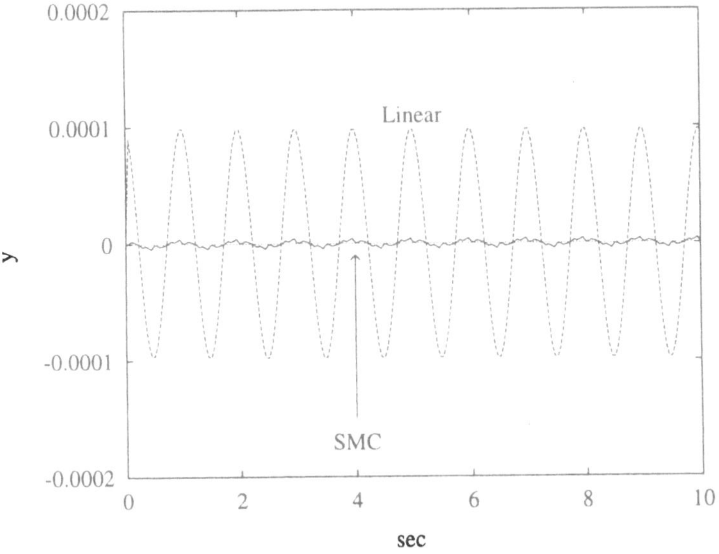

Fig. 3. z **(Plant with relative degree one)**

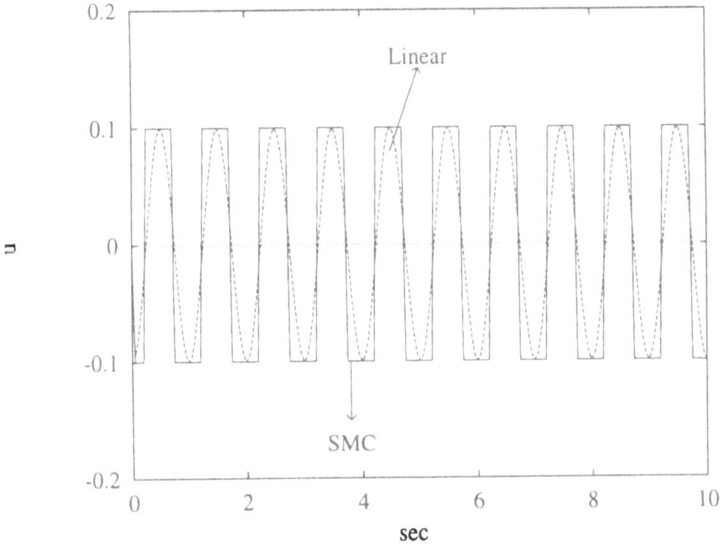

Fig. 4. u **(Plant with relative degree one)**

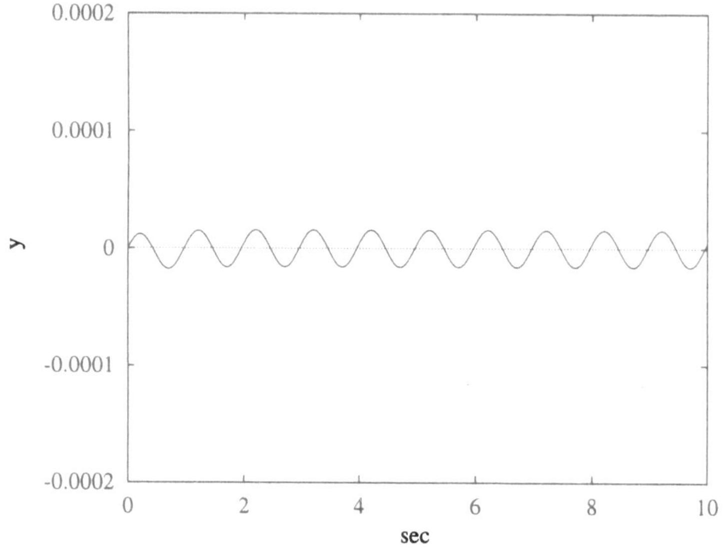

Fig. 5. z **(Plant with relative degree two)**

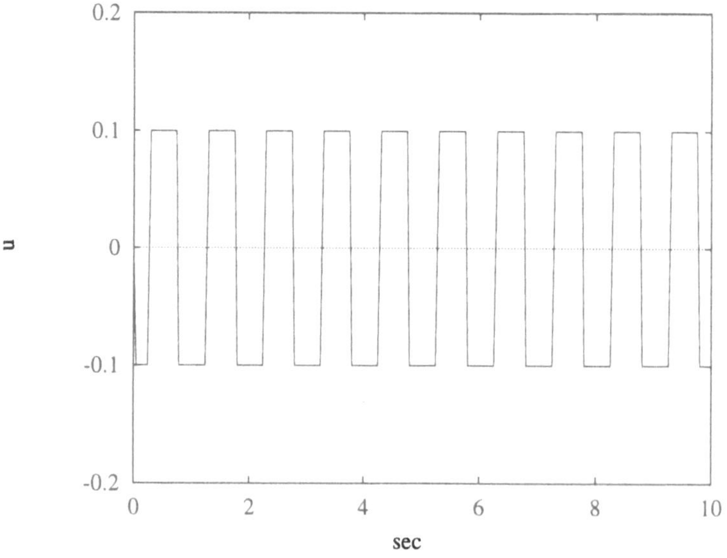

Fig. 6. u **(Plant with relative degree two)**

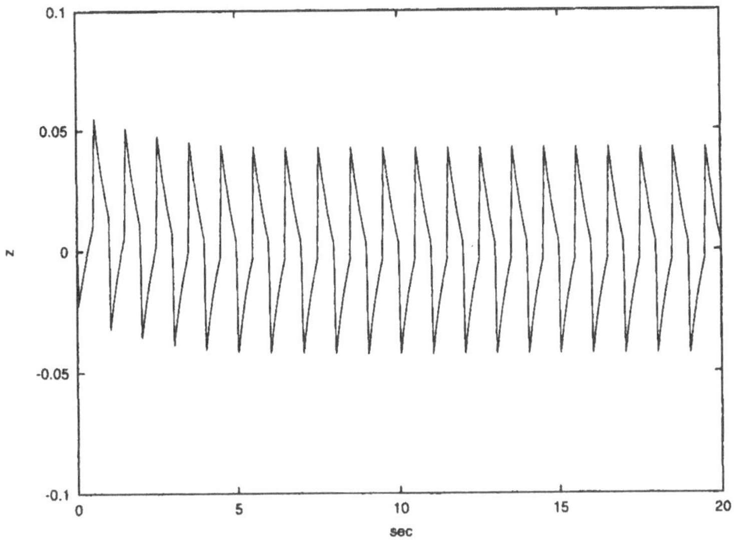

Fig. 7. z **(Plant with an unstable zero)**

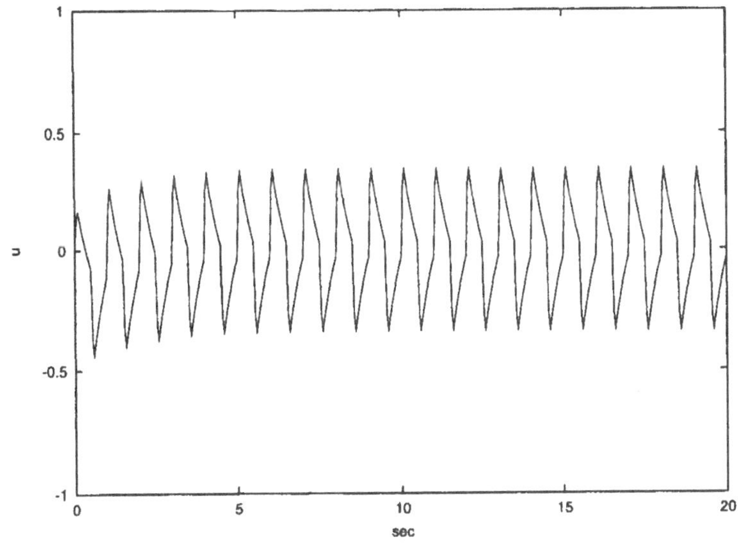

Fig. 8. u **(Plant with an unstable zero)**

Sliding Mode Tracking Control of Systems with Unstable Zero Dynamics

Heon-Sul Jeong[1] and Vadim I. Utkin[2]

Kunsan National University, Kunsan, Chonbuk, Korea,
hsjeong@knusun1.kunsan.ac.kr
The Ohio State University, Columbus, OH43210-1272,
utkin@ee.eng.ohio-state.edu

Abstract. In this paper, a sliding mode tracking control methodology, based on the block control principle, for an arbitrary reference signal of multivariable systems with unstable zero dynamics is developed. The proposed approach does not require an exosystem and is applicable to any relative degree systems. It is shown that the two step design procedure of the sliding mode theory is applicable to provide tracking without an error. Nonlinear systems reducable to a regular form can also be incorporated in this approach. As an illustrative example, tracking controller for EGR/VGT diesel engine was designed which is a 7th order 2-input 2-output nonminimum phase system.

1 Introduction

Output tracking of reference signals produced by some external generator (exosystem) has been investigated in depth for linear systems in general framework [1], [2]. Solvability of the tracking problem is seen to be related to the transmission zeros of a combination of the system and the exosystem. The presence of right-half plane transmission zeros impose limitations on tracking.

Many previous researchers have worked on nonminimum phase tracking systems. The work of Byrnes and Isidori [3] identified the dynamics on a particular center manifold which satisfies the tracking requirement. Gopalswamy and Hedrick [4] extended the center manifold concept to obtain the desirable internal dynamics by redefining output variables and applying sliding surface approach to track a modified trajectory which leads to a system with a stable zero dynamics. This technique requires that at least one of eigenvalues of the given system or the exosystem should be on the imaginary axis. Shtessel and Tournes combined sliding mode controllers with dynamic sliding manifolds to overcome the nonminimum phase nature of a plant and to reject disturbances [5]. The dynamic sliding manifold is designed in the form of a transfer function to satisfy certain requirements under the given types of references and disturbances. However, when the original system has relative degree one, a dynamic manifold can not be found stabilizing the closed loop system which is one of the requirements.

Recently an inversion-based tracking control is introduced to treat nonlinear nonminimum phase system [6], [7]. These results generalize the earlier results of

Lanari and Wen [8] for linear time invariant systems. The inversion algorithm is applied to exact output tracking and to derive a feedforward type controller for robotics application, respectively. The stable inversion process is similar to the Pickard iteration method to solve nonlinear differential equations.

The sliding mode control theory gains wide interest due to its attractiveness such as order reduction, easier implementation, design simplification, decoupling property of the design procedure and the robustness against modeling error. The design procedure consists of two steps: selection of a sliding surface and design of a discontinuous control enforcing sliding mode.

In this paper, we seek the tracking control of multivariable systems with unstable zero dynamics in the framework of the sliding mode theory and to control the system to follow the predefined smooth arbitrary reference signal without an exosystem. It is based on the two step design procedure of the sliding mode theory. One of the key idea is that the bounded solution for unstable zero dynamics can be obtained if noncausality is allowed. The other one is that tracking problem can be reduce to stabilization problem via neutralization with respect to deviation from the desired trajectory.

2 Outline of Design Methodology

Consider the following n-th order nonlinear systems with the same number m of inputs and outputs

$$\begin{aligned} \dot{\mathbf{x}} &= f(\mathbf{x}) + g(\mathbf{x})u\,, \quad \mathbf{x} \in \mathbf{R}^n,\ u \in \mathbf{R}^m \\ y &= h(\mathbf{x}) \end{aligned} \tag{1}$$

where rank $(g(\mathbf{x})) = m \leq n$, $u = \mathrm{col}(u_1, ..., u_m)$, $y = \mathrm{col}(y_1, ..., y_m) = \mathrm{col}(h_1(\mathbf{x}), ..., h_m(\mathbf{x}))$, $g(\mathbf{x}) = [g_1(\mathbf{x}), ..., g_m(\mathbf{x})]$ in which $f(\mathbf{x})$, $g_1(\mathbf{x}), ..., g_m(\mathbf{x})$ are smooth vector fields and $h_1(\mathbf{x}), ..., h_m(\mathbf{x})$ are smooth functions defined on an open set of $\mathbf{R}^n$. Assume that the linear approximation of the system (1) is stabilizable.

Let us first briefly review the sliding mode theory and block control principle utilizing some of the components of the state vector as a fictitious control [16], [17]. Suppose the above system can be decomposed into two subsystems of the following form, so-called *regular form* through a nonlinear transformation

$$\dot{z} = q(z, x), \quad z \in \mathbf{R}^{n-m} \tag{2a}$$

$$\dot{x} = b(z, x) + a(z, x)u, \quad x \in \mathbf{R}^m \tag{2b}$$

where $a(z, x)$ has rank m and the upper equation (2a) does not depend on the control input u. Necessary and sufficient condition for the transformation may be obtained based on the theory of Phaffian's form which solves the matrix partial differential equation for the nonlinear vector function. The reader is referred to [10] for a complete review of this approach.

Following the regular form design approach, one assumes that u is a discontinuous control enforcing sliding mode in the manifold $s(\mathbf{x}) = 0$ with m selected switching surfaces denoted by the vector $s(\mathbf{x}) = [s_1(\mathbf{x}), ..., s_m(\mathbf{x})]^T$. Generally speaking, after sliding mode occurs on $s(x) = 0$, m components of the state vector may be found as functions of the remaining $(n-m)$ ones, $x = s_o(z)$. As a result, the sliding mode equation on the manifold $s(\mathbf{x}) = x - s_o(z) = 0$ is

$$\dot{z} = q(z, s_o(z)) \tag{3}$$

The sliding dynamics (3) are determined only by the state z. The desired behavior may be achieved by a suitable choice of the function $x = s_o(z)$, with state x regarded as a fictitious control for the upper subsystem (2a). To confine the state trajectory to the selected manifold $s(\mathbf{x}) = 0$, the discontinuous control

$$u = -M\ sign(s(\mathbf{x})) \tag{4}$$

may be employed with gain matrix M satisfying the existence condition [1] of sliding mode [12]. Then, the state is steered to the manifold $s(\mathbf{x}) = 0$ within finite time.
Thus, the design reduces to an $(n-m)$ dimensional problem of a proper selection of the function $x = s_o(z)$ and the design of a control u enforcing the sliding mode in the manifold $s(\mathbf{x}) = 0$.

If the state x is reduced to zero by some control u, then the equation (2a) on the surface $x = 0$, $\dot{z} = q(z, 0)$ are referred to *zero dynamics* with respect to the state x. When the zero dynamics are unstable, the system is called nonminimum phase. We concentrate on the nonminimum phase system, i.e, the case that (2a) contains unstable zero dynamics.

Now let us turn to the tracking problem. The block control principle [9] introduced in sliding mode approach greatly simplifies the stabilization problem especially when it is transformed into a regular form. This principle may be applied for tracking of minimum phase systems but it can not be directly applied to nonminimum phase systems since a tracking problem implies stabilization of unstable zero dynamics as well as signal following.
It is well known that a tracking problem may be handled as a stabilization problem when the system is represented by the deviation of the system outputs and reference points. However, when the reference signal varies with time, tracking error appears because the derivative terms of the reference signal exists in the mismatch equation. Let us explain this facts using the following nonminimum phase linear example.

$$\begin{aligned} \dot{z} &= z + x \\ \dot{x} &= -z + x + u \\ e &= x - r \end{aligned} \tag{5}$$

[1] Sliding mode exists on a surface $s(\mathbf{x})$=0 whenever the distance to this surface and the rate of its change $\dot{s}$ are of opposite signs as a results of control u, i.e. when $lim_{s \to -0} \dot{s} > 0$ and $lim_{s \to +0} \dot{s} < 0$

Defining the deviation $\Delta z = z + r$ with respect to the steady state $e = 0$ and $\dot{z} = 0$ i.e, $x = r$ and $z = -r$, it becomes

$$\begin{aligned} \dot{\Delta z} &= \Delta z + e + \dot{r} \\ \dot{e} &= -\Delta z + e + u + 2r - \dot{r} \end{aligned} \tag{6}$$

One may take the fictitious control $e = -k\Delta z$. It may stabilize the unstable zero dynamics of Δz, but it causes tracking error due to the presence of the term $\dot{r}$ which may vanish in case of set point regulation.

Let us assume that $x^o \in R^m$ is the desired trajectory. Suppose the *bounded solution* [2] $z^o(t)$ of unstable zero dynamics with respect to x is available on the trajectory $x = x^o$. Then, tracking can be accomplished within finite time without an error as follows. Define the deviation $\Delta z = z - z^o$ and $\Delta x = x - x^o$, where Δx means the tracking error. Mismatch dynamics may be obtained from eq. (2)

$$\dot{\Delta z} = [q(z^o + \Delta z, x^o + \Delta x) - q(z^o, x^o)] \tag{7a}$$
$$\dot{\Delta x} = b(z, x) + a(z, x)u - \dot{x}^o \tag{7b}$$

Now it can be seen from (7) that the tracking problem making $x \to x^o$ yields the stabilization problem $\Delta z \to 0$ and $\Delta x \to 0$. Applying the sliding mode technique described above, the stabilization of the upper mismatch dynamics (7a) can be achieved by the fictitious control $\Delta x = f_s(\cdot)$. Accordingly, define the sliding surface

$$s = \Delta x - f_s(\Delta z, z^o, x^o) \tag{8}$$

with $f_s(0, z^o, x^o) = 0$. The real control enforcing sliding mode along the sliding surface $s = 0$ may be selected as discontinuous function of type (4). Note that, when the mismatch dynamics are stabilized, the term $[\cdot]$ in (7a) vanishes. Thus, the mismatch $\Delta z \to 0$ and no steady state error can be expected from the surface (8). Since the bounded solution $z^o(t)$ of (2a) is available, by assumption, on the trajectory $x = x^o$, an arbitrary smooth signal can be tracked even though the upper (zero) dynamics [3] are unstable. This is a natural extension of the concept of the set point neutralization.

The variable x of (2b) coincides with the system outputs y only when each output y_i has *relative degree* [4] 1. Generally, they are not identical. Even if it is the case, we can calculate the desired trajectory x^o for the desired reference

[2] In this paper, the *bounded solution* means that the solution of a (possibly unstable) system is bounded in the sense of BIBS (bounded-input and bounded-state) stability, i.e., the state is L_∞ signal for L_∞ input.

[3] Calling the upper subsystems of a regular form as *zero dynamics*, we may sometimes abuse the terminology of zero dynamics whenever there is no confusion, in the point that the zero dynamics are an autonomous one but the upper dynamics have input variables.

[4] The system (1) has local *relative degree* r at a point $\mathbf{x}_o$ if (i) $L_g L_f^k h(\mathbf{x}) = 0$ for all x in a neighborhood of $\mathbf{x}_o$ and all $k < r - 1$ (ii) $L_g L_f^{r-1} h(\mathbf{x}) \neq 0$. Where the Lie

signal r of the system output y. Let us suppose $p(p < m)$ sets of the system outputs have higher relative degree than 1. Then the corresponding output, say y_z, may be contained in the state vector z, i.e, $z^T = [z_1^T \; y_z^T]^T$ and the remaining $(m-p)$ sets of system outputs, say y_x, should be contained in the state vector x, i.e, $x^T = [x_1^T \; y_x^T]^T$. From the equations corresponding to y_z among (2a), one may derive the following relation

$$x_1 = w(z_1, y_x, y_z, \dot{y}_z) \tag{9}$$

Substituting (9) to z_1 dynamics among (2a) may result in

$$\dot{z}_1 = q_1(z_1, y_x, y_z, \dot{y}_z) \tag{10}$$

The bounded solution z_1^o of (10) gives z^o for the upper system (2a) since the desired state for y_z is readily available. Substituting z_1^o to (9) gives us x_1^o and in turn will provide the desired trajectory x^o. Thus without loss of generality, we will frequently consider x^o as the reference signal r of the system output y. Note that the relation (10) with $y = 0$ is nothing but the zero dynamics with respect to the system outputs y.

Consider the example with the output x.

$$\begin{aligned} \dot{z} &= 2z + z^2 + x \\ \dot{x} &= x^2 - z + u \end{aligned} \tag{11}$$

It is in the form of (2). The zero dynamics with respect to the output is unstable. Let us pick $x^o = 1$ as reference signal. Then, one may find that bounded solution of the unstable upper dynamics for the selected reference is simply $z^o = -1$. The mismatch dynamics of (11) are

$$\begin{aligned} \dot{\Delta z} &= \Delta z^2 + \Delta x \\ \dot{\Delta x} &= (x^o + \Delta x)^2 - z - \dot{x}^o + u \end{aligned} \tag{12}$$

Fortunately, the upper dynamics of (12) can be stabilized simply by $\Delta x = -k\Delta z$ with large gain k. Thus, choose the sliding surface $s = \Delta x + k\Delta z$ and design a control like (4) to enforce the surface to zero. Figure 1 shows the simulation results.

Fig. 1 Nonlinear Example

3 Sliding Mode Tracking of Linear Systems

Consider nth order stabilizable linear systems in a regular form

$$\dot{z} = A_{11}z + A_{12}x, \quad z \in R^{n-m} \tag{13a}$$

derivative $L_f\lambda(\mathbf{x})$ is the derivative of λ along f. If λ differentiated k times along f, the notation $L_f^k\lambda(\mathbf{x})$ is used satisfying the recursion $L_f^k\lambda(\mathbf{x}) = (\partial L_f^{k-1}\lambda(\mathbf{x})/\partial\mathbf{x})f(\mathbf{x})$ with $L_f^0\lambda(\mathbf{x}) = \lambda(\mathbf{x})$.

$$\dot{x} = A_{21}z + A_{22}x + Bu + d(t), \quad x, u \in R^m \tag{13b}$$

$$e = x - x^o = x - r, \quad e, r \in R^m \tag{13c}$$

where B is a full rank matrix, $d(t)$ is a matched disturbance [5] and e is the tracking error. Let's suppose bounded solution $z^o(t)$ of (13a) on the trajectory $e = 0$ i.e. $x = x^o$ is available. Then, define the mismatch as new variables

$$\Delta x = x - x^o, \ \Delta z = z - z^o, \ \Delta u = u - u^o \tag{14a}$$

$$u^o = B^{-1}(-A_{21}z^o - A_{22}r + \dot{r}) \tag{14b}$$

where $u^o(t)$ is the nominal control for the desired state trajectory $x = r$. Note that B is nonsingular. Then the mismatch dynamics are

$$\begin{aligned} \dot{\Delta z} &= A_{11}\Delta z + A_{12}\Delta x \\ \dot{\Delta x} &= A_{21}\Delta z + A_{22}\Delta x + B\Delta u + d(t) \end{aligned} \tag{15}$$

It is known that if the original linear system is controllable (stabilizable) then the pair (A_{11}, A_{12}) is controllable (stabilizable) as well [12]. The sliding surface (16a) based on the block control principle may be selected to stabilize the upper dynamics of (15) and then the sliding mode control (16b), the gain of which satisfy the existence condition of sliding mode, can be chosen to enforce the sliding mode on the selected sliding surface.

$$s = \Delta x + k\Delta z \tag{16a}$$

$$\Delta u = -M \ sign(s) \tag{16b}$$

Again, equation (15) means that the original tracking problem turns out to be stabilization problem noting that the mismatch Δx is nothing but the error signal. Since the nominal solution z^o is bounded, the nominal control u^o is also bounded which can be seen from (14b). Also note that there is no derivative terms of the reference signal in the dynamics of Δz, from which one can anticipate a tracking possibility without errors.
In this section, the bounded solution will be found even if the system is nonminimum phase. As the result, stabilization of mismatch dynamics achieved by block control principle may accomplish the output tracking of an arbitrary reference signal with no exosystem at the presence of matched disturbances.

3.1 Linear Systems with Reference Model

Suppose the model of a reference signal of dimension l is known as follows

$$\begin{aligned} \dot{\tau} &= R\tau, \quad \tau \in R^l \\ r &= F\tau, \quad r \in R^m \end{aligned} \tag{17}$$

[5] Suppose the system (1) is additionally introduced by disturbance $d(\mathbf{x}, t)$. If $d(\cdot)$ satisfies the following

$$d(\mathbf{x}, t) \in G(\mathbf{x}, t)$$

where $G(\mathbf{x}, t)$ is a subspace formed for each t by the vector $g(\mathbf{x})$, the disturbance is called *matched.* This relation, *matching condition*, means that there exists scalar function $\lambda(\mathbf{x}, t)$ such that $d(\mathbf{x}, t) = g(\mathbf{x})\lambda(\mathbf{x}, t)$.

where τ will be called as a *reference vector* since it consists of reference signal and its derivatives. Then the bounded solution of (13a) can be given in compact form $z^o = Z\tau$.

Theorem (Bounded Solution Driven by Reference Model) : The bounded solution of zero dynamics (13a) with respect to x driven by the reference signal $x^o = r$ of model (17) is given by $z^o = Z\tau$ if $[(n-m)\times l]\times[(n-m)\times l]$ dimensional matrix $R^T \bigotimes I_{(n-m)} - I_l \bigotimes A_{11}]$ is nonsingular where $\bigotimes$ denotes Kronecker products. [6]

Proof) Suppose the driven dynamics solution is given by $z^o = Z\tau$. Substituting z^o to (13a) and from (17)

$$\begin{aligned} \dot{z}^o &= Z\dot{\tau} = ZR\tau \\ &= A_{11}z^o + A_{12}r = A_{11}Z\tau + A_{12}F\tau \\ ZR - A_{11}Z &= A_{12}F \end{aligned} \tag{18}$$

Utilizing Kronecker products, the above algebraic matrix equation may yields

$$[R^T \bigotimes I_{(n-m)} - I_l \bigotimes A_{11}]Z_c = (A_{12}F)_c \tag{19}$$

where the subscript c denotes the vector formed by stacking up the columns in the matrix [13]. As the result the conclusion follows if the matrix in $[\,\cdot\,]$ is nonsingular. $\diamondsuit$

Thus, if the model of reference signal is known, e.g. finite-times differentiable or harmonic signal, the bounded solution of the (possibly unstable) upper dynamics reduces to find $(n-m)\times l$ matrix Z of the algebraic equation (18).

3.2 Tracking of an Arbitrary Reference Signal

Let us consider the linear system

$$\dot{z}(t) = Az(t) + Bu(t) \tag{20}$$

where matrix A has no eigenvalues on imaginary axis. When A is stable, the solution satisfying the initial condition is in the following form

$$z(t) = e^{At}z(0) + \int_0^t e^{A(t-\tau)}Bu(\tau)d\tau \tag{21}$$

which is bounded if input u is bounded. If all eigenvalues of A have positive real part, the bounded solution satisfying the following boundary condition can also be obtained

$$z(T) = 0, \quad t \le T \le \infty \tag{22}$$

[6] The Kronecker product of matrices $M(p\times q)$ and $N(m\times n)$ is a $(pm\times qn)$ matrix defined by $M\otimes N = \begin{bmatrix} M_{11}N & \dots & M_{1q}N \\ \vdots & \dots & \vdots \\ M_{p1}N & \dots & M_{pq}N \end{bmatrix}$ where M_{ij} is (i,j)-th element of matrix M.
And the Kronecker sum of $M(m\times m)$ and $N(n\times n)$ is a $(mn\times mn)$ matrix defined by $M\oplus N = N\otimes I_n + I_m\otimes N$.

Theorem (Bounded Solution of Unstable System) : The bounded solution of system (20) all modes of which are unstable, satisfying boundary condition (22) with T=constant is

$$z^o(t) = -\int_t^T e^{A(t-\tau)} Bu(\tau)d\tau \tag{23}$$

Proof: Differentiating (23) shows that it is indeed the solution of the unstable system (20).

$$\begin{aligned}\dot{z}^o(t) &= -Ae^{At}\int_t^T e^{-A\tau}Bu(\tau)d\tau - e^{At}[0 - e^{-At}Bu(t)] \\ &= Az^o(t) + Bu(t)\end{aligned}$$

The boundedness can be seen from the facts that A is unstable and $(t-\tau) < 0$. $\diamond$

Generally the system matrix A has both stable and unstable modes. Without loss of generality, one can transform the system (20) into a block diagonal (modal) form via a suitable similarity transformation $z_m = Wz$

$$\begin{aligned}\dot{z}_m &= A_m z_m + B_m u \\ &= \begin{pmatrix} A_s & 0 \\ 0 & A_u \end{pmatrix} \begin{bmatrix} z_s \\ z_u \end{bmatrix} + \begin{pmatrix} B_s \\ B_u \end{pmatrix} u\end{aligned} \tag{24}$$

where $z_s(z_u)$ is the state corresponding to the stable (unstable) mode, i.e, the matrix $A_s(A_u)$ has only open left (right) half plane eigenvalues, respectively.

According to the theorem, the bounded solution of linear systems (20) can always be obtained even in the presence of unstable matrix A_u since the corresponding one may be associated with bounded boundary condition rather than initial condition. Also note that the solutions (21) and (23) corresponding to A_s and A_u are causal and non-causal respectively.

When the zero dynamics of the system (13) are unstable i.e, A_{11} is a unstable matrix, the block control principle is not applicable in the case of output tracking. But if noncausality is allowed, in other words if the future signal of the input x^o of upper system (13a) is available, one can obtain the corresponding bounded solution z^o. Tracking problem is one of the case because x^o is the desired trajectory we want to trace. Thus our approach described before may be applicable regardless of the stability property of the zero dynamics.

Let us note here that Desavia et. al. [6] applied the similar concept to get a bounded solution for unstable dynamics in nonlinear setting. The solution was represented by integral form which is also non-causal when the system is unstable. At the beginning, applying the trial solution to the integral form, it may converge to the true solution iteratively, which is similar to the Pickard process for solving nonlinear differential equation numerically. Lanari and Wen [8] tried to choose a non-zero initial condition as

$$z(0) = -\int_0^\infty e^{A\tau}Bu(\tau)d\tau$$

so that the unstable dynamics are not excited. If one choose the initial condition as above, this may be explained from the fact that (21) reduces to (23) with $T = \infty$. They used this idea for feedforward type controller which may be highly sensitive to numerical conditions.

The bounded solution of the linear system (20), either stable or unstable, can also be represented with the derivatives of input signal.

Theorem (Bounded Solution in Derivative Form) : Suppose the system matrix A of linear systems (20) has no imaginary axis eigenvalues, then bounded solution $z^o(t)$ of the system with boundary condition $z(\infty) = 0$ can be given in the following derivative form

$$z^o(t) = -\sum_{n=0}^{\infty} A^{-(n+1)} B u^{(n)}(t) \tag{25}$$

Proof: Let us assume that the system (20) is transformed into modal form (24). At first, consider the unstable mode A_u of (24). From (23) with $T = \infty$, integrating by parts

$$\begin{aligned} z_u^o(t) &= -\int_t^{\infty} e^{A_u(t-\tau)} B_u u(\tau) d\tau \\ &= A_u^{-1} e^{A_u(t-\tau)} B_u u(\tau)|_t^{\infty} - \int_t^{\infty} A_u^{-1} e^{A_u(t-\tau)} B_u \dot{u}(\tau) d\tau \\ &= -A_u^{-1} B_u u(t) - A_u^{-2} B_u \dot{u}(t) - \cdots \\ &= -\sum_{n=0}^{\infty} A_u^{-(n+1)} B_u u^{(n)}(t) \end{aligned}$$

For the stable mode A_s, it may be easily checked by substitution that

$$z_s^o(t) = \int_0^t e^{A_s(t-\tau)} B_s u(\tau) d\tau$$

is one of the bounded solution which can be regarded as zero-state solution. Similarly integrating by parts, we can get the following under the causality assumption, i.e, $u^{(n)}(t) = 0$ for all $t \leq 0$

$$z_s^o(t) = -\sum_{n=0}^{\infty} A_s^{-(n+1)} B_s u^{(n)}(t)$$

From above derivations, the solution of (24) can be obtained as

$$\begin{aligned} z_m^o(t) &= \begin{bmatrix} z_s^o(t) \\ z_u^o(t) \end{bmatrix} \\ &= -\sum_{n=0}^{\infty} \begin{pmatrix} A_s^{-(n+1)} & 0 \\ 0 & A_u^{-(n+1)} \end{pmatrix} \begin{bmatrix} B_s \\ B_u \end{bmatrix} u^{(n)} \end{aligned}$$

$$= \quad -\sum_{n=0}^{\infty} \begin{pmatrix} A_s & 0 \\ 0 & A_u \end{pmatrix}^{-(n+1)} \begin{bmatrix} B_s \\ B_u \end{bmatrix} u^{(n)}$$

$$= \quad -\sum_{n=0}^{\infty} A_m^{-(n+1)} B_m u^{(n)}$$

Since (24) is transformed from (20) with $z_m = Wz$

$$z_m^o(t) = -\sum_{n=0}^{\infty} (WAW^{-1})^{-(n+1)} (WB) u^{(n)}$$

$$= -\sum_{n=0}^{\infty} (WA^{-(n+1)} W^{-1})(WB) u^{(n)}$$

$$= -W \sum_{n=0}^{\infty} A^{-(n+1)} B u^{(n)}$$

Thus, the bounded solution of (20) can be represented as (25) due to similarity transformation $z_m = Wz$. ◇

Even though (25) consists of infinite numbers of derivative terms, this theorem is very important in the point that the solution does not require the future input signal unlikely to the solution of convolution integral form. Also the compact form $z^o = Z\tau$ of Theorem (Bounded Solution Driven by Reference Model) may be expected from the derivative form (25) since a higher order derivative than l can be described as a combination of less order derivatives than l due to model (17).

3.3 Design Aspect

For the non-minimum phase case, the nominal trajectory $z^o(t)$ is noncausal and can be obtained through a convolution integral (23). However, in many cases a reference signal may be pre-defined so that the corresponding trajectory may be provided priori. Moreover if reference signal has finite-time derivatives or has its own model, it can be given in derivative form (25) and it can also be in compact form $z^o = Z\tau$. This fact validates the applicability of our design methodology.

In the case of a high order MIMO system, the dimensionality may cause some difficulty in practical application. If, fortunately, some parts of the zero dynamics are stable, we have a chance to limit our design only to the unstable part. Without loss of generality, upper dynamics (13a) can be decomposed into stable and unstable parts via modal form coordinate transformation as (24). On the desired trajectory $x = r$, the stable part $z_s(t)$ is already bounded and approaches to zero. One need only the unstable trajectory $z_u(t)$ to be bounded. Thus, the design may omit the stable part as free dynamics and reduce the order of problem. These will be illustrated in section 5.

4 Relation with Output Regulation

The approach proposed here is somewhat related with output regulator of the following linear system written in error coordinate system.

$$\begin{aligned} \dot{x} &= Ax + Bu + H\tau \quad && x \in R^n,\ u \in R^m \\ e &= Cx && e \in R^m \end{aligned} \tag{26}$$

Now output tracking requires that the output variables e should be regulated to zero. This may be achieved by static state feedback as stated in the following theorem.

Theorem (Stabilization and Output Regulation) : Tracking control for the system (26), possibly nonminimum phase, can be achieved by a state and reference feedback control of the form (27) if the reference signal has its model (17)

$$u = Kx + L\tau \tag{27}$$

where τ is a l dimensional reference vector and its elements consist of reference and its derivatives.

Proof: Suppose state feedback gain K is chosen such that $A^* = A + BK$ is Huruwitz. Also suppose the state x is composed of the stable solution x_s of a autonomous system (28) and a reference vector τ as (29).

$$\dot{x}_s = (A + BK)x_s = A^* x_s \tag{28}$$

$$x = x_s + M\tau \tag{29}$$

where M should be determined in the sequel. Differentiating (29) and from reference model (17)

$$\dot{x} = \dot{x}_s + M\dot{\tau} = A^* x_s + MR\tau$$

And the system dynamics and output yield the followings by substituting (27)-(29) into (26)

$$\begin{aligned} \dot{x} &= Ax + B(Kx + L\tau) + (BL + H)\tau \\ &= A^*(x_s + M\tau) + (BL + H)\tau \\ &= A^* x_s + (A^* M + BL + H)\tau \\ e &= Cx = Cx_s + CM\tau \end{aligned}$$

Investigating above equations and noting that x_s is a autonomous stable trajectory, output regulation is accomplished if one can find matrices M and L based on the stabilizing control gain K and reference model R such that

$$\begin{aligned} CM &= 0 \\ MR &= A^* M + BL + H \end{aligned} \tag{30}$$

Now let's think about the existence condition for matrices M and L. Without loss of generality, one can assume the following structure about matrices B and C

$$B = \begin{bmatrix} 0 \\ B_m \end{bmatrix}, \quad C = \begin{bmatrix} e_1 \\ \cdots \\ e_m \end{bmatrix}$$

where B_m is full rank and e_i is a elementary vector which has only one non-zero element 1. From (30) it may be easily seen that M has $(n-m)\times l$ free parameters to be chosen as follows

$$M = [m_1 \cdots m_l] = \begin{bmatrix} \bar{m}_1 \\ \cdots \\ \bar{m}_n \end{bmatrix}$$

$$m_i \in Ker(C) \ \in \ R^n$$

$$m \ \text{ sets of vectors : } \ \bar{m}_i = 0 \ \in \ R^l$$

and (30) may be arranged as follows

$$MR - A^* M - \begin{bmatrix} 0 \\ B_m L \end{bmatrix} = H$$

$(n-m) \times l$ sets of unknown m_{ij}, element of matrix M, and $m \times l$ sets of unknown elements of matrix L are same in number with $n \times l$ equations. Moreover A^* matrix can be arbitrarily chosen through the choice of gain K if the pair (A, B) is controllable. Thus, adjusting the gain K the matrices M and L can always be found. These close the proof. ◇

According to theorem, tracking problem can be achieved by stabilizing control Kx, treating the nature of unstable transmission zero, and by reference feedback L regulating the output. In the context of the proof, one can see the similarity with the former design concept. The difference is that sliding mode tracking deals the problem with two distinctive steps, while the latter mixes the stabilization and output regulation in one step. Also it is explicitly seen that the state trajectory consists of components of an autonomous stable system and reference vector.

Above explanation can also be interpreted in frequency domain. Consider the following linear system and control

$$\begin{aligned} \dot{x} &= Ax + Bu, \quad x \in R^n \\ y &= Cx, \quad y \in R^m \\ u &= Kx + D(s)Y_d, \quad u \in R^m \end{aligned} \tag{31}$$

where Y_d is a desired output, $D(s)$ is $m \times m$ polynomial matrix of order l in s which is a differential operator. Taking Laplace transformation, error $E(s)$ may be straight forwardly given as follows

$$\begin{aligned} E(s) &= Y(s) - Y_d(s) \\ Y(s) &= C(sI - A - BK)^{-1}BD(s)Y_d(s) \\ E(s) &= C(sI - A - BK)^{-1}BD(s)Y_d - Y_d(s) \\ &= -\{I_m - C(sI - A - BK)^{-1}BD(s)\}Y_d \end{aligned}$$

Multiplying both sides by $det(sI - A - BK)$

$$\begin{aligned} det(sI - A - BK)E(s) = det(sI - A - BK) \\ \{C(sI - A - BK)^{-1}BD(s) - I_m\}Y_d(s) \end{aligned} \tag{32}$$

Applying matrix inversion lemma to the above equation, it yields

$$\begin{aligned} det(sI - A - BK - BD(s)C)Y_d(s) = \\ -det(sI - A - BK - BD(s)C)E(s) \\ \{I_m + C(sI - A - BK - BD(s)C)^{-1}BD(s)\} \end{aligned} \tag{33}$$

From left hand side of both (32) and (33), it can be seen that the n-th order polynomial $det(sI - A - BK)$ represents the closed loop stability and the polynomial $det(sI - A - BK - BD(s)C)$ decides the tracking performance of the closed loop system. Tracking with no error means this polynomial becomes zero at the corresponding frequency components of reference signal. In other words, tracking problem is nothing but the problem of how to match the polynomial $det(sI - A - BK - BD(s)C)$ with reference, so called signal matching problem.

One more things to be mentioned in the same spirit is that every state trajectories of a linear system (26), with additional disturbance input term Dd, are composed of three elements, i.e. autonomous element, known modeled reference vector and a component due to unknown disturbances. To see these, define the state variable $x = x^* + M\tau = x_s + x_d + M\tau$ and take the control $u = Kx + L\tau + v$, similarly as before, where x_s is defined as (28), x_d is for disturbance and v is an additional control. If we take M and L as (30), it is straightforward to see

$$\begin{aligned} \dot{x}_d &= A^* x_d + Bv + Dd \\ \dot{x}^* &= A^* x^* + Bv + Dd \\ y &= Cx^* + CM\tau = Cx_s + Cx_d + CM\tau \end{aligned} \tag{34}$$

From (34), it may be seen that the known reference vector and unknown disturbances can be completely decoupled and state trajectories are composed of three components, i.e. x_s, x_d, τ. Until now, linear system theory tends to solve three major goals of control system design, stabilization, output tracking and disturbance rejection, by one controller which inevitably requires trade-off between design specifications. However, our discussion shows that system design can be decoupled into three separate steps such that state feedback Kx, reference tracking L and disturbance rejection through an additional control v may achieve its own design purpose with no interference with other one.

5 Case Study (EGR/VGT Diesel Engine)

5.1 Background of EGR/VGT diesel engine

For production of passenger car internal combustion engines to meet strict emission regulations and customer demands of improved fuel economy and driveability, advanced hardware components are increasingly being considered. However, the basic requirements in terms of peak power, transient response, fuel economy and emissions are often contradictory and require judicious tradeoffs at every stage of the design process. Diesel (compression ignition) engines have a significant advantage over gasoline spark ignition (SI) engines in fuel economy. Moreover, diesel engines have lower feed gas emissions of the regulated exhaust gases, but the after-treatment devices for diesel engines are far less efficient than the conventional three way catalysts for SI engines.

An effective way to reduce the formation of NO_x during combustion is to re-circulate the exhaust gas through the exhaust gas recirculation (EGR) valve into the intake manifold. The fraction of EGR recirculated flow must be scheduled on the operating condition. To improve its relative low power density, a diesel engine can be equipped with a turbocharger which consist of a turbine and a compressor attached to the common shaft. A conventional turbocharger faces a tradeoff between fast transient response at low engine speeds and high power without over-speeding at high engine speeds. This tradeoff can be adjusted by employing a variable geometry turbocharger (VGT) in which the amount of exhaust gas flowing through the turbine and the power transferred is controlled by changing the position of the guide vane [18].

For diesel engines equipped with EGR and VGT actuators, the design objective is to supply an amount of air and a fraction of EGR appropriate for a given operating condition (engine speed and fueling rate demanded by the driver). An insufficient amount of air leads to an increase in particulate emissions and possibly visible smoke, while an insufficient EGR fraction leads to an increase in NOx emissions. In addition, the system has to provide fast increase of engine air intake at driver tip-ins to allow fast increase in fueling rate and engine torque. Stabilizing controller for this system based on block control principle was designed in [20].

Fig. 2 Input-output of diesel engine

The input-output representation of diesel engine is shown in figure 2. A nonlinear model for this system has been described in [19]. The dynamical model is developed using the conservation of energy and mass and the ideal gas law together with several experimentally derived maps. The model involves seven states: mass (m_1), burnt gas density fraction (F_1) and pressure (p_1) in intake manifold: mass (m_2), burnt gas density fraction (F_2) and pressure (p_2) in exhaust manifold and turbocharger rotational speed (N_tc). The control inputs are the EGR valve position χ_{EGR} and the VGT actuator position χ_{VGT}.

5.2 Tracking controller design

Our design objective is to regulate the air-fuel ratio and the EGR flow fraction to the set points determined from a static engine data based on engine speed N and the desired fueling rate W_f^d.

$$AF_{ref} = AF_{ref}(N, W_f^d),\ EGR_{ref} = EGR_{ref}(N, W_f^d)$$

Typically, diesel engines are equipped with sensors for intake manifold pressure p_1 and compressor mass flow rate W_{c1}. Thus, the set points for AF and EGR are transformed into the set points for the intake manifold pressure (p_1^d) and the compressor mass flow rate (W_{c1}^d) from the static engine map.

We restrict our attention to the linearization of the model at a medium engine speed operating point (N=2000 rpm and W_f=6 kg/hr).

$$\dot{x} = A_{7\times 7}\begin{bmatrix} m_1 \\ F_1 \\ p_1 \\ m_2 \\ F_2 \\ p_2 \\ N_{tc} \end{bmatrix} + B_{7\times 2}\begin{bmatrix} \chi_{EGR} \\ \chi_{VGT} \end{bmatrix} = Ax + Bu \tag{35}$$
$$y = \begin{bmatrix} p_1 \\ W_{c1} \end{bmatrix} = Cx$$

Thus all state variables as well as control inputs are deviations from the corresponding nominal values. The system matrices given by maker are shown in Table 1 and the states used are described in Table 2. System outputs have both relative degree 1 and the zero dynamics with respect to the outputs has one unstable mode i.e. non-minimum phase system.

According to our design procedure, system (35) is transformed into the following form

$$\begin{aligned}
\dot{Z}_s &= A_s Z_s && +B_{s1}y_1 + B_{s2}y_2 \\
\dot{z}_u &= \qquad\quad A_u z_u && +B_{u1}y_1 + B_{u2}y_2 \\
\dot{y}_1 &= A_{11}Z_s + A_{12}z_u && +A_{13}y_1 + A_{14}y_2 + B_{11}u_1 \\
\dot{y}_2 &= A_{21}Z_s + A_{22}z_u && +A_{23}y_1 + A_{24}y_2 + B_{21}u_1 + B_{22}u_2
\end{aligned} \tag{36}$$

where Z_s is a stable mode vector and z_u is a unstable mode. It is obtained in the following steps. First, state p_1 is moved to the sixth position and then states p_1 and N_{tc} is replaced with system outputs p_1 and W_{c1} since the relative degree of system outputs are 1. Second, it is transformed into regular form using the nonsingular matrix

$$T = \begin{pmatrix} I_5 & -B_1 B_2^- 1 \\ 0 & I_2 \end{pmatrix} \tag{37}$$

where B_2 is the last two rows and B_1 is the remaining matrix of B. Third, the zero dynamics are decomposed into stable / unstable modal form. Finally, unstable mode of zero dynamics

$$\dot{z}_u = A_u z_u + B_{u1} y_1 + B_{u2} y_2 \tag{38}$$

is moved to the lowest position of zero dynamics. The transformed system matrices is given in table 3.

To simplify the design, the bounded solution of only unstable mode is numerically calculated via convolution integral

$$z_u^o(t) = -\int_t^T e^{A_u(t-\tau)}(B_{u1} y_1^d + B_{u2} y_2^d) d\tau \tag{39}$$

Our reference signals, y_1^d and y_2^d, are chosen as in figure 3. The curves in first 3 seconds represents feasible practical commands change from one points to another and the remainders are designed to show the performance of tracking arbitrary signals. To calculate the integral (39), one needs data $e^{A_u t}$, $t \in [-T, 0]$ and $y_1^d(t), y_2^d(t), t \in [t, T]$. It may be obtained in advance for the given desired trajectory.

Fig. 3 Reference signals

The corresponding mismatch dynamics are

$$\begin{aligned}
\dot{Z}_s &= A_s Z_s & &+B_{s1} y_1 + B_{s2} y_2 \\
\dot{\Delta z}_u &= & A_u \Delta z_u &+B_{u1} \Delta y_1 + B_{u2} \Delta y_2 \\
\dot{\Delta y}_1 &= A_{11} Z_s + & A_{12} \Delta z_u &+A_{13} \Delta y_1 + A_{14} \Delta y_2 + B_{11} u_1 \\
& & +(A_{12} z_u^o &+A_{13} y_1^d + A_{14} y_2^d - \dot{y}_1^d) \\
\dot{\Delta y}_2 &= A_{21} Z_s + & A_{22} \Delta z_u &+A_{23} \Delta y_1 + A_{24} \Delta y_2 + B_{21} u_1 + B_{22} u_2 \\
& & +(A_{22} z_u^o &+A_{23} y_1^d + A_{24} y_2^d - \dot{y}_2^d)
\end{aligned} \tag{40}$$

where mismatch variables are $\Delta z_u = z_u - z_u^o$, $\Delta y_1 = y_1 - y_1^d$ and $\Delta y_2 = y_2 - y_2^d$. Utilizing Δy_1 as a fictitious control for the unstable Δz_u zero dynamics, sliding surfaces

$$\begin{aligned}
s_1 &= \Delta y_1 + k_1 \Delta z_u \\
s_2 &= \Delta y_2
\end{aligned} \tag{41}$$

and control enforcing the surfaces s_1 and s_2 to zero may stabilize the mismatch dynamics. Note that stable Z_s dynamics are left as free motions and Z_s in Δy_1 and Δy_2 dynamics may be regarded as a kind of disturbance, which can be rejected because it is located in the input channel, i.e, it satisfies the so-called matching conditions.

5.3 Simulation results

The bounded solution z_u^o is shown in figure 4. From the figures 3 and 4, it may be expected that noncausal actuation about 0.15 sec is necessary to track the desired signal accuratly, i.e. control action must applied before the output needs to move. The required time of noncausal actuation depends on the eigenvalue of unstable mode. The time may decrease fast as the unstable pole moves far from the imaginary axis.

Fig. 4 Bounded solution z_u^o

Considering practical application to alleviate the chattering of mechanical actuators, a sigmoid function, $tanh(as)$, will be used in simulation study instead of a signum function in (4), which is a kind of continuous approximation of discontinuous function. The benefit of this function is that it may smoothen the chattering and the magnitude of its derivative can be limited by the parameter a.

Control gains used for each control input are $M_1 = -0.8$ and $M_2 = -0.8$ and the sliding surface gain in (41) is $k_1 = -0.2$. Simulation results of non-causal actuation, the time of which was 1 sec to enhance visibility, are shown in figures 5, 6 and 7 and those of causal actuation are in figures 8, 9 and 10.
Non-causal actuation result, Fig. 5 show that outputs keep tracking arbitrary reference signals exactly, they are almost indistinguishable in the figure. And unstable mode z_u is enforced to follow the bounded trajectory z_u^o. Even though noncausal actuation is not allowed, figures 8 - 10 show similar results after short initial transient undershoot which is the typical phenomenon of a non-minimum phase system. Control inputs shows smooth curve on behalf of sigmoid function and the state Z_s remains bounded as expected. It must be approaching to the curve $-A_s^{-1}(B_{s1}y_1^d + B_{s2}y_2^d)$ which can be obtained from (40).

6 Tracking of Nonlinear Systems

Some nonlinear systems may be given in the form of (2), but in general we need coordinates transformation. As described in section 2, if a regular form and the relation (9) is obtained, one needs only the bounded solution z_1^o of (10) to get the desired state trajectory x^o and the associated mismatch dynamics (7). Then, one can apply our design procedure to stabilize the mismatch dynamics by the usage of fictitious control $\Delta x = f_s(\Delta z, z^o, x^o)$, $f_s(0, z^o, x^o) = 0$ and sliding mode control.

Even if (9) is unstable, the bounded solution can be found from an elegant theory of stable inversion which is a Pickard-like iterative numerical methods. If the subsystem (10) is stable, our approach may leave it as free dynamics and one may concentrate to stabilize the mismatch dynamics associated with y_z in (7a) by a control of the form $\Delta x = f_s(\Delta y_z, y_z^o, x^o)$, $f_s(0, y_z^o, x^o) = 0$.

If the nonlinear control $\Delta x = f_s(\cdot)$ is difficult to find for some reasons, one can resort to the principle of stability in the first approximation [15]. The principle tell us that the local asymptotic stability of the equilibrium point of nonlinear systems can be determined by the behavior of the linear approximation of the nonlinear system at the equilibrium point. To this end, take the linear approximation form of the upper dynamics (2a) $q(z,x) = Qz + Px + q_o(z,x)$ where $q_o(z,x)$ is a higher order nonlinear term. Then, (7a) becomes

$$\begin{aligned} \dot{\Delta z} &= Q\Delta z + P\Delta x \\ &\quad + [q_o(z^o + \Delta z, x^o + \Delta x) - q_o(z^o, x^o)] \\ &= Q'\Delta z + P'\Delta x + q_o'(z^o, \Delta z, x^o, \Delta x) \end{aligned} \tag{42}$$

where $q_o'(\cdot)$ includes resulting nonlinear terms after re-arrangement. If we choose the fictitious control as $\Delta x = f_s(\cdot) = K\Delta z$ such that the matrix $Q' + P'K$ is stable, then Δz tends to zero by virtue of the principle of stability in the first approximation. Enforcing the relationship $\Delta x = K\Delta z$ by a discontinuous control (4) satisfying the existence condition of sliding mode, $\Delta x = K\Delta z \to 0$ on the sliding surface (8) as well. Note that the term $[\cdot]$ in the above equation also vanish.

Now let us turn our attention to the nonlinear system representation in *normal form* [14] which is developed from the differential geometric system theory connected with the concept of relative degree. Suppose the system (1) has an well-defined (vector) relative degree $r_1, ..., r_m$, $m \leq r = r_1 + ... + r_m \leq n$ at a point $\mathbf{x}_o$. By setting a new local coordinate in a neighborhood of $\mathbf{x}_o$

$$\Phi(\mathbf{x}) = \mathrm{col}[\phi^1{}_1(\mathbf{x}), ..., \phi^1{}_{r_1}(\mathbf{x}), ..., \phi^m{}_1(\mathbf{x}), ..., \phi^m{}_{r_m}(\mathbf{x}), \phi_{r+1}(\mathbf{x}), ..., \phi_n(\mathbf{x})]$$

where $\phi_1^i(\mathbf{x}) = h_i(\mathbf{x})$, $\phi_2^i(\mathbf{x}) = L_f h_i(\mathbf{x})$, ..., $\phi_{r_i}^i(\mathbf{x}) = L_f^{r_i - 1} h_i(\mathbf{x})$ and $(n - r)$ functions $\phi_{r+1}(\mathbf{x})$, ... , $\phi_n(\mathbf{x})$ are chosen arbitrarily such that the jacobian matrix of $\Phi(\mathbf{x})$ is nonsingular at $\mathbf{x}_o$. Then, the system can be transformed into the following

$$\dot{\eta} = q'(\eta, \xi) + p'(\eta, \xi)u, \quad \eta \in R^{n-r} \tag{43a}$$
$$\dot{\xi} = b'(\eta, \xi) + a'(\eta, \xi)u, \quad \xi \in R^r \tag{43b}$$

where $\eta = \mathrm{col}(\phi_{r+1}(\mathbf{x}), ..., \phi_n(\mathbf{x}))$ and ξ is the vector which has the first r elements of the vector $\Phi(\mathbf{x})$. If the distribution $G = \mathrm{span}\{g_1, ..., g_m\}$ is involutive near $\mathbf{x}_o$, it is always possible to choose η in such a way that $L_g^j \phi_i(\mathbf{x}) = 0$ for all $r + 1 \leq i \leq n$, for all $1 \leq j \leq m$ and for all $\mathbf{x}$ around $\mathbf{x}_o$. By choosing η in this way, (43a) yields

$$\dot{\eta} = q(\eta, \xi), \quad \eta \in R^{n-r} \tag{44}$$

Equation (43b) can be further decomposed into

$$\dot{\rho} = c\rho + dx, \quad \rho \in R^{r-m} \tag{45a}$$
$$\dot{x} = b(\eta, \xi) + a(\eta, \xi)u, \quad x \in R^m \tag{45b}$$

where $\xi = \mathrm{col}(\rho, x), x = \mathrm{col}(\phi_{r_1}^1, ..., \phi_{r_m}^m), \rho = (\phi_1^1, ..., \phi_{r_1 - 1}^1, ..., \phi_1^m, ..., \phi_{r_m - 1}^m)$ and c, d are constant matrices as follows

$$c = diag(c_1, .., c_m), \; c_i = \begin{bmatrix} 0 & I_{r_i - 2} \\ 0 & 0 \end{bmatrix}, \; c_i \in R^{(r_i - 1)\times(r_i - 1)}$$

$$d = diag(d_1, .., d_m), \; d_i = \mathrm{col}(0, .., 0, 1), \; d_i \in R^{r_i - 1}$$

The zero dynamics of the system with respect to the system output y, i.e. the internal dynamics when initial condition and input have been chosen to constrain the output to remain identically zero are (44) with $\xi = 0$. If the zero dynamics are stable, stabilization and tracking is solvable via a feedback linearization technique [14].

Now, the nonlinear system (44)-(45) has the form of (2). Note that (45a) consists of series of integrators of input x and its element x_i is $(r_i - 1)$-th derivative of the system output $y_i = h_i(\mathbf{x})$. Thus, if the bounded solution η^o of unstable zero dynamics on the desired trajectory ξ^o=col$(\rho^o\ x^o)$ are available, one can derive the corresponding mismatch dynamics. We can now follow the design procedure outlined in section 2 in two steps. If the zero dynamics (44) are stable, our approach may leave it as free dynamics and concentrate on the stabilization of mismatch dynamics corresponding to (45a). Note that, while the feedback linearization technique has no measure to disturbances, the approach proposed here can reject the disturbances injected through the equation (45b) i.e. satisfying the matching conditions. Moreover, even if some parts of zero dynamics are unstable, our approach enables the tracking as proposed.

A normal form may be viewed as a special kind of regular form. While a regular form is suitable for two step design procedure of sliding mode control, a normal form exhibits the internal dynamics structure of the system associated with the system output and is tightly connected with the notion of relative degree.

7 Concluding Remarks

Tracking control of an arbitrary reference without an exosystem has been discussed for nonminimum phase systems which is applicable to any relative degree systems. If the bounded solution of the upper subsystem of a regular form is available, tracking problem is shown to be reduced to stabilization problem, when the system is described in the form of mismatch dynamics via neutralization on the desired reference trajectory.
The bounded solution for a linear system has been provided in a convolution integral form which is causal for minimum phase and non- causal for nonminimum phase. Also it is given in a derivative form which does not need a future input signal even in the case of nonminimum phase system. If the model of a reference signal is known, it can be in a compact form $z^o = Z\tau$ of reference vector.
Then, the sliding mode control technique utilizing some states as a fictitious control was applied to solve the stabilization problem effectively rejecting the matched disturbances. The design could be further simplified if the zero dynamics are decomposed into stable and unstable parts.

Our approach has also interpreted in conventional linear system theory. It shows that two step design may be mixed in one step. And our spirit summarizes that three design goals of control system design such as stabilization, tracking and disturbance rejection, can be treated in separate steps not sacrificing the other's design specification, while linear system theory until now tends to solve these by one controller. For a nonlinear system, comparison of design aspects such as design of fictitious control, linear approximation and matched disturbance rejection etc, based on regular form and normal form were discussed. As an illustrative example, tracking controller for EGR/VGT diesel engine was designed which is a 7th order 2-input 2-output nonminimum phase system.

References

1. Francis B. A., Wonham W. M., "The internal model principle of control theory", Automatica, vol. 12, pp.457-465, 1976
2. Wonham W. M., Linear Multivariable Control: A Geometric Approach, 3rd ed., Springer-Verlag, 1985
3. Byrnes C. I., Isidori A., "Output regulation of nonlinear systems", IEEE Transactions on Automatic Control, Vol. 35, pp.131-140, 1990
4. Gopalswamy S., Hedrick J. K., "Tracking nonlinear non-minimum phase systems using sliding control", Int. J. Control, vol. 57, no.5, pp.1141-1158, 1993
5. Shtessel Y. B., Tournes C., "Nonminimum phase output tracking in dynamic sliding manifolds with application to aircraft control", Proc. of the 35th Conference on Decision and Control, Kobe, Japan, Dec., 1996
6. Devasia S., Chen D., Paden B., "Nonlinear inversion-based output tracking", IEEE Transactions on Automatic Control, Vol. 41, no. 7, pp. 930-942, 1996
7. Hunt L. R., Meyer G., "Stable inversion for nonlinear systems", Automatica, vol. 33, no. 8, pp. 1549-1554, 1997
8. Lanari L., Wen J., "Feedforward calculation in tracking control of flexible robots", Proc. of IEEE Conference on Decision and Control, pp.1403-1408, Brighton, England, 1991
9. Drakunov S. V., Izosimov D. B., Luk'yanov A. G., Utkin V. A., Utkin V. I., "Block control principle I & II", Avtomatika i Telemekhanika, No. 5, pp. 38-47, May, 1990, No. 6, pp. 20-31, June, 1990
10. Luk'yanov A. G., Utkin V. I., "Methods of reducing equations for dynamic systems to a regular form", Avtomatika i Telemekhanika, No. 4, pp. 5-13, April, 1981
11. Hale J. K., Ordinary Differential Equations, vol. 21, New York: Wiley- Interscience, 1980
12. Utkin V. I., Sliding Modes in Control and Optimization, Springer-Verlag, 1992
13. Brewer J. W., "Kronecker products and Matrix Calculus in System Theory", IEEE Transactions on Circuits and Systems, vol. CAS-25, no. 9, pp.772-781, Sep. 1978
14. Isidori A., Nonlinear Control Systems, 2nd Ed., Springer-Verlag, 1989
15. Hahn W., Stability of Motion, Spinger-Verlag, 1967
16. Luk'yanov A. G., "Optimal Nonlinear Block-Control Method", Proc. of the 2nd European Control Conference, Groningen, 1853-1855, 1993
17. Luk'yanov A. G., Dodds S. J., "Sliding Mode Block Control of Uncertain Nonlinear Plants", Proc. IFAC World Congr.
18. Moody J., "Variable Geometry Turbocharging with Electronic Control", SAE Paper 860107, 1986

19. Kolmanovsky I., Moraal P., van Nieuwstadt M., and Stefanopoulou A., "Issues in Modelling and Control of Intake Flow in Variable Geometry Turbocharged Engines", in System Modelling and Optimization, Addison Wesley Longman, to appear
20. Utkin, V. I., Chang, H., Kolmanovsky, I., and Chen, D., 1998, "Sliding Mode Control Design based on Block Control Principle," *The Fourth International Conference on Motion and Vibration Control (MOVIC)*, Zurich, Switzerland, August 25–28.

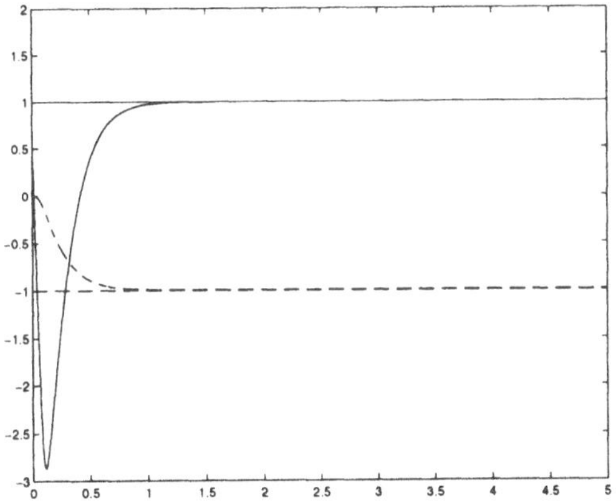

Fig. 1. Nonlinear example

Table 1. Original system matrices

A	x_1	-2.10	0	-3.7e-3	24.60	0	1.0e-4	4.03e-6
	x_2	-6.9e-13	-6.41	0.046	377.39	1.48	0.01	-6.87e-5
	x_3	1.18e5	0	-99.01	4.48e5	0	7.09	0.097
	x_4	2.09	0	4.0e-4	-47.67	0	-3.0e-4	0
	x_5	-1067.5	57.67	-0.12	9130.7	-59.48	-0.022	0
	x_6	-7.57e5	-2308.7	143.45	5.79e6	0	-202.13	0
	x_7	0	0	9.14e4	-8.71e8	0	3.16e4	-142.86
B^T	u_1	0.024	1.394	935.79	-0.025	0	-7.17e3	0
	u_2	0	0	0	-0.064	0	-1.86e4	3.97e6
C	y_1	0	0	1.00	0	0	0	0
	y_2	0	0	-3.3ee-3	0	0	0	4.03e-6

Table 2. Physical states and operating range

	physical meaning	operating range	nominal point
x_1	m_1	0.001 - 0.01 [kg]	8.3000e-3
x_2	F_1	0 - 1	1.4160e-1
x_3	p_1	100 - 300 [kPa]	1.6332e2
x_4	m_2	0.0001 - 0.001 [kg]	9.2365e-4
x_5	F_2	0 - 1	6.1180e-1
x_6	p_2	100 - 300 [kPa]	1.9191e2
x_7	N_{tc}	10e3 - 200e3 [rpm]	1.1664e5
y_1	p_1	100 - 300 [kPa]	
y_2	W_{c1}	0.001 - 0.1 [kg/sec]	

Table 3. Transformed system matrices

A	z_{s1}	-72.5	0	0	0	0	-28.77	-5479.1
	z_{s2}	0	-4.6718	0.5712	0	0	-0.3205	8968.1
	z_{s3}	0	-0.5712	-4.6718	0	0	-0.0155	-5251.9
	z_{s4}	0	0	0	-61.245	0	-0.5968	-32.464
	z_u	0	0	0	0	22.434	-356.11	3502.2
	y_1	-1.0671	-0.6465	5.939	-3.124	9.027	-115.22	14256
	y_2	-0.1746	-0.0712	0.0701	0.1395	0.0830	-1.049	-323.78
B^T	u_1	0	0	0	0	0	935.79	-3.077
	u_2	0	0	0	0	0	0	16.044

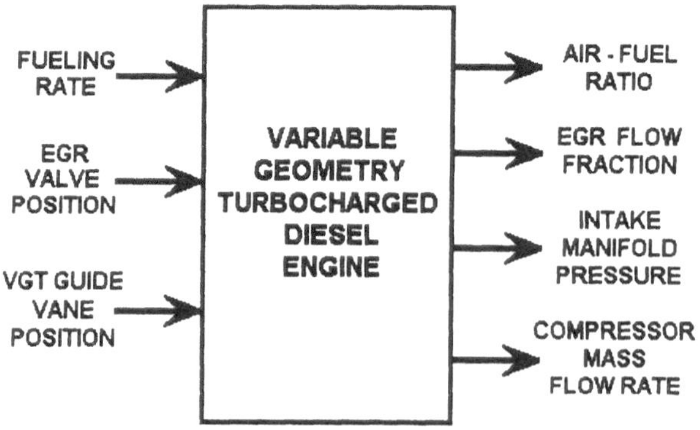

Fig. 2. Input-output of diesel engine

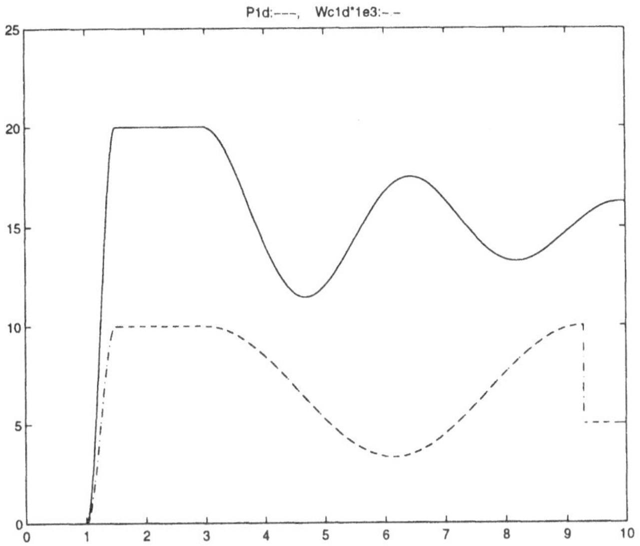

Fig. 3. Reference signals

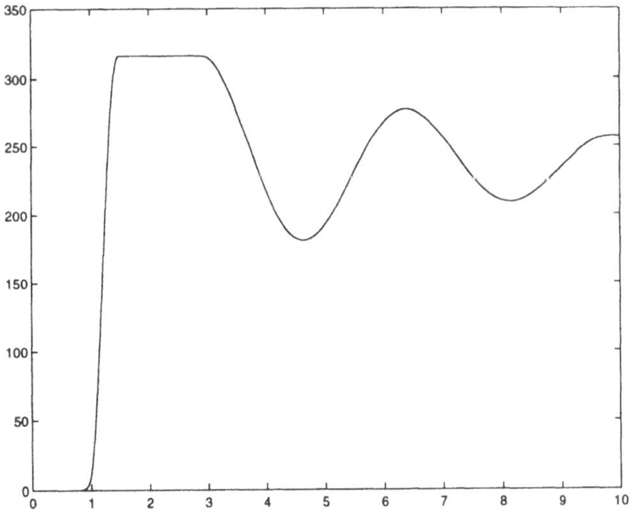

Fig. 4. Bounded solution z_u^o

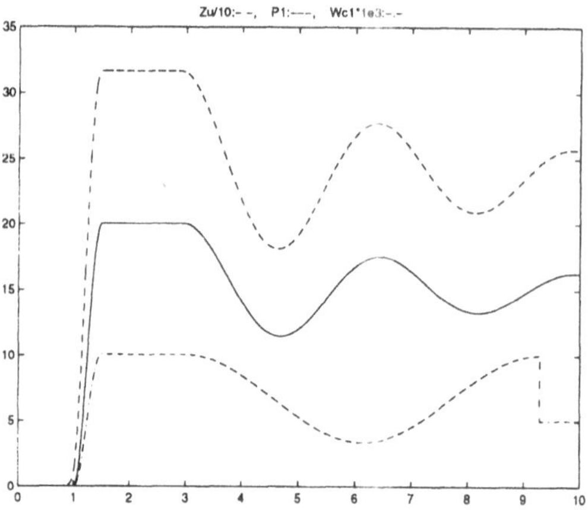

Fig. 5. Non-causal actuation: outputs

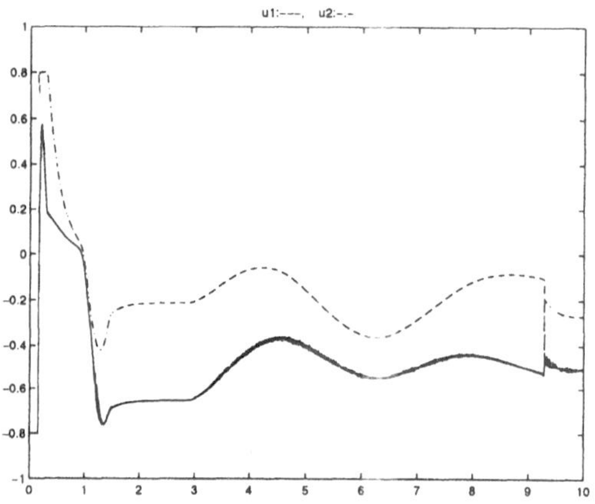

Fig. 6. Non-causal actuation: control inputs

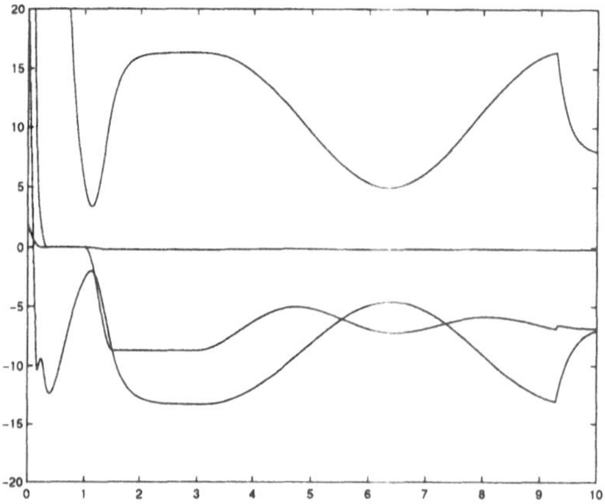

Fig. 7. Non-causal actuation: state Z_s

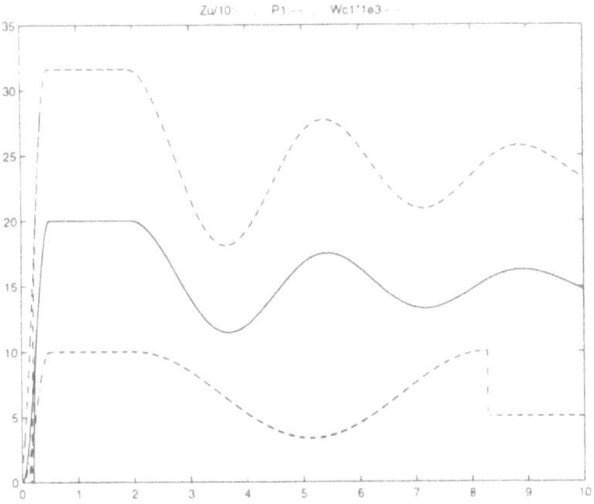

Fig. 8. Causal actuation: outputs

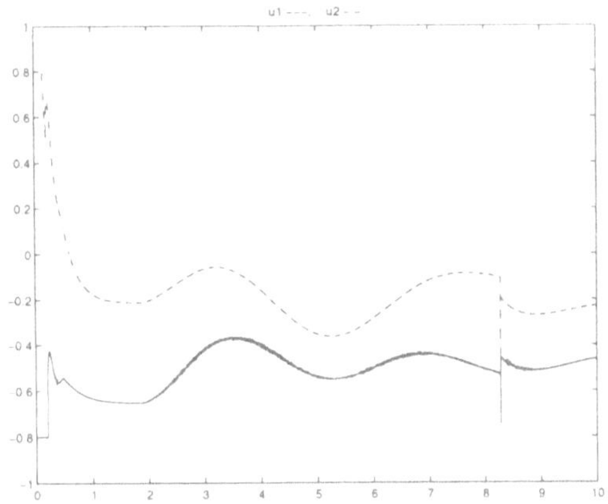

Fig. 9. Causal actuation: control inputs

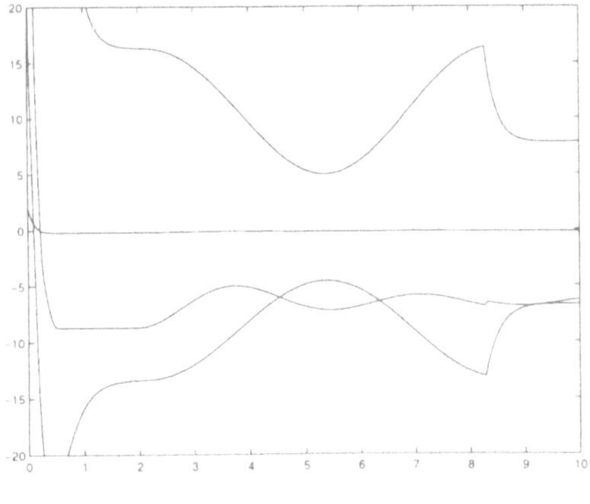

Fig. 10. Causal actuation: state Z_s

On Second Order Sliding Mode Controllers

Giorgio Bartolini[1], Antonella Ferrara[2], Arie Levant[3], Elio Usai[1]

1 Department of Electrical and Electronic Engineering, University of Cagliari, Piazza d' Armi, 09123 Cagliari, Italy

2 Department of Communication, Computer and System Sciences, University of Genova - Via Opera Pia 13, genova, Italy

3 Institute for Industrial Mathematics, 4/24 Yehuda Ha-Nachtom St., Beer-Sheva 84249, Israel

1 Introduction

The ability to manage systems in an uncertain context is one of the most intriguing desires of humans; as an example many popular games, such as chess and bridge, are strictly related to such a topic. This expectation can be reduced to a system theory context; indeed the development of the control theory has given rise to a number of sophisticated control techniques devoted to solving the control problem for some classes of uncertain systems. Most of them are based on adaptation methods [24], relying both on identification and observation, and on absolute stability methods [11, 12], which often lead to very complicated control algorithms whose implementation can imply a relevant computational cost and/or the use of very expensive devices.

In the last years an increasing interest for sliding modes led to an almost complete formalization of the mathematical background and of the robustness properties, with respect to system uncertainties and external disturbances, of this control technique [26].

Variable Structure Systems (VSS) in which the control is able to constrain the uncertain system behaviour on an "a priori" specified manifold (the sliding manifold) by "brutal force" seem to be the most obvious and heuristic way to withstand the uncertainty. In such systems the control immediately reacts to any deviation of the system, steering it back to the constraint by means of a sufficiently energetic control effort. Any strictly satisfied equality removes one "uncertainty dimension".

Furthermore, VSS may be considered as a general term for any dynamical system with discontinuous feedback control which can be defined by means of suitable sliding mode techniques.

Classical sliding mode control is based on the possibility of making and keeping identically null an auxiliary output variable (the sliding variable), which represents the deviation from the constraint, by means of a discontinuous control acting on the first time derivative of the sliding variable, and switching between high amplitude opposite values with theoretically infinite frequency. Moreover, due to its regularity properties, any system evolving in a Δ boundary layer of the sliding manifold because of the nonideal realization, both of the system and of the control devices, has the same trajectories as the ideal one apart from some perturbing terms whose influence grows with the size Δ of the boundary layer [23, 26]. Nevertheless, the implementation of sliding mode control techniques is troublesome because of the large control effort usually needed to assure robustness properties, and the possibility that the so-called chattering phenomenon can arise [26]. In particular the latter may lead to large undesired oscillations which can damage the controlled system.

Recently invented, higher order sliding modes generalize the basic sliding mode idea. They are characterized by a discontinuous control acting on the higher order time derivatives of the sliding variable instead of influencing its first time derivative, as it happens in standard sliding modes. Preserving the main advantages of the original approach with respect to robustness and easiness of implementation, at the same time, they totally remove the chattering effect and guarantee even higher accuracy in presence of plant and/or control devices imperfections [20, 19].

In particular, the sliding order characterizes the smoothness degree of the system dynamics in the vicinity of the sliding mode. Generally speaking, the task in sliding mode control is to keep the system on the sliding manifold defined by the equality of the sliding variable to zero. The sliding order is defined as the number of continuous, and of course null when in sliding mode, total time derivatives of the sliding variable, the zero one included. Furthermore an r-th order real sliding mode provides a sliding precision, i.e., the size of the boundary layer of the sliding manifold, up to the r-th order with respect to plant imperfections which result in delays in the the switchings [20, 19].

Higher order sliding modes can appear naturally when fast dynamic actuators are used in VSS applications [19]. Indeed, when some dynamic actuator is present between a relay and the controlled process, the switching is moved to higher order derivatives of the actual plant input. As a result some new modes appear providing for exact satisfaction of the constraint, and actually being higher order sliding modes. This phenomenon reveals itself by the spontaneous disappearance of chattering in VSS.

Some controllers which are able to induce asymptotically stable sliding behaviours of any order have been presented in the literature [13, 10, 25, 1]. In this contribution we deal with the first generation of specifically designed controllers which give rise to finite time second order sliding behaviours in VSS [13, 14, 20, 3, 4]. In particular, the most effective second order sliding algorithms presented by the authors are described.

2 Second Order Sliding Modes

VSS dynamics is characterized by differential equations with a discontinuous right-hand side. According to the definition by Filippov, any discontinuous differential equation $\dot{\mathbf{x}} = v(\mathbf{x})$, where $\mathbf{x} \in I\!R^n$ and v is a locally bounded measurable vector function, is replaced by an equivalent differential inclusion $\dot{\mathbf{x}} \in V(\mathbf{x})$ [15]. In the simplest case, when v is continuous almost everywhere, $V(\mathbf{x})$ is the convex closure of the set of all possible limits of $v(z)$ as $\mathbf{z} \to \mathbf{x}$, while $\{\mathbf{z}\}$ are continuity points of $v(\mathbf{z})$. Any solution of the differential equation is defined as an absolute continuous function $\mathbf{x}(t)$ satisfying the differential inclusion almost everywhere. The extension to the non–autonomous case is straightforward by considering time t as an element of vector $\mathbf{x}$.

Consider an uncertain single–input nonlinear system whose dynamics is defined by the differential system

$$\dot{\mathbf{x}}(t) = f(\mathbf{x}(t), t, u(t)) \tag{1}$$

where $\mathbf{x} \in X \subset I\!R^n$ is the state vector, $u \in U \subset I\!R$ is the bounded input, t is the independent variable time, and $f : I\!R^{n+2} \rightarrow I\!R^n$ is a sufficiently smooth uncertain vector function. Assume that the control task is fulfilled by constraining the state trajectory on a proper sliding manifold in the state space defined by the vanishing of a corresponding sliding variable $s(t)$, i.e.,

$$s(t) = s(\mathbf{x}(t), t) = 0 \tag{2}$$

where $s : I\!R^{n+1} \rightarrow I\!R$ is a known single valued function such that its total time derivatives $s^{(k)}$, $k = 0, 1, \ldots, r-1$, along the system trajectories exist and are single valued functions of the system state $\mathbf{x}$. The latter assumption means that discontinuity does not appear in the first $r-1$ total time derivatives of the sliding variable s.

Definition 1. Given the constraint function (2), its r–th order sliding set is defined by the r equalities

$$s = \dot{s} = \ddot{s} = \ldots = s^{(r-1)} = 0 \tag{3}$$

which constitute an r–dimensional condition on the system dynamics [19]. □

Definition 2. Let the r–th order sliding set (3) be not empty, and assume that it is locally an integral set in the Filippov sense, i.e., it consists of Filippov's trajectories of the discontinuous dynamic system. The corresponding motion of system (1) satisfying (3) is called r–th order sliding mode with respect to the constraint function s. □

For shortening purposes, the words "r–th order sliding" will be abridged below to "r–sliding".

On the basis of Definition 2 system (1) evolves featuring a 2–sliding mode on the sliding manifold (2) iff its state trajectories lie on the intersection of the two manifolds $s = 0$ and $\dot{s} = 0$ in the state space. It is easy to see that at 2–sliding points Filippov's set of admissible velocities lies in the tangential space to $s = 0$ (Fig.1).

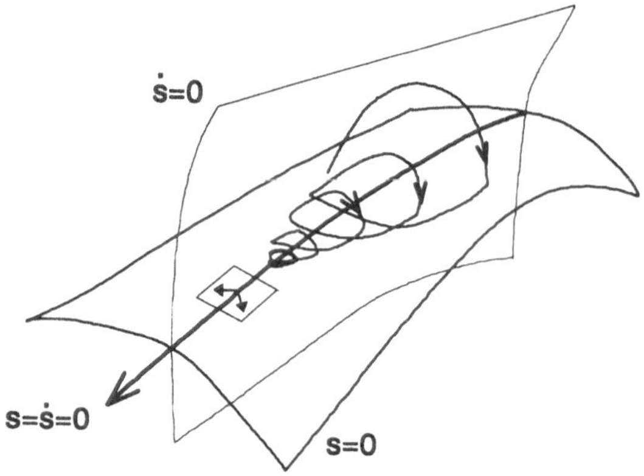

Fig. 1. Second order sliding mode trajectory

2.1 The Sliding Variable Dynamics

Consider system (1) and assume that the control task is fulfilled by its zero dynamics [16] with respect to a properly defined output variable $s(\mathbf{x}, t)$ as in (2). By differentiating the sliding variable s twice, the following relationships are derived

$$\dot{s}(t) = \dot{s}(\mathbf{x}(t), t, u(t)) = \frac{\partial}{\partial t} s(\mathbf{x}, t) + \frac{\partial}{\partial \mathbf{x}} s(\mathbf{x}, t) f(\mathbf{x}, t, u) \tag{4}$$

$$\begin{aligned} \ddot{s}(t) &= \ddot{s}(\mathbf{x}(t), t, u(t), \dot{u}(t)) = \\ &= \tfrac{\partial}{\partial t} \dot{s}(\mathbf{x}, t, u) + \tfrac{\partial}{\partial \mathbf{x}} \dot{s}(\mathbf{x}, t, u) f(\mathbf{x}, t, u) + \tfrac{\partial}{\partial u} \dot{s}(\mathbf{x}, t, u) \dot{u}(t) \end{aligned} \tag{5}$$

Depending on the relative degree [16] of the nonlinear SISO system (1), (2), different cases should be considered

a) relative degree $p = 1$, i.e., $\frac{\partial}{\partial u} \dot{s} \neq 0$;
b) relative degree $p \geq 2$, i.e., $\frac{\partial}{\partial u} s^{(i)} = 0$ $(i = 1, 2, \ldots, p-1)$, $\frac{\partial}{\partial u} s^{(p)} \neq 0$.

In case a) the classical approach to VSS by means of 1–sliding mode control solves the control problem, nevertheless 2–sliding mode control can be used in order to avoid chattering as well. In fact, if the time derivative of the plant control, $\dot{u}(t)$, is considered as the actual control variable, the 2–sliding mode control approach allows the definition of a discontinuous control $\dot{u}$ steering both the sliding variable s and its time derivative $\dot{s}$ to zero, so that the plant control u is continuous and chattering is avoided [20, 6].

In case b), because of the system uncertainties and the possible not complete availability of the system state, the r–sliding mode approach, with $r \geq p$, is the most appropriate control technique. In this contribution we limit our interest to 2–sliding mode control problems, and, therefore, to systems with $p = 2$. These class of control problems can arise when the output control problem of systems with relative degree two is faced [4], or when the differentiation of a smooth signal is considered [21, 7].

Chattering avoidance; the generalized constraint fulfillment problem
When considering classical VSS the control variable $u(t)$ is a feedback-designed relay output. The most direct application of 2–sliding mode control is that of attaining the sliding motion on the sliding manifold (2) by means of a continuous bounded input $u(t)$. This means that $u(t)$ can be considered as a continuous output of a suitable first–order dynamical system which can be driven by a proper discontinuous signal. Such first–order dynamics can be either inherent to the control device [19] or specially introduced for chattering elimination purposes [6], and the feedback control signal generated by the 2–sliding control algorithm is mostly the time derivative of the plant input $u(t)$.

Consider system (1) and the constraint function (2); assume that f and s are respectively $\mathcal{C}^1$ and $\mathcal{C}^2$ functions, and that the only available information consists of the current values of t, $u(t)$, $s(\mathbf{x},t)$ and, possibly, of the sign of the time derivative of the latter. The control goal for a 2–sliding mode controller is that of steering s to zero in a finite time by means of a control $u(t)$ continuously dependent on time.

In order to define the control problem the following conditions must be assumed:

1) $U = \{u : |u| \leq U_{\mathrm{M}}\}$, where $U_{\mathrm{M}} > 1$ is a real constant; furthermore the solution of (1) is well defined for all t, provided that $u(t)$ is continuous and $u(t) \in U \; \forall t$.
2) There exists $u_1 \in (0,1)$ such that for any continuous function $u(t)$ with $|u(t)| > u_1$, there is t_1 such that $s(t)u(t) > 0$ for each $t > t_1$. Hence, the control $u(t) = -\mathrm{sign}[s(t_0)]$, where t_0 is the initial value of time, provides hitting the manifold (2) in finite time.
3) Let $\dot{s}(\mathbf{x},t,u)$ be the total time derivative of the sliding variable $s(\mathbf{x},t)$ as defined in (4). There are positive constants s_0, $u_0 < 1$, Γ_{m}, Γ_{M} such that if $|s(\mathbf{x},t)| < s_0$ then

$$\Gamma_{\mathrm{m}} \leq \frac{\partial}{\partial u}\dot{s}(\mathbf{x},t,u) \leq \Gamma_{\mathrm{M}} \quad ,\forall u \in U, \mathbf{x} \in X \tag{6}$$

and the inequality $|u| > u_0$ entails $\dot{s}u > 0$.
4) There is a positive constant Φ such that within the region $|s| < s_0$ the following inequality holds $\forall t, \mathbf{x} \in X, u \in U$

$$\left|\frac{\partial}{\partial t}\dot{s}(\mathbf{x},t,u) + \frac{\partial}{\partial \mathbf{x}}\dot{s}(\mathbf{x},t,u)f(\mathbf{x},t,u)\right| \leq \Phi \tag{7}$$

Condition 2 means that starting from any point of the state space it is possible to define a proper control $u(t)$ steering the sliding variable within a set such that the boundedness conditions on the sliding dynamics defined by conditions 3 and 4 are satisfied. In particular such conditions state that the second time derivative of the sliding variable s, evaluated with fixed values of the control u, is uniformly bounded in a bounded domain.

It follows from the theorem on implicit function that there is a function $u_{\mathrm{eq}}(t,\mathbf{x})$, which can be viewed as Utkin's equivalent control [26], satisfying the equation $\dot{s} = 0$. Once $s = 0$ is achieved, the control $u = u_{\text{eq}}(t, \mathbf{x})$ would provide for the exact constraint fulfillment. Conditions 3 and 4 mean that $|s| < s_0$ implies $|u_{\text{eq}}| < u_0 < 1$, and that the velocity of the u_{eq} changes is bounded. This opens the possibility to approximate u_{eq} by a Lipschitzian control.

Note that the unit upper bound for u_0 and u_1 can be considered as a scaling factor, and somewhere in the following it is not explicitly considered. Note also that linear dependence on control u is not required and that the usual form of the uncertain systems dealt with by the VSS theory, i.e., systems affine in the control, is a special case of the considered systems (1), (2).

Systems with relative degree two The conditions to define the control problem for system (1), (2) in case of relative degree two could be derived from those above by considering the variable u as a state variable and $\dot{u}$ as the actual control. Nevertheless, they will be re–stated in case the system dynamics is affine in the control law, i.e.,

$$f(\mathbf{x}, t, u) = a(\mathbf{x}, t) + b(\mathbf{x}, t)u(t) \tag{8}$$

where $a : \mathbb{R}^{n+1} \rightarrow \mathbb{R}^n$ and $b : \mathbb{R}^{n+1} \rightarrow \mathbb{R}^n$ are sufficiently smooth uncertain vector functions.

By substituting (8) in (4) and (5) the following relationships are derived

$$\dot{s}(t) = \frac{\partial}{\partial t}s(\mathbf{x}, t) + \frac{\partial}{\partial \mathbf{x}}s(\mathbf{x}, t)a(\mathbf{x}, t) + \frac{\partial}{\partial \mathbf{x}}s(\mathbf{x}, t)b(\mathbf{x}, t)u(t) \tag{9}$$

$$\begin{aligned} \ddot{s}(t) = {} & \tfrac{\partial^2}{\partial t^2}s(\mathbf{x}, t) + \left[\tfrac{\partial^2}{\partial t \partial \mathbf{x}}s(\mathbf{x}, t) + a^T(\mathbf{x}, t)\tfrac{\partial^2}{\partial \mathbf{x}^2}s(\mathbf{x}, t)\right. \\ & \left. + \tfrac{\partial}{\partial \mathbf{x}}s(\mathbf{x}, t)\tfrac{\partial}{\partial \mathbf{x}}a(\mathbf{x}, t)\right]\left[a(\mathbf{x}, t,) + b(\mathbf{x}, t)u(t)\right] \end{aligned} \tag{10}$$

In order to define the control problem the following conditions must be assumed:

I) Let (9) and (10) be, respectively, the first and second total time derivative of the sliding variable $s(\mathbf{x}, t)$, such that $\dot{s} = \dot{s}(\mathbf{x}, t)$, $\ddot{s} = \ddot{s}(\mathbf{x}, t, u)$, i.e.,

$$\begin{gathered} \tfrac{\partial}{\partial \mathbf{x}}s(\mathbf{x}, t)b(\mathbf{x}, t) \equiv 0 \\ \left[\tfrac{\partial^2}{\partial t \partial \mathbf{x}}s(\mathbf{x}, t) + a^T(\mathbf{x}, t)\tfrac{\partial^2}{\partial \mathbf{x}^2}s(\mathbf{x}, t) + \tfrac{\partial}{\partial \mathbf{x}}s(\mathbf{x}, t)\tfrac{\partial}{\partial \mathbf{x}}a(\mathbf{x}, t)\right] b(\mathbf{x}, t) \neq 0 \\ \forall t, u \in U, \mathbf{x} \in X \end{gathered} \tag{11}$$

$U = \{u : |u| \leq U_{\text{M}}\}$, where U_{M} is a real constant, so that $u(t)$ is a bounded discontinuous function of time; furthermore, the differential equation with discontinuous right–hand side (1) , (8) admits solutions in the Filippov sense on the 2–sliding manifold $s = \dot{s} = 0$ for all t.

II) There exists $u_1 \in (0, U_{\text{M}})$ such that for any continuous function $u(t) \in U$ with $|u(t)| > u_1$, there is t_1 such that $\dot{s}u(t) > 0$ for each $t > t_1$. Hence, the control $u(t) = -U_{\text{M}}\text{sign}[\dot{s}(t_0)]$, where t_0 is the time initial value, provides hitting the manifold $\dot{s} = 0$ in finite time.

III) There are positive constants s_0, Γ_m, Γ_M such that if $|s(\mathbf{x},t)| < s_0$ then

$$\Gamma_m \le \left[\frac{\partial^2}{\partial t \partial \mathbf{x}} s(\mathbf{x},t) + a^T(\mathbf{x},t)\frac{\partial^2}{\partial \mathbf{x}^2} s(\mathbf{x},t) + \frac{\partial}{\partial \mathbf{x}} s(\mathbf{x},t)\frac{\partial}{\partial \mathbf{x}} a(\mathbf{x},t)\right] b(\mathbf{x},t) \le \Gamma_M \qquad \forall t > t_1, \mathbf{x} \in X \tag{12}$$

IV) There is a positive constant Φ such that within the region $|s| < s_0$ the following inequality holds $\forall t > t_1, \mathbf{x} \in X$

$$\left|\frac{\partial^2}{\partial t^2} s(\mathbf{x},t) + \left[\frac{\partial^2}{\partial t \partial \mathbf{x}} s(\mathbf{x},t) + a^T(\mathbf{x},t)\frac{\partial^2}{\partial \mathbf{x}^2} s(\mathbf{x},t) + \frac{\partial}{\partial \mathbf{x}} s(\mathbf{x},t)\frac{\partial}{\partial \mathbf{x}} a(\mathbf{x},t)\right] a(\mathbf{x},t,)\right| \le \Phi \tag{13}$$

Condition I states that the differential equation with discontinuous right–hand side (1), (8) admits solution in the Filippov sense on a 2–sliding manifold, while condition II means that starting from any point of the state space it is possible to define a proper control $u(t)$ steering the sliding variable within a set such that the boundedness conditions on the sliding dynamics defined by III and IV are satisfied. In particular, they state that the second time derivative of the sliding variable s is uniformly bounded in a bounded domain, for all $u \in U$.

The auxiliary problem The sliding variable s can be considered as a suitable output variable of the uncertain system (1), and the control aim is that of steering this output to zero in a finite time. The 2–sliding mode approach allows for the finite time stabilization of both the output variable s and its time derivative $\dot{s}$ by defining a suitable discontinuous control function which can be either the actual control plant or its time derivative, depending on the system relative degree.

Consider the local coordinates (y_1, y_2), where $y_1 \equiv s$, $y_2 \equiv \dot{s}$, on the base of the previous definitions and conditions, apart from a proper initialization phase, the 2–sliding mode control problem is equivalent to the finite time stabilization problem for the following uncertain second order system

$$\begin{cases} \dot{y}_1(t) = y_2(t) \\ \dot{y}_2(t) = \varphi(\mathbf{y}(t),t) + \gamma(\mathbf{y}(t),t)v(t) \end{cases} \tag{14}$$

with $y_2(t)$ unmeasurable but with a possibly known sign, and $\varphi(\mathbf{y}(t),t)$ and $\gamma(\mathbf{y}(t),t)$ uncertain functions such that

$$\begin{array}{c} |\varphi(\mathbf{y}(t),t)| \le \Phi \\ 0 < \Gamma_m \le \gamma(\mathbf{y}(t),t) \le \Gamma_M \end{array} \quad \mathbf{y} \in Y \subset \mathbb{R}^2 \tag{15}$$

in which, referring to the previous notation, v is the actual control plant if system (1) has relative degree $p = 2$, with respect to y_1, or its time derivative if $p = 1$, and Y is a bounded region within which the boundedness of the uncertain sliding dynamics is assured, i.e., $|y_1| \le s_0$.

Since y_2 is not available and $\varphi(\mathbf{y}(t),t)$, $\gamma(\mathbf{y}(t),t)$ are uncertain, this problem is not easily solvable by consolidated theory. It has been solved, recently, in previous papers by the authors [20, 3, 4]. Hereafter, a synthetic and qualitative presentation of the solution procedure is provided for the readers' convenience.

Consider a double integrator $\dot{y}_1 = y_2$, $\dot{y}_2 = v$, that is $\varphi(\mathbf{y}(t), t)$, $\gamma(\mathbf{y}(t), t)$ are perfectly known and evaluable functions. In this case, there are algorithms capable of causing the finite time reaching of the origin. One, in particular, can be derived by the well-known time optimal bang-bang control approach according to the following propositions.

Proposition 3. *The time optimal switching line [17, 18]*

$$y_1(t) + \frac{y_2(t)|y_2(t)|}{2V_{\mathrm{M}}} = 0$$

when $y_1(0)y_2(0) \geq 0$, can be replaced by

$$y_1(t) - \frac{1}{2}y_1(t_{\mathrm{M}_1}) = 0$$

with t_{M_1} such that $y_2(t_{\mathrm{M}_1}) = 0$. □

Proposition 4. *When $y_1(t)y_2(t) < 0$, the initialization control*

$$v(t) = -V_{\mathrm{M}}\mathrm{sign}\left(y_1(t) - \frac{1}{2}y_1(0)\right) = 0$$

guarantees that the condition $y_1(t)y_2(t) > 0$ is achieved in finite time. □

The combination of the initialization phase with the use of the modified optimal switching line steers the state trajectory to the origin of the state plane with at most two commutations, instead of the single commutation characterizing the optimal bang-bang control (Fig.2). In this sense, such a combination gives rise to a suboptimal strategy.

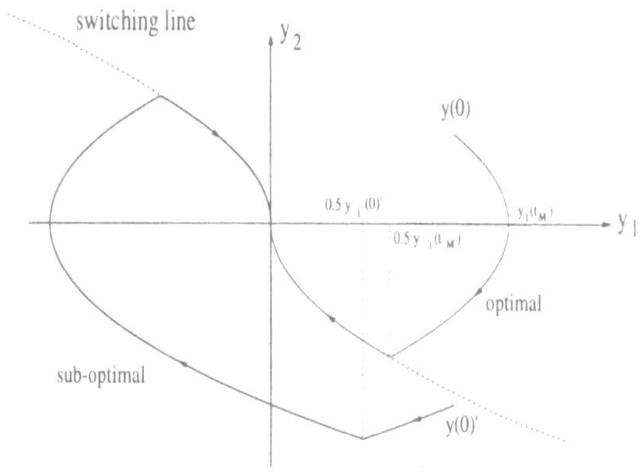

Fig. 2. Modified time optimal switchings

Another way to steer the state of a double integrator to zero in finite time is that produced by the so-called twisting algorithm by Levant [20], i.e.

$$v(t) = -\alpha(t) V_{\mathrm{M}} \mathrm{sign}(y_1(t))$$

with

$$\alpha(t) = \begin{cases} 1 & \text{if } y_1(t)y_2(t) > 0 \\ \alpha^* \in (0,1) & \text{if } y_1(t)y_2(t) < 0 \end{cases}$$

With reference to this algorithm the convergence to zero in finite time can be proved following a procedure which can be used also to deal with the case of perturbed couple of integrators. More specifically, it can be verified that the sequence of values $y_1(t_{\mathrm{M}_i})$ is contractive, that is

$$\frac{y_1(t_{\mathrm{M}_{i+1}})}{y_1(t_{\mathrm{M}_i})} \leq q < 1$$

and that $\lim_{i \to \infty} y_1(t_{\mathrm{M}_i}) = 0$. Moreover, the reaching time is a series of positive elements upperbounded by a geometric series with ratio strictly less than one. Therefore, $\lim_{i \to \infty} t_{\mathrm{M}_i} = T < \infty$.

Now consider system (14): it can be viewed as a double integrator with perturbation uncertain terms. The solution to the perturbed case proposed in [4] is based on the suboptimal algorithm indicated through Proposition 1 and 2. The point is to prove that a contractive behaviour between two successive singular points is achieved despite the uncertainties φ and γ. The analysis has been carried out by considering, as the worst case, that in which uncertainties act always against the attainment of the contraction.

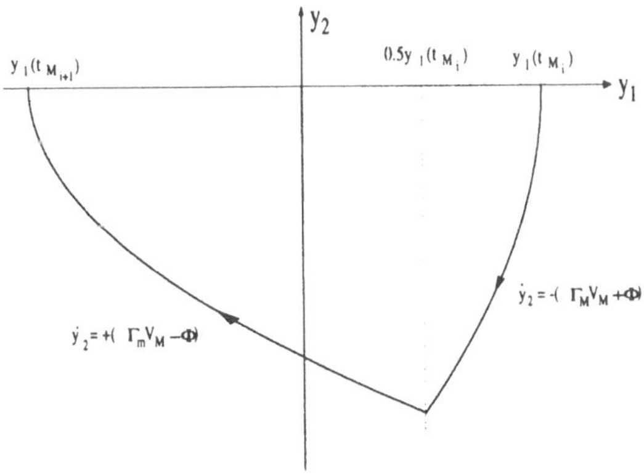

Fig. 3. Perturbed sub time optimal trajectory (worst case)

Assume that, for a certain choice of V_{M}, the worst case behaviour, sketched in Fig.3, is not characterized by the contraction effect. There are three ways to achieve the prefixed control objective:

a) to increase the control amplitude V_M, that is to reduce $\frac{\Gamma_M V_M + \Phi}{\Gamma_m V_M - \Phi}$ (Fig.4), but this is effective only if $3\Gamma_m > \Gamma_M$ [3];
b) to use an asymmetric commutation logic as in the twisting algorithm, (Fig.5);
c) to anticipate the commutation at the moment when $y_1 = \beta y_1(t_{M_i})$, with $\beta \in [\frac{1}{2}, 1)$ (Fig.6).

With each of the above actions, the worst case turns out to be characterized by a finite time damped oscillatory time response. A particular possible behaviour which has to be further considered is that which corresponds to uncertainties producing series of two successive singular points with the same sign, (Fig.7). In this case, $y_1(t_{M_i}) y_1(t_{M_{i+1}}) > 0$, and the strategy is to change the sign of the control not only when $y_1(t) = \beta y_1(t_{M_i})$, but also at $t_{M_{i+1}}$.

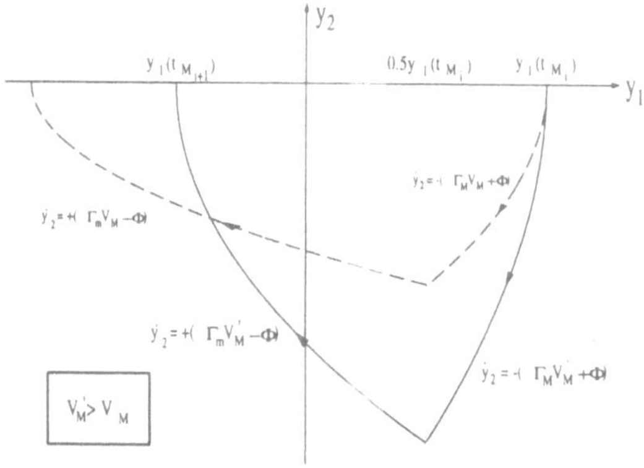

Fig. 4. Perturbed sub time optimal trajectory with increased control authority (worst case)

2.2 The Initialization Phase

In most cases the bounds (15) are not global even if the uncertain functions φ and γ are bounded in any bounded domain, i.e., $Y \subset \mathbb{R}^2$. Furthermore it can be assumed, without loss of generality, that function s is such that

$$\|\mathbf{x}\| \leq N \Longleftrightarrow \|[s, \dot{s}]\| \leq M$$

where N and M are real nonegative constants, so that bounded domains of the state variables correspond to bounded domains of the sliding variable and its time derivative, and vice–versa.

Usually 2–sliding control algorithms are defined with respect to constant bounds of the uncertain dynamics (14) and are such that y_1 and y_2 converge to the origin of the phase state plane, and, therefore, once set Y is reached, the y_1 and y_2 trajectories never leave it. The above considerations imply that, apart

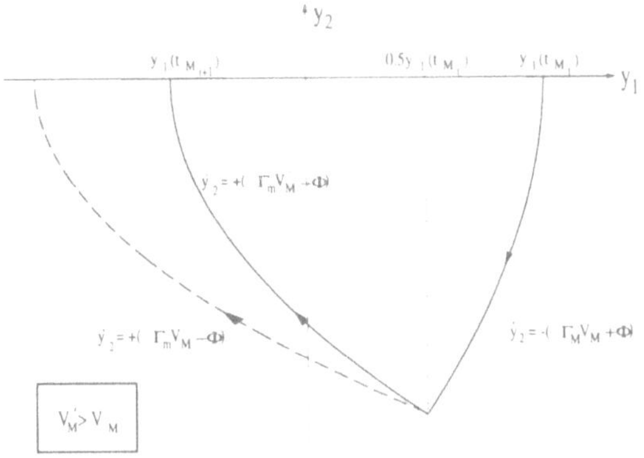

Fig. 5. Perturbed sub time optimal trajectory with control magnitude modulation (worst case)

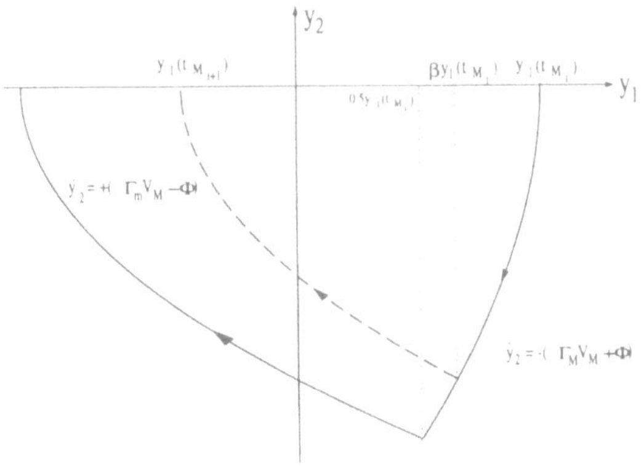

Fig. 6. Perturbed sub time optimal trajectory with switching antecipation (worst case)

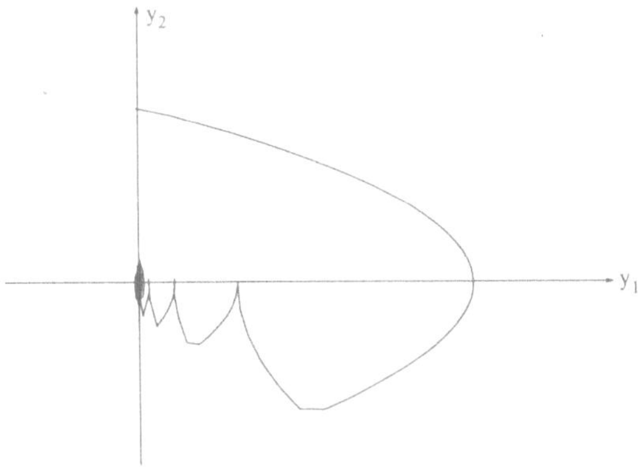

Fig. 7. Perturbed sub time optimal trajectory with constant signs of y_1 and y_2

from the case in wich bounds (15) are gobal, an initialization phase in which the domain Y is reached in finite time must be implemented.

The control law during the initialization phase depends on the class of system to which a 2-sliding control strategy is applied:

a) systems with relative degree one;
b) systems with relative degree two.

Indeed, in case a) the possibility of using the plant control u directly allows to assign the sign of the y_2 variable, and not only that of its time derivative $\dot{y}_2$ as in case b).

In case of relative degree one systems, 2–sliding mode control can be used in order to avoid chattering in VSS; this means that the plant control u is computed by integration of the discontinuous control generated by a 2–sliding controller. Due to the availability of both the control and its time derivative two different strategies are possible in order to reach the domain Y.

If condition 2 is verified then control u is able to assign the sign of the y_2 variable. By choosing u such that $y_1 y_2 \leq -k^2$, a point of the ordinate axis of the $y_1 O y_2$ plane, i.e., $y_1 = 0$, is reached in finite time, at least by using

$$u(t) = -U_{\mathrm{M}} \mathrm{sign}(y_1(t_0)) \; t \geq t_0$$

During the initialization phase a time varying control magnitude can be used if some knowledge about a function of the available signals which upperbounds the uncertain dynamics is available. As an example, assume that the sliding surface is a linear manifold in the state space

$$s(\mathbf{x}) = C\mathbf{x} \tag{16}$$

where C is a n–dimensional row vector whose elements define a Hurwitz polinomial. Furthermore the uncertain system dynamics (1) (8) is such that

$$\begin{aligned} &||a(\mathbf{x},t)|| \leq A + K_{\mathrm{a}}||\mathbf{x}|| \\ &0 < B_{\mathrm{m}} \leq b_i(\mathbf{x},t) \leq B + K_{\mathrm{b}}||\mathbf{x}|| \; i = 1, 2, \ldots, n \end{aligned} \tag{17}$$

where A, B, B_{m}, K_{a} and K_{b} are known positive constants. In this case an initialization phase with the following control law guarantees the finite time reaching of the ordinate axis of the $y_1 O y_2$ plane

$$u(\mathbf{x}, y_1) = -\frac{A + K_{\mathrm{a}}||\mathbf{x}|| + h^2}{B_{\mathrm{m}}} \mathrm{sign}(y_1)$$

When systems with relative degree two are considered, the plant control v acts on the y_2 time derivative so that the previous approach cannot be applied. If condition II is verified then control v is able to assign the sign of the time derivative of the y_2 variable. By choosing v such that $y_2 \dot{y}_2 \leq -k^2$, a point of the abscissa axis of the $y_1 O y_2$ plane, i.e., $y_2 = 0$, is reached in finite time, at least by using

$$v(t) = -V_{\mathrm{M}} \mathrm{sign}(y_2(t_0)) \; t \geq t_0$$

Also in this case the knowledge of bound functions both for the uncertain dynamics and its partial derivatives can allow to design time varying control magnitudes during the initialization phase. Assume that the uncertain dynamics (14) is such that the following bounds are satisfied

$$\begin{array}{l} |\varphi(\mathbf{y}(t),t)| \leq P + \mathcal{P}(y_1) + (Q + \mathcal{Q}(y_1))\,||\mathbf{y}|| \\ 0 < \Gamma_{\mathrm{m}} \leq \gamma(\mathbf{y}(t),t)v(t) \leq R + \mathcal{R}(y_1)||\mathbf{y}|| \end{array} \tag{18}$$

where P, Q, and R are nonegative constants, and $\mathcal{P}$, $\mathcal{Q}$ and $\mathcal{R}$ are positive semi-definite functions. In this case an initialization phase with the following control law guarantees the finite time reaching of the abscissa axis of the y_1Oy_2 plane [2]

$$\begin{array}{c} v(y_1) = -V_{in}(y_1)\mathrm{sign}(y_2(t_0)) \\ V_{in}(y_1) = \frac{1+h^2}{\Gamma_{\mathrm{m}}}\left(P + \mathcal{P}(y_1) + (Q + \mathcal{Q}(y_1))\left\|[y_1; y_2(t_0)]^T\right\|\right) \end{array}$$

Often $\mathrm{sign}(y_2(t_0))$ is unknown but it can be evaluated by the first difference of the available quantity y_1, that is $\mathrm{sign}(y_2(t_0))$ is estimated by $\mathrm{sign}(y_1(t_0+\delta) - y_1(t_0))$, where δ is an arbitray small time interval.

The last initialization procedure is effective if the chattering avoidance is dealt with as well. Furthermore, in this case, it could be possible to assign the initial sign of the y_2 variable by a proper choice of the initial value of the available plant control u [6].

Once the abscissa or the ordinate axis of the y_1Oy_2 plane is reached, the constant bounds (15) exist and can be evaluated if the global uncertainty bounds have a first order dependence on the not available signal y_2 [2, 3, 6]. If $||\mathbf{y}||$ in (18) is an usual m–order norm defined in an Euclidean space, i.e., $||\mathbf{y}|| = [|y_1|^m + |y_2|^m]^{\frac{1}{m}}$ $(m \geq 1)$, the previous condition is satisfied, so that after a finite time transient a subspace, in which any of the controllers described in the sequel can be applied, is reached.

3 Second Order Sliding Controllers

Each of the following controllers is characterized by few constant parameters. These parameters have to be tuned in order to achieve the control goal for the considered class of processes and sliding functions which will be defined in terms of the constants Φ, Γ_{m}, Γ_{M} and s_0. By increasing the constants Φ, Γ_{m}, Γ_{M} and reducing s_0 at the same time, it is possible to enlarge the class of controlled systems too. Such algorithms are obviously insensitive to any model perturbations and external disturbances which do not move the dynamic system from the given class.

It must be pointed out that, given the system to be controlled and the desired sliding manifold, it is possible to define the above constants by uncertainty maximization. Nevertheless this evaluation procedure usually defines very large parameters, and, as a consequence, very large control signals as well, which are not really needed for controlling the real system. In practice, the convergence conditions for the control algorithms are only sufficient but not necessary and the tuning of the controller is often better made heuristically.

3.1 Twisting Algorithm

This algorithm is characterized by a twisting around the origin of the y_1Oy_2 2–sliding plane (Fig.8). The finite time convergence to the origin of the plane is due to the switching of the control amplitude between two different values such that the abscissas and ordinates axes are crossed nearer and nearer to the origin. The control amplitude commutes at each axis crossing, and the sign of the sliding variable time derivative y_2 is needed.

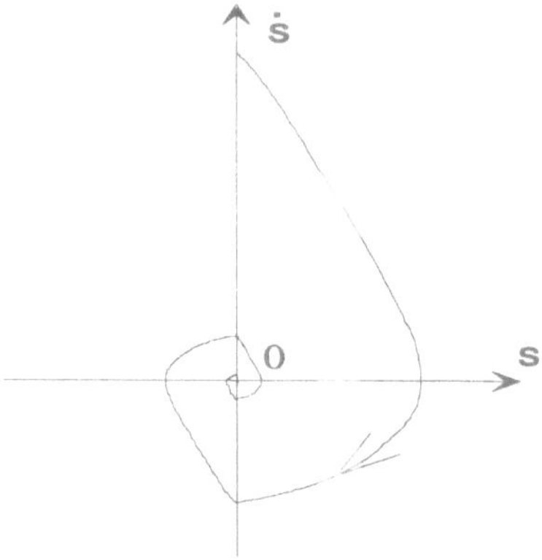

Fig. 8. Twisting algorithm phase trajectory

The control algorithm is defined by the following control law [20], in which the condition on $|u|$ must be taken into account when considering the chattering avoidance problem,

$$v(t) = \begin{cases} -u & if \ \ |u| > 1 \\ -V_{\mathrm{m}}\mathrm{sign}(y_1) & if \ \ y_1 y_2 \leq 0; \ |u| \leq 1 \\ -V_{\mathrm{M}}\mathrm{sign}(y_1) & if \ \ y_1 y_2 > 0; \ |u| \leq 1 \end{cases} \tag{19}$$

and the corresponding sufficient conditions for the finite time convergence to the sliding manifold are [20]

$$\begin{cases} V_{\mathrm{M}} > V_{\mathrm{m}} \\ V_{\mathrm{m}} > \frac{4\Gamma_{\mathrm{M}}}{s_0} \\ V_{\mathrm{m}} > \frac{\Phi}{\Gamma_{\mathrm{m}}} \\ \Gamma_{\mathrm{m}} V_{\mathrm{M}} - \Phi > \Gamma_{\mathrm{M}} V_{\mathrm{m}} + \Phi \end{cases} \tag{20}$$

By taking into account the different limit trajectories arising from the uncertain dynamics (14) and evaluating the time intervals between subsequent crossings of the abscissa axis, it is possible to define the following upper bound for the convergence time

$$t_{\mathrm{tw}_\infty} \leq t_{M_1} + \Theta_{\mathrm{tw}} \frac{1}{1 - \theta_{\mathrm{tw}}} \sqrt{|y_{1_{M_1}}|} \tag{21}$$

where y_{1M_1} is the value of the y_1 variable at the first abscissa crossing in the y_1Oy_2 plane, t_{M_1} the corresponding time instant and

$$\begin{aligned}\Theta_{\mathrm{tw}} &= \sqrt{2}\frac{\Gamma_{\mathrm{m}}V_{\mathrm{M}}+\Gamma_{\mathrm{M}}V_{\mathrm{m}}}{(\Gamma_{\mathrm{m}}V_{\mathrm{M}}-\Phi)\sqrt{\Gamma_{\mathrm{M}}V_{\mathrm{m}}+\Phi}}\\ \theta_{\mathrm{tw}} &= \sqrt{\frac{\Gamma_{\mathrm{M}}V_{\mathrm{m}}+\Phi}{\Gamma_{\mathrm{m}}V_{\mathrm{M}}-\Phi}}\end{aligned}$$

In many practical cases the y_2 variable is completely unmeasurable, then its sign can be estimated by the sign of the first difference of the available sliding variable y_1 in a time interval δ, i.e., $\mathrm{sign}[y_2(t)]$ is estimated by $\mathrm{sign}[y_1(t)-y_1(t-\delta)]$. In this case only 2–sliding precision with respect to the measurement time interval is provided, and the size of the boundary layer of the sliding manifold is $\Delta \sim \mathcal{O}(\delta^2)$ [20].

3.2 Sub–Optimal Algorithm

This 2–sliding control algorithm derives from a sub–optimal feedback implementation of the classical time optimal control for a double integrator. The trajectories on the y_1Oy_2 plane are confined within limit parabolic arcs which include the origin, so that both twisting and jumping (in which y_1 and y_2 do not change sign) behaviours are allowed (Fig.9)

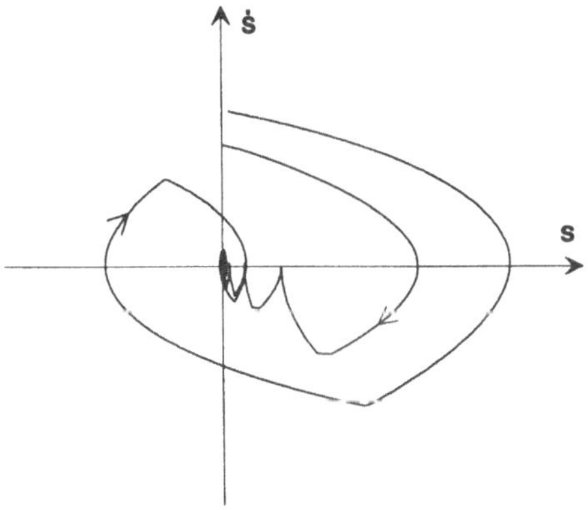

Fig. 9. Sub–Optimal algorithm phase trajectories

Apart from a possible initialization phase [2, 3, 6], the control algorithm is defined by the following control law

$$\begin{aligned}v(t) &= -\alpha(t)V_{\mathrm{M}}\mathrm{sign}(y_1(t)-\tfrac{1}{2}y_{1\mathrm{M}})\\ \alpha(t) &= \begin{cases}\alpha^* & if\ \ [y_1(t)-\frac{1}{2}y_{1\mathrm{M}}][y_{1\mathrm{M}}-y_1(t)]>0\\ 1 & if\ \ [y_1(t)-\frac{1}{2}y_{1\mathrm{M}}][y_{1\mathrm{M}}-y_1(t)]\le 0\end{cases}\end{aligned} \tag{22}$$

where y_{1_M} represents the last extremal value of the $y_1(t)$ function, i.e., the last local maximum, local minimum or horizontal flex point of $y_1(t)$. The corresponding sufficient conditions for the finite time convergence to the sliding manifold are [3]

$$\begin{cases} \alpha^* \in (0,1] \cap (0, \frac{3\Gamma_m}{\Gamma_M}) \\ V_M > \max\left(\frac{\Phi}{\alpha^*\Gamma_m}, \frac{4\Phi}{3\Gamma_m - \alpha^*\Gamma_M}\right) \end{cases} \tag{23}$$

Also in this case an upper bound for the convergence time can be defined [3]

$$t_{\text{opt}_\infty} \leq t_{M_1} + \Theta_{\text{opt}} \frac{1}{1-\theta_{\text{opt}}} \sqrt{|y_{1_{M_1}}|} \tag{24}$$

where $y_{1_{M_1}}$ and t_{M_1} are those defined previously, and

$$\Theta_{\text{opt}} = \frac{(\Gamma_m + \alpha^*\Gamma_M)V_M}{(\Gamma_m V_M - \Phi)\sqrt{\alpha^*\Gamma_M V_M + \Phi}}$$
$$\theta_{\text{opt}} = \sqrt{\frac{\alpha^*\Gamma_M - \Gamma_m)V_M + 2\Phi}{2(\Gamma_m V_M - \Phi)}}$$

In [2, 3, 8] the effectiveness of the above algorithm was extended to larger classes of uncertain systems, while in [4] it was proved that in case of unit gain function the control law (22) can be simplified by setting $\alpha = 1$ and choosing $V_M > 2\Phi$.

The sub-optimal algorithm needs a device capable of detecting local maxima, local minima and horizontal flexes of the available sliding variable. In practical cases y_{1_M} can be estimated by checking the sign of the quantity $D(t) = [y_1(t-\delta) - y_1(t)]\, y_1(t)$, in which $\frac{\delta}{2}$ is the estimation delay. In this case the control amplitude V_M needs to belong to a finite set instead of a semi-infinite one, so that the second of (23) is modified into the following [5]

$$V_M \in \left(\max\left(\frac{\Phi}{\alpha^*\Gamma_m}, V_{M_1}(\delta, y_{1_M})\right), V_{M_2}(\delta; y_{1_M})\right) \tag{25}$$

where $V_{M_1} < V_{M_2}$ are the solutions of the second order algebraic equation

$$\left[(3\Gamma_m - \alpha^*\Gamma_M)\frac{V_{M_i}}{\Phi} - 4\right]\frac{y_{1_M}}{\Phi\delta^2} - \frac{V_{M_i}}{8\Phi}\left[\Gamma_m + \Gamma_M(2-\alpha^*)\right]\left(\Gamma_M\frac{V_{M_i}}{\Phi} + 1\right) = 0$$

In accordance with the definition of real higher order sliding mode, in the case of approximated evaluation of the y_{1_M} values the size of the boundary layer of the sliding manifold is $\Delta \sim \mathcal{O}(\delta^2)$, and it can be minimized by choosing V_M as follows [5]

$$V_M = \frac{4\Phi}{3\Gamma_m - \alpha^*\Gamma_M}\left[1 + \sqrt{1 + \frac{3\Gamma_m - \alpha^*\Gamma_M}{4\Gamma_M}}\right]$$

An extension of the real sub-optimal 2-sliding control algorithm to a class of sampled data systems characterized by a constant gain function in (14), i.e., $\gamma = 1$, was recently presented in [9]. A development of the cited digital controller is presented in the next chapter.

3.3 Super–Twisting Algorithm

This algorithm has been developed for the case of systems with relative degree one in order to avoid chattering in VSS. Also in this case the trajectories on the 2–sliding plane are characterized by twistings around the origin (Fig.10), but the continuous control law $u(t)$ is constituted by two terms. The first is defined by means of its discontinuous time derivative, while the other, which is present during the reaching phase only, is a continuous function of the available sliding variable.

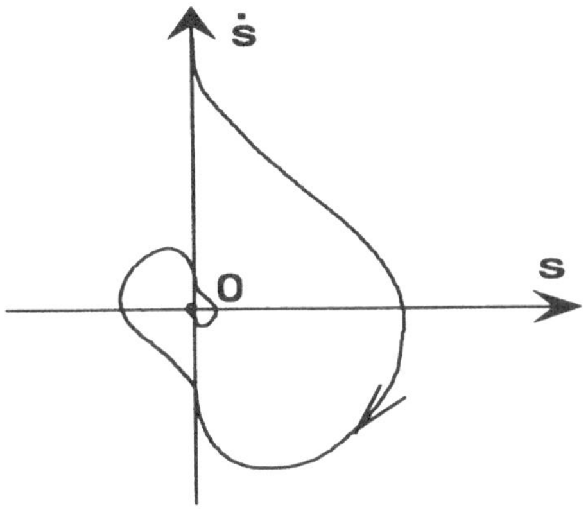

Fig. 10. Super–Twisting algorithm phase trajectory

The control algorithm is defined by the following control law [20]

$$\begin{aligned} u(t) &= u_1(t) + u_2(t) \\ \dot{u}_1(t) &= \begin{cases} -u & if \ \ |u| > 1 \\ -W \mathrm{sign}(y_1) & if \ \ |u| \leq 1 \end{cases} \\ u_2(t) &= \begin{cases} -\lambda |s_0|^\rho \mathrm{sign}(y_1) & if \ \ |y_1| > s_0 \\ -\lambda |y_1|^\rho \mathrm{sign}(y_1) & if \ \ |y_1| \leq s_0 \end{cases} \end{aligned} \tag{26}$$

and the corresponding sufficient conditions for the finite time convergence to the sliding manifold are [20]

$$\begin{cases} W > \frac{\Phi}{\Gamma_m} \\ \lambda^2 \geq \frac{4\Phi}{\Gamma_m^2} \frac{\Gamma_M (W + \Phi)}{\Gamma_m (W - \Phi)} \\ 0 < \rho \leq 0.5 \end{cases} \tag{27}$$

The above algorithm does not need the evaluation of the sign of the time derivative of the sliding variable. An exponentially stable 2–sliding mode appears if the control law (26) with $\rho = 1$ is used. The choice $\rho = 0.5$ assures that the maximum real sliding order for 2–sliding realization is achieved.

3.4 Drift Algorithm

When using the drift algorithm the phase trajectories on the 2-sliding plane are characterized by loops with constant sign of the sliding variable y_1 (Fig.11), furthermore it is characterized by the use of sampled values of the available signal y_1 with sampling period δ.

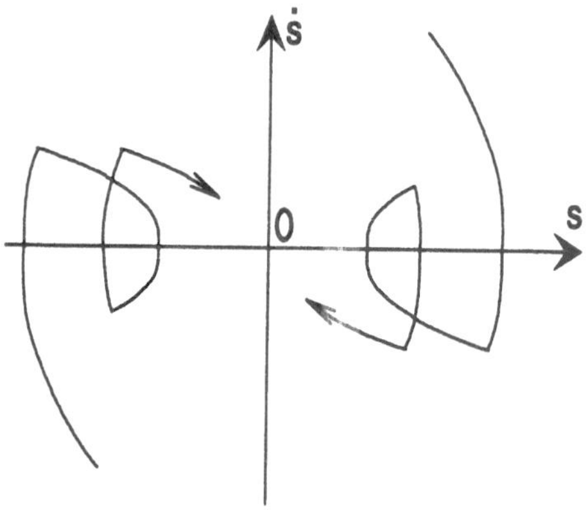

Fig. 11. Drift algorithm phase trajectories

The control algorithm is defined by the following control law, in which the condition on $|u|$ must be considered when dealing with the chattering avoidance problem. [14, 20]

$$v(t) = \begin{cases} -u & if \ \ |u| > 1 \\ -V_{\mathrm{m}}\mathrm{sign}(\Delta y_{1_i}) & if \ \ y_1 \Delta y_{1_i} \le 0; \ |u| \le 1 \\ -V_{\mathrm{M}}\mathrm{sign}(\Delta y_{1_i}) & if \ \ y_1 \Delta y_{1_i} > 0; \ |u| \le 1 \end{cases} \tag{28}$$

where V_{m} and V_{M} are suitable positive constants such that $V_{\mathrm{m}} < V_{\mathrm{M}}$ and $\frac{V_{\mathrm{M}}}{V_{\mathrm{m}}}$ sufficiently large, and $\Delta y_{1_i} = y_1(t_i) - y_1(t_i - \delta)$, $t \in [t_i, t_{i+1})$. The corresponding sufficient conditions for the convergence to the sliding manifold are rather cumbersome [14] and are omitted here for the sake of simplicity.

After substituting y_2 for Δy_{1_i} a first order sliding mode on $y_2 = 0$ would be achieved. This implies $y_1 = const.$, but, since an artificial switching time delay appears, we ensure a real sliding on y_2 with most of time spent in the set $y_1 y_2 < 0$, and therefore, $y_1 \to 0$. The accuracy of the real sliding on $y_2 = 0$ is proportional to the sampling time interval δ; hence the duration of the transient process is proportional to δ^{-1}.

Such an algorithm does not satisfy the definition of a real sliding algorithm [20] requiring the convergence time to be uniformly bounded with respect to δ. Let us consider a variable sampling time $\delta_{i+1}[y_1(t_i)] = t_{i+1} - t_i$, $i = 0, 1, 2, \ldots$ with $\delta = \max(\delta_{\mathrm{M}}, \min(\delta_{\mathrm{m}}, \eta|y_1(t_i)|^{\rho}))$ himself where $0.5 \le \rho \le 1$, $\delta_{\mathrm{M}} > \delta_{\mathrm{m}} > 0$, $\eta > 0$. Then with η, $\frac{V_{\mathrm{m}}}{V_{\mathrm{M}}}$ sufficiently small and V_{m} sufficiently large the drift algorithm constitutes a second order real sliding algorithm with respect to $\delta \to 0$. This algorithm has no overshoot if parameters are chosen properly [14].

3.5 Algorithm with a Prescribed Law of Variation of

This class of sliding control algorithms is characterized by the fact that the switchings of the control time derivative depend on suitable functions of the sliding variable. Therefore the convergence properties are strictly related to the considered function (Fig.12).

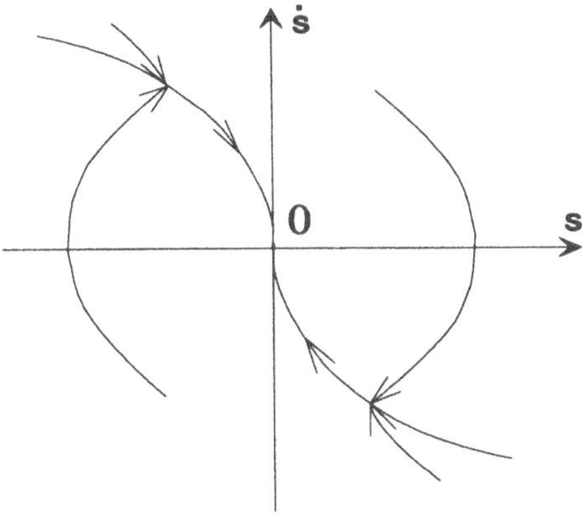

Fig. 12. Phase trajectories for the algorithm with prescribed law of variation of s

The general formulation of such a class of 2–sliding control algorithms is the following

$$v(t) = \begin{cases} -u & if \ |u| > 1 \\ -V_{\mathrm{M}}\mathrm{sign}(y_2 - g(y_1)) & if \ |u| \leq 1 \end{cases} \tag{29}$$

where V_{M} is a positive constant and the continuous function $g(y_1)$ is smooth everywhere but in $y_1 = 0$.

Function g must be chosen so that all the solutions of the equation $\dot{y}_1 = g(y_1)$ vanish in a finite time, and function $g' \cdot g$ is bounded. For example, the following function can be used

$$g(y_1) = -\lambda |y_1|^{\rho} \mathrm{sign}(y_1)$$
$$\lambda > 0 \ , 0.5 \leq \rho < 1$$

The sufficient condition for the finite time convergence to the sliding manifold is defined by the following inequality

$$V_{\mathrm{M}} > \frac{\Phi + \sup(g'(y_1) g(y_1))}{\Gamma_{\mathrm{m}}} \tag{30}$$

and the convergence time depends on the function g [13, 20, 27].

This algorithm needs the variable y_2 to be known and that is not always the case. The substitution of y_2 with the first difference of the available y_1, i.e.,

$\text{sign}[y_2 - g(y_1)] \rightarrow \text{sign}[\Delta y_{1_i} - \delta_i g(y_1)]$ $(t \in [t_i\ ; t_{i+1}),\ \ \delta_i = t_i - t_{i-1})$, turns this algorithm into a real sliding algorithm, and its order equals two if g is chosen as in the above example with $\rho = 0.5$ [20].

An extension of the control algorithm with prescribed law of variation of s to arbitrary order sliding mode control of uncertain system was recently presented in [22].

4 Conclusions

In this paper a collection of control algorithms which are able to give rise to 2–sliding modes have been presented, and for each of them the sufficient convergence conditions are given. Furthermore, the real sliding behaviour is briefly considered, and, in some cases, the upper bound of the convergence time is given.

2–sliding mode control seems to be an effective tool for the control of uncertain nonlinear systems since it overcomes the main drawbacks of the classical sliding mode control, i.e., the chattering phenomenon and the large control effort. Its real implementation implies very simple control laws and assures an improvement of the sliding accuracy with respect to real 1–sliding mode control.

The main difficulty in using 2–sliding mode controllers is the tuning of the parameters which characterize the various algorithms. Their values depend on the bounds of the uncertain dynamics and on the chosen sliding manifold, and only sufficient conditions for the convergence to the sliding behaviour are known. These conditions are very conservative, and, in practice, the parameters are heuristically tuned. It depends on the engineer's experience to define which of the presented algorithms is more suitable for the specific control problem, even if, in the authors' opinion, the super–twisting and the sub–optimal ones seem to be able to cover a large class of control problems with a remarkable implementation easiness.

Acknowledgments

This work has been partially supported by the contract MAS3CT95-0024-AMADEUS of the European Community, by Regione Autonoma della Sardegna, and by MURST Project "Identification and control of industrial systems".

References

1. G. Bartolini, P. Pydynowski, "An improved chattering free VSC scheme for uncertain dynamical systems," *IEEE Trans. Automatic Control*, vol. 41, pp. 1220–1226, 1996.
2. G. Bartolini, A. Ferrara, E. Usai, "Second order VSC for non linear systems subjected to a wide class of uncertainty conditions", *Proc. of the 1996 IEEE Int. Workshop on Variable Structure Systems - VSS'96, Seiken Symposium No. 19*, pp. 49–54, Tokyo, Japan, December 1996.

3. G. Bartolini, A. Ferrara, E. Usai, "Applications of a sub-optimal discontinuous control algorithm for uncertain second order systems", *Int. J. of Robust and Nonlinear Control*, vol. 7, no. 4, pp. 299–319, 1997.
4. G. Bartolini, A. Ferrara, E. Usai, "Output Tracking Control of Uncertain Nonlinear Second-Order Systems", *Automatica*, vol. 33, no. 12, pp. 2203–2212, 1997.
5. G. Bartolini, A. Ferrara, A. Pisano, E. Usai, "Adaptive reduction of the control effort in chattering free sliding mode control of uncertain nonlinear systems", *Applied Mathematics and Computer Science*, vol. 8, no. 1, pp. 51–71, 1998.
6. G. Bartolini, A. Ferrara, E. Usai, "Chattering Avoidance by Second Order Sliding Mode Control", *IEEE Trans. Automatic Control*, vol. 43, no. 2, pp. 241–246, 1998.
7. G. Bartolini, A. Ferrara, E. Usai, "Real–time output derivatives estimation by means of higher order sliding modes", *Proc. of CESA'98 IMACS Multiconference*, pp. 782–787, Nabeul–Hammamet, Tunisia, April 1998.
8. G. Bartolini, A. Ferrara, E. Usai, V.I. Utkin, "Second order chattering–free sliding mode control for some classes of multi–input uncertain nonlinear systems", *Proc. of the 6th IEEE Mediterranean Conf. on Control and Systems*, Alghero, Italy, June 1998, to be published.
9. G. Bartolini, A. Pisano, E. Usai, "Digital Second Order Sliding Mode Control of SISO Uncertain Nonlinear Systems", *Proc. of the 1998 American Control Conference ACC'98*, vol. 1, pp. 119–124, Philadelfia, Pensilvania, June 1998.
10. ML.W. Chang, "A MIMO sliding control with second order sliding condition," *ASME W.A.M.*, paper no. 90-WA/DSC-5, Dallas, Texas, 1990.
11. M. Corless, and G. Leitmann, "Continuous state feedback guaranteeing uniform ultimate boundedness for uncertain dynamic systems," *IEEE Trans. Automatic Control*, vol. 26, pp. 1139–1141, 1981.
12. M. Corless, G. Leitmann, "Exponential Convergence for Uncertain Systems with Component–Wise Bounded Controllers", *Robust Control via variable structure and Lyapunov techniques*. F.Garofalo and L.Glielmo eds., Lecture Notes in Control and Information Science no. 217, pp. 175–196, Springer-Verlag, London, 1996.
13. S.V. Emel'yanov, S.K. Korovin, L.V. Levantovsky, "Higher–Order Sliding Modes in the Binary Control Systems", *Soviet Physics*, Dokady, vol. 31, no. 4, pp. 291–293, 1986.
14. S.V. Emel'yanov, S.K. Korovin, L.V. Levantovsky, "Drift Algorithm in Control of Uncertain Processes", *Problems of Control and Information Theory*, vol. 15, no. 6, pp. 425–438, 1986.
15. A.F. Filippov,*Differential Equations with Discontinuous Righthand Side*, Kluwer Academic Publishers, Dordrecht, 1988.
16. A. Isidori, *Nonlinear control systems*, Springer-Verlag, New York, 1989.
17. D.E. Kirk, *Optimal control theory*, Prentice Hall, Englewood Cliffs, NJ, 1970.
18. G. Leitmann, *The calculus of variations and optimal control*, Plenum Press, New York, 1981.
19. L. Fridman, A. Levant, "Higher Order Sliding Modes as a Natural Phenomenon in Control Theory", in *Robust Control via variable structure and Lyapunov techniques*. F.Garofalo and L.Glielmo eds., Lecture Notes in Control and Information Science no. 217, pp. 107–133, Springer-Verlag, London, 1996.
20. A. Levant (Levantovsky L.V.), "Sliding order and sliding accuracy in sliding mode control", *Int. J. of Control*, vol. 58, no. 6, pp. 1247–1263, 1993.
21. A. Levant, "Robust exact differentiation via sliding mode technique", *Automatica*, vol.34, no. 3, pp. 379–384, 1998.

22. A. Levant, "Arbitrary–order sliding modes with finite time convergence", *Proc. of the 6th IEEE Mediterranean Conf. on Control and Systems*, Alghero, Italy, June 1998, to be published.
23. C. Milosavljevic, "General conditions for the existence of a quasisliding mode on the switching hyperplane in discrete variable systems", *Automation Remote Control*, vol. 43, no. 1, pp. 307–314, 1985.
24. K.S. Narendra, A.M. Annaswamy, *Stable adaptive systems*, Prentice–Hall, Englewood Cliffs, NJ, 1989.
25. H. Sira–Ramirez, "On the sliding mode control of nonlinear systems," *System and Control Letters*, vol. 19, pp. 303–312, 1992.
26. V.I. Utkin, *Sliding modes in control and optimization*, Springer Verlag, Berlin, 1992.
27. M. Zhihong, A.P. Paplinski, H.R. Wu, "A robust MIMO terminal sliding mode control for rigid robotic manipulators", *IEEE Trans. Automatic Control*, vol. 39, no. 12, pp. 2464–2468, 1994.

A Tale of Two Discontinuities

K. David Young

YKK Systems, Mountain View, California 94040-4470 USA

Abstract. A class of systems which has two discontinuous functions on the right hand side of the governing differential equation is being examined in this paper. These systems are motivated by mechanical systems in which Columb friction is a major contributing factor to the limitations on repeatibility in precision positioning applications. We shall provide analytical results showing the existence of a new class of dynamic behavior which may be described as an extension of sliding mode for continuous time dynamic systems.

1 Introduction

Control of mechanical systems is typically challenged by the presence of friction. The most familiar friction model perhaps is the so-called Columb friction in which the friction force is discontinuous with respect to the relative velocity of the two body surfaces in contact. At zero relative velocity, the force is undefined. The stick or stagnation zone is a set of equilibrium points in simple spring mass systems in which the relative motion stops, and the final position in this stuck zone is uncertain. When the relative velocity is nonzero, that is when the two surfaces is moving at different speeds, friction force is viscous in nature, thus providing viscous damping.

A great deal of research have been devoted to improve feedback control system design to mitigate the effects of friction, and to improve the quality of mechanical motion. The task of accomplishing precision positioning in the presence of static and Columb friction remains a challenge in the design of positioning servo. Many design approaches have been proposed. Designs which employ high gain position and velocity feedback have been reportedly applied with great success in industry, whereas adaptive control techniques in which the friction force is estimated and subsequently used in a friction compensation scheme have also been proposed [1].

Southward *et al* [2] introduced a bang-bang friction compensation scheme for stick-slip friction which employed a feedback control which is discontinuous with respect to position. By evoking Lyapunov stability and additional analyses to show sliding mode does not exist, it was concluded that the resulting discontinuous feedback system is globally asymptotically stable.

This mechanical system with stick-slip friction and a feedback control which is discsontinuous with respect to position is represented by a differential equation with two discontinuity functions on the right hand side. In this paper we shall analyze the dynamic behavior of this class of systems and show the existence of system trajectories which are not sliding mode in the classical sense, yet they exhibit behavior which may be described as an extension of sliding mode for continuous time dynamic systems.

2 Nonlinear Friction Compensation

We first give a description of the mechanical system control problem and the associated discontinous feedback control solution as proposed by Southward [2]. This provides the physical connections to this class of systems which involves two discontinous functions, a problem which we refer to as "a tale of two discontinuities".

2.1 Stick-Slip Friction - the first discontinuity

The mechanical system considered is modeled by a single degree-of-freedom system,

$$m\ddot{x} = u - f_d\,, \tag{1}$$

with stick-slip friction, f_d, developed between the lumped mass and the support surface, and u represents all other forces acting on the mass, which includes the control force. For simplicity, we assume u is the control force. The stick-slip friction model is partitioned into two regimes. For nonzero velocity, it represents the slip friction

$$f_d(\dot{x}) = \begin{cases} f_{sp}(\dot{x}) > 0\,, \text{ if } & \dot{x} > 0 \\ -f_{sp}(\dot{x})\,, \quad \text{if } & \dot{x} < 0 \end{cases}, \tag{2}$$

and for zero velocity, stick friction is given by

$$f_d(u) = \begin{cases} \bar{f}_{sk} > 0\,, \text{ if } & u \geq \bar{f}_{sk} \\ u\,, \quad\quad \text{ if } & -\bar{f}_{sk} < u < \bar{f}_{sk} \\ -\bar{f}_{sk}\,, \quad \text{if } & u \leq -\bar{f}_{sk} \end{cases}. \tag{3}$$

We have simplified these models by assuming equal magnitudes for the upper and lower limits in the friction force. The slip friction function in its most general form includes a viscous friction term and an exponential term in the velocity [3],

$$f_{sp}(\dot{x}) = \bar{f}_{sp} + (\bar{f}_{sk} - \bar{f}_{sp})e^{-(\dot{x}/v_o)^2} + b\dot{x}\,, \tag{4}$$

in which $\bar{f}_{sk}$ is the constant Columb friction force, and v_o is a parameter that characterizes the rate of the Stribeck effect, *i.e.*, the decrease in friction force with increasing velocity in a low velocity region.

2.2 The second discontinuity: Bang-Bang Compensation

If the control force has a spring force component $-k_p x$, the equilibrium position of the system is a set:

$$X_E = \{ x : |x| \leq x_l \triangleq \bar{f}_{sk}/k_p \} . \tag{5}$$

Without any additional compensation control forces, the system is stuck with a maximum steady state positioning error when the equilibrium point is at the limits of this set. The actual error depends on the initial conditions. Southward et al proposed a bang bang compensation control scheme which is applied when the position x reaches an ϵ-neighborhood of the upper and lower bounds of the equilibrium set X_E:

$$x_{lc} \triangleq x_l + \epsilon \, , \; \epsilon > 0 \, , \tag{6}$$

$$\bar{u} \triangleq k_p x_{lc} \, , \tag{7}$$

$$u = \begin{cases} -k_v \dot{x} - \bar{u}\mathrm{sgn}(x) \, , & \text{if } |x| \leq x_{lc} \\ -k_v \dot{x} \, , & \text{if } x = 0 \end{cases} . \tag{8}$$

The resulting control action is discontinuous on $x = 0$, as indicated by the signum function sgn(·). [1] Various variable structure control approaches to this class of problems using standard sliding mode techniques have been proposed by Young [4, 5].

3 A Second Order System with Two Discontinuities

The essence of the tale of discontinuities can be illustrated with a simplified version of the mechanical system with nonlinear friction compensation described above:

$$\begin{aligned} \dot{x}_1 &= x_2 \\ \dot{x}_2 &= u_1 + u_2 \end{aligned} \tag{9}$$

where $u_1(\cdot)$ and $u_2(\cdot)$ are two functions which are discontinuous on the manifold $x_1 = 0$ and $x_2 = 0$ respectively:

$$u_1 = -\bar{u}_1 \mathrm{sgn}(x_1) \tag{10}$$

$$u_2 = -\bar{u}_2 \mathrm{sgn}(x_2) \, , \tag{11}$$

with $\bar{u}_1$ and $\bar{u}_2$ being positive scalars.

We shall examine the behavior of this system under two different conditions. First, let

$$\bar{u}_2 > \bar{u}_1 > 0 \tag{12}$$

[1] $\mathrm{sgn}(\xi) = 1, \; \forall \xi > 0 \, ; \mathrm{sgn}(\xi) = -1, \; \forall \xi < 0.$

From the reaching condition for sliding mode on the manifold $x_2 = 0$,

$$\dot{x}_2 x_2 = [-\bar{u}_2 + u_1 \text{sgn}(x_2)]|x_2| < 0 \tag{13}$$

it is concluded that $x_2 = 0$ is a sliding manifold with $x_1 = x_{1,\text{stick}}$. Clearly, no further motion on this manifold is possible. This is the so-called stick zone, and the phenomenon is explained superbly by Andronov [6].

The second condition in which the relative magnitudes of the two inputs are reversed:

$$\bar{u}_1 > \bar{u}_2 > 0 \tag{14}$$

leads to a much less obvious and more intriguing dynamic behavior. It is possible to construct a Lyapunov function $V(x_1, x_2)$ and show that

$$\dot{V} \leq -\alpha x_2^2 \leq 0 \tag{15}$$

is negative semi-definite; and by evoking LaSalle's Theorem to determine if the origin of the system is indeed asymtotically stable. This however requires knowing the solution properties of this system with two discontinuities on the right hand side.

We first provide a conjecture of how this system behaves under this condition. This provides a framework to introduce the more rigorous arguments later. Despite the fact that the state variable x_2 cannot be discontinuous, suppose x_2 is a substituted control variable which forces "sliding mode" to occur on the manfiold $x_1 = 0$. In addition, sliding mode exists as in the hierarchy of control [7, ?] sense, $x_1 = 0 \rightarrow x_2 = 0$, *i.e.*, sliding mode exists on $x_1 = 0$, and then on the intersections of $x_1 = 0$ and $x_2 = 0$.

This argument is obviously flawed given the continuous time nature of the underlying system dynamics. Nevertheless, we are reminded of sliding mode in sampled data systems [9] – the control variable does not have to be discontinuous in order for discrete time sliding mode to occur. The "sliding mode" which occcurs for this class of systems with two discontinuities is a special class of system trajectories that is encompassed by a broader definition of the classical sliding mode.

4 Non-uniform Sampling Sliding Mode

The second order system with two discontinuities given by Eqn.(9)–(11) can be wriiten in a sector form:

$$\text{For } x_1 > 0 : \dot{x}_2 = \begin{cases} -\epsilon_2 \, , \, x_2 > 0 \\ -\epsilon_1 \, , \, x_2 < 0 \end{cases} \tag{16}$$

$$\text{For } x_1 < 0 : \dot{x}_2 = \begin{cases} \epsilon_1 \, , \, x_2 > 0 \\ \epsilon_2 \, , \, x_2 < 0 \end{cases} \tag{17}$$

where in view of the second condition stated in Eqn.(14),

$$\epsilon_1 \stackrel{\Delta}{=} \bar{u}_1 - \bar{u}_2 > 0, \quad \epsilon_2 \stackrel{\Delta}{=} \bar{u}_1 + \bar{u}_2, \qquad \epsilon_2 > \epsilon_1 > 0 \tag{18}$$

The following theorem describes the sampled data sliding mode characteristics of the system trajectories.

Theorem 1 *Suppose $x_1(t_o) = 0$, then there exists $t_o < t_1 < t_2 < t_3 < \ldots < t_j \ldots$ such that*

$$x_1(t_j) = 0\,, \quad j = 0, 1, 2, \ldots \tag{19}$$

$$x_2(t_{j+1}) \;\; = -\left(\tfrac{\epsilon_1}{\epsilon_2}\right)^{\frac{1}{2}} x_2(t_j) \tag{20}$$

Furthermore,

$$\Delta t_j \triangleq t_{j+1} - t_j = \left(\frac{1}{\epsilon_2} + \frac{1}{\sqrt{\epsilon_1 \epsilon_2}}\right) |x_2(t_j)| \tag{21}$$

If we interpret the time points $t_1\,, t_2\,, \ldots$ as the sampling points of the continuous time trajectories $x_1(t)$ and $x_2(t)$, then this theorem shows that a discrete time sliding mode exists on the manifold $x_1 = 0$. Figure 1 illustrates the convergence of $x_2(t_j)$ at the sampled time points to the origin which is given in a Corollary to the above theorem:

Corollary 1 *As $t_j \to \infty$, the phase trajectory of the system at the sampled time points $t_1\,, t_2\,, \ldots\,, t_j\, \ldots$ remains on the manifold $x_1 = 0$ and approaches the origin while the time difference between each successive sampled time points also tends to zero:*

$$\lim\nolimits_{j \to \infty} x_2(t_j) = 0\,, \quad x_1(t_j) = 0 \tag{22}$$

$$\lim\nolimits_{j \to \infty} \Delta t_j = 0 \tag{23}$$

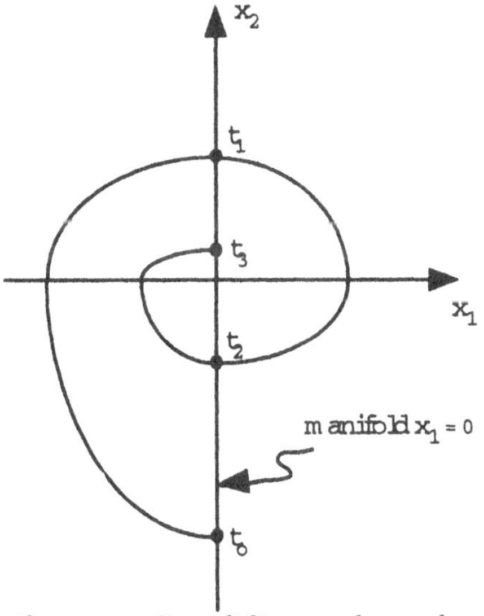

Figure 1: Non-uniform sampling sliding mode on the manifold $x_1 = 0$

The dynamic behavior of this class of systems with two discontinuous functions when the second condition for the relative magnitudes holds (Eqn.(14)) can be summarized as follows:

- A discrete time sliding mode occurs on the manifold $x_1 = 0$; sampling is however non-uniform, and the sampling period also tends to zero with increasing time
- While on the manfiold $x_1 = 0$, $x_2(t)$ converges to the manifold $x_2 = 0$. However, discrete time sliding mode does not occur since $x_2(t_j)$ is not zero after a finite time (or finite number of samples).
- The system is kept on the manifold $x_1 = 0$ by a substituted "control" x_2 which is not discontinuous, but nevertheless, it is changing sign.
- The system's trajectories belong to a new class of sliding mode which involves multiple manifolds in a hierarchy of control sense.

5 Robustness of systems with two discontinuities

For more interesting and realistic extensions of the simple system introduced in the last section, consider the system with uncertain nonlinearities:

$$\begin{aligned} \dot{x}_1 &= x_2 \\ \dot{x}_2 &= f(x_1, x_2) + u_1 + u_2 \end{aligned} \tag{24}$$

and let the magnitudes of the discontinuous functions be nonlinear functionals of the states,

$$u_1(x_1, x_2) = -u_1^*(x_1, x_2)\text{sgn}(x_1) \tag{25}$$
$$u_2(x_1, x_2) = -u_2^*(x_1, x_2)\text{sgn}(x_2) \tag{26}$$

The analysis of the system's behavior is simplied by taking lower and upper bound estimates of the nonlinearities and consider the upper and lower bounds of the right hand sides of the system rewritten in the sector form introduced earlier. The following bounds are introduced:

$$|f(\cdot,\cdot)| \leq \bar{f} \tag{27}$$
$$\bar{u}_1 \geq u_1^*(\cdot,\cdot) \geq \tilde{u}_1 > 0 \tag{28}$$
$$\bar{u}_2 \geq u_2^*(\cdot,\cdot) \geq \tilde{u}_2 > 0 \tag{29}$$

Given $\bar{f}$, there exists conditions for the various bounds in Eqn.(28-29) such that the system's phase trajectories are bounded within finite intervals on the manifold $x_1 = 0$ at the non-uniform sampling time points. Furthermore, these interval bounds tend to zero with increasing time. Thus the origin of the system is asymptotically stable. The following Theorem provides a characterization of the system trajectories:

Theorem 2 *Suppose $x_1(t_o) = 0$ and $x_2(t_o) \neq 0$, then there exists $t_o < t_1 < t_2 < t_3 < \ldots < t_j \ldots$ and $0 < r^- < r^+ < 1$ such that*

$$x_1(t_j) = 0, \quad j = 0, 1, 2, \ldots \tag{30}$$
$$r^-|x_2(t_j)| \leq |x_2(t_{j+1})| \leq r^+|x_2(t_j)| \tag{31}$$

Furthermore, the time difference between two consecutive sampling points on the manifold $x_1 = 0$, $\Delta t_j \triangleq t_{j+1} - t_j$ *satisfies*

$$\epsilon_l |x_2(t_j)| \leq \Delta t_j \leq \epsilon_u |x_2(t_j)| \tag{32}$$

$$\epsilon_l \triangleq \frac{1}{\epsilon_2^+} + \frac{1}{\sqrt{\epsilon_1^+ \epsilon_2^+}} \tag{33}$$

$$\epsilon_u \triangleq \frac{1}{\epsilon_2^-} + \frac{1}{\sqrt{\epsilon_1^- \epsilon_2^-}} \tag{34}$$

Proof. We provide a proof for this Theorem by considering the following system in sector form:

$$\dot{y}_1 = y_2 \tag{35}$$

$$\text{For } y_1 > 0 : \begin{cases} -\epsilon_2^+ < \dot{y}_2 < -\epsilon_2^- \,, \, y_2 > 0 \\ -\epsilon_1^+ < \dot{y}_2 < -\epsilon_1^- \,, \, y_2 < 0 \end{cases} \tag{36}$$

$$\text{For } y_1 < 0 : \quad \begin{cases} \epsilon_1^- < \dot{y}_2 < \epsilon_1^+ \,, \, y_2 > 0 \\ \epsilon_2^- < \dot{y}_2 < \epsilon_2^+ \,, \, y_2 < 0 \end{cases} \tag{37}$$

where $\epsilon_1^-, \epsilon_1^+, \epsilon_2^-, \epsilon_2^+$ are positive constants satisfying

$$0 < \epsilon_1^- < \epsilon_1^+ < \epsilon_2^- < \epsilon_2^+ \tag{38}$$

Suppose $y_1(t_o) = 0$ and $y_2(t_o) = y_{2o}$ is negative. The phase trajectory which starts at this initial point and ends on the manifold $y_1 = 0$ at a later time t_1 is composed of two segments; each of which is governed by one of the two cases in Eqn.(37). The first segment is characterized by an end point on $y_2 = 0$. Let this intermediate phase trajectory point be $(y_1(t_1'), 0)$. We have

$$\frac{-y_{2o}^2}{2\epsilon_2^+} \leq y_1(t_1') \leq \frac{-y_{2o}^2}{2\epsilon_2^-} \tag{39}$$

The second segment starts at this end point of the first segment and ends at the point $(0, y_2(t_1))$ on the manifold $y_1 = 0$. For the lower bound of $y_1(t_1')$ in Eqn.(39),

$$\left(\frac{\epsilon_1^-}{\epsilon_2^+}\right)^{\frac{1}{2}} |y_{2o}| \leq y_2(t_1) \leq \left(\frac{\epsilon_1^+}{\epsilon_2^+}\right)^{\frac{1}{2}} |y_{2o}| \tag{40}$$

Similarly for the upperbound,

$$\left(\frac{\epsilon_1^-}{\epsilon_2^-}\right)^{\frac{1}{2}} |y_{2o}| \leq y_2(t_1) \leq \left(\frac{\epsilon_1^+}{\epsilon_2^-}\right)^{\frac{1}{2}} |y_{2o}| \tag{41}$$

In view of the assumption in Eqn.(38), we have

$$r^- |y_2(t_o)| \leq |y_2(t_1)| \leq r^+ |y_2(t_o)| \tag{42}$$

$$r^- \triangleq \left(\frac{\epsilon_1^-}{\epsilon_2^+}\right)^{\frac{1}{2}} < 1, \quad r^+ \triangleq \left(\frac{\epsilon_1^+}{\epsilon_2^-}\right)^{\frac{1}{2}} < 1 \tag{43}$$

For positive $y_2(t_o)$, we can apply similar arguments and obtain the same bounds for $y_2(t_1)$ on the manifold $y_1 = 0$. By applying Eqn.(42) repeatedly to successive transitions from an initial value of y_2 on $y_1 = 0$ to a final value of y_2 returning onto $y_1 = 0$ again, we obtain the bounds given by Eqn.(31). Figure 2 illustrates the idea behind these derivations.

The estimate for the time of travel starting from a point on the manifold $y_1 = 0$ and returning to the manifold is computed also by considering the time spent in the first segment and for the second segment. Let Δt_j^1 be the time of the phase trajectory from $(0, y_2(t_j)$ to $(y_1(t_j + \Delta t_j^1), 0)$. We have

$$\frac{|y_2(t_j)|}{\epsilon_2^+} < \Delta t_j^1 < \frac{|y_2(t_j)|}{\epsilon_2^-} \tag{44}$$

For the second segment, suppose Δt_j^2 is the time from $((y_1(t_j + \Delta t_j^1), 0)$ to $(0, y_2(t_{j+1})$ which is on $y_1 = 0$, then

$$\frac{|y_2(t_j)|}{\sqrt{\epsilon_1^+ \epsilon_2^+}} < \Delta t_j^2 < \frac{|y_2(t_j)|}{\sqrt{\epsilon_1^- \epsilon_2^-}} \tag{45}$$

The total time of travel which is $t_{j+1} - t_j$ is the sum of Δt_j^1 and Δt_j^2.

It remains to be shown that the assumption (38) is valid and relevant to the system in question. Comparing the y-system in sector form and the system given by Eqns.(24) and (26) and taking into the bounds in Eqns.(27-29), we have the following corresponding relationships:

$$\epsilon_2^+ \Leftrightarrow \bar{u}_1 + \bar{u}_2 + \bar{f}\,,\; \epsilon_2^- \Leftrightarrow \tilde{u}_1 + \tilde{u}_2 - \bar{f} \tag{46}$$

$$\epsilon_1^+ \Leftrightarrow \bar{u}_1 - \tilde{u}_2 + \bar{f}\,,\; \epsilon_1^- \Leftrightarrow \tilde{u}_1 - \bar{u}_2 - \bar{f} \tag{47}$$

Given $\bar{f}$, the conditions for the $\tilde{u}_i$'s and $\bar{u}_i$'s satisfying the assumption ((38) are given as follows. In order for these four constants to be positive, we need

$$\tilde{u}_1 > \bar{u}_2 + \bar{f}\,, \tag{48}$$

and if

$$\tilde{u}_2 > \frac{1}{2}(\bar{u}_1 - \tilde{u}_1) + \bar{f} \tag{49}$$

then $\epsilon_2^- > \epsilon_1^+$. The remaining assumptions are then all satisfied.

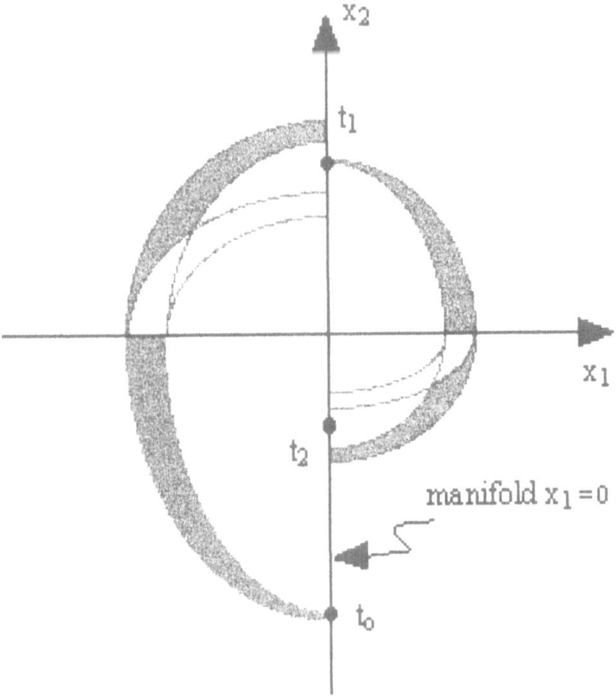

Figure 2: An Illustration of the effects of uncertainty bounds on the two solution segments of the sector system.

Remarks: Corollary 1 follows equally from Theorem 1 and Theorem 2. Theorem 1 is in fact a special case of Theorem 2. Indeed if $\bar{f} = 0$ and $\tilde{u}_1 = \bar{u}_1$ and $\tilde{u}_2 = \bar{u}_2$, then $\epsilon_1^+ = \epsilon_1^- = \epsilon_1$ and $\epsilon_2^+ = \epsilon_2^- = \epsilon_2$. Eqn.(31) and Eqn.(32) in Theorem 2 are reduced to Eqn.(20) and Eqn.(21) in Theorem 1 respectively.

6 Simulation Study

We demonstrate the dynamic behavior of this class of systems with two discontinuties by simulating thc solutions of a number of different systems of the form given by Eqn.(24). In these systems, the uncertain nonlinearity $f(\cdot,\cdot)$ takes on different forms, including discontinuous functions, and time varying functions. The only common constraint is that they are all bounded from above by $\bar{f} = 0.05$. The feedback functionals in Eqn.(26) are assumed to be of constant magnitudes with $\bar{u}_1 = \tilde{u}_1 = 1$ and $\bar{u}_2 = \tilde{u}_2 = 1.1$. For these values, we have $r^- = 0.1525$ and $r^+ = 0.2705$. Figure 3 shows the phase trajectories corresponding to the different choices of $f(\cdot,\cdot)$ and the two trajectories representing the system in sector form taking the constant upperbound pair $(\epsilon_1^+ = 0.15 \epsilon_2^- = 2.05$ and the lowewrbound pair $(\epsilon_1^- = 0.05, \epsilon_2^+ = 2.15)$. All the systems start on $x_1 = 0$ at t_o and the initial value $x_2(t_o) = -0.1$ is assumed. The interval bounds on the manifold $x_1 = 0$ are indicated by "+" and "o" markers which correspond the the crossing points of the upper nd lowebound sector systems. On the manifold

$x_1 = 0$, the phase trajectory points of the uncertain nonlinear systems are inside the bounding intervals. For $x_2(t_1)$, the bounds as computed from Eqn.(43) are $[.01525, .02705]$. The computed bounds for $x_2(t_2)$ and $x_2(t_3)$ are $[.0023, .0073]$ and $[.00036, .0020]$ respectively. These bounds agree with the simulation results.

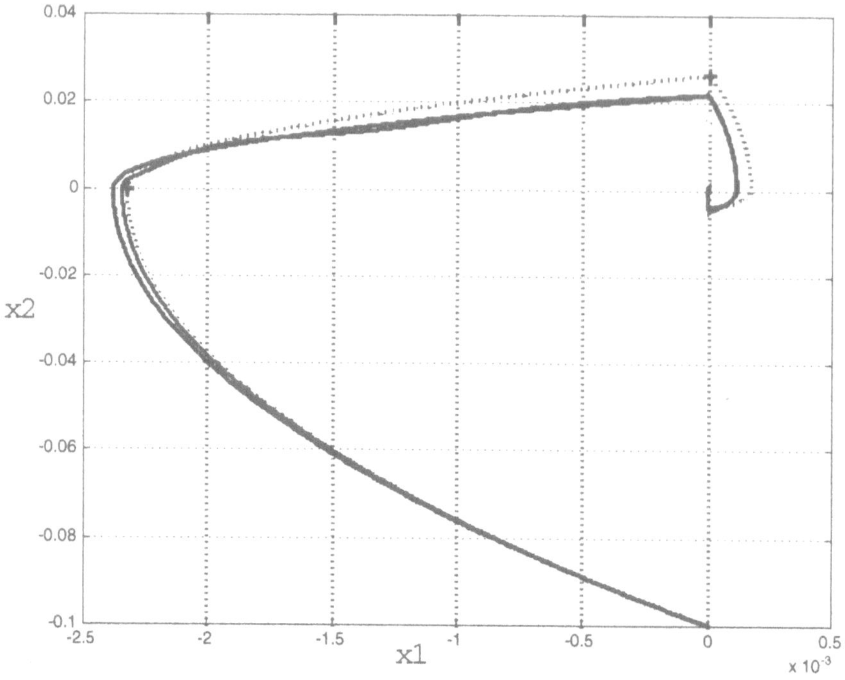

Figure 3: Phase trajectories of uncertain nonlinear systems with two discontinuities.

7 Conclusions

We have identified motion in a system with two discontinuities which may be considered as an alternative form of existence for sliding mode. This discovery adds to our knowledge of variable structure system behavior: two discontinuous functions co-exist in a co-operative manner and create a new class of system trajectories which is closely related to sliding mode. Indeed the two discontinuities give rise to a peculiar combinations of sector dynamics in the variable structure context which produce the unique system behavior. From an application standpoint, this information enhances our understanding of switching feedback control as a mechanism for overcoming the limitations of positioning repeatibility due to Columb friction. Discontinuous control works co-operatively with Columb friction and drives the system to the desired equilibrium point, thus avoiding the sticking zone which otherwise limits repeatibility. Generalizations of these results for single DOF mechanical systems to multiple DOFs mechanical systems are also feasible once it is recognized that the non-uniform sampling sliding mode and the associated convergence of the generalized velocity to zero fundamentally occur within each DOF. This decoupling effect which reduces multiple DOFs mechanical system dynamics to a set of uncoupled sliding mode dynamics is a well documented charactersitic of sliding mode control [10].

References

1. Armstrong-Helouvry, B., P. Dupont and C. Canudas De Wit,"A Survey of Models, Analysis Tools and Compensation Methods for the Control of Machines with Friction," *Automatica*, Vol.30, No.7, pp. 1083-1138, 1994.
2. Southward, S. C., C. J. Radcliffe, and C. R. MacCluer, "Robust Nonlinear Stick-Slip Friction Compensation," *ASME J. of Dynamic Systems, Measurement and Control*, Vol.113, No.4, December 1991, pp. 639-645.
3. Canudas de Wit, C., C. H. Olsson, K. J. Astrom and P. Lischinsky,"Dynamic Friction Models and Control Design," *Procs. 1993 American Control Conference*, San Franscisco, California, pp. 1920-1926.
4. Young, K. D.,"Discontinuous Control of Sliding Base Isolated Structures under Earthquakes," *Robotics, Mechatronics and Manufacturing Systems*, T. Takamori and T. Tsuchiya (Editors), pp. 485-490, Elsevier Science Publishers B.V., (North-Holland), 1993.
5. Young, K. D., "A Variable Structure Control Approach to Friction Force Compensation," *Proc. 1998 American Control Conference, Philadelphia, PA, June 1998.*
6. A. A. Andronov, A. A. Vitt, nd S. E. Khaikin, *Theory of Oscillators*, pp. 151-157, Pergamon Press, 1966.
7. V.I. Utkin, *Sliding Mode and Its Applications*, Mir, Moscow, (1978).
8. V. I. Utkin, *Sliding Modes in Control Optimization*, Springer-Verlag, (1992).
9. Young, K. D., V. I. Utkin, and U. Ozguner,"A Control Engineer's Guide to Sliding Mode Control," *Procs. 1996 IEEE International Workshop on Variable Structure Systems*, Tokyo, Japan, pp. 1-14; to appear in *IEEE Transactions on Control Systems Technology*, 1999.
10. Young, K. D. (editor), *Variable Structure Control for Robotics and Aerospace Applications*, Elsevier Science Publishers, 1993.

The Problem of Chattering: an Averaging Approach

L. Fridman

Department of Mathamatics
Samara Architecture and Civil Engineering Academy
194, Molodogvardeiscaja str., Samara, 443001, Russia

Abstract. The singularly perturbed relay control systems (SPRCS) are examined. The mathematical apparatus for investigation of the fast periodic oscillations of SPRCS is developed. The theorem about existence of fast periodic solution of SPRCS is proved. The theorem about averaging is given. It is proved that the slow motions in SPRCS with fast periodic solutions are approximately described by equations obtained from the equations for the slow variables of SPRCS by averaging along fast periodic motions. The algorithm of asymptotic representation for the fast periodic solution of SPRCS is suggested. The algorithm for correction of the averaged equation is given. The stability of the fast periodic solution is investigated.

It is shown that in the case when the original SPRCS contains the relay control linearly the averaged equations and equations which describe the motions of the reduced system in the sliding mode are coincide. The example is given which shows that in the general case when the original SPCSC contains the relay control nonlinearly, the averaging equations do not coincide with the equivalent control equations or the Filippov extension definition which describe the motions in the sliding mode in the reduced system. The algorithm is proposed which allows to solve the problem of eigenvalue assignment for averaged equations using the additional dynamics of fast actuator.

1 Introduction

The chattering phenomena is one of the actual problems in modern sliding mode control theory. The presence of actuators and measuring devices is one of the basic reasons for chattering in sliding mode control systems ([17],[4],[2]). The behaviour of such systems is described by singularly perturbed relay control systems (SPRCS). Moreover, for such systems the conditions of dynamic uncertainty hold, which means that for the original SPRCS there are no stable first order sliding modes, but for the reduced system the sufficient conditions for existence of a stable first order sliding mode hold ([4],[3],[9]).

If the original SPRCS contains either a sliding mode of 3rd order and greater or positive feedback, then the sliding modes are unstable ([1],[10]). In such systems fast periodic oscillations can occur ([17],[9]).

The general model of sliding mode control systems with fast actuators and measuring devices has the following form (see [4])

$$\mu dz/dt = g(z, s, x, u(s)), \tag{I.1}$$

$$ds/dt = h_1(z, s, x, u(s)), \quad dx/dt = h_2(z, s, x, u(s)),$$

where $z \in \mathbf{R}^m$, $s \in \mathbf{R}$, $x \in \mathbf{R}^n$, $u(s) = sign(s)$ and g, h_1, h_2 are smooth functions of their arguments. System (I.1) under such assumptions can describe for instance the behaviour of control systems in which variables x, s describe plant behaviour, and the vector z describes the behaviour of the fast actuator.

Conditions of dynamic uncertainty for system (I.1) mean that letting $\mu = 0$ and expressing $z_0 = \varphi(s, x, u(s))$ from the equation

$$g(z_0, s, x, u(s)) = 0$$

according to the formula $z_0 = \varphi(s, x, u(s))$, we obtain the reduced system

$$ds/dt = h_1(\varphi(s, x, u(s)), s, x, u(s)) = H_1(s, x, u(s)), \tag{I.2}$$

$$dx/dt = h_2(\varphi(s, x, u(s)), s, x, u(s)) = H_2(s, x, u(s)).$$

It is assumed that

(i) almost everywhere on $s = 0$

$$h_1(z, 0, x, 1)h_1(z, 0, x, -1) > 0 \tag{I.3}$$

or

$$h_1(z, 0, x, 1) > 0, \; h_1(z, 0, x, -1) < 0;$$

(ii) the measure of the domain

$$S = \{x : H_1(0, x, 1) < 0, \; H_1(0, x, -1) > 0\} \in \mathbf{R}^n$$

is nonzero and consequently S is the domain of stable first order sliding for system (I.2).

This chapter is devoted to the investigation of chattering problem in (I.2) from the viewpoint of averaging and specific features of fast periodic solutions of system (I.1). The chapter consists of four sections. Section 2 is devoted to the development of mathematical tools for the investigation of periodic solutions of SPRCS. In section 3 these tools are used for the investigation of the behaviour of sliding mode control systems with fast actuators. The proposed new approach is used in section 4 for the design of desired averaged equation in sliding mode control system with fast actuators.

2 Mathematical Tools

2.1 Problem Formulation

In this section we will consider the existence and stability of the fast periodic solutions for singularly perturbed relay control systems (SPRCS) of the form

$$\mu dz/dt = g(z, \xi, x, u(\xi)), \tag{1}$$

$$\mu d\xi/dt = h_1(z, \xi, x, u(\xi)),\ dx/dt = h_2(z, \xi, x, u(\xi)),$$

where $z \in \mathbf{R}^m$, $\xi \in \mathbf{R}$, $x \in \mathbf{R}^n, u(\xi) = sign\,(\xi)$ and g, h_1, h_2 are smooth functions of their arguments. Introducing the 'fast time' $\tau = t/\mu$ into (1), we obtain

$$dz/d\tau = g(z, \xi, x, u(\xi)), \tag{2}$$

$$d\xi/d\tau = h_1(z, \xi, x, u(\xi)), \quad dx/d\tau = \mu h_2(z, \xi, x, u(\xi)).$$

For smoothly singularly perturbed system the existence and stability of the first approximation of the fast periodic solution was investigated by [14]. The existence and stability of the first approximation of fast periodic solution of (1) was investigated in [9].

It turns out that for the investigation of the fast periodic solutions of the singularly perturbed system (2) it's impossible to use standard small parameter methods [7] for autonomous systems because setting $\mu = 0$ in (2) we will obtain a degenerate equation for the slow variables x.

In this section we develop the mathematical tools for investigation of the fast periodic oscillations (1),(2). To this end we employ the point mapping method (see [12],[13]).

2.2 Some Properties of the Point Mapping which Made by SPRCS

Let us denote the variation domain as Z, X with variables (z, s, x) and x.

Definition 1. We shall call the surface $\xi = 0$ the surface without stable sliding towards trajectories of the system

$$dz/d\tau = g(z, \xi, x, u(\xi)), \tag{3}$$

$$d\xi/d\tau = h_1(z, \xi, x, u(\xi)),$$

if all the trajectories of (3) which start outside the surface $\xi = 0$ cross it at the point $(z, 0, x)$ where the conditions

$$h_1(z, 0, x, 1)h_1(z, 0, x, -1) > 0$$

are fulfilled.

Suppose that the following conditions are true:

1^0 $h_1, h_2, g \in \mathbf{C}^2[\bar{Z} \times [-1,1]]$;

2^0 the surface $\xi = 0$ for all $x \in \bar{X}$ is a surface without stable sliding towards trajectories of system (3);

3^0 system (3) for all $x \in \bar{X}$ has an isolated orbitally asymptotically stable solution $(z_0(\tau, x), \xi_0(\tau, x))$ with period $T(x)$;

4^0 let $R(z, x)$ be a point mapping of the set $V = \{(z, x) : h_1(z, 0, x, 1) > 0\}$ on the surface $\xi = 0$ into itself performed by system (3), which has a fixed point $z^*(x)$ corresponding to $(z_0(\tau, x), \xi_0(\tau, x))$;

5^0 suppose that for $\lambda_i(x_0)$ $(i = 1, ..., m)$, the eigenvalues of the matrix

$$\frac{\partial R}{\partial z}(z^*(x_0), x_0),$$

the inequalities $|\lambda_i(x_0)| \neq 1$ are true;

6^0 the averaged system $dx/dt = p(x)$, where

$$p(x) = \frac{1}{T(x)} \int_0^{T(x)} h_2(z_0(\tau, x), \xi_0(\tau, x), x, u(\xi_0(\tau, x))d\tau, \tag{4}$$

has an isolated equilibrium point x_0 such that

$$p(x_0) = 0, \quad det|\frac{dp}{dx}(x_0)| \neq 0.$$

Let us denote as $z^{\pm}(\tau, z, x, \mu)$ and $\xi^{\pm}(\tau, z, x, \mu)$, the solutions of system (2) with the initial conditions $z^{\pm}(0, z, x, \mu) = z, \xi^{\pm}(0, z, x, \mu) = 0$ for $\xi > 0$ and $\xi < 0$.

The point mapping of the domain V of the surface $\xi = 0$, associated with system (2), has the following form

$$\Phi(z, x, \mu) = (\Phi_1(z, x, \mu), \Phi_2(z, x, \mu)) =$$

$$(z^-(\Theta, z^+(\theta, z, x, \mu), x^+(\theta, z, x, \mu), \mu), x^-(\Theta, z^+(\theta, z, x, \mu), x^+(\theta, z, x, \mu), \mu)),$$

where the functions $\theta(z, x, \mu), \Theta(z, x, \mu)$ are determined by equations

$$\xi^+(\theta, z, x, \mu) = 0,$$

$$\xi^-(\Theta, z^+(\theta, z, x, \mu), x^+(\theta, z, x, \mu), \mu) = 0.$$

This means that $\Phi_1(z, x, 0) = R(z, x)$.

The surface $\xi = 0$ is the surface without stable sliding for system (3). This means that there exists a neighbourhood of the point $(z^*(x_0), x_0)$ on the surface $\xi = 0$ for which

$$max\{|d\xi^+/d\theta|, |d\xi^-/d\Theta|\} > 0.$$

It follows from condition 1^0 and the implicit function theorem that for some small μ_0, the functions Φ, θ, Θ have continuous derivatives in some set $U \times [0, \mu_0]$ on the surface $\xi = 0$. This means that we can consider the function Φ as the point mapping of the set $U \times [0, \mu_0]$ on the surface $\xi = 0$ into itself. Moreover we can rewrite $\Phi(z, x, \mu)$ in the form

$$\Phi(z, x, \mu) = (\bar{R}(z, x, \mu), x + \mu\bar{Q}(z, x, \mu)),$$

where $\bar{R}(z, x, \mu), \bar{Q}(z, x, \mu)$ are sufficiently smooth functions and

$$\bar{Q}(z^*(x_0), x_0, 0) = 0, \bar{R}(z^*(x), x, 0) = z^*(x).$$

Let's introduce into the function Φ new variables using the formula $\eta = z - z^*(x)$. Then the point mapping Φ takes the form

$$\Psi(\eta, x, \mu) = (\Psi_1(\eta, x, \mu), \Psi_2(\eta, x, \mu)) =$$

$$= (\bar{R}(\eta + z^*(x), x, \mu) - z^*(x), x + \mu\bar{Q}(\eta + z^*(x), x, \mu)), \qquad (5)$$

and consequently $\Psi(0, x, 0) = (0, x)$.

2.3 Existence of the Fast Periodic Solution

Theorem 2. *Under conditions* $1^0 - 6^0$, *system* (1) *has an isolated periodic solution with the period* $\mu(T(x_0) + O(\mu))$ *near to the circle*

$$(z_0(t/\mu, x_0), \xi_0(t/\mu, x_0), x_0).$$

Proof. We will prove the existence of the periodic solution as the existence of the fixed point $(\eta^*(\mu), x^*(\mu))$ of the point mapping Ψ. Let's rewrite the conditions of existence of this fixed point in the form

$$G(\eta^*, x^*, \mu) = \begin{pmatrix} G_1(\eta^*, x^*, \mu) \\ G_2(\eta^*, x^*, \mu) \end{pmatrix} = \begin{pmatrix} \eta^* - \Psi_1(\eta^*, x^*, \mu) \\ \frac{1}{\mu}[x^* - \Psi_2(\eta^*, x^*, \mu)] \end{pmatrix} = 0. \qquad (6)$$

It is necessary to take into account that for $\mu = 0$ $\eta^*(0) = 0, x^*(0) = x_0$ and $G_2(0, x_0, 0) = -T(x_0)p(x_0) = 0$ and consequently for $\mu = 0$ conditions (6) are fulfilled. Moreover, taking into account that $G_1(0, x, 0) = 0$ for all $x \in \bar{X}$ we can conclude that $\frac{\partial G_1}{\partial x}(0, x_0, 0) = 0$. Let us compute the Jacobian of function G with respect to variables η, x at $\mu = 0$.

$$\frac{\partial G}{\partial(\eta, x)}(0, x_0, 0) =$$

$$= \begin{vmatrix} I_m - \frac{\partial R}{\partial z}(z^*(x_0), x_0) & 0 \\ \frac{\partial G_2}{\partial \eta}(0, x_0, 0) & -T(x_0)\frac{\partial p}{\partial x}(x_0) \end{vmatrix} \neq 0.$$

This means that there exists an isolated fixed point $(z^*(\mu), x^*(\mu))$ of the point mapping G which corresponds to the periodic solution of systems (1) and (2) and in this case $z^*(\mu) = z^*(x_0) + O(\mu), x^*(\mu) = x_0 + O(\mu)$. □

2.4 Stability in the First Approximation

Assume that

7^0 the eigenvalues $\lambda_i(x_0)$ of the matrix $\frac{\partial R}{\partial z}(z(x_0), x_0)$ satisfy the inequalities $|\lambda_i(x_0)| < 1\,(i = 1, ..., m)$;

8^0 the eigenvalues $\nu_j(x_0)$, $j = 1, ..., n$ of the matrix $\frac{dp}{dx}(x_0)$ satisfy the inequalities

$$\mathbf{Re}\,\nu_j(x_0) < 0.$$

Theorem 3. *Under conditions* $1^0 - 8^0$ *the periodic solution of* (1), (2) *is orbitally asymptotically stable.*

Proof. Let's find the derivatives Ψ by variables η, x

$$\frac{\partial \Psi}{\partial(\eta, x, \mu)} = \Gamma(\eta, x, \mu) = \begin{bmatrix} I_m - \frac{\partial R}{\partial z}(x_0) + O(\mu) & O(\mu) \\ \frac{\partial \Psi_2}{\partial \eta}(0, x_0, 0) + O(\mu) & I_m + \mu T(x_0)\frac{\partial p}{\partial x}(x_0) + O(\mu) \end{bmatrix}.$$

Consequently the matrix $\Gamma(\eta, x, \mu)$ has at the small vicinity of $(0, x_0, 0)$ two groups of eigenvalues

$$\lambda_i(x_0) + O(\mu),\ i = 1, ..., m,$$
$$1 + \mu T(x_0)\nu_j(x_0) + o(\mu),\ j = 1, ..., n.$$

This means that under conditions of theorem 3 there exists some neighbourhood of $(0, x_0, 0)$ for which Ψ is contraction mapping and corresponding fast periodic solution of systems (1),(2) is orbitally asymptotically stable. □

2.5 Some Auxiliary Theorems about Decomposition of Two - Speed Point Mappings

It is obvious that the problem of stability of fast periodic solution of system (1) is equivalent to the problem of stability of the fixed point $\eta^*(\mu), x^*(\mu)$ of $\Psi(\mu)$. Let's introduce into Ψ new variables according to the formulae $\kappa = \eta - \eta^*(\mu), \chi = x - x^*(\mu)$. Then, taking into account that $\partial\Psi(0, x_0, 0)/\partial x = 0$, we have

$$\Lambda_1(\kappa, \chi, \mu) = P\kappa + Q(\kappa, \chi, \mu),$$
$$\Lambda_2(\kappa, \chi, \mu) = \chi + \mu R(\kappa, \chi, \mu), \qquad (D.1)$$

where Q and R are smooth functions and under conditions $1^0 - 7^0$

$$P = \partial\bar{R}/\partial z(z(x_0), x_0),\ ||P|| < 1;$$
$$Q(\kappa, \chi, \mu) = O(\mu)O(|\kappa| + |\chi|) + O(|\kappa|^2 + |\chi|^2),$$
$$R(\kappa, \chi, \mu) = O(|\kappa| + |\chi|).$$

Thus we can reduce Cauchy's problem for system (1) with initial conditions

$$z(0, \mu) = z^0,\ s(0, \mu) = 0,\ x(0, \mu) = x^0. \qquad (IC)$$

to the investigation of the two-speed discrete system

$$\kappa_{k+1} = P\kappa_k + Q(\kappa_k, \chi_k, \mu),\ \chi_{k+1} = \chi_k + \mu R(\kappa_k, \mu), \qquad (D.2)$$
$$\kappa_0 = z^0 - z^*(x^*(\mu)),\ \chi_0 = x^0 - x^*(\mu).$$

In the sequel we will use the following theorems about decomposition of point mappings (D.1),(D.2) (see [16]).

Proposition D.1. *Assume that for system (D.1) conditions (D.2) hold. Then system (D.1) has a slow motions integral manifold of the form $\kappa = V(\chi, \mu)$ for small μ. Then there exist C_1, C_2 such that*

$$||V(\chi, \mu)|| < C_1,$$

$$||V(\chi, \mu) - V(\bar{\chi}, \mu)|| < C_2||\chi - \bar{\chi}||.$$

The motion on the manifold $\kappa = V(\chi, \mu)$ is described by the equation

$$\Lambda_1(V(\chi, \mu), \chi, \mu) = \chi + \mu R(V(\chi, \mu), \chi, \mu). \tag{D.3}$$

For the slow coordinate solution of (D.2) $\chi_k(\chi_0)$, and $\bar{\chi}_k(\tilde{\chi})$ the solution of system (D.3) with initial condition $\bar{\chi}_0 = \tilde{\chi}$, there exist $c > 0$, $0 < q < 1$ and $\tilde{\chi} \in \mathbf{R}$ for which the inequality

$$|\chi(\chi_0) - \bar{\chi}(\tilde{\chi})| < cq^k$$

is true.

Proposition D.2 (reduction principle). *If*

$$Q(0, 0, \mu) = 0; \quad R(0, 0, \mu) = 0,$$

then the problem of stability of the zero solutions of systems (D.1) and (D.3) are equivalent. This means that the zero solution of (D.1) is stable (asymptotically stable, unstable) if and only if the zero solution of (D.3) is stable (asymptotically stable, unstable).

The function $V(\chi, \mu)$ may be found from the equation

$$PV(\chi, \mu) + Q(V(\chi, \mu), \chi, \mu) = V(\chi + \mu R(V(\chi, \mu), \chi, \mu), \mu)$$

with any level of precision in form

$$V(\chi, \mu) = V_0(\chi) + \mu V_1(\chi) + \mu^2 V_2(\chi) + \cdots.$$

The function $V_0(\chi)$ is a solution of the equation

$$PV_0(\chi) + Q(V_0(\chi), \chi, 0) = V_0(\chi).$$

Function $V_1(\chi)$ can be found from equation

$$PV_1(\chi) + Q'_\mu(V_0(\chi), \chi, 0) = V_1(\chi).$$

An equation describing the flow on the slow motions integral manifold has the form

$$\Lambda_2(V(\chi, \mu), \chi, \mu) = \chi + \mu R(V_0(\chi), \chi, \mu) + \tag{D.4}$$
$$+ \mu^2 (R'_\kappa(V_0(\chi), \chi, 0) V_1(\chi) + R'_\mu(V_0(\chi), \chi, 0)) + O(\mu^3).$$

2.6 Theorem about Averaging

Assume that

9^0 The solution $\bar{x}(t)$ of the averaged system (3) with initial conditions $\bar{x}(0) = x^0$ for $t \in [0, L]$ is situated in the closed subdomain $\bar{X} \in X$.

Theorem 4. *Under conditions* $1^0 - 7^0$ *and* 9^0, *the slow coordinate* $x(t, \mu)$ *of solution* (1), (IC) *and* $\bar{x}(t)$ *satisfy the inequality*

$$\sup_{t \in [0,L]} |x(t, \mu) - \bar{x}(t)| = O(\mu).$$

2.7 Searching for the Periodic Solution

Assume now that

$1A^0$ $h_1, h_2, g \in \mathbf{C}^{k+2}[\bar{Z} \times [-1, 1]]$.

We will find the period of the desired periodic solution of (2) in the form

$$T(\mu) = T_0 + \mu T_1 + \mu^2 T_2 + \ldots, \tag{7}$$

where $T_0 = T(x_0)$ and the time interval for which $u = 1$ and $u = -1$ in form

$$\theta^{\pm}(\mu) = \theta_0^{\pm} + \mu\theta_1^{\pm} + \mu^2\theta_2^{\pm} + \cdots + \mu^k\theta_k^{\pm} + \cdots,$$

where $\theta_0^{\pm} = \theta^{\pm}(x_0)$. Then the asymptotic representation of the desired periodic solution on $[0, T(\mu)]$ takes the form

$$z(\tau, \mu) = z_0(\tau) + \mu z_1(\tau) + \mu^2 z_2(\tau) + \ldots + \mu^k z_k(\tau) + \ldots,$$

$$x(\tau, \mu) = \xi_0(\tau) + \mu\xi_1(\tau) + \mu^2\xi_2(\tau) + \ldots + \mu^k\xi_k(\tau) + \ldots,$$

$$x(\tau, \mu) = x_0 + \mu x_1(\tau) + \mu^2 x_2(\tau) + \ldots + \mu^k x_k(\tau) + \ldots \quad .$$

Denote

$$\tilde{T}_k(\mu) = T_0 + \mu T_1 + \mu^2 T_2 + \ldots + \mu^k T_k,$$

$$\tilde{\theta}_k^{\pm}(\mu) = \theta_0^{\pm} + \mu\theta_0^{\pm} + \mu^2\theta_2^{\pm} + \ldots + \mu^k\theta_k^{\pm}.$$

Let's find the $k - th$ approximation of the asymptotic representation for the desired periodic solution for $\tau \in [0, \tilde{T}_k(\mu)]$ in the form

$$Z_k(\tau, \mu) = z_0(\tau) + \mu z_1(\tau) + \mu^2 z_2(\tau) + \ldots + \mu^k z_k(\tau),$$

$$\Xi_k(\tau, \mu) = \xi_0(\tau) + \mu\xi_1(\tau) + \mu^2\xi_2(\tau) + \ldots + \mu^k\xi_k(\tau),$$

$$X_k(\tau, \mu) = x_0 + \mu x_1(\tau) + \mu^2 x_2(\tau) + \ldots + \mu^k x_k(\tau),$$

where the continuous functions z_i, ξ_i, x_i are smooth on $[0, \tilde{\theta}_k^{+}(\mu)) \cup (\theta_k^{+}(\mu), T_k(\mu)]$ but have jumps in the derivative at $\tau = \tilde{\theta}_k^{+}(\mu)$. Let's show that under the conditions of theorem 2 the functions $z_i^{\pm}, \xi_i^{\pm}, x_i^{\pm}$ and the constants θ_i, Θ_i for every $i = 1, \ldots, k$ can be uniquely found.

Let's introduce in system (2) two 'new times' according to the formulae

$$\tau^{+} = \tau/\tilde{\theta}_k^{+}(\mu); \tau^{-} = (\tau - \tilde{\theta}_k^{+}(\mu))/\tilde{\theta}_k^{-}(\mu), \tau^{\pm} \in [0, 1],$$

and the auxiliary functions $z_0^{\pm}(\tau^{\pm}), \xi_0^{\pm}(\tau^{\pm})$ as the solutions to

$$dz_0^{\pm}/d\tau^{\pm} = \theta_0^{\pm} g(z_0^{\pm}, \xi_0^{\pm}, x_0, \pm 1), \tag{8}$$

$$d\xi_0^{\pm}/d\tau = \theta_0^{\pm} h_1(z_0^{\pm}, \xi_0^{\pm}, x_0, \pm 1)$$

with initial and periodicity conditions

$$z_0^{+}(0) = z^*(x_0) = z_0^*, \quad \xi_0^{+}(0) = 0; \tag{9}$$

$$z_0^{-}(0) = z_0^{+}(1), \quad \xi_0^{-}(0) = \xi_0^{+}(1) = 0;$$

$$z_0^{-}(1) = z^*(x_0), \quad \xi_0^{-}(1) = 0.$$

From the periodicity of functions $z_0(\tau), \xi_0(\tau)$ it follows that the system (8),(9) has a unique solution.

The functions $x_1^{\pm}(\tau)$ are described by the equations

$$dx_1^{\pm}/d\tau^{\pm} = \theta_0^{\pm} h_2(z_0^{\pm}(\tau^{\pm}, x_0), \xi_0^{\pm}(\tau^{\pm}, x_0), x_0, \pm 1), \tag{10}$$

with initial conditions and periodicity conditions given by

$$x_1^{+}(0) = x_1^*, \quad x_0^{-}(0) = x_1^{+}(1), \quad x_1^{-}(1) = x_1^*, \tag{11}$$

Moreover

$$[h_{20}](x_0) = \int_0^1 h_2(z_0(\tau^{+}, x_0), \xi_0^{+}(\tau^{+}, x_0), x_0, 1) d\tau^{+} +$$

$$+ \int_0^1 h_2(z_0(\tau^{-}, x_0), \xi_0^{-}(\tau^{-}, x_0), x_0, -1) d\tau^{-} = 0, \tag{12}$$

$$det|\frac{d[h_{20}]}{dx}(x_0)| = T(x_0)\frac{dp}{dx}(x_0) \neq 0.$$

This means that for every x_1^* there exists a unique solution of (10) and (11) for which $\int_0^1 \tilde{x}_1^{+}(\tau^{+})d\tau^{+} + \int_0^1 \tilde{x}_1^{-}(\tau^{-})d\tau^{-} = 0$ and we can define the function $x_1(\tau)$ in the form

$$x_1(\tau) = x_1^* + \tilde{x}_1(\tau) =$$

$$= \begin{cases} x_1^* + \tilde{x}_1^{+}(\tau/\tilde{\theta}_k^{+}(\mu)) & \text{for} \quad \tau \in [0, \tilde{\theta}_k^{+}(\mu)] \\ x_1^* + \tilde{x}_1^{-}((\tau - \tilde{\theta}_k^{+}(\mu))/\tilde{\theta}_k^{-}(\mu)) & \text{for} \quad \tau \in [\tilde{\theta}_k^{+}(\mu), \tilde{T}_k(\mu)]. \end{cases}$$

The functions $z_1^{\pm}(\tau^{\pm}, x_1^*), \xi_1^{\pm}(\tau^{\pm}, x_1^*)$ are defined by equations

$$dz_1^{\pm}/d\tau = \theta_0^{\pm}(g_z'^{\pm} z_1^{\pm} + g_\xi'^{\pm}\xi_1^{\pm} + g_x'^{\pm}x_1^{\pm}) + \theta_1^{\pm} g^{\pm}; \tag{13}$$

$$d\xi_1/d\tau = \theta_0^{\pm}(h_{1z}'^{\pm} z_1^{\pm} + h_{1\xi}'^{\pm}\xi_1^{\pm} + h_{1x}'^{\pm}x_1^{\pm}) + \theta_1^{\pm} h_1^{\pm},$$

where the values of $g^{\pm}, h_1^{\pm}$ and their derivatives are calculated at the points

$$(z_0^{\pm}(\tau^{\pm}, x_0), \xi_0^{\pm}(\tau^{\pm}, x_0), x_0, \pm 1).$$

Initial and periodicity conditions for system (13) are defined by equations

$$z_1^+(0, x_1^*) = z_1^-(1, x_1^*) = z_1^*, z_1^-(0, x_1^*) = z_1^+(1, x_1^*); \tag{14}$$

$$\xi_1^+(0, x_1^*) = \xi_1^+(1, x_1^*) = \xi_1^-(0, x_1^*) = \xi_1^-(1, x_1^*) = 0.$$

Equations (13) depend linearly on $z_1^{\pm}, \xi_1^{\pm}, \theta_1^{\pm}$ and consequently their solutions $z_1^{\pm}(\tau, x_1^*)$, $\xi_1^{\pm}(\tau, x_1^*)$, $\theta_1^{\pm}(x_1^*)$ are linearly dependent on the initial conditions $z_1^{\pm}(0, x_1^*)$. Expressing $z_1^{\pm}(\tau, x_1^*), \xi_1^{\pm}(\tau, x_1^*), \theta_1^{\pm}(x_1^*)$ through $z_1^+(0, x_1^*)$ and substituting the results in the first equation of (14) we have a system of algebraic equations linear in $z_i^+(0, x_1^*)$ whose determinant coincides with $det|I_m - \partial R(z^*(x_0), x_0)/\partial z| \neq 0$.

The functions $x_2^{\pm}(\tau)$ are described by the equations

$$dx_2^{\pm}/d\tau = \theta_0^{\pm}(h'_{2\,z} z_1^{\pm} +$$

$$+ h'_{2\,\xi}\xi_1^{\pm} + h'_{2\,x} x_1^{\pm}) + \theta_1^{\pm} h_2, \tag{15}$$

where the values of $h_2^{\pm}$ are calculated at the points

$$(z_0^{\pm}(\tau^{\pm}, x_0), \xi_0^{\pm}(\tau^{\pm}, x_0), x_0, \pm 1).$$

Initial and periodicity conditions are

$$x_2^+(0) = x_2^*, \quad x_2^-(0) = x_2^+(1), \quad x_2^-(1) = x_2^*. \tag{16}$$

The condition under which system (16) has periodic solutions for every x_2^* takes the form

$$\int_0^1 [\theta_0^+(h'^+_{2\,z} z_1^+(\tau^+, x_1^*) + h'^+_{2\,\xi}\xi_1^+(\tau^+, x_1^*) +$$

$$+ h'^+_{2\,x} x_1^+(\tau^+, x_1^*)) + \theta_1^+(\tau^+, x_1^*) h_2^+] d\tau^+ +$$

$$+ \int_0^1 [\theta_0^-(h'^-_{2\,z} z_1^-(\tau^-, x_1^*) + h'^-_{2\,\xi}\xi_1^-(\tau^-, x_1^*) +$$

$$+ h'^-_{2\,x} x_1^-(\tau^-, x_1^*)) + +\theta_1^-(x_1^*) h_2^-] d\tau^- = 0. \tag{17}$$

Conditions (17) are a system of linear equations for obtaining x_1^*, whose determinant coincides with $\frac{dp}{dx}(x_0) \neq 0$. This means that we can find uniquely the function $x_2(\tau)$ in form $x_2(\tau) = \tilde{x}_2^* + \bar{x}_2(\tau)$, where $\bar{x}_2(\tau)$ is the function with zero averaged value.

Now suppose that the functions $z_j(\tau), \xi_j(\tau), x_j(\tau)$ and the constants $x_j^*, \theta_j^{\pm}$, $j = 1, ..., i-1$ have been found. Moreover, the periodic function $x_i(\tau)$ for every x_i^* can be represented in form of the sum of x_i^* and the function $\tilde{x}_i(\tau)$ with zero averaged value.

Then the functions $z_i^{\pm}(\tau^{\pm}, x_i^*), \xi_i^{\pm}(\tau^{\pm}, x_i^*), x_i^{\pm}(\tau^{\pm}, x_i^*)$ are defined by equations

$$dz_i/d\tau^{\pm} = \theta_0^{\pm}(g_z'^{\pm} z_i^{\pm} + g_\xi'^{\pm}\xi_i^{\pm} + g_x'^{\pm}x_i^{\pm}) + \\ + \theta_i^{\pm}(x_i^*)g^{\pm} + \Pi_{1i}^{\pm}(\tau^{\pm}); \tag{18}$$

$$d\xi_i/d\tau^{\pm} = \theta_0^{\pm}(h_{1z}'^{\pm} z_i^{\pm} + h_{1\xi}'^{\pm}\xi_i^{\pm} + h_{1x}'^{\pm}x_i^{\pm}) + \\ \theta_i^{\pm}(x_i^*)h_1^{\pm} + \Pi_{2i}^{\pm}(\tau^{\pm}),$$

where the values of $g^{\pm}, h_1^{\pm}$ and their derivatives are calculated at the points

$$(z_0^{\pm}(\tau^{\pm}, x_0), \xi_0^{\pm}(\tau^{\pm}, x_0), x_0, \pm 1),$$

and $\Pi_{ji}^{\pm}, j = 1, 2$ are uniquely defined functions containing the terms of order μ^i in the asymptotic representations of $g^{\pm}, h_1^{\pm}$ depending on $z_j^{\pm}, \xi^{\pm}, x_j^{\pm}, x_j^*, j \leq i-1$. Initial and periodicity conditions for system (18) are defined by the equations

$$z_i^+(0, x_1^*) = z_i^-(1, x_i^*) = z_i^*, \; z_i^-(0, x_i^*) = z_i^+(1, x_i^*) \tag{19}$$

$$\xi_i^+(0, x_i^*) = \xi_i^+(1, x_i^*) = \xi_i^-(0, x_i^*) = \xi_i^-(1, x_i^*) = 0.$$

Equations (18) depend linearly on $z_i^{\pm}, \xi_i^{\pm}, \theta_i^{\pm}$ and consequently their solutions $z_i^{\pm}(\tau, x_i^*)$, $\xi_i^{\pm}(\tau, x_i^*)$, $\theta_i^{\pm}(x_i^*)$ are linearly dependent on the initial conditions $z_i^{\pm}(0, x_i^*)$. Expressing $z_i^{\pm}(\tau, x_i^*), \xi_i^{\pm}(\tau, x_i^*), \theta_i^{\pm}(x_i^*)$ through $z_i^+(0, x_i^*)$ and substituting the results in the first equation of (19) we have a system of algebraic equations linear in $z_i^+(0, x_i^*)$ whose determinant coincides with $det|I_m - \partial R(z^*(x_0), x_0)/\partial z| \neq 0$.

The functions $x_{i+1}(\tau)$ are described by the equations

$$dx_{i+1}^{\pm}/d\tau = \theta_0^{\pm}(h_{2z}' z_1^{\pm} + h_{2\xi}'\xi_1^{\pm} + h_{2x}' x_1^{\pm}) + \\ + \theta_1^{\pm} h_{i+1} + \pi_{3i}^{\pm}(\tau),$$

where the values of $h_2^{\pm}$ and its derivatives are calculated at the

$$(z_0^{\pm}(\tau^{\pm}, x_0), \xi_0^{\pm}(\tau^{\pm}, x_0), x_0, \pm 1)$$

and $\pi_{3i}^{\pm}$ are uniquely defined functions containing the terms of order μ^i in the asymptotic representations of $h_2^{\pm}$ depending on $z_j^{\pm}, \xi_j^{\pm}, x_j^{\pm}, x_j^*, j \leq i-1$. Initial and periodicity conditions are

$$x_{i+1}^+(0) = x_{i+1}^*, \; x_{i+1}^-(0) = x_{i+1}^+(1), \; x_{i+1}^-(1) = x_{i+1}^*. \tag{20}$$

The conditions under which system (20) has a periodic solution with zero averaged value for every x_{i+1}^* takes the form

$$\int_0^1 [\theta_0^+(h_{2z}'^+ z_i^+(\tau^+, x_i^*) + h_{2\xi}'^+\xi_i^+(\tau^+, x_i^*) + \\ + h_{2x}'^+ x_i^+(\tau^+, x_i^*)) + \theta_i^+(\tau^+, x_i^*)h_2^+]d\tau^+ + \tag{21}$$

$$+\int_0^1 [\theta_0^-(h'^-_{2z} z_i^-(\tau^-, x_i^*) + h'^-_{2\xi}\xi_i^-(\tau^-, x_i^*) +$$

$$+ h'^-_{2x} x_i^-(\tau^-, x_i^*)) + \theta_i^-(x_i^*) h_2^-] d\tau^- = 0.$$

Conditions (21) are a system of linear equations for obtaining of x_i^*, whose determinant coincides with $\frac{dp}{dx}(x_0) \neq 0$. This means that for every x_{i+1}^* we can uniquely find the function $x_{i+1}(\tau)$ in the form $x_{i+1}(\tau) = \tilde{x}_{i+1}^* + \bar{x}_{i+1}(\tau)$, where $\bar{x}_{i+1}(\tau)$ is a function with zero averaged value. To finish the algorithm for the design of the desired asymptotic representation, it is necessary to define

$$(z_i(\tau), \xi_i(\tau)) =$$

$$= \begin{cases} (z_i^+(\tau/\tilde{\theta}_k^+(\mu), x_i^*), \xi_i^+(\tau/\tilde{\theta}_k^+(\mu), x_i^*)) \text{ for } \quad \tau \in [0, \tilde{\theta}_k^+(\mu)], \\ (z_i^-((\tau - \theta_k^+(\mu))/\tilde{\theta}_k^-(\mu), x_i^*), \xi_i^+((\tau - \theta_k^+(\mu))/\tilde{\theta}_k^-(\mu), x_i^*)) \\ \text{for} \quad \tau \in [\tilde{\theta}_k^+(\mu), \tilde{T}_k(\mu)], i = 0, ..., k. \end{cases}$$

$$x_j(\tau) = \begin{cases} x_j^* + \tilde{x}_j^+(\tau/\tilde{\theta}_k^+(\mu)) \quad \text{for} \quad \tau \in [0, \tilde{\theta}_k^+(\mu)], \\ x_j^* + \tilde{x}_j^-((\tau - \tilde{\theta}_k^+(\mu))/\tilde{\theta}_k^-(\mu)) \\ \text{for} \quad \tau \in [\tilde{\theta}_k^+(\mu), \tilde{T}_k(\mu)], j = 1, ..., k. \end{cases}$$

2.8 Correction of Averaged Equations

Let us now show how we can use knowledge about the fast periodic solution for the correction of averaged equations with any precision level according to the small parameter degrees. The knowledge of such equations is necessary the case when the linear part of the averaged equations (3) has spectral points on the imaginary axis.

Assume that we have found the functions

$$\theta^{\pm}(x, \mu) = \theta_0^{\pm}(x) + \sum_{i=1}^{\infty} \mu^i \theta_i^{\pm}(x),$$

$$T(x, \mu) = \theta^+(x, \mu) + \theta^-(x, \mu)$$

and

$$(z_i(\tau, x), \xi_i(\tau, x)) =$$

$$= \begin{cases} (z_i^+(\tau/\theta^+(\mu, x), x), \xi_i^+(\tau/\theta^+(\mu, x), x)) \\ \text{for} \quad [0, \theta^+(\mu, x)], \\ (z_i^-((\tau - \bar{\theta}^+(\mu))/\theta^-(\mu, x), x), \xi_i^-((\tau - \theta^+(\mu, x))/\theta^-(\mu, x), x)) \\ \text{for} \quad [\theta^+(\mu, x), T(\mu, x)], \end{cases}$$

then

$$z(\tau, x, \mu) = z_0(\tau, x) + \mu z_1(\tau, x) + ... + \mu^i z_i(\tau, x) + ...,$$

$$\xi(\tau, x, \mu) = \xi_0(\tau, x) + \mu \xi_1(\tau, x) + ... + \mu^i \xi_i(\tau, x) + ...,$$

$$x(\tau, x, \mu) = \mu x_1(\tau, x) + ... + \mu^i x_i(\tau, x) + ... \quad .$$

Then the precise averaged equation has the form

$$dx/dt = \frac{1}{T(x\mu)} \int_0^{T(x,\mu)} h_2(z(\tau, x, \mu), \xi(\tau, x, \mu), x+$$

$$+\tilde{x}(\tau, \mu), u(\xi(\tau, x, \mu)))d\tau. \qquad (PAE)$$

Equation (PAE) correspond to the system (D.4) which describes a flow on the slow motion manifold in system (D.1). In this case the first order approximation of (PAE) has the form

$$dx/dt = \frac{1}{T_0(x)} \Bigg\{ (1 - \mu T_1(x)) \int_0^{T_0(x)} h_2 d\tau +$$

$$+\mu \Bigg[\int_0^{T_0(x)} \Bigg(h'_{2_z} z_1(\tau, x) + h'_{2\xi} \xi_1(\tau, x) + h'_{2x} \tilde{x}_1(\tau) \Bigg) d\tau +$$

$$+\theta_1^+(x) h_2(z_0(\theta_0(x), x), \xi_0(\theta_0(x), x), x, 1) +$$

$$+\theta_1^-(x) h_2(z_0(T_0(x), x), \xi_0(T_0(x), x), x, -1) \Bigg] \Bigg\}, \qquad (FAAE)$$

where the values of the functions h_2 and its derivatives in the integral terms are calculated at the points $(z_0(\tau, x), \xi_0(\tau, x), x, u(\xi_0(\tau, x)))$. Analogously we can obtain the averaged equations with any precision level expanding in powers of the small parameter.

2.9 Investigation of stability in critical Case

Theorem 5 Reduction Principle. *Under conditions* $1^0 - 7^0$ *the periodic solution for original system* (1) *is stable (asymptotically stable, unstable) if and only if the equilibrium point of system* (*PAE*) *is stable (asymptotically stable, unstable).*

Corollary 6. *Assume that for system* (1) *conditions* $1^0 - 7^0$ *are true. If the equilibrium point of system (FAAE) is asymptotically stable (unstable) in the first approximation then the periodic solution for original system* (1) *is asymptotically stable (unstable).*

3 Analysis of Averaged Equations in Sliding Mode with Fast Actuators

3.1 Averaged Equations of Systems which Linearly Depend on Relay Control

In this section we will consider the SPRCS which linearly depend on relay control. We will show that the averaged equations which describe the slow motions in such SPRCS and the equations which describe the sliding motion in the reduced systems coincide.

Let's consider the system

$$\mu dz/dt = A(s,x)z + f_1(s,x) + K_1(s,x)u(s),$$

$$ds/dt = B(s,x)z + f_2(s,x) + K_2(s,x)u(s),$$

$$dx/dt = D(s,x)z + f_3(s,x) + K_3(s,x)u(s), \qquad (22)$$

where $z \in \mathbf{R}^m$, $s \in \mathbf{R}$, $x \in \mathbf{R}^n$, $u(s) = sign\,(s)$ and $f_i, K_i\ (i = 1,2,3)$ are smooth functions of their arguments. Setting $\mu = 0$ and expressing z_0 from the first equation of system (22) according to the formula $z_0 = -A^{-1}(s,x)[f_1(s,x) + K_1(s,x)u(s)]$ we obtain the reduced system

$$ds/dt = -B(s,x)A^{-1}(s,x)f_1(s,x) + f_2(s,x) -$$

$$-[B(s,x)A^{-1}(s,x)K_1(s,x) - K_2(s,x)]u(s),$$

$$dx/dt = D(s,x)A^{-1}(s,x)f_1(s,x) + f_3(s,x) -$$

$$-[D(s,x)A^{-1}(s,x)K_1(s,x) - K_3(s,x)]u(s).$$

Suppose that for the original system (22) the conditions of dynamic uncertainty hold which means that

$$K_2(0,x) \geq 0, \quad B(0,x)A^{-1}(0,x)K_1(0,x) - K_2(0,x) > 0. \qquad (CDU)$$

The equations which describe the motion in sliding modes in the reduced system have the form

$$dx/dt = -D(0,x)A^{-1}(0,x)f_1(0,x) + f_3(0,x) -$$

$$-\,[D(0,x)A^{-1}(0,x)K_1(0,x) - K_3(0,x)](u(s) - u_{eq}(x)), \qquad (23)$$

$$u_{eq}(x) = [B(0,x)A^{-1}(0,x)K_1(0,x) - K_2(0,x)]^{-1} \times$$

$$\times[-B(0,x)A^{-1}(0,x)f_1(0,x) + f_2(0,x)].$$

Let's show that the averaged equations for the original system (22) coincide with system (23).

Suppose that for all $x \in \bar{X}$ the following conditions are true:

(*) $Re \quad Spec\, A(0,x) < 0$;

(**) $|u_{eq}(x)| < 1$.

It is obvious that if the conditions of dynamical uncertainty are true it is reasonable to consider only solutions of system (22) with initial conditions

$$z(0,\mu)=z^0,\ s(0,\mu)=\mu s^0,\ x(0,\mu)=x^0,$$

which are situated in the $O(\mu)$ vicinity of the switching surface. Following [3],[9] let us increase $1/\mu$ times the neighbourhood of the discontinuity surface $s=0$ in system (22) and let the variable $\xi=s/\mu$. Then we will rewrite system (22) in the form

$$\begin{aligned}\mu dz/dt &= A(\mu\xi,x)z+f_1(\mu\xi,x)+K_1(\mu\xi,x)u(\xi),\\ \mu d\xi/dt &= B(\mu\xi,x)z+f_2(\mu\xi,x)+K_2(\mu\xi,x)u(\xi),\\ dx/dt &= D(\mu\xi,x)z+f_3(\mu\xi,x)+K_3(\mu\xi,x)u(\xi).\end{aligned} \tag{24}$$

In this case the system which describes the fast motions in (24) has, analogous to (3), the form

$$\begin{aligned}dz/d\tau &= A(0,x)z+f_1(0,x)+K_1(0,x)u(\xi),\\ d\xi/d\tau &= B(0,x)z+f_2(0,x)+K_2(0,x)u(\xi),\quad (x-parameter).\end{aligned} \tag{25}$$

Introducing into system (25) the new variables $\eta=z+A^{-1}(0,x)[f_1(0,x)+K_1(0,x)u_{eq}(x)]$ we will have

$$\begin{aligned}d\eta/d\tau &= A(0,x)\eta+K_1(0,x)\bar{u}(\xi,x),\\ d\xi/d\tau &= B(0,x)\eta+K_2(0,x)\bar{u}(\xi,x),\\ \bar{u}(\xi,x) &= u(\xi)-u_{eq}(x).\end{aligned} \tag{26}$$

Let's consider the point mapping of the surface $\xi=0$ into itself which is made by system (26). The solution of system (26) with initial conditions

$$\eta^+(0,\mu)=\eta,\quad \xi^+=0;$$

$$\eta\in\Omega^+=\{(\eta,\mu):\ D(0,x)\eta+K_2^+(0,x)\bar{u}(\xi,x)>0\},$$

$$K_i^+=K_i(1-u_{eq}),i=1,2$$

has the form

$$\begin{aligned}\eta^+(\tau,\eta,x) &= e^{A\tau}(\eta+A^{-1}K_1^+)-A^{-1}K_1^+,\\ \xi^+(\tau,\eta,\mu) &= BA^{-1}(e^{A\tau}-I)(\eta+A^{-1}K_1^+)-\\ &\quad -(BA^{-1}K_1^+-K_2^+)\tau.\end{aligned}$$

Here and in the sequel the functions A,B,K_1,K_2 are computed at the point $(0,x)$. For $\tau=0$ $d\xi/d\tau=B(0,x)\eta+K_2^+(0,x)\bar{u}(\xi,x)$ and consequently

$$\xi^+(\tau,\eta,\mu)>0$$

at least for small $\tau>0$. On the other hand from condition (i) it follows that

$$\lim_{\tau\to\infty}\xi^+(\tau,\eta,\mu)=-\infty.$$

This means that there exists $\theta(\tau, \eta, \mu)$, the smallest root of equation

$$\xi^+(\theta(\tau, \eta, \mu), \eta, \mu) = 0.$$

Let's rewrite this equation in form

$$BA^{-1}(e^{A\theta} - I)(\eta + A^{-1}K_1^+) = (BA^{-1}K_1^+ - K_2^+)\theta.$$

It follows from the definition of θ that $d\xi^+/d\tau(\theta) \leq 0$. This means that we can define the point mapping of the set Ω^+ into the set

$$\Omega^- = \{(\eta, \mu) : B(0, x)\eta^+(\theta, \eta, x) - K_2^-(0, x)\bar{u}_{eq}(\xi, x) < 0\},$$

where $K_i^- = K_i(1 + u_{eq}), i = 1, 2$. Analogously the point mapping

$$\eta^-(\Theta, \eta, x) = e^{A\Theta}(\eta - A^{-1}K_1^-) + A^{-1}K_1^-,$$

$$BA^{-1}(e^{A\Theta} - I)(\eta - A^1K_1^-) = -(BA^{-1}K_1^- - K_2^-)\Theta$$

transforms the set Ω^+ into the set

$$\Omega^* = \{(\eta, \mu) : B(0, x)\eta - K_2^-(0, x) > 0, \ x \in \bar{X}\}.$$

This means that the point mapping

$$\Phi(\eta, x) = \eta^-(\Theta, \eta^+(\theta, \eta, x), x)$$

given by formula

$$\Phi(\eta, x) = e^{A\Theta}(e^{A\theta}(\eta + A^{-1}K_1^+) - 2A^{-1}K_1) + A^{-1}K_1^-$$

transforms the set Ω^* into itself. Let's denote $\eta^*(x)$ as the fixed point of the point mapping $\Phi(\eta, x)$ which corresponds to the periodic solution

$$(z_0(\tau, x), \xi_0(\tau, x)).$$

For $\eta^*(x)$ we have the formula

$$\eta^*(x) = \{2[I - e^{A(\theta+\Theta)}]^{-1}[I - e^{A\Theta}] - (1 - u_{eq})\}A^{-1}K_1.$$

Let's study the properties of averaged values of the $T(x) = \theta(x) + \Theta(x)$ - periodic solutions $\eta_0(\tau, x)$ and $\xi_0(\tau, x)$.

$$I(x) = \int_0^{T(x)} \eta_0(\tau, x)d\tau = [(1 + u_{eq})\Theta - (1 - u_{eq})\theta]A^{-1}K_1.$$

Taking into account that $\xi_0(T(x), x) = 0$ we have

$$\xi_0(T(x), x) =$$

$$= \int_0^{T(x)} B\eta_0(\tau, x)d\tau - K_2[(1 + u_{eq})\Theta - (1 - u_{eq})\theta] = 0.$$

This means that $(1 + u_{eq})\Theta = (1 - u_{eq})\theta$. The following lemma is true.

Lemma 7. *If there exists a solution of system (26) of period $T(x)$ then*

$$\int_0^{T(x)} \eta_0(\tau, x) d\tau = 0,$$

$$\int_0^{T(x)} u(\eta_0(\tau, x)) d\tau = \frac{\Theta(x) - \theta(x)}{T(x)} = u_{eq}(x).$$

Remark. This lemma was obtained for the first time in [11] by using transfer function methods.

Let's turn to system (22). If for system (24) the conditions of theorem 3 are true there exists an isolated periodic solution $(z(\tau, \mu), \xi(\tau, \mu), x(\tau, \mu))$ which corresponds to the periodic solution $(\eta_0(\tau, x), \xi_0(\tau, x))$ of system (26). Moreover

$$\int_0^{T(x)} z_0(\tau, x) d\tau = A^{-1}(0, x)(f_1(0, x) + K_1(0, x) u_{eq}(x)).$$

This means that the averaged equations which approximately describe the behaviour of the slow motions in system (22) coincide with equations (23) for the sliding motions in the reduced system .

3.2 Example

Suppose that a mathematical model of a control system taking account of actuator behaviour has the following form

$$\mu dz/dt = -z - u, ds/dt = z + (\alpha + x)u, \ \alpha > 0 \tag{27}$$

$$dx/dt = -z + x - u, \tag{28}$$

$z, s, x \in \mathbf{R}, u(s) = sign(s), \mu$ - actuator time constant. The fast motions taking place in (27), (28) are described by the system

$$dz/d\tau = -z - u, \ d\xi/d\tau = z + (\alpha + x)u, \ u = sign(\xi). \tag{29}$$

System (29) is symmetric relative to the point $z = \xi = 0$ so we shall consider it as a point mapping $R(z, x)$ of the domain $z + \alpha > 0$ on the switching line $\xi = 0$ into the domain $z + \alpha < 0$ with $\xi > 0$. Then $\Psi(z) = -1 + e^{-T}(z + 1)$ where τ is the smallest root of equation

$$(1 - e^{-T})(z + 1) = (1 - \alpha - x)\tau.$$

The fixed point $z^* = \Psi(z^*(x), x)$ corresponding to the periodic solution (29) is determined by the equation $\Psi(z^*(x), x) = -z^*(x)$. Then the fixed point $z^*(x)$ (amplitude) and the semiperiod $T(x)$ of the periodic solution are determined by equations

$$2th(T/2) = (1 - \alpha - x)T, \ z^*(x) = th(T/2). \tag{30}$$

Equations (30) with $0 < \alpha + x < 1$ have positive solutions which corresponds to the existence of a $2T$ periodic solution in system (29). The slow motions averaged equation for system (28) assumes the form

$$dx/dt = -x.$$

This equation has the asymptotically stable equilibrium point $x = 0$. It follows from (30) that $T \approx 3.83$, $\lambda \approx -0.07$, and so system (27), (28) has an orbitally asymptotically stable periodic solution which lies in the $O(\mu)$ neighbourhood of the switching surface.

3.3 The Systems Containing The Relay Control Nonlinearly

Consider the control system which is described by the equations

$$\mu dz/dt = -z - u, ds/dt = z + \alpha u,$$

$$dx/dt = (z^4 - z^2 + \beta)x, \tag{31}$$

where $x, s, z \in \mathbf{R}$, $u(s) = sign(s)$, $0 < \alpha, \beta < 1$ and μ is the actuator time constant. If we take $\mu = 0$ in system (31) we will have

$$ds/dt = (\alpha - 1)u, \; dx/dt = (u^4 - u^2 + \beta)x. \tag{32}$$

In system (32) a stable sliding mode exists. Both the classical extension definition of solutions according Filippov [8] and the equivalent control method [17] coincide. These motions are described by the equation $dx/dt = \beta x$. The zero solution of this equation is unstable for $\beta > 0$.

At the same time if $0 < \alpha < 1$ in system (31) fast periodic solutions occur. Let us denote $z(\tau)$ as the first coordinate of the periodic solution (29) for $x = 0$. If α and β are selected so that

$$-\gamma = \int_0^{T(x_0)} [z^4(\tau) - z^2(\tau)]d\tau < -\beta < 0$$

the averaged equation has the form $dx/dt = -(\gamma - \beta)x$. The zero solution of this equation is asymptotically stable. This means that system (31) has an asymptotically orbitally stable periodic solution in the $O(\mu)$ neighbourhood of the point $s = x = 0$. The averaged equation does not coincide with the equations of the equivalent control method and Filippov determination of solution. Moreover the introduction of positive feedback was used for transition from one vector of convex closure of the right hand part to the other one and for giving the system desired dynamic properties.

4 Eigenvalue Assignment in Averaged Equations using Dynamics of Actuators

4.1 Problem Formulation

Let us suppose that the behaviour of control system is described by the state vector (s, x) $(s \in R, x \in R^n)$ with equations

$$\dot{s} = A_1 s + A_2 x + b_1 u(s), \dot{x} = A_3 s + A_4 x + b_2 u(s), \tag{33}$$

where $s \in \mathbf{R}, x \in \mathbf{R}^n$ the discontinuous control law has been designed in the form $u(s) = sign(s)$. Let us suppose that this control law ensures a stable sliding mode on the surface $s = 0$.

Then the motions in the sliding mode in system (33) are described by the equations

$$\dot{x} = (A_4 - b_2 b_1^{-1} A_3)x. \tag{34}$$

In [17] two methods were proposed to solve the problem of eigenvalue assignment in (34):

(i)to extend the state space by using additional dynamics and to solve the problem of eigenvalue assignment in the extended state space;

(ii) to include the derivatives of the variable s into the equation of the switching surface.

In [15] the fast variable describing the behaviour of the fast actuator was introduced in the equation for the switching surface for motion control in singularly perturbed discontinuous control systems. This approach ensures the existence of a first order sliding mode in overall system. For such systems the composite control method (see [6]) was used [5].

We suggest using the dynamics of the actuators which are present in the original system to solve the problem of eigenvalue assignment in (34). The proposed algorithm is based on theorems 2,3. It is necessary to note that in this case we can use only the slow coordinates of the state-vector for control design and we can solve the eigenvalue assignment problem in the space of the sliding mode equations. On the other hand, the proposed algorithm is useful only in case when the actuator is a MIMO system.

4.2 The Eigenvalues Assignment in Averaged Equations

Let us suppose that the complete model of the control system taking into account the fast actuator dynamics has the form

$$\mu \dot{z} = B_1 z + B_2 s + B_3 x + d_1 v$$

$$\dot{s} = B_4 z + B_5 s + B_6 x + d_2 v, \quad \dot{x} = B_7 z + B_8 s + B_9 x + d_3 v, \tag{35}$$

where $z \in \mathbf{R}^m$, $v \in \mathbf{R}^l$ and μ is actuator time constant. Now we suppose that the conditions

$$rank \begin{pmatrix} B_4 \\ B_7 \end{pmatrix} \geq 2,$$

$$rank\, d_1 \geq 2, \quad m \geq l \geq 2 \tag{36}$$

are satisfied. The conditions (36) mean that the discontinuous control is transmitted to the plant through the actuators with the number of inputs and outputs no less than two. Ignoring fast dynamics, having accepted that $\mu = 0$ and expressing z according to the formula $z = -B_1^{-1}(B_2 s + B_3 x + d_1 v)$ we obtain

$$\begin{aligned} \dot{s} &= (B_5 - B_4 B_1^{-1} B_2)s + (B_6 - B_4 B_1^{-1} B_3)x + (d_2 - B_4 B_1^{-1} d_1)v, \\ \dot{x} &= (B_8 - B_7 B_1^{-1} B_2)s + (B_9 - B_7 B_1^{-1} B_3)x + (d_3 - B_7 B_1^{-1} d_1)v. \end{aligned} \tag{37}$$

Let us suppose that in the case when the control law has been designed in the form $v = Ku(s)$ $(u(s) = sign\,(s), K$ is constant vector), systems (37) and (33) coincide.

The proposed algorithm uses theorem 3. We propose to use the control law in the form

$$v = Ku(s) + w. \tag{38}$$

From theorem 3 it follows that the slow motions in (35) are described by the equations

$$\dot{x} = (A_4 - b_2 b_1^{-1} A_2)x - [d_3 - B_7 b_1^{-1}(d_2 - B_4 B_1^{-1} d_1)]w \tag{39}$$

Assume that

D.1.*Matrices*

$$(A_4 - b_2 b_1^{-1} A_2)\, and\, [d_3 - B_7 b_1^{-1}(d_2 - B_4 B_1^{-1} d_1)]$$

are controllable.

Then, choosing the control vector in the form $w = Lx$, we can solve the eigenvalue assignment problem for (39). This means, that an algorithm for design of the control vector w is proposed which allows us to solve the eigenvalue assignment problem for system (39). If the conditions of lemma 7 are fulfilled, the equations (39) coincide with the averaged equations which approximately describe the slow motions in a small neighbourhood of the switching surface of system (35). To use this algorithm it is necessary to ensure existence and stability in the first approximation of periodic solutions of the system

$$dz/d\tau = B_1 z + d_1 Ku(s),\; ds/d\tau = B_4 z + d_2 Ku(s), \tag{40}$$

describing the fast motions in (35) in the small neighbourhood of the equilibrium point $x = 0$ of the averaged equation (39).

To formulate the sufficient conditions consider the point mapping $R(z)$ of the domain

$$\Omega^* = \{z : B_4 z - d_2 K > 0, z \in \mathbf{R}^m\}$$

on the surface $s = 0$ into itself, given by the formulae

$$R^+(z) = e^{B_1 \tau_1}(z + B_1^{-1} d_1 K) - B_1^{-1} d_1 K,$$

$$R(z) = e^{B_1 \tau_2}(R^+(z) - B_1^{-1} d_1 K) + B_1^{-1} d_1 K,$$

where $\tau_1, \tau_2$the smallest positive roots of the equations

$$B_4B_1^{-1}(e^{B_1\tau_1} - I)(z + B_1^{-1}d_1K) = (B_4B_1^{-1}d_1 - d_2)K\tau_1,$$

$$B_4B_1^{-1}(e^{B_1\tau_2} - I)[e^{B_1\tau_1}(z + B_1^{-1}d_1K) - $$
$$-2B_1^{-1}d_1K] = -(B_4B_1^{-1}d_1 - d_2)K\tau_2.$$

Taking into account the symmetry of system (33) for $u(s) = sign\,(s)$ we can rewrite the conditions of existence of the fixed point in the form $R^+(z^*) = -z^*$. Then

$$z^* = [I + e^{B_1T}]^{-1}(I - e^{B_1T})B_1^{-1}d_1K,$$

where the semiperiod of the desired periodic solution $T > 0$ is the smallest root of the equation

$$B_4B_1^{-1}(e^{B_1T} - I)(z^* + B_1^{-1}d_1K) = (B_4B_1^{-1}d_1 - d_2)KT.$$

Then from theorems 2,3 it follows

Theorem 8. *Assume that condition D.1 is true and B_1 and $R(z)$ satisfy the conditions*

D.2 Re Spec $B_1 < 0$.

D.3 The point mapping $R(z)$ has an isolated fixed point $z^* \in \Omega^*$.

D.4. For $\lambda_i(x_0)$ $(i = 1, ..., m)$, the eigenvalues of the matrix $\frac{\partial R}{\partial z}(z^*)$, the inequalities $|\lambda_i| < 1$ hold.

Then there exists a matrix L which ensures that the characteristic polynomial of the matrix

$$(A_4 - b_2b_1^{-1}A_2) - [d_3 - B_7b_1^{-1}(d_2 - B_4B_1^{-1}d_1)]L$$

has the desired form and the averaged equations for system (35) *has the form* (39) *and for systems* (35) *and* (39) *theorems* 2, 3 *are true.*

4.3 Example

Let us suppose that the state vector of the control system is described by the equations

$$\dot{s} = u(s)/2, \ \dot{x}_1 = x_2, \ \dot{x}_2 = -x_1, \ s, x_1, x_2 \in \mathbf{R} \qquad (41)$$

and the discontinuous control $u(s) = -sign(s)$ has been designed. The motions in the sliding mode in (41) are described by the equations

$$\dot{x}_1 = x_2, \quad \dot{x}_2 = -x_1. \qquad (42)$$

The spectrum of matrix in (42) is situated on the imaginary axis. Let us suppose that the discontinuous control $u(s)$ is transmitted to the plant with the help of actuators whose behaviour is described by the variables z_1, z_2 and the overall model of the system has the following form

$$\mu\dot{z}_1 = -z_1 + v_1 - x_1, \mu\dot{z}_2 = -z_2 + v_2,$$

$$\dot{s} = z_2 + v_2/2,\ \dot{x}_1 = x_2,\ \dot{x}_2 = z_1. \tag{43}$$

It can be easily seen that in the case where we suppose that $v_1 = v_2 = u(s) = -sign(s)$, system (43) takes the form

$$\mu\dot{z}_1 = -z_1 - sign\,(s) - x_1, \quad \mu\dot{z}_2 = -z_2 - sign\,(s),$$

$$\dot{s} = z_2 + 1/2sign\,(s),\ \dot{x}_1 = x_2,\ \dot{x}_2 = z_1$$

and the slow motions in it are described by system (42) with an accuracy of $O(\mu)$.

Let's show that for system (43) the conditions of theorem 8 are fulfilled. Denote $z = (z_1, z_2)$. Consider the point mapping $R^+(z)$ of the domain

$$\Omega^* = \{(z, x) \,:\, z_2 - 1/2 > 0\}$$

on the surface $s = 0$ into the domain $\Omega^- = \{(z, x) \,:\, z_2 + 1/2 < 0\}$ made by the system

$$dz_1/d\tau = -z_1 - sign\,(s), \quad dz_2/d\tau = -z_2 - sign\,(s),$$

$$d\xi/d\tau = z_2 + 1/2sign\,(s). \tag{44}$$

The point mapping $R^+(z)$ of the domain Ω^* into the domain $\Omega^- = \{z : z_2 + 1/2 < 0\}$ made by the system (44) has the form

$$R^+(z) = \{R_1^+(z), R_2^+(z)\} =$$

$$= \{-1 + e^{-\tau}(z_1 + 1), -1 + e^{-\tau}(z_2 + 1)\},$$

where $\tau > 0$ is the smallest root of the equation $(1 - e^{-\tau})(z_2 + 1) = \tau/2$. System (44) is symmetric with respect to the point $(0, 0)$ and consequently the condition of existence of a fixed point z^* corresponding to the desired periodic solution of (44) takes the form $R^+(z^*) = -z^*$. Then z^* and the semiperiod T satisfy the equations $z_2^* = th(T/2)$, $4th(T/2) = T$ with the solution $z_1^* = z_2^* = 0,95$, $T \approx 3,83$. Moreover

$$\frac{\partial R^+}{\partial(z_1, z_2)}(z^*) = \begin{pmatrix} -0,07 & 0 \\ 0 & -0,07 \end{pmatrix}.$$

This means that for system (44) the conditions of theorem 3 are fulfilled and the slow motions in (44) are described by the averaged equations (42). This means that for eigenvalue assignment in the system we can use a control law of the form

$$v_1 = -sign\,(s) + l_1x_1 + l_2x_2, v_2 = -sign\,(s).$$

Assume that for our goal the desired characteristic polynomial of averaged equations

$$\dot{x}_1 = x_2,\ \dot{x}_2 = (l_1 - 1)x_1 + l_2x_2 \tag{45}$$

has the form

$$\lambda^2 + \alpha\lambda + \beta,\ \alpha, \beta \,\text{are constants.} \tag{46}$$

This means that choosing $l_1 = 1 - \beta, l_2 = -\alpha$, we can ensure that the characteristic polynomial of the averaged system (45) has the form (46).

Conclusion

As a mathematical model of chattering in the small neighbourhood of the switching surface in sliding mode systems the singularly perturbed relay control systems was examined.

The following mathematical apparatus for the investigation of the SPRCS has been designed

- the sufficient conditions for existence of fast periodic solutions;
- the averaging theorem;
- the algorithm for asymptotic representation of fast periodic solutions;
- the sufficient conditions and reduction principle theorem for the investigation of the stability of fast periodic solutions.

It was shown that the slow motions in such SPRCS are approximately described by equations obtained from the equations for the slow variables of SPRCS by averaging along fast periodic motions.

The analysis of oscillations of sliding mode systems in the small neighbourhood of the sliding mode control systems has shown that

- in the general case when the original SPCSC contains the relay control nonlinearly, the averaging equations do not coincide with the equivalent control equations or the Filippov extension definition which describe the motions in the sliding mode in the reduced system;
- in the case when the original SPCSC contain the relay control linearly, the averaging equations and equations which describe the motions in the sliding mode in the reduced system coincide.

The results obtained were used for eigenvalue assignment for slow motions using the dynamics of the actuators in the case when the actuator is a MIMO system.

References

1. Anosov D.V.: On stability of equilibrium points of relay systems. Automatica i telemechanica (Automation and Remote Control) (1959) 135–149 (Russian).
2. Astrom K. , Barabanov A. and Johanson K.H.: Limit cycles with chattering in relay feedbacks systems. Proc. of 36th CDC: San-Diego (1997).
3. Bogatyrev S.V., Fridman L.M.: Singular correction of the equivalent control method. Differentialnye uravnenija (Differential equations) 28 (1992) 740–751.
4. Bondarev A.G., Bondarev S.A., Kostylyeva N.Ye. and Utkin V.I.: Sliding Modes in Systems with Asymptotic State Observers. Automatica i telemechanica (Automation and Remote Control) 46 (1985) 679–684.
5. Heck B.: Sliding mode control for singularly perturbed system. International J. of Control 50 (1991) 985 - 1001.
6. Kokotovic P.V., Khalil H.K. and O'Reilly J.: Singular Perturbation Methods in Control: Analysis and Design. Academic Press (1986).
7. Coddington E.A., Levinson N: Theory of ordinary differential equations. McGraw-Hill: New-York (1955).

8. FilippovA.F.: Differential equations with discontinuous right hand side. Kluwer Publishers: Dodrecht (1988).
9. Fridman L.M.: Singular extension definition of discontinuous control systems and stability. Differentialnye uravnenija (Differential equation) 26 (1990) 1307–1312.
10. Fridman L., Levant A.: Higher order sliding modes as the natural phenomenon in control theory. In: F. Garafalo and L. Glielmo,Eds. Robust Control via variable structure and Lyapunov techniques. Springer - Verlag: Berlin (1996) 103–130.
11. Neimark Y.I.: About periodic solutions of relay systems. In.: Memorize of A.A. Andronov. Nauka: Moscow (1955) (Russian).
12. Neimark Y.I.: The method of averaging from the viewpoint of the point mapping method. Izvestija Vusov: Radiophysika. VI (1963) 1023–1032 (Russian).
13. Neimark Y.I.: The point mapping method in the theory of nonlinear oscillations. Nauka: Moscow (1972) (Russian).
14. Pontriagin L.S., Rodygin L.V.: Periodic solution of one system of differential equation with small parameter near the derivative. Doklady Academii Nauk 132 (1960) 537–540 (Russian).
15. Sira-Ramires H.: Sliding Regimes on Slow Manifolds of Systems with fast Actuators. International J. of System Science 37 (1988) 875–887.
16. Varapaeva N.V.: Decomposition of discret system with fast and slow motions. Transaction of RANS, Mathematics, Mathematical Modelling, Informatics and Control II 4 (1998) 123–132 (Russian).
17. Utkin V.I.: Sliding Modes in Optimization Control. Springer Verlag: Berlin (1992).

Lecture Notes in Control and Information Sciences

Edited by M. Thoma

1993–1999 Published Titles:

Vol. 186: Sreenath, N.
Systems Representation of Global Climate Change Models. Foundation for a Systems Science Approach.
288 pp. 1993 [3-540-19824-5]

Vol. 187: Morecki, A.; Bianchi, G.; Jaworeck, K. (Eds)
RoManSy 9: Proceedings of the Ninth CISM-IFToMM Symposium on Theory and Practice of Robots and Manipulators.
476 pp. 1993 [3-540-19834-2]

Vol. 188: Naidu, D. Subbaram
Aeroassisted Orbital Transfer: Guidance and Control Strategies
192 pp. 1993 [3-540-19819-9]

Vol. 189: Ilchmann, A.
Non-Identifier-Based High-Gain Adaptive Control
220 pp. 1993 [3-540-19845-8]

Vol. 190: Chatila, R.; Hirzinger, G. (Eds)
Experimental Robotics II: The 2nd International Symposium, Toulouse, France, June 25-27 1991
580 pp. 1993 [3-540-19851-2]

Vol. 191: Blondel, V.
Simultaneous Stabilization of Linear Systems
212 pp. 1993 [3-540-19862-8]

Vol. 192: Smith, R.S.; Dahleh, M. (Eds)
The Modeling of Uncertainty in Control Systems
412 pp. 1993 [3-540-19870-9]

Vol. 193: Zinober, A.S.I. (Ed.)
Variable Structure and Lyapunov Control
428 pp. 1993 [3-540-19869-5]

Vol. 194: Cao, Xi-Ren
Realization Probabilities: The Dynamics of Queuing Systems
336 pp. 1993 [3-540-19872-5]

Vol. 195: Liu, D.; Michel, A.N.
Dynamical Systems with Saturation Nonlinearities: Analysis and Design
212 pp. 1994 [3-540-19888-1]

Vol. 196: Battilotti, S.
Noninteracting Control with Stability for Nonlinear Systems
196 pp. 1994 [3-540-19891-1]

Vol. 197: Henry, J.; Yvon, J.P. (Eds)
System Modelling and Optimization
975 pp approx. 1994 [3-540-19893-8]

Vol. 198: Winter, H.; Nüßer, H.-G. (Eds)
Advanced Technologies for Air Traffic Flow Management
225 pp approx. 1994 [3-540-19895-4]

Vol. 199: Cohen, G.; Quadrat, J.-P. (Eds)
11th International Conference on Analysis and Optimization of Systems – Discrete Event Systems: Sophia-Antipolis, June 15–16–17, 1994
648 pp. 1994 [3-540-19896-2]

Vol. 200: Yoshikawa, T.; Miyazaki, F. (Eds)
Experimental Robotics III: The 3rd International Symposium, Kyoto, Japan, October 28-30, 1993
624 pp. 1994 [3-540-19905-5]

Vol. 201: Kogan, J.
Robust Stability and Convexity
192 pp. 1994 [3-540-19919-5]

Vol. 202: Francis, B.A.; Tannenbaum, A.R. (Eds)
Feedback Control, Nonlinear Systems, and Complexity
288 pp. 1995 [3-540-19943-8]
Vol. 203: Popkov, Y.S.

Macrosystems Theory and its Applications: Equilibrium Models
344 pp. 1995 [3-540-19955-1]

Vol. 204: Takahashi, S.; Takahara, Y.
Logical Approach to Systems Theory
192 pp. 1995 [3-540-19956-X]

Vol. 205: Kotta, U.
Inversion Method in the Discrete-time Nonlinear Control Systems Synthesis Problems
168 pp. 1995 [3-540-19966-7]

Vol. 206: Aganovic, Z.; Gajic, Z.
Linear Optimal Control of Bilinear Systems with Applications to Singular Perturbations and Weak Coupling
133 pp. 1995 [3-540-19976-4]

Vol. 207: Gabasov, R.; Kirillova, F.M.; Prischepova, S.V.
Optimal Feedback Control
224 pp. 1995 [3-540-19991-8]

Vol. 208: Khalil, H.K.; Chow, J.H.; Ioannou, P.A. (Eds)
Proceedings of Workshop on Advances inControl and its Applications
300 pp. 1995 [3-540-19993-4]

Vol. 209: Foias, C.; Özbay, H.; Tannenbaum, A.
Robust Control of Infinite Dimensional Systems: Frequency Domain Methods
230 pp. 1995 [3-540-19994-2]

Vol. 210: De Wilde, P.
Neural Network Models: An Analysis
164 pp. 1996 [3-540-19995-0]

Vol. 211: Gawronski, W.
Balanced Control of Flexible Structures
280 pp. 1996 [3-540-76017-2]

Vol. 212: Sanchez, A.
Formal Specification and Synthesis of Procedural Controllers for Process Systems
248 pp. 1996 [3-540-76021-0]

Vol. 213: Patra, A.; Rao, G.P.
General Hybrid Orthogonal Functions and their Applications in Systems and Control
144 pp. 1996 [3-540-76039-3]

Vol. 214: Yin, G.; Zhang, Q. (Eds)
Recent Advances in Control and Optimization of Manufacturing Systems
240 pp. 1996 [3-540-76055-5]

Vol. 215: Bonivento, C.; Marro, G.; Zanasi, R. (Eds)
Colloquium on Automatic Control
240 pp. 1996 [3-540-76060-1]

Vol. 216: Kulhavý, R.
Recursive Nonlinear Estimation: A Geometric Approach
244 pp. 1996 [3-540-76063-6]

Vol. 217: Garofalo, F.; Glielmo, L. (Eds)
Robust Control via Variable Structure and Lyapunov Techniques
336 pp. 1996 [3-540-76067-9]

Vol. 218: van der Schaft, A.
L_2 Gain and Passivity Techniques in Nonlinear Control
176 pp. 1996 [3-540-76074-1]

Vol. 219: Berger, M.-O.; Deriche, R.; Herlin, I.; Jaffré, J.; Morel, J.-M. (Eds)
ICAOS '96: 12th International Conference on Analysis and Optimization of Systems - Images, Wavelets and PDEs:
Paris, June 26-28 1996
378 pp. 1996 [3-540-76076-8]

Vol. 220: Brogliato, B.
Nonsmooth Impact Mechanics: Models, Dynamics and Control
420 pp. 1996 [3-540-76079-2]

Vol. 221: Kelkar, A.; Joshi, S.
Control of Nonlinear Multibody Flexible Space Structures
160 pp. 1996 [3-540-76093-8]

Vol. 222: Morse, A.S.
Control Using Logic-Based Switching
288 pp. 1997 [3-540-76097-0]

Vol. 223: Khatib, O.; Salisbury, J.K.
Experimental Robotics IV: The 4th International Symposium, Stanford, California, June 30 - July 2, 1995
596 pp. 1997 [3-540-76133-0]

Vol. 224: Magni, J.-F.; Bennani, S.; Terlouw, J. (Eds)
Robust Flight Control: A Design Challenge
664 pp. 1997 [3-540-76151-9]

Vol. 225: Poznyak, A.S.; Najim, K.
Learning Automata and Stochastic Optimization
219 pp. 1997 [3-540-76154-3]

Vol. 226: Cooperman, G.; Michler, G.; Vinck, H. (Eds)
Workshop on High Performance Computing and Gigabit Local Area Networks
248 pp. 1997 [3-540-76169-1]

Vol. 227: Tarbouriech, S.; Garcia, G. (Eds)
Control of Uncertain Systems with Bounded Inputs
203 pp. 1997 [3-540-76183-7]

Vol. 228: Dugard, L.; Verriest, E.I. (Eds)
Stability and Control of Time-delay Systems
344 pp. 1998 [3-540-76193-4]

Vol. 229: Laumond, J.-P. (Ed.)
Robot Motion Planning and Control
360 pp. 1998 [3-540-76219-1]

Vol. 230: Siciliano, B.; Valavanis, K.P. (Eds)
Control Problems in Robotics and Automation
328 pp. 1998 [3-540-76220-5]

Vol. 231: Emel'yanov, S.V.; Burovoi, I.A.; Levada, F.Yu.
Control of Indefinite Nonlinear Dynamic Systems
196 pp. 1998 [3-540-76245-0]

Vol. 232: Casals, A.; de Almeida, A.T. (Eds)
Experimental Robotics V: The Fifth International Symposium Barcelona, Catalonia, June 15-18, 1997
190 pp. 1998 [3-540-76218-3]

Vol. 233: Chiacchio, P.; Chiaverini, S. (Eds)
Complex Robotic Systems
189 pp. 1998 [3-540-76265-5]

Vol. 234: Arena, P.; Fortuna, L.; Muscato, G.; Xibilia, M.G.
Neural Networks in Multidimensional Domains: Fundamentals and New Trends in Modelling and Control
179 pp. 1998 [1-85233-006-6]

Vol. 235: Chen, B.M.
H∞ Control and Its Applications
361 pp. 1998 [1-85233-026-0]

Vol. 236: de Almeida, A.T.; Khatib, O. (Eds)
Autonomous Robotic Systems
283 pp. 1998 [1-85233-036-8]

Vol. 237: Kreigman, D.J.; Hagar, G.D.; Morse, A.S. (Eds)
The Confluence of Vision and Control
304 pp. 1998 [1-85233-025-2]

Vol. 238: Elia, N. ; Dahleh, M.A.
Computational Methods for Controller Design
200 pp. 1998 [1-85233-075-9]

Vol. 239: Wang, Q.G.; Lee, T.H.; Tan, K.K.
Finite Spectrum Assignment for Time-Delay Systems
200 pp. 1998 [1-85233-065-1]

Vol. 240: Lin, Z.
Low Gain Feedback
376 pp. 1999 [1-85233-081-3]

Vol. 241: Yamamoto, Y.; Hara S.
Learning, Control and Hybrid Systems
472 pp. 1999 [1-85233-076-7]

Vol. 242: Conte, G.; Moog, C.H.; Perdon A.M.
Nonlinear Control Systems
192 pp. 1999 [1-85233-151-8]

Vol. 243: Tzafestas, S.G.; Schmidt, G. (Eds)
Progress in Systems and Robot Analysis and Control Design
624 pp. 1999 [1-85233-123-2]

Vol. 244: Nijmeijer, H.; Fossen, T.I. (Eds)
New Directions in Nonlinear Observer Design
552pp: 1999 [1-85233-134-8]

Vol. 245: Garulli, A.; Tesi, A.; Vicino, A. (Eds)
Robustness in Identification and Control
448pp: 1999 [1-85233-179-8]

Vol. 246: Aeyels, D.;
Lamnabhi-Lagarrigue, F.; van der Schaft, A. (Eds)
Stability and Stabilization of Nonlinear Systems
408pp: 1999 [1-85233-638-2]